Methods in Enzymology

Volume 391
LIPOSOMES
Part E

METHODS IN ENZYMOLOGY

EDITORS-IN-CHIEF

John N. Abelson   Melvin I. Simon

DIVISION OF BIOLOGY
CALIFORNIA INSTITUTE OF TECHNOLOGY
PASADENA, CALIFORNIA

FOUNDING EDITORS

Sidney P. Colowick and Nathan O. Kaplan

Methods in Enzymology

Volume 391

Liposomes

Part E

EDITED BY

Nejat Düzgüneş

ELSEVIER
ACADEMIC
PRESS

AMSTERDAM • BOSTON • HEIDELBERG • LONDON
NEW YORK • OXFORD • PARIS • SAN DIEGO
SAN FRANCISCO • SINGAPORE • SYDNEY • TOKYO

Table of Contents

Section I. Liposomal Anticancer Agents

Section II. Liposomal Antibacterial, Antifungal, and Antiviral Agents

Section III. Miscellaneous Liposomal Therapies

Section IV. Electron Microscopy of Liposomes

Contributors to Volume 391

Article numbers are in parentheses and following the names of contributors.
Affiliations listed are current.

SHEELA A. ABRAHAM (4), *Cambridge Institute for Medical Research, Addenbrooke's Hospital, Cambridge, United Kingdom CB2 2XY*

ATEEQ AHMAD (10), *NeoPharm, Inc., Research and Development, Waukegan, Illinois 60085*

IMRAN AHMAD (10), *NeoPharm, Inc., Research and Development, Waukegan, Illinois 60085*

TOMOHIRO ASAI (9), *Department of Medical Biochemistry, School of Pharmaceutical Sciences, University of Shizuoka, Yada, Shizuoka 422-8526, Japan*

IRMA A. J. M. BAKKER-WOUNDENBERG (14), *Department of Medical Microbiology and Infectious Diseases, Erasmus University Medical Center Rotterdam, 3000 DR Rotterdam, The Netherlands*

SATHYAMANGALAM V. BALASUBRAMANIAN (5), *Department of Pharmaceutical Sciences, University of Buffalo, State University of New York, Amherst, New York 14260-1200*

MARCEL B. BALLY (2, 4), *Advanced Therapeutics—Medical Oncology, British Columbia Cancer Agency, Vancouver, V5Z 1L3*

MARTIN J. BECKER (14), *Department of Medical Microbiology and Infectious Diseases, Erasmus University Medical Center, Rotterdam, 3000 DR Rotterdam, The Netherlands*

MICHEL G. BERGERON (19), *Centre de Recherche en Infectiologie, RC 709 Centre Hospitalier Universitaire de Quebec, Pavillon CHUL, Quebec, Canada G1V 4G2*

KOERT N. J. BURGER (6), *Department of Molecular Cell Biology, Institute of Biomembranes, Utrecht University, 3584 CH Utrecht, The Netherlands*

PIETER R. CULLIS (1, 2, 4), *Inex Pharmaceuticals Corporation, Burnaby, British Columbia, Canada V5J 5L8*

M. LUÍSA CORVO (22), *Instituto Nacional de Engenharia e Technologia Industrial (INETI), Departmento de Biotecnologia, Unidade Novas Formas de Agentes Bioactivos, Estrada do Paco do Lumiar, 22, 1649-038 Lisbon, Portugal*

M. EUGÉNIA M. CRUZ (22), *Instituto Nacional de Engenharia e Technologia Industrial (INETI), Departmento de Biotecnologia, Unidade Novas Formas de Agentes Bioactivos, Estrada do Paco do Lumiar, 22, 1649-038 Lisbon, Portugal*

ANDRÉ DÉSORMEAUX (19), *Centre de Recherche en Infectiologie, RC 709 Centre Hospitalier Universitaire de Quebec, Pavillon CHUI, Quebec, Canada G1V 4G2*

NEJAT DÜZGÜNEŞ (Introduction, 15, 20, 23), *Department of Microbiology, Arthur A. Dugoni School of Dentistry, University of the Pacific, San Francisco, California 94115*

DAVID B. FENSKE (1), *Department of Biochemistry and Molecular Biology, University of British Columbia, Vancouver, British Columbia, Canada V6T 1Z3*

DIANA FLASHER (15), *Department of Microbiology, Arthur A. Dugoni School of Dentistry, University of the Pacific, San Francisco, California 94115*

PETER M. FREDERIK (24), *EM-Unit, Universiteit Maastricht, 6200 MD Maastricht, The Netherlands*

PATTISAPU R. J. GANGADHARAM[†] (23), *Section of Infectious Diseases, Department of Medicine, College of Medicine, University of Illinois at Chicago, Chicago, Illinois 60612*

M. MANUELA GASPAR (22), *Instituto Nacional de Engenharia e Technologia Industrial (INETI), Departmento de Biotecnologia, Unidade Novas Formas de Agentes Bioactivos, Estrada do Paco do Lumiar, 22, 1649–038 Lisbon, Portugal*

GREGORY GREGORIADIS (Introduction) *School of Pharmacy, University of London, London, WC1N 1AX, United Kingdom*

LUKE GUO (14), *Alza Corporation, 1900 Charleston Road, Mountain View, California*

C. M. GUPTA (16), *Director, Central Drug Research Institute, Lucknow 226 001 (India)*

TIMO L. M. TEN HAGEN (7), *Department of Surgical Oncology, Laboratory of Experimental Surgical Oncology, Erasmus University Rotterdam, 3000 DR Rotterdam, The Netherlands*

W. HAQ (16), *Central Drug Research Institute, Lucknow 226 001 (India)*

TIMOTHY D. HEATH (11), *University of Wisconsin, School of Pharmacy, Madison, Wisconsin 53705-2222*

D. H. W. HUBERT (24), *FEI Electron Optics, 5600 MD Eindhoven, The Netherlands*

ANDREAS JACOBS (12), *Max-Planck Institute for Neurological Research, 50866 Cologne, Germany*

MALCOLM N. JONES (13), *School of Biological Sciences, University of Manchester, Manchester M33 2EU, United Kingdom*

KAMESWARI S. KONDURI (23), *Departments of Pediatrics and Physiology/Surgery, Zablocki Veterans Administration Medical Center, and Medical College of Wisconsin, Milwaukee, Wisconsin 53201*

KRYSTYNA KONOPKA (20), *Department of Microbiology, Arthur A. Dugoni School of Dentistry, University of the Pacific, San Francisco, California 94115*

ANTON I. P. M. DE KROON (6), *Department of Biochemistry of Membranes, Centre for Biomembranes and Lipid Enzymology, Institute of Biomembranes, Utrecht University, 3584 CH Utrecht, The Netherlands*

BEN DE KRUIJFF (6), *Department of Biochemistry of Membranes, Centre for Biomembranes and Lipid Enzymology, Institute of Biomembranes, Utrecht University, 3584 CH Utrecht, The Netherlands*

GABRIEL LOPEZ-BERESTEIN (17), *Section of Immunobiology and Drug Carriers, Department of Bioimmunotherapy, Division of Medicine, The University of Texas MD, Houston, Texas 77030*

[†]Deceased

THOMAS D. MADDEN (2, 4), *Inex Pharmaceuticals Corporation, Burnaby, British Columbia, Canada V5J 5L8*

M. BÁRBARA F. MARTINS (22), *Instituto Nacional de Engenharia e Technologia Industrial (INETI), Departmento de Biotecnologia, Unidade Novas Formas de Agentes Bioactivos, Estrada do Paco do Lumiar, 22, 1649-038 Lisbon, Portugal*

LAWRENCE MAYER (4), *Advanced Therapeutics—Medical Oncology, British Columbia Cancer Agency, Vancouver, British Columbia, Canada V5Z 1L3*

LAWRENCE D. MAYER (2), *Celator Technologies, Vancouver, British Columbia, Canada V5Z 1G1*

YUKIHIRO NAMBA (8), *Nippon Fine Chemical Co. Ltd., Takasago, Hyogo 676-0074, Japan*

SANDHYA NANDEDKAR (23), *Departments of Pediatrics and Physiology/Surgery, Zablocki Veterans Administration Medical Center, and Medical College of Wisconsin, Milwaukee, Wisconsin 53201*

AGATHA W. K. NG (17), *Division of Pharmaceutics and Biopharmaceutics, Faculty of Pharmaceutical Sciences, The University of British Columbia Vancouver British Columbia, Canada, V6T 1Z3*

NAOTO OKU (8, 9), *Department of Medical Biochemistry, School of Pharmaceutical Sciences, University of Shizuoka, Yada, Shizuoka 422-8526, Japan*

HAYAT ÖNYÜKSEL (21), *Departments of Pharmaceutics and Pharmacodynamics, University of Illinois at Chicago, Chicago, Illinois 60612*

MARIA C. PEDROSO DE LIMA (20), *Department of Biochemistry and Center for Neurosciences and Cell Biology, University of Coimbra, 3001-401 Coimbra, Portugal*

ELIZABETH PRETZER (20), *Department of Microbiology, University of the Pacific, 2155 Webster Street, San Francisco, California 94115*

REGINA C. RESZKA (12), *Max-Delbrück-Center for Molecular Medicine, G.O.T. Therapeutics GmbH, 13125 Berlin, Germany*

ISRAEL RUBINSTEIN (21), *Departments of Medicine, University of Illinois at Chicago, Chicago, Illinois 60612*

ISAM I. SALEM (15, 20), *Department of Pharmacy and Pharmaceutical Technology, Faculty of Pharmacy, University of Granada, 18071 Granada, Spain**

RAYMOND M. SCHIFFELERS (14), *Department of Pharmaceutics, Utrecht Institute for Pharmaceutical Sciences, Utrecht, The Netherlands*

HERBERT SCHOTT (3), *Institute of Organic Chemistry, Eberhard-Karls University, D-72076 Tuebingen, Germany*

RETO SCHWENDENER (3), *Molecular Cell Biology, Paul Scherrer Insitute, CH-5232 Villigen, Switzerland*

VARUN SETHI (21), *Departments of Pharmaceutics and Pharmacodynamics, University of Illinois at Chicago, Chicago, Illinois 60612*

SERGIO SIMÕES (20), *Laboratory of Pharmaceutical Technology, Faculty of Pharmacy, and Center for Neurosciences and Cell Biology, University of Coimbra, 3001-517 Coimbra, Portugal*

VLADIMIR SLEPUSHKIN (20), *Department of Microbiology, University of the Pacific, San Francisco, California 94115†*

*Present address: Department of Microbiology, Arthur A. Dugoni School of Dentistry, University of the Pacific, San Francisco, California 94115
†Present address: VIRxSYS, 200 Perry Parkway, suite 1A, Gaithersburg, Maryland 20877

RUTGER W. H. M. STAFFHORST (6), Department of Biochemistry of Membranes, Centre for Biomembranes and Lipid Enzymology, Institute of Biomembranes, Utrecht University, 3584 CH Utrecht, The Netherlands

GERHARD STEFFAN (20), Department of Microbiology, University of the Pacific, San Francisco, California 94545

GERT STORM (14), Department of Pharmaceutics, Utrecht Institute for Pharmaceutical Sciences, Utrecht, The Netherlands

ROBERT M. STRAUBINGER (5), Department of Pharmaceutical Sciences, University of Buffalo, State University of New York, Amherst, New York 14260-1200

JÜRGEN VOGES (12), Department of Stereotaxy and Functional Neurosurgery, University of Cologne, 50924 Cologne, Germany

YUE-FEN WANG (10), NeoPharm, Inc., Research and Development, Waukegan, Illinois 60085

KISHOR M. WASAN (17), Division of Pharmaceutics and Biopharmaceutics, Faculty of Pharmaceutical Sciences, The University of British Columbia, Vancouver British Columbia, Canada V6T 1Z3

DAWN W. WATERHOUSE (4), Advanced Therapeutics—Medical Oncology, British Columbia Cancer Agency, Vancouver, British Columbia, Canada V5Z 1L3

DAWN N. WATERHOUSE (2), Department of Advanced Therapeutics, British Columbia Cancer Agency, Vancouver, British Columbia, Canada V5Z 4E6

MURRAY S. WEBB (2), Director, Clinical Development, Panagin Pharmaceuticals Inc., Richmond, British Columbia, Canada V6Z 177

LEILA ZARIF (18), 1400 Route de Cannes, Villa Layalina, Valbonne 06560, France

Preface

The origins of liposome research can be traced to the work of Alec Bangham and colleagues in the mid 1960s. The description of lecithin dispersions as containing "spherulites composed of concentric lamellae" (A. D. Bangham and R. W. Horne, *J. Mol. Biol.* **8**, 660, 1964) was followed by the observation that "the diffusion of univalent cations and anions out of spontaneously formed liquid crystals of lecithin is remarkably similar to the diffusion of such ions across biological membranes" (A. D. Bangham, M. M. Standish, and J. C. Watkins, *J. Mol. Biol.* **13**, 238, 1965). Following the early studies on the biophysical characterization of multilamellar and unilamellar liposomes, investigators began to utilize liposomes as a well-defined model to understand the structure and function of biological membranes. It was also recognized by pioneers including Gregory Gregoriadis and Demetrios Papahadjopoulos that liposomes could be used as drug delivery vehicles. It is gratifying that their efforts and the work of those inspired by them have lead to the development of liposomal formulations of doxorubicin, daunorubicin and amphotericin B now utilized in the clinic.

The first chapter of this volume is a tribute to Alec Bangham. The rest of the volume focuses on therapeutic applications of liposomes, and comprises sections on liposomal anti-cancer agents, liposomal antibacterial, antifungal and antiviral agents, miscellaneous liposomal therapies, and electron microscopy of liposomes. I hope that these chapters will facilitate the work of graduate students, post-doctoral fellows, as well as established scientists entering liposome research or shifting their focus in the field. The previous volumes in this series have covered additional aspects of liposomology: Preparation and physicochemical characterization of liposomes (volume 367), the use of liposomes in biochemistry and in molecular cell biology (volume 372), liposomes in immunology, diagnostics, gene delivery and gene therapy (volume 373), antibody- or ligand-targeted liposomes, environment-sensitive liposomes, liposomal oligonucleotides, and liposomes *in vivo* (volume 387).

The areas represented in this volume are not exhaustive, but represent the major therapeutic applications of liposomes. I have tried to identify the experts in each area of liposome research, particularly those who have contributed to the field over some time. It is unfortunate that I was unable to convince some prominent investigators to contribute to the volume. Some invited contributors were not able to prepare their chapters, despite generous extensions of time. In some cases I may have inadvertently overlooked some experts in a particular area, and to these individuals, I extend my apologies. Their primary contributions to the field will, nevertheless, not go unnoticed in the citations in these volumes and in the hearts and minds of the many investigators in liposome research.

I wish to express my gratitude to all the colleagues who graciously contributed to these volumes. I also thank Shirley Light of Academic Press for her encouragement and initiation of this project and Cindy Minor of Elsevier Science and Christine Brandt of Kolam USA for their help and patience during the preparation of this volume. I am especially grateful to my wife Diana Flasher for her support and love during the seemingly endless editing process and to my dear children Avery and Maxine for their companionship, creativity, and love.

I would like to dedicate this volume to the mentors who have helped me during my early education in science, Drs. Cengiz Yalçın, Feriha Erman, Richard Burger, and Marcel Bastin at the Middle East Technical University in Ankara, and Drs. Shinpei Ohki and Robert Spangler at the State University of New York at Buffalo.

NEJAT DÜZGÜNEŞ

METHODS IN ENZYMOLOGY

VOLUME 72. Lipids (Part D)
Edited by JOHN M. LOWENSTEIN

VOLUME 73. Immunochemical Techniques (Part B)
Edited by JOHN J. LANGONE AND HELEN VAN VUNAKIS

VOLUME 74. Immunochemical Techniques (Part C)
Edited by JOHN J. LANGONE AND HELEN VAN VUNAKIS

VOLUME 75. Cumulative Subject Index Volumes XXXI, XXXII, XXXIV–LX
Edited by EDWARD A. DENNIS AND MARTHA G. DENNIS

VOLUME 76. Hemoglobins
Edited by ERALDO ANTONINI, LUIGI ROSSI-BERNARDI, AND EMILIA CHIANCONE

VOLUME 77. Detoxication and Drug Metabolism
Edited by WILLIAM B. JAKOBY

VOLUME 78. Interferons (Part A)
Edited by SIDNEY PESTKA

VOLUME 79. Interferons (Part B)
Edited by SIDNEY PESTKA

VOLUME 80. Proteolytic Enzymes (Part C)
Edited by LASZLO LORAND

VOLUME 81. Biomembranes (Part H: Visual Pigments and Purple Membranes, I)
Edited by LESTER PACKER

VOLUME 82. Structural and Contractile Proteins (Part A: Extracellular Matrix)
Edited by LEON W. CUNNINGHAM AND DIXIE W. FREDERIKSEN

VOLUME 83. Complex Carbohydrates (Part D)
Edited by VICTOR GINSBURG

VOLUME 84. Immunochemical Techniques (Part D: Selected Immunoassays)
Edited by JOHN J. LANGONE AND HELEN VAN VUNAKIS

VOLUME 85. Structural and Contractile Proteins (Part B: The Contractile Apparatus and the Cytoskeleton)
Edited by DIXIE W. FREDERIKSEN AND LEON W. CUNNINGHAM

VOLUME 86. Prostaglandins and Arachidonate Metabolites
Edited by WILLIAM E. M. LANDS AND WILLIAM L. SMITH

VOLUME 87. Enzyme Kinetics and Mechanism (Part C: Intermediates, Stereo-chemistry, and Rate Studies)
Edited by DANIEL L. PURICH

VOLUME 88. Biomembranes (Part I: Visual Pigments and Purple Membranes, II)
Edited by LESTER PACKER

VOLUME 89. Carbohydrate Metabolism (Part D)
Edited by WILLIS A. WOOD

Introduction: The Origins of Liposomes: Alec Bangham at Babraham

By NEJAT DÜZGÜNEŞ and GREGORY GREGORIADIS

The field of liposomology has its origins in the work of Alec Bangham and colleagues at the Agricultural Research Council Institute of Animal Physiology at Babraham, Cambridge in the mid-1960s.

In 1964, Bangham and Horne described electron microscopic observations of phosphatidylcholine (lecithin) or its mixtures with cholesterol dispersed in water and stained with a 2% solution of potassium phosphotungstate. The dispersions produced large and small "spherulites," whether produced by hand shaking or sonication, with the latter preparation having fewer large spherulites. The spherulites had a lamellar structure, with a lipid layer of 44.2 Å and a water layer of 25.6 Å. Treatment of lecithin spherulites with lysolecithin resulted in the beading of the lamellae to produce micelles of 70–80 Å.

The next year, Bangham *et al.* (1965) reported that "the diffusion of univalent cations and anions out of spontaneously formed liquid crystals of lecithin is remarkably similar to the diffusion of such ions across biological membranes." The bimolecular leaflet structure of the liquid crystal was shown to be orders of magnitude more permeable to anions than cations, and the diffusion rate of the cations was controlled by the sign and magnitude of the surface charge. The authors argued that "if the cation is sequestered in aqueous compartments between the bimolecular leaflets, and if the thickness of the aqueous compartments is determined by the surface charge density of the lipid head groups and by the ionic strength of the aqueous phase in accordance with the double-layer theory, the amount of cation trapped would also be expected to vary." Bangham *et al.* (1974) pointed out that it was not until the ionophore valinomycin was shown to facilitate selectively the diffusion of K^+ over Na^+ from liposomes containing equal concentrations of the ions (Bangham *et al.*, 1967) that the closed membrane theory (i.e., a lipid bilayer completely enclosing an aqueous space) could be claimed.

Bangham *et al.* (1974) have described some of the earlier work on ordered phase systems involving lipids and water that were not recognized as closed membranes, including a British patent application by J. Y. Johnson in 1932 for pharmaceutical preparations for injection combining "medicaments with liquids, such as fats or fatty oils, if necessary together with waxes or wax-like substances, with water, or other liquids,

METHODS IN ENZYMOLOGY, VOL. 391
0076-6879/05 $35.00

and a dispersing agent." The resulting "depot" would be "capable of holding any desired doses of the medicament but releasing it over any desired space of time only gradually ... without the slightest detriment to the organism."

Earlier, Bangham and Dawson had described the interaction of phospholipases with lecithin dispersions in buffer but termed the resulting bodies "emulsion particles" (Bangham and Dawson, 1958, 1959) in 1959 and "micelles" (Bangham and Dawson, 1962) in 1962, and were able to perform electrophoretic mobility measurements on these particles by direct observation. According to Bangham, "starting in November 1961, shortly after the arrival of the first EM at Babraham, Horne and I attempted to visualize dispersion of phospholipids in aqueous negative stain. Toward the end of 1962 we had persuaded ourselves that we were seeing minute sacs of approximately 50-nm diameter, the first 'lipid somes,' as we have come to know them" (Bangham, 1983).

Following the initial characterization of multilamellar liposomes, Papahadjopoulos and Miller (1967) described the structure of sonicated microvesicles, later to be known as "small unilamellar vesicles" (SUVs). Papahadjopoulos and Watkins (1967) showed the differential permeability of anions and cations through the SUV membrane, depending on the phospholipid composition. An excellent account of the biophysical characterization of liposomes can be found in Bangham *et al.* (1974). The ability of liposomes to sequester solutes formed the basis for the initiation in 1970 of work by Gregoriadis and Ryman on the use of the system in drug delivery (Gregoriadis, 1976a,b; Gregoriadis and Ryman, 1972). The evolution of liposomes, from the early years to the emergence of biotechnology companies producing therapeutic liposomes, has been reviewed by Bangham (1993, 1995).

The contributors to the *Methods in Enzymology* volumes on "Liposomes," as well as many others working in this field, are grateful to Alec Bangham for his discoveries, his subsequent work on the development and characterization of liposomes, and his inspiration to those who knew him.

References

Bangham, A. D. (1983). *In* "Liposome Letters" (A. D. Bangham, ed.), p. xiii. Academic Press, London.

Bangham, A. D. (1993). Liposomes: The Babraham connection. *Chem. Phys. Lipids* **64,** 275–285.

Bangham, A. D. (1995). Surrogate cells of Trojan horses. The discovery of liposomes. *Bioessays* **17,** 1081–1088.

Bangham, A. D., and Dawson, R. M. C. (1958). Control of lecithinase activity by the electrophoretic charge on its substrate surface. *Nature* **182,** 1292–1293.

Bangham, A. D., and Dawson, R. M. C. (1959). The relation between the activity of a lecithinase and the electrophoretic charge of the substrate. *Biochem. J.* **72,** 486–492.

Bangham, A. D., and Dawson, R. M. C. (1962). Electrokinetic requirements for the reaction between Cl. perfringens α-toxin (phospholipase C) and phospholipid substrates. *Biochim. Biophys. Acta* **59,** 103–115.

Bangham, A. D., and Horne, R. W. (1964). Negative staining of phospholipids and their structural modification by surface-active agents as observed in the electron microscope. *J. Mol. Biol.* **8,** 660–668.

Bangham, A. D., Hill, M. W., and Miller, N. G. A. (1974). Preparation and use of liposomes as models of biological membranes. *Methods Membr. Biol.* **1,** 1–68.

Bangham, A. D., Standish, M. M., and Watkins, J. C. (1965). Diffusion of univalent ions across the lamellae of swollen phospholipids. *J. Mol. Biol.* **13,** 238–252.

Bangham, A. D., Standish, M. M., and Watkins, J. C. (1967). The diffusion of ions from a phospholipid model membrane system. *Protoplasm* **63,** 183–187.

Gregoriadis, G. (1976a). The carrier potential of liposomes in biology and medicine (first of two parts). *N. Engl. J. Med.* **295,** 704–710.

Gregoriadis, G. (1976b). The carrier potential of liposomes in biology and medicine (second of two parts). *N. Engl. J. Med.* **295,** 765–770.

Gregoriadis, G., and Ryman, B. E. (1972). Fate of protein-containing liposomes injected into rats. An approach to the treatment of storage diseases. *Eur. J. Biochem.* **24,** 485–491.

Papahadjopoulos, D., and Miller, N. (1967). Phospholipid model membranes. I. Structural characteristics of hydrated liquid crystals. *Biochim. Biophys. Acta* **135,** 624–638.

Papahadjopoulos, D., and Watkins, J. C. (1967). Phospholipid model membranes. II. Permeability properties of hydrated liquid crystals *Biochim.. Biophys. Acta* **135,** 639–652.

Section I

Liposomal Anticancer Agents

[1] Entrapment of Small Molecules and Nucleic Acid–Based Drugs in Liposomes

By DAVID B. FENSKE and PIETER R. CULLIS

Abstract

In the past two decades there have been major advances in the development of liposomal drug delivery systems suitable for applications ranging from cancer chemotherapy to gene therapy. In general, an optimized system consists of liposomes with a diameter of ~100 nm that possess a long circulation lifetime (half-life >5 h). Such liposomes will circulate sufficiently long to take advantage of a phenomenon known as disease site targeting, wherein liposomes accumulate at sites of disease, such as tumors, as a result of the leaky vasculature and reduced blood flow exhibited by the diseased tissue. The extended circulation lifetime is achieved by the use of saturated lipids and cholesterol or by the presence of PEG-containing lipids. This chapter will focus on the methodology required for the generation of two very different classes of liposomal carrier systems: those containing conventional small molecular weight (usually anticancer) drugs and those containing larger genetic (oligonucleotide and plasmid DNA) drugs. Initially, we will examine the encapsulation of small, weakly basic drugs within liposomes in response to transmembrane pH and ion gradients. Procedures will be described for the formation of large unilamellar vesicles (LUVs) by extrusion methods and for loading anticancer drugs into LUVs in response to transmembrane pH gradients. Three methods for generating transmembrane pH gradients will be discussed: (1) the use of intravesicular citrate buffer, (2) the use of transmembrane ammonia gradients, and (3) ionophore-mediated generation of pH gradients via transmembrane ion gradients. We will also discuss the loading of doxorubicin into LUVs by formation of drug–metal ion complexes. Different approaches are required for encapsulating macromolecules within LUVs. Plasmid DNA can be encapsulated by a detergent-dialysis approach, giving rise to stabilized plasmid–lipid particles, vectors with potential for systemic gene delivery. Antisense oligonucleotides can be spontaneously entrapped upon electrostatic interaction with ethanol-destabilized cationic liposomes, giving rise to small multilamellar systems known as stabilized antisense–lipid particles (SALP). These vectors have the potential to regulate gene expression.

Introduction

It has now been over 35 years since it was discovered that vigorous dispersal of purified phospholipids in water resulted in the formation of microscopic closed membrane spheres (Bangham, 1968). These artificial membranes, referred to as liposomes, were found to consist of one or more lipid bilayers arranged concentrically about a central aqueous core. Studies on the membrane permeability of small molecules demonstrated that polar and charged molecules could be retained within liposomes, an observation that immediately suggested their potential as systems for the systemic delivery of drugs (Sessa and Weissmann, 1968). Unfortunately, a significant amount of technological development was required before this potential could be realized. In addition to a better understanding of the physical properties of membranes and their lipid components, techniques were required for the generation of unilamellar vesicles and encapsulation of drugs and macromolecules within them. Although a wide variety of methods were developed for the formation of liposomes (Hope *et al.*, 1986; Lichtenberg and Barenholz, 1988), many of them did not generate liposomes of optimal size and polydispersity and often were technically demanding and time consuming. Furthermore, the drug-loading technology at the time was based on passive entrapment methods, which resulted in low encapsulation levels (<30%) and poor retention of drugs (Mayer *et al.*, 1990a). Nevertheless, early animal studies using liposomal drug carriers were encouraging enough to warrant further development (see Mayer *et al.*, 1990a and references therein).

The development of extrusion technology for the rapid generation of monodisperse populations of unilamellar vesicles (Hope *et al.*, 1985; Mayer *et al.*, 1986b; Olson *et al.*, 1979) allowed characterization of the physical properties and *in vivo* characteristics of a wide variety of liposomal systems. This information revealed that optimized drug delivery systems would possess two key parameters: a small size (on the order of 100 nm) and long circulation lifetimes (half-life >5 h in mice). The basic structural framework on which most delivery systems are based is the large unilamellar vesicle (LUV) with a diameter close to 100 nm. These systems possess internal volumes large enough to carry adequate quantities of encapsulated material but are small enough to circulate for a time sufficient to reach sites of disease, such as tumors or sites of inflammation. Vesicles that are much larger or smaller are rapidly cleared from the circulation. However, circulation lifetime is determined by factors other than size. Both circulation lifetimes and drug retention are dependent on lipid composition and were found to be greatly enhanced in systems made from phosphatidylcholine (or sphingomyelin) and cholesterol (Mayer *et al.*,

1989, 1993; Webb *et al.*, 1995, 1998a). Further improvements in circulation longevity were achieved by the inclusion of ganglioside G_{M1} in the vesicle formulation (Boman *et al.*, 1994; Gabizon and Papahadjopoulos, 1988; Woodle *et al.*, 1994) or by grafting water-soluble polymers such as poly (ethylene glycol) (PEG) onto the vesicle surface, thereby generating what have come to be known as "stealth" liposomes (Allen, 1994, 1998; Allen *et al.*, 1991; Woodle *et al.*, 1994).

A major advance in the design of the first generation of drug transport systems came with the development of methods for achieving the encapsulation and retention of large quantities of drug within liposomal systems. Perhaps the most important insight in this area was the recognition that many chemotherapeutic drugs could be accumulated within vesicles in response to transmembrane pH gradients (ΔpH) (Cullis *et al.*, 1997; Madden *et al.*, 1990; Mayer *et al.*, 1986a). The ability of ΔpH to influence transmembrane distributions of certain weak acids and bases had long been recognized (see Cullis *et al.*, 1997 and references therein). The fact that many chemotherapeutics were weak bases led us to investigate the transport of these substances into liposomes in response to membrane potentials and ΔpH. Subsequent studies led to considerably broader applications involving the transport and accumulation of a wide variety of drugs, biogenic amines, amino acids, peptides, lipids, and ions in LUVs exhibiting a ΔpH (for a review, see Cullis *et al.*, 1997). Application of this technology has led to the development of several liposomal anticancer systems that exhibit improved therapeutic properties over free drug. Early studies (see Mayer *et al.*, 1990a and references therein) had shown that reduced side effects with equal or enhanced efficacy could be obtained in liposomal systems, despite low encapsulation levels and poor drug retention. This led to our initial efforts to develop a liposomal version of doxorubicin, the most commonly employed chemotherapeutic agent, which is active against a variety of ascitic and solid tumors, yet exhibits a variety of toxic side effects. The pH gradient approach (Mayer *et al.*, 1989, 1990a–c, 1993) was expected to provide significant improvements in overall efficacy due to high drug-to-lipid ratios and excellent retention observed both *in vitro* and *in vivo*. This has been realized in liposomal doxorubicin preparations that are currently either in advanced clinical trials (Cheung *et al.*, 1999; Chonn and Cullis, 1995) or have been approved by the U.S. FDA for clinical use (Muggia, 2001). Other liposomal doxorubicin formulations (Burstein *et al.*, 1999; Campos *et al.*, 2001; Coukell and Spencer, 1997; Gokhale *et al.*, 1996; Gordon *et al.*, 2000; Grunaug *et al.*, 1998; Israel *et al.*, 2000; Judson *et al.*, 2001; Northfelt *et al.*, 1998; Shields *et al.*, 2001) are in various Phase I or II clinical trials, often with promising results. A variety of other liposomal drugs are currently in preclinical or clinical development; these include vincristine (Gelmon *et al.*, 1999; Millar *et al.*, 1998; Tokudome *et al.*, 1996; Webb *et al.*, 1995, 1998a), mitoxantrone (Adlakha-Hutcheon *et al.*, 1999; Chang *et al.*, 1997; Lim *et al.*, 1997, 2000; Madden *et al.*, 1990), daunorubicin (Gill *et al.*, 1996; Madden *et al.*, 1990; Muggia, 2001; Pratt *et al.*, 1998), ciprofloxacin (Bakker-Woudenberg *et al.*, 2001; Webb *et al.*, 1998b), topotecan (Tardi *et al.*, 2000), and vinorelbine, to name a few. Of these, our group has been prominent in devising methods for the encapsulation of doxorubicin, vincristine, and ciprofloxacin.

Liposomal delivery systems are finally reaching a stage of development where significant advances can reasonably be expected in the short term. The first of the conventional drug carriers are reaching the market while new liposomal drugs are being developed and entered into clinical trials. These advances stem from the fact that the design features required of drug delivery systems that have systemic utility are becoming better defined. Based on the studies indicated above, we now know that liposomal systems that are small (diameter $\leq$ 100 nm) and that exhibit long circulation life-times (half-life $\geq$ 5 h in mice) following intravenous (iv) injection exhibit a remarkable property termed "disease site targeting" or "passive targeting" that results in large improvements in the amounts of drug arriving at the disease site. For example, liposomal vincristine formulations can deliver 50- to 100-fold higher amounts of drug to a tumor site relative to the free drug (Boman *et al.*, 1994; Mayer *et al.*, 1993; Webb *et al.*, 1995, 1998a). This can result in large increases in efficacy (Boman *et al.*, 1994). These improvements stem from the increased permeability of the vasculature at tumor sites (Brown and Giaccia, 1998; Dvorak *et al.*, 1988) or sites of inflammation, which results in preferential extravasation of small, long-circulating carriers in these regions.

The insights gleaned from conventional drug carriers have implications for the design of liposomal systems for the delivery of larger macromolecules. There is currently much interest in developing systemic vectors for the delivery of the therapeutic genetic drugs such as antisense oligonucleotides or plasmid DNA. To obtain appreciable amounts of a vector containing the antisense oligonucleotides or therapeutic gene to the site of disease, the vector must be stable, small, and long-circulating. Of course, the vector must also be accumulated by target cells, escape the endocytotic pathway, and be delivered to the nucleus.

Over the past 20 years, our laboratory has played a major role in the development of liposomal systems optimized for the delivery of both conventional drugs and, more recently, genetic drugs. Our early studies on the production of LUVs by extrusion led to the characterization of several liposomal drug delivery systems (Bally *et al.*, 1988; Boman *et al.*, 1993, 1994; Chonn and Cullis, 1995; Cullis *et al.*, 1997; Fenske *et al.*, 1998;

Hope and Wong, 1995; Madden *et al.*, 1990; Maurer-Spurej *et al.*, 1999; Mayer *et al.*, 1986a), the development of new approaches for the loading of drugs via generation of ΔpH (Fenske *et al.*, 1998; Maurer-Spurej *et al.*, 1999) or other ion gradients (Cheung *et al.*, 1998), and finally new methods for the encapsulation of antisense oligonucleotides (Maurer *et al.*, 2001; Semple *et al.*, 2000, 2001) and plasmid DNA (Fenske *et al.*, 2002; Maurer *et al.*, 2001; Mok *et al.*, 1999; Tam *et al.*, 2000; Wheeler *et al.*, 1999) within liposomes. In this chapter we will provide an overview of these methods, along with detailed descriptions of procedures for the encapsulation of both conventional and genetic drugs within liposomes.

Encapsulation of Small, Weakly Basic Drugs within LUVs in Response to Transmembrane pH and Ion Gradients

The Formation of LUVs by Extrusion Methods

Many research questions in membrane science, specifically those involving the dynamic properties of lipid bilayers, can be addressed using very basic model membrane systems, such as the multilamellar vesicle (MLV) formed spontaneously upon vigorous agitation of lipid–water mixtures. These large (1–10 μm) multilamellar liposomes are ideal for biophysical investigations of lipid dynamics and order using techniques such as flourescence, electron spin resonance (ESR), or broadband (^{2}H and ^{31}P) nuclear magnetic resonance (NMR). However, many properties of biological membranes, such as the presence of pH or ion gradients, cannot be adequately modeled using large, multilamellar systems. These kinds of studies require the use of unilamellar vesicles in the nanometer size range. In our case, investigations relating ion and pH gradients to lipid asymmetry (Cullis *et al.*, 1997, 2000) were the driving force for the development of extrusion technology. While it was clear that MLVs were not appropriate for such topics, it was also apparent that the methods available for the generation of unilamellar vesicles, which included dispersion of lipids from organic solvents (Batzri and Korn, 1973), sonication (Huang, 1969), detergent dialysis (Mimms *et al.*, 1981), and reverse-phase evaporation (Szoka and Papahadjopoulos, 1978), had serious drawbacks (Cullis, 2000). However, Papahadjopolous, Szoka, and co-workers (Olson *et al.*, 1979) had observed that sequential extrusion of MLVs through a series of filters of reducing pore size under low pressure gave rise to LUV systems. Further development of this method in our laboratory led to an approach involving direct extrusion of MLVs, at relatively high pressures (200–400 psi), through polycarbonate filters with a pore size ranging from

30 to 400 nm. This allowed generation of narrow, monodisperse vesicle populations with a narrow size distribution and diameters close to the chosen pore size (Fig. 1) (Hope *et al.*, 1985; Mayer *et al.*, 1986). The method is rapid and simple and can be performed for a wide variety of lipid compositions and temperatures. As it is necessary to extrude the lipid emulsions at temperatures 5–10° above the gel-to-liquid crystalline phase transition temperature, the system is manufactured so that it may be attached to a variable-temperature circulating water bath.

Initially, we will describe in some detail the formation of a 30 mM solution of 100 nm LUVs composed of distearoylphosphatidylcholine (DSPC)/cholesterol (Chol), a highly ordered lipid mixture that is frequently chosen for drug delivery applications due to its good circulation lifetime and drug retention properties.

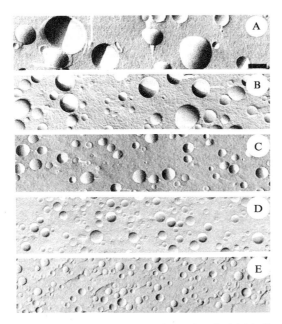

Fig. 1. Freeze-fracture electron micrographs of egg phosphatidylcholine LUVs prepared by extrusion through polycarbonate filters with pore sizes of (A) 400 nm, (B) 200 nm, (C) 100 nm, (D) 50 nm, and (E) 30 nm. The bar in (A) represents 150 nm. [Reprinted from Hope, M. J., Bally, M. B., Mayer, L. D., Janoff, A. S., and Cullis, P. R. (1986). *Chem. Phys. Lipids* **40**, 89–107, with permission.]

Experimental Procedure: Preparation of DSPC/Chol (55:45)
   LUVs by Extrusion

Stock solutions of DSPC and Chol in $CHCl_3$ should be prepared with concentrations ranging from 100 to 200 mM. Whenever possible, lipid concentrations should be verified by an appropriate assay. Phospholipid concentrations are verified using the assay of Fiske and Subbarow (1925) for the quantitative determination of inorganic phosphate, which is still in use today. The assay, which combines simplicity, accuracy, and reproducibility, is generally used for determining the concentration of phospholipid stock solutions or of LUV preparations. The assay is based on the ability of perchloric acid to liberate and oxidize phosphorus, forming phosphate ions, which form a colored complex that is quantified by absorbance spectrophotometry. The entire assay takes about 2 h, including preparation of standard curve and clean up. Detailed protocols for this assay have been described recently by Fenske *et al.* (2003) and will not be repeated here.

To prepare an LUV preparation with a final lipid concentration of 30 mM and a DSPC/Chol ratio of 55:45, DSPC (33 μmol) (Avanti Polar Lipids, Alabaster, AL; Northern Lipids, Vancouver, BC, Canada) and cholesterol (27 μmol) (Sigma-Aldrich, St. Louis, MO; Northern Lipids) are codissolved in a 13 × 100-mm test tube. If required, a trace of a radioactive tracer such as [^{3}H]cholesterol hexadecyl ether ([^{3}H]CHE) (PerkinElmer Life Sciences, Woodbridget, Ontario, Canada) is added to give a specific activity of 10–30 dpm/nmol lipid. The $CHCl_3$ is then mostly removed under a gentle stream of nitrogen gas while immersing the tube in hot water (50–60°), following which the lipid film is placed under high vacuum for a minimum of 1 h. The dry lipid film can be used immediately or stored in the freezer for later use.

For the formation of vesicles, the lipid film is hydrated with 2 ml of appropriate buffer (internal buffer or hydration buffer), such as 300 mM citrate pH 4.0 (pH gradient loading), 300 mM ammonium sulfate (for amine loading), or 300 mM $MgSO_4$, pH 6.5 (for ionophore loading) (these will all be discussed below). The lipid–water mixture is vortexed at 65° until a lipid emulsion is obtained and the lipid film can no longer be seen on the glass tube. This step must always be performed at a temperature approximately 10° higher than the gel-to-liquid crystalline phase transition temperature (T_m) of the phospholipid being used.

The lipid emulsion is then transferred to cryovial tubes for five cycles of freeze-thawing. The cryovials are immersed in liquid nitrogen for 3–5 min, then transferred to lukewarm water (for 1 min), and then to a water bath at 65°. After thawing completely, the emulsions are vortexed vigorously, and the freeze–thaw cycle is repeated four more times.

The extruder (Northern Lipids) is assembled with two polycarbonate filters (Nuclepore polycarbonate membranes; Whatman, Clifton, NJ) with a pore size of 0.1 μm and a diameter of 25 mm, and connected to a circulating waterbath equilibrated at 65°. The lipid emulsion is extruded 10 times through the filters under a pressure of approximately 400 psi. For larger LUVs (200–400 nm), lower pressures will be adequate (100–200 psi). After each pass, the sample is cycled back to the extruder. It is important to start at a low pressure and gradually increase until each pass takes less than 1 min.

The lipid concentration of the final LUVs is determined by a phosphate assay (Fenske *et al.*, 2003; Fiske and Subbarow, 1925) or by liquid scintillation counting.

Generation of pH Gradients via Internal Citrate Buffer

Early studies in our laboratory on membrane potentials and the uptake of weak bases used for the measurement of ΔpH led to the recognition that a variety of chemotherapeutic drugs could be accumulated within LUVs exhibiting transmembrane pH gradients (Cullis, 2000). This "remote-loading" technique, so named because drug is loaded into preformed vesicles, is based on the membrane permeability of the neutral form of weakly basic drugs such as doxorubicin. When doxorubicin ($pK_a = 8.6$) is incubated at neutral pH in the presence of LUVs exhibiting a ΔpH (interior acidic), the neutral form of the drug will diffuse down its concentration gradient into the LUV interior, where it will be subsequently protonated and trapped (the charged form is membrane impermeable). As long as the internal buffer (300 mM citrate pH 4) is able to maintain the ΔpH, diffusion of neutral drug will continue until either all the drug has been taken up or the buffering capacity of the vesicle interior has been overwhelmed. This process is illustrated in Fig. 2 for the uptake of doxorubicin into egg phosphatidylcholine (EPC)/Chol and EPC LUVs, where it is seen that uptake is dependent on time, temperature, and lipid composition (Mayer *et al.*, 1986a). If conditions are chosen correctly, high drug-to-lipid ratios can be achieved (D/L = 0.2 mol:mol) with high trapping efficiencies (98% and higher) and excellent drug retention. A diagrammatic illustration of this process is given in Fig. 3A (also see inset). Interestingly, much higher levels of doxorubicin can be loaded than would be predicted on the basis of the magnitude of ΔpH (Cullis *et al.*, 1997; Harrigan *et al.*, 1993). This would appear to result from the formation of doxorubicin precipitates within the LUV interior, which provides an additional driving force for accumulation (Lasic *et al.*, 1992, 1995). Doxorubicin forms fibrous precipitates that are aggregated into bundles by citrate (Li *et al.*, 1998) or sulfate (Lasic *et al.*,

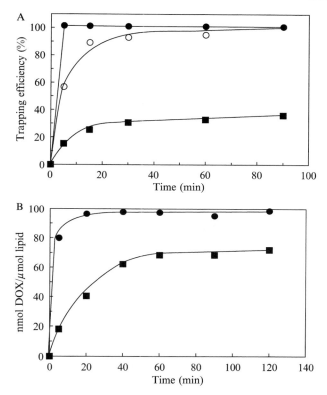

Fig. 2. (A) Effect of incubation temperature on uptake of doxorubicin into 200 nm EPC/cholesterol (55:45 mol/mol) LUVs exhibiting a transmembrane pH gradient (pH 4 inside, 7.8 outside). Doxorubicin was added to LUVs (D/L = 0.3 wt:wt) equilibrated at 21(■), 37(○), and 60°. [Reprinted from Mayer, L. D., Tai, L. C., Ko, D. S., Masin, D., Ginsberg, R. S., Cullis, P. R., and Bally, M. B. (1989). *Cancer Res.* **49**, 5922–5930, with permission.] (B) Effect of cholesterol on the uptake of doxorubicin at 20 into 100 nm LUVs exhibiting a transmembrane pH gradient (pH 4.6 inside, 7.5 outside). Lipid compositions were EPC (■) and EPC/cholesterol (1:1 mol/mol) (●). The initial D/L ratio was 100 nmol/μmol. [Reprinted from Mayer, L. D., Bally, M. B., and Cullis, P. R. (1986a). *Biochim. Biophys. Acta* **857**, 123–126, with permission.]

1992, 1995) counteranions, which affect the rate of doxorubicin release from LUVs (Li *et al.*, 2000). These precipitates can be visualized by cryoelectron microscopy, where they are seen to give the LUVs a "coffee-bean" appearance (Fig. 4). This has been corroborated by recent observations that very high levels of uptake can be achieved in the absence of a pH gradient by the formation of doxorubicin–Mn^{2+} complexes (Cheung *et al.*, 1998), as will be described below.

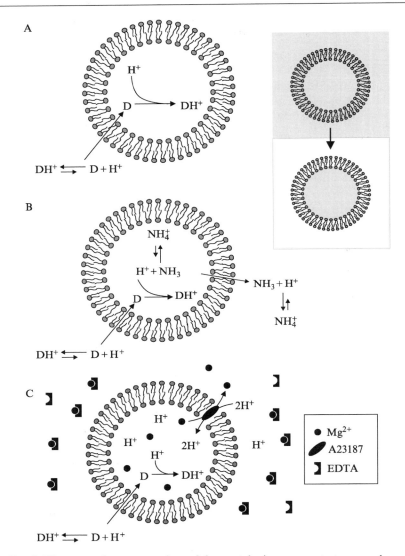

Fig. 3. Diagrammatic representations of drug uptake in response to transmembrane pH gradients. Prior to drug loading, it is necessary to establish the primary pH gradient or the primary ion gradient that will generate a ΔpH. Lipid films or powders are initially hydrated and then extruded in the internal (or hydration) buffer, giving rise to a vesicle solution in which both the external and internal solutions are the same, as indicated by the gray shading in the upper frame of the insert (top right). The vesicles are then passed down a gel exclusion column (Sephadex G-50) hydrated in the external buffer, giving rise to vesicles with a pH or ion gradient (lower frame of insert). (A) The standard pH gradient method. The internal

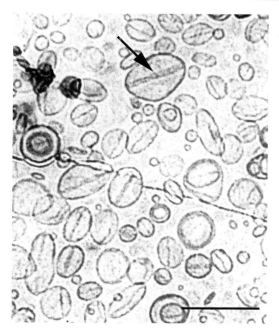

FIG. 4. Cryoelectron micrograph of 100 nm EPC/cholesterol LUVs containing doxorubicin (D/L = 0.3 mol/mol) loaded in response to a transmembrane pH gradient (inside acidic). The arrow indicates the presence of precipitated doxorubicin fibers within the LUV, which gives the vesicle the appearance of a "coffee-bean." [Reprinted from Fenske, D. B., Maurer, N., and Cullis, P. R. (2003). *In* "Liposomes: A Practical Approach" (V. P. Torchilin and V. Weissig, eds.). Oxford University Press, Oxford, with permission.]

buffer is 300 m*M* citrate, pH 4, and the external buffer is 20 m*M* HEPES, 150 m*M* NaCl, pH 7.5. The precipitation of certain drugs such as doxorubicin, which provides an addition driving force for uptake, is not indicated in the figure. (B) A second method for generating ΔpH involves the initial formation of a transmembrane gradient of ammonium sulfate, which leads to an acidified vesicle interior as neutral ammonia leaks from the vesicles. Here, the internal buffer is 300 m*M* ammonium sulfate, and the external buffer is 150 m*M* NaCl. Possible drug precipitation is not indicated. (C) Transmembrane pH gradients can also be established by ionophores (such as A23187) in response to transmembrane ion gradients (e.g., Mg^{2+}, represented as solid circles). A23187 couples the external transport of one Mg^{2+} ion (down its concentration gradient) to the internal transport of two protons, resulting in acidification of the vesicle interior. An external chelator such as EDTA is required to bind Mg^{2+} ions as they are transported out of the vesicle. Other divalent cations such as Mn^{2+} can also be used. See text for further details.

The experimental procedure described below for the accumulation of doxorubicin within DSPC/Chol LUVs represents our "basic" pH gradient method for drug loading. This basic system can be used for the uptake of a wide variety of drugs (Madden *et al.*, 1990), and all the remote-loading methods that follow are based on similar principles involving the generation of ΔpH, even though this may not always be immediately obvious.

Experimental Procedure: Remote Loading of Doxorubicin into DSPC/Chol (55:45) LUVs

DSPC/Chol (55:45) LUVs (diameter = 100 nm) are prepared as described previously ([Lipid] = 30 mM, volume = 2 ml), using 300 mM citrate, pH 4.0, as the hydration buffer. There are two common methods for formation of the pH gradient, the first of which is passing the LUVs down a column of Sephadex G-50 (Amersham Pharmacia Biotech) equilibrated in HEPES-buffered saline (HBS) (20 mM HEPES, 150 mM NaCl, pH 7.5) (the method described below), and the second of which involves the use of spin columns.

Spin columns permit the rapid separation of LUVs from their hydration buffer (or from unencapsulated drug) on a hydrated gel at low centrifugation speeds. They are particularly useful for monitoring drug uptake with time (as described below). On the day prior to drug loading, a slurry of Sephadex G-50 in HBS is prepared by adding a small volume (2–3 ml) of dry G-50 powder to 200–300 ml HBS with frequent swirling. Small quantities of gel are added as necessary until the settled G-50 occupies about half the aqueous volume. The hydrated gel is allowed to swell overnight. To prepare the spin columns, a tiny plug of glass wool is packed into the end of a 1-ml disposable syringe (without the needle), which is then placed in a 13 × 100-mm glass test tube. The G-50 slurry is swirled, and the syringes are immediately filled using a Pasteur pipette. The syringes (in test tubes) are then placed in a desktop centrifuge, and the gel is packed (to 0.6 or 0.7 ml) by bringing the speed to 2000 rpm (670g) momentarily. More G-50 slurry is added, and the centrifugation is repeated. When finished, the moist G-50 bed should be 0.9–1.0 ml. The spin columns are covered with parafilm to prevent drying and are used within an 8-h period.

Transmembrane pH gradients are formed by running an aliquot (200 μl) of the LUVs down a column (1.5 × 15 cm) of Sephadex G-50 eluted in HBS. The LUV fractions, which will elute at the void volume and are visible to the eye, are collected and pooled. The final volume will be ~2 ml, and the lipid concentration will be around 15 mM. Alternatively, the pH gradient can be formed using spin columns prepared in HBS (spin 4 × 100 μl) and pooling the fractions.

Doxorubicin (Sigma-Aldrich) is often loaded at a drug-to-lipid (D/L) ratio of 0.2 mol:mol. A doxorubicin standard solution is prepared by dissolving 1.0 mg of drug in 0.5 ml of saline (150 mM NaCl). The concentration is verified on the spectrophotometer using the doxorubicin extinction coefficient $\varepsilon = 1.06 \times 10^4\ M^{-1}\ cm^{-1}$ (Rottenberg, 1979). Aliquots of lipid (5 μmol) and doxorubicin (1 μmol, approximately 0.5 mg) are combined in a glass test tube (or plastic Eppendorf tube) with HBS to give a final volume of 1 ml (5 mM lipid concentration). Drug uptake occurs during a 30-min incubation at 65°. This is verified at appropriate time points (0, 5, 15, 30 min) by applying an aliquot (50–100 μl) to a spin column and centrifuging at 2000 rpm for 2 min. LUVs containing entrapped drug will elute off the column, while free doxorubicin will be trapped in the gel. An aliquot (50 μl) of the initial lipid–drug mixture is saved for determination of the initial D/L ratio.

The initial mixture and each time point are then assayed for doxorubicin and lipid. Lipid concentrations can be quantified by the phosphate assay (see above), by liquid scintillation counting of an appropriate radiolabel, or by measuring the fluorescence of LUVs containing 0.5–1 mol% of rhodamine-PE (see below). Doxorubicin is quantified by an absorbance assay (see below). The percent uptake at any time point (e.g., $t = 30$ min) is determined by % uptake = $[(D/L)_{t\ =\ 30\ min}] \times 100 / [(D/L)_{initial}]$. Doxorubicin can be assayed by both a fluorescence assay and an absorbance assay, but we find the latter to be more accurate. The standard curve consists of four or five cuvettes containing 0–150 nmol doxorubicin in a volume of 0.1 ml; samples to be assayed are the same volume. To each tube is added 0.9 ml of 1% (v/v in water) Triton X-100 solution. For saturated lipid systems such as DSPC/Chol, the tubes should be heated in a boiling waterbath for 10–15 s, until the detergent turns cloudy. The samples are allowed to cool, and the absorbance is read at 480 nm on a UV/visible spectrophotometer.

Generation of pH Gradients via Transmembrane Ammonia Gradients

Despite its successful application in several drug delivery systems (Madden et al., 1990), the pH gradient approach utilizing internal citrate buffer does not provide adequate uptake of all weakly basic drugs. A case in point is the antibiotic ciprofloxacin, a commercially successful, quinolone antibiotic widely used in the treatment of respiratory and urinary tract infections (Hope and Wong, 1995). Ciprofloxacin is a zwitterionic compound that is charged and soluble under acidic and alkaline conditions but is neutral and poorly soluble in the physiological pH range, precisely the external conditions of most drug-loading techniques. This low solubility results in low levels of uptake (<20%) when the drug is loaded using the standard citrate technique. However, high levels of encapsulation can be achieved using an alternate ΔpH-loading method that is based on transmembrane gradients of ammonium sulfate (Haran *et al.*, 1993; Hope and Wong, 1995; Lasic *et al.*, 1995). If a given drug is incubated with LUVs containing internal ammonium sulfate in an unbuffered external saline solution, a small quantity of neutral ammonia will diffuse out of the vesicle, creating an unbuffered acidic interior with a pH ~2.7 (Hope and Wong, 1995). Any neutral drug that diffuses into the vesicle interior will consume a proton and become charged and therefore trapped. If continued drug uptake leads to consumption of the available protons, the diffusion of additional neutral ammonia from the vesicles will create more protons to drive drug uptake. This will continue until all the drug has been loaded or until the internal proton supply is depleted, leaving a final internal pH of about 5.1 (Hope and Wong, 1995). The technique is ideal for ciprofloxacin as the drug is supplied as an HCl salt, and thus is acidic and soluble when dissolved in water. In addition, the amine method results in higher uptake levels for other drugs such as doxorubicin. A diagrammatic scheme of the method is given in Fig. 3B.

This technique has been applied to a variety of drugs including doxorubicin (Haran *et al.*, 1993; Lasic *et al.*, 1992, 1995; Maurer-Spurej *et al.*, 1999), epirubicin (Maurer-Spurej *et al.*, 1999), ciprofloxacin (Hope and Wong, 1995; Lasic *et al.*, 1995; Maurer *et al.*, 1998; Maurer-Spurej *et al.*, 1999), and vincristine (Maurer-Spurej *et al.*, 1999). The method is equally effective using a range of alkylammonium salts (e.g., methylammonium sulfate, propylammonium sulfate, or amylammonium sulfate) to drive uptake (Maurer-Spurej *et al.*, 1999). Some drugs, such as doxorubicin, precipitate and form a gel in the vesicle interior (Lasic *et al.*, 1992, 1995; Li *et al.*, 2000), while others, such as ciprofloxacin, do not (Lasic *et al.*, 1995; Maurer *et al.*, 1998). Ciprofloxacin can reach intraliposomal concentrations as high as 300 mM, and while the drug does form small stacks as shown by ^{1}H-NMR, it does not form large precipitates (Maurer *et al.*, 1998), even though its solubility in buffer cannot exceed 5 mM. As a result of the lack of precipitation and rapid exchange properties, ciprofloxacin can respond rapidly to changes in electrochemical equilibria, such as depletion of the pH gradient. This explains the observed rapid leakage of ciprofloxacin from LUVs in response to serum destabilization or loss of pH gradient. In contrast, doxorubicin, which is known to form insoluble precipitates within LUVs in the presence of both citrate (Li *et al.*, 1998, 2000) and ammonium sulfate (Lasic *et al.*, 1992, 1995), is retained within vesicles in the presence of serum. This is a clear illustration of how the physical state of encapsulated drugs will affect retention and therefore may impact efficacy.

The experimental procedure below describes the uptake of ciprofloxacin into sphingomyelin (SPM)/Chol LUVs. Drug delivery vehicles prepared from SPM/Chol often exhibit greater efficacy than those prepared from DSPC/Chol (Webb *et al.*, 1995). Included is a description of the Bligh–Dyer extraction procedure (Bligh and Dyer, 1959), which involves partitioning the lipid and water-soluble drug into organic solvent and aqueous layers, respectively. This is necessary as lipid interferes with the ciprofloxacin assay.

Experimental Procedure: Remote Loading of Ciprofloxacin into SPM/Chol LUVs

When preparing radiolabeled lipid stock solutions, it is often convenient to prepare a large batch for determination of specific activity and then divide that batch into smaller aliquots for later use. For the preparation of 10 lipid samples of 50 μmol each, 275 μmol egg sphingomyelin (193 mg) (Avanti Polar Lipids; Northern Lipids; Sigma Aldrich) is codissolved with 225 μmol Chol (87 mg) in 7 ml t-butanol. If desired, 12 μCi of [^{3}H]cholesteryl hexadecyl ether (CHE) or [^{14}C]CHE is added. The specific activity (in dpm/μmol lipid) of the lipid mixture is easily obtained by measuring the activity of an aliquot by liquid scintillation counting, and measuring SPM concentration via the phosphate assay (Fenske *et al.*, 2003; Fiske and Subbarow, 1925), removing t-BuOH by lyophilization prior to assay (freeze tubes in liquid nitrogen and place under high vacuum for 30 min). The lipid mixture is aliquoted into 10 glass Pyrex tubes (50 μmol lipid per tube), frozen in liquid nitrogen, and lyophilized for at least 4 h. SPM/Chol (55:45) LUVs (diameter = 100 nm) are prepared using 50 μmol lipid as described above ([Lipid] = 25 mM, volume = 2 ml). The hydration solution is 300 mM ammonium sulfate [(NH$_4$)$_2$SO$_4$]. Spin columns are prepared (using saline rather than HBS) for monitoring drug uptake with time.

The amine gradient is formed by running an aliquot (200 μl) of the LUVs down a column (1.5 $\times$ 15 cm) of Sephadex G-50 eluted in saline (150 mM NaCl), as described above. Alternatively, the gradient could be formed using spin columns (spin 2 $\times$ 100 μl) and pooling the fractions.

Ciprofloxacin (Bayer Corporation) is often loaded at a drug-to-lipid (D/L) ratio of 0.3 mol:mol. After preparation of a ciprofloxacin standard solution (4 mM in water), 5 μmol of lipid and 1.5 μmol of ciprofloxacin are pipetted into a glass test tube (or plastic Eppendorf tube), adding saline to give a final volume of 1 ml (5 mM lipid concentration). This solution is incubated at 65° for 30 min, with aliquots (50–100 μl) withdrawn at appropriate time points (0, 5, 15, 30 min) and applied to a spin column [centrifuge at 2000 rpm (670g) for 2 min]. An aliquot (50 μl) of the initial lipid–drug mixture is saved for determination of initial D/L.

The initial mixture and each time point are then assayed for ciprofloxa-cin and lipid. Lipid can be quantified using the phosphate assay (Fenske *et al.*, 2003; Fiske and Subbarow, 1925) or by liquid scintillation counting. Ciprofloxacin is quantified by an absorbance assay following removal of drug from lipid by a Bligh–Dyer extraction procedure (Bligh and Dyer, 1959) (see below). The percent uptake is determined as previously described.

To perform the ciprofloxacin assay, a standard curve is prepared con-sisting of six glass test tubes containing 0, 50, 100, 150, 200, and 250 nmol ciprofloxacin (in water). The volume is made up to 1 ml with 200 mM NaOH. For the blank, 1 ml of 200 mM NaOH is used. Each LUV sample to be assayed should contain <250 nmol ciprofloxacin in a volume of 1 ml.

To each standard and assay sample, 2.1 ml of methanol and 1 ml of chloroform are added and vortexed gently. Only one phase should be present (if two phases form, 0.1 ml methanol is added and the solution is vortexed again). One milliliter of 200 mM NaOH and 1 ml chloroform are then added to each tube, which is then vortexed at high speed. Two phases should form, an aqueous phase containing the ciprofloxacin with a volume of 4.1 ml (top) and an organic phase containing the lipid with a volume of 2 ml (bottom). If a clean separation is not obtained, the tubes can be centrifuged at 2000 rpm for 2 min in a desktop centrifuge.

After carefully removing the aqueous phase, the absorbance at 273.5 nm is read to obtain nmol ciprofloxacin present in the original sample volume (in μl), thereby yielding the sample drug concentration (mM).

Ionophore-Mediated Generation of pH Gradients via Transmembrane Ion Gradients

The observation that improved remote loading of ciprofloxacin could be achieved using ammonium sulfate solutions rather than sodium citrate buffers highlighted the need for further investigation and development of drug-loading methodologies. In this section and the next we examine two approaches that utilize transmembrane ion gradients to generate the driving force for uptake. In the first method, uptake is driven by a second-ary pH gradient that is generated through the action of ionophores re-sponding to a primary ion gradient (involving Na^+, Mn^{2+}, or Mg^{2+}). In the second method, drug uptake is driven by the formation of intravesicular drug–ion complexes and does not involve a ΔpH at all. These will be discussed in turn below.

Recently, we have developed a new method of remote loading that is based on the ionophore-mediated generation of a secondary pH gradient in response to transmembrane gradients of monovalent and divalent cations

(Fenske *et al.*, 1998). The process is diagrammed in Fig. 3C. A primary ion gradient is generated when LUVs formed by extrusion in K_2SO_4, $MnSO_4$, or $MgSO_4$ solutions are passed down a column equilibrated in a sucrose-containing buffer. The use of sulfate salts is important, as chloride ion can dissipate pH gradients by forming neutral HCl that can diffuse out of the vesicle. Likewise, sucrose is chosen as a component of the external buffer rather than saline, as chloride ion can interfere with some ionophores (Wheeler *et al.*, 1994). After establishing the primary ion gradient, the drug (which to date includes doxorubicin, mitoxantrone, ciprofloxacin, and vincristine) is added. If the LUVs contain a potassium salt, the ionophore nigericin is added, whereas if the LUVs contain either Mn^{2+} or Mg^{2+}, the ionophore A23187 and the chelator ethylenediaminetetraacetic acid (EDTA) are used. Under the current conditions, nigericin couples the outward flow of a potassium ion (down its concentration gradient) to the inward flow of a proton. Likewise, A23187 couples the outward flow of a single divalent cation to the inward flow of a pair of protons. In both cases, ionophore-mediated ion transport is electrically neutral and results in acidification of the vesicle interior, thereby creating a pH gradient that drives drug uptake. For systems containing divalent cations, EDTA chelates manganese and magnesium as they are transported out of the vesicles and is required to drive drug uptake. Both ionophore methods result in high levels of encapsulation for the drugs ciprofloxacin and vincristine (80–90%) and excellent *in vitro* retention (Fenske *et al.*, 1998). However, the A23187-loaded systems exhibit excellent *in vivo* circulation and drug retention properties that are comparable to systems loaded by the citrate or amine methods, whereas the nigericin-loaded systems do not.

Experimental Procedure: Ionophore-Mediated Loading of Vincristine into SPM/Chol LUVs

LUVs (diameter = 100 nm) are prepared from SPM/Chol (55:45) using 50 μmol lipid as described above ([Lipid] = 25 mM). It has been shown that liposomal vincristine prepared from SPM/Chol exhibits greater efficacy than systems prepared from DSPC/Chol (Webb *et al.*, 1995). The hydration solution is 300 mM $MnSO_4$ (2 ml).

Spin columns for monitoring drug uptake with time are hydrated in 300 mM sucrose. A Mn^{2+} gradient is formed by running an aliquot (200 μl) of the LUVs down a column (1.5 × 15 cm) of Sephadex G-50 eluted in HEPES-buffered sucrose (20 mM HEPES, 300 mM sucrose, pH 7.5) containing 3 mM Na_2EDTA. Alternatively, the gradient can be formed using spin columns prepared in the same buffer (spin 2 × 100 μl) and pooling the fractions.

Vincristine sulfate (commercially available from Eli Lilly Canada at 1 mg/ml) is added to give a D/L ratio of 0.03 mol:mol. Five micromoles of lipid and 0.15 μmol of vincristine are combined in a glass test tube, with addition of HEPES-buffered sucrose containing 3 mM Na$_2$EDTA to give a final volume of 1 ml (5 mM lipid concentration). The mixture is incubated at 60° for 10 min, saving an aliquot (50 μl) for determination of the initial D/L ratio. An aliquot (50–100 μl) is applied to a spin column to assess any drug uptake prior to addition of ionophore. There may be a small amount of uptake if the salt solution is acidic (as MnSO$_4$ solutions tend to be), but the majority of uptake will begin with addition of the ionophore.

At time $t = 0$, the A23187 (Sigma-Aldrich) (in ethanol) is added in a volume of approximately 5 μl to give a concentration of 0.1 μg/μmol lipid. If necessary, the concentration of A23187 can be increased 10 fold. Aliquots (100 μl) are withdrawn at appropriate time points (0, 5, 15, 30 min) and applied to spin columns (2000 rpm for 2 min). The initial mixture and each time point are then assayed for vincristine and lipid. Lipid can be quantified as described above. Vincristine is quantified by an absorbance assay. A standard curve is prepared consisting of four or five tubes containing 0–150 nmol vincristine in a volume of 0.2 ml. After addition of 0.8 ml of 95% ethanol, the absorbance of the standard curve and assay samples are read at 295 nm.

A representative uptake experiment is shown in Fig. 5, where the effects of ionophore, external pH, and EDTA are apparent. Under the

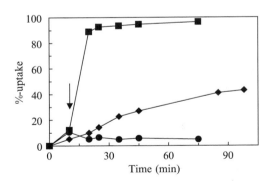

Fig. 5. Effect of external pH and EDTA on the uptake of vincristine in 100 nm SPM/Chol LUVs containing 300 mM MnSO$_4$. The external medium was 20 mM HEPES, 300 mM sucrose, 3 mM EDTA, pH 7.5 (■), 300 mM sucrose, 3 mM EDTA, pH 6 (♦), or 20 mM HEPES, 300 mM sucrose, pH 7.5 (●). The addition of A23187 (0.1 μg/μmol lipid) is indicated by the arrow. The uptake temperature was 60°, and the initial D/L ratio was 0.03 (mol:mol). [Reprinted from Fenske, D. B., Wong, K. F., Maurer, E., Maurer, N., Leenhouts, J. M., Boman, N., Amankwa, L., and Cullis, P. R. (1998). *Biochim. Biophys. Acta* **1414**, 188–204, with permission.]

conditions described above, >95% uptake is observed within 20 min of addition of ionophore. However, if the external pH is reduced to 6.0, the uptake reaches 40% only after 90 min of incubation. If EDTA is absent, no uptake occurs.

If the toxicity of Mn^{2+} is of concern, equally effective loading can be achieved using $MgSO_4$ solutions. In this case, the lipids are hydrated with 300 mM $MgSO_4$, pH 6.5, and the Mg^{2+} gradient is formed on a Sephadex column eluted with 20 mM HEPES, 300 mM sucrose, pH 6.0, containing 15 mM Na_2EDTA. Due to the lower affinity of Mg^{2+} for the ionophore relative to Mn^{2+}, a higher concentration of A23187 is required, usually in the range of 0.5–1 $\mu g/\mu mol$ lipid.

Encapsulation of Doxorubicin within LUVs in Response to Transmembrane Manganese Gradients

During our studies on ionophore loading, it was observed that doxorubicin uptake could occur in the presence of a $MnSO_4$ gradient but in the *absence* of a pH gradient. The driving force for drug uptake was found to be the formation of a doxorubicin–Mn^{2+} complex, which can be observed by eye as the characteristic red color of doxorubicin changes to an intense purple and which is monitored by absorption spectroscopy by a shift in absorbance maximum from 480 to 580 nm (Cheung *et al.*, 1998). The method was very efficient, allowing D/L ratios of 0.5 (mol:mol) to be achieved with 100% efficiency and ratios as high as 0.8 at lower loading efficiencies. Interestingly, the formation of Mn^{2+}–drug complexes does not appear to play a role in A23187-mediated uptake of doxorubicin, as no evidence of complex formation was observed (Cheung *et al.*, 1998). This formulation may have pharmaceutical potential in that complexes of doxorubicin with other cations (such as Fe^{3+} and Cu^{2+}) have been found to increase drug cytotoxicity (Gutteridge, 1984; Gutteridge and Quinlan, 1985; Hasinoff, 1989a,b; Hasinoff *et al.*, 1989; Muindi *et al.*, 1984). In addition, the high D/L ratios achievable will permit a higher payload of drug to be delivered to cells. As the methodology is similar to that of ionophore loading, only a brief description will be provided here.

Experimental Procedure: Manganese-Mediated Loading of Doxorubicin into SPM/Chol LUVs

LUVs composed of SPM/Chol (55:45 mol:mol), containing a trace of [^{3}H]cholesterol hexadecyl ether, are prepared as described above using lipid mixtures lyophilized from *t*-butanol (lyophilization guards against phase separation during solvent evaporation). The hydration solution is 300 mM $MnSO_4$ prepared in 30 mM HEPES, pH 7.4. A transmembrane

Mn^{2+} gradient is formed using spin columns equilibrated in 300 mM sucrose, 30 mM HEPES, pH 7.4, and drug uptake is accomplished by addition of doxorubicin (D/L ratio = 0.5 mol:mol) followed by incubation at 60° for 70 min. Drug uptake is monitored as described above.

Encapsulation of Genetic Drugs within LUVs: Long-Circulating Vectors for the Systemic Delivery of Genes and Antisense Oligonucleotides

The development of genetic drugs, such as antisense oligonucleotides or plasmid DNA carrying a therapeutic gene, holds great promise in the treatment of acquired diseases such as cancer and inflammation. Much current effort is directed at development of gene delivery vehicles capable of accessing distal disease sites following systemic (intravenous) administration. While numerous methods exist for effective *in vitro* gene delivery, the situation *in vivo* is very different. Many current systems, such as viral vectors, lipoplexes, and lipopolyplexes, have limited utility for systemic applications (Fenske *et al.*, 2001). Viral vectors cannot carry plasmids that exceed a certain size and often elicit strong immune responses. Lipoplexes and lipopolyplexes tend to be cleared rapidly from the circulation due to their large size and positive charge characteristics and suffer from toxicity issues (see Fenske *et al.*, 2001, and references therein). Unfortunately, liposomal carriers of genetic drugs that possess the optimized characteristics of conventional drug carriers (small size, serum stability, and long circulation lifetimes) have been difficult to achieve (Maurer *et al.*, 1999) due to the technical challenges involved in encapsulating large, highly charged molecules within relatively small vesicles. Recently, though, we have developed two very different methods for the generation of liposomes capable of carrying either antisense oligonucleotides or plasmid DNA. The first of these employs a detergent-dialysis approach for the formation of liposomal DNA carriers known as stabilized plasmid–lipid particles (SPLP) (Fenske *et al.*, 2002; Tam *et al.*, 2000; Wheeler *et al.*, 1999). The second method involves entrapping polynucleotides, via electrostatic interactions, within preformed ethanol-destabilized cationic liposomes (Maurer *et al.*, 2001; Semple *et al.*, 2000, 2001). Both methods are discussed below.

Stabilized Plasmid–Lipid Particles (SPLP): Vectors for Systemic Gene Delivery

SPLP are small (∼70 nm), monodisperse particles consisting of a single plasmid encapsulated in a unilamellar lipid vesicle. The lipid composition can be varied and can include additional components, but the basic system is composed of DOPE, a cationic lipid [usually *N,N*-dioleoyl-*N*,

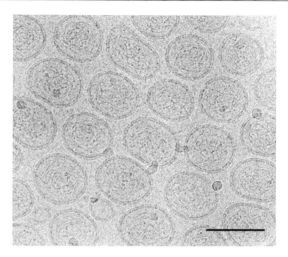

FIG. 6. Cryoelectron micrographs of purified SPLP prepared from DOPE:DODAC:PEG-CerC$_{20}$ (83:7:10; mol:mol:mol) and pCMVluc as described in the text. The bar indicates 100 nm. [Reprinted from Tam, P., Monck, M., Lee, D., Ludkovski, O., Leng, E. C., Clow, K., Stark, H., Scherrer, P., Graham, R. W., and Cullis, P. R. (2000). *Gene Ther.* **7,** 1867–1874, with permission.]

N-dimethylammonium chloride (DODAC)] and PEG-ceramide (PEG-Cer) (Tam *et al.*, 2000; Wheeler *et al.*, 1999). The plasmid DNA can be visualized within the particles by cryo-EM (Fig. 6). SPLP are formed from mixtures of plasmid and lipid by a detergent-dialysis procedure involving octyl-glucopyranoside (OGP). SPLP protect plasmid DNA from DNase I and serum nucleases (Wheeler *et al.*, 1999), possess extended circulation half-lives (6–7 h) (Tam *et al.*, 2000), and have been shown to accumulate in distal tumor sites with subsequent gene expression in mouse tumor models following iv injection (Fenske *et al.*, 2001). These vectors are clearly capable of disease-site targeting, making them promising vectors for *in vivo* gene transfer. The experimental procedures discussed below outline the production of SPLP for smaller-scale *in vitro* or *in vivo* studies. A detailed protocol for the production of a large-scale SPLP batch, suitable for animal studies, has been published elsewhere (Fenske *et al.*, 2002).

Experimental Procedure: A Detergent Dialysis Approach to the Encapsulation of Plasmid DNA within LUVs

The basic procedure involves solubilizing lipid and plasmid (200 μg) in a detergent solution of appropriate ionic strength and removing the detergent by dialysis to form SPLP and empty vesicles. Unencapsulated DNA is removed by ion-exchange chromatography and empty vesicles by sucrose density gradient centrifugation.

Due to the relatively large quantities of buffer required, it is convenient to initially prepare 2 liters of $10\times$ HBS ($1\times = 20$ mM HEPES, 150 mM NaCl, pH 7.4) and to dilute as required. Lipid stock solutions can be prepared in chloroform or in methanol, at (approximate) concentrations of 100 mM for DOPE (Avanti Polar Lipids; Northern Lipids), 25 mM for DODAC (Inex Pharmaceuticals), 10 mM for PEGCerC$_{20}$ (Northern Lipids; Inex Pharmaceuticals), and 0.78 mM (1 mg/ml) for rhodamine-PE [1,2-dioleoyl-sn-glycero-3-phosphoethanolamine-N-(Lissamine Rhodamine B Sulfonyl)] (Avanti Polar Lipids). These are combined in the proportions DOPE:DODAC:PEGCerC$_{20}$:Rho-PE (81.5:8:10:0.5) for a total of 10 μmol lipid. Chloroform is dried under a stream of nitrogen gas and then under high vacuum as described above.

The 10 μmol of dry lipid is solubilized by addition of 200 μl of OGP (1 M in water) (Sigma-Aldrich), with heating (in a 60° waterbath) and vortexing until the lipid is mostly dissolved and no longer sticking to the glass tube. The addition of 600 μl HBS is followed by vortexing with heating until clear. The addition of 200 μg plasmid DNA (from a 1 mg/ml stock solution), with gentle vortexing, will give a 1 ml solution with 200 μg plasmid DNA, [lipid] = 10 mM, and [OGP] = 200 mM.

Encapsulation of plasmid and formation of SPLP is highly sensitive to the cationic lipid content and the salt concentration of the dialysis buffer. In general, a salt concentration is selected that gives high levels of encapsulation (50–70%), a small diameter (80–100 nm), and a monodisperse particle distribution (a χ^2 value of 3 or less obtained from QELS). In practice, it is often necessary to test a range of salt concentrations to achieve optimal encapsulation. This is particularly critical when preparing SPLP with high DODAC content (14–24 mol%), which requires inclusion of a polyanionic salt such as citrate or phosphate in addition to NaCl (Saravolac et $al.$, 2000; Zhang et $al.$, 1999). For SPLP containing 8 mol% DODAC, particle formation should occur for 20 mM HEPES, pH 7.5, containing 140–150 mM NaCl. As it may be necessary to test several salt concentrations for a given lipid formulation, it is advisable to prepare more SPLP than will be required, taking into account final particle yields.

Optimal encapsulation conditions can be assessed by dialyzing 250-μl aliquots of the above preparation in 1-cm-diameter dialysis tubing overnight against 1 liter of 20 mM HEPES, pH 7.5, containing 140, 145, and 150 mM NaCl. For dialysis, Spectrum Spectra/Por Molecular porous 10-mm membrane tubing with 12,000–14,000 MWCO (Spectrum Laboratories, Rancho Dominguez, CA) gives good results. The aliquots are analyzed for size (via QELS analysis using a Nicomp Model 370 Submicron Particle Sizer or equivalent unit) and encapsulation efficiency (picogreen assay). The picogreen assay involves pipetting 5-μl aliquots of each formulation into disposable 1-cm fluorescence cuvettes and adding 1.8 ml of picogreen buffer [containing 1 μl of picogreen (Molecular Probes, Eugene, OR) per 2 ml HBS]. The fluorescence is read in the absence and presence of Triton X-100 (Sigma-Aldrich), using excitation and emission wavelengths of 480 and 520 nm, respectively (with slit widths set to 5 nm). After reading the fluorescence in the absence of detergent ($-$ Triton value), 30 μl of 10% (v/v) Triton X-100 (in water) is added to the cuvette with thorough mixing, after which the fluorescence is read again ($+$ Triton value). The % encapsulation = (([+ Triton]$-$[$-$ Triton]) $\times$ 100)/[+ Triton].

The optimal NaCl concentration is determined by the percent of plasmid encapsulation and particle size. The optimal NaCl provides for 50–80% encapsulation and a single population with a particle size of $\leq$100 nm. Particle sizes greater than 120 nm indicate the presence of vesicle aggregation. Once the optimal NaCl concentration has been selected, 4 $\times$ 2 liters of dialysis buffer is prepared for a 48-h dialysis. The dialysis bags should be transferred to fresh dialysis buffer (2 liters) every 12 h. Following dialysis, both a picogreen assay and QELS analysis are performed on the resulting material. The percent encapsulation should be in the range of 50–80%. If this is not achieved, another salt curve should be performed before proceeding to the next step (the lipid–DNA mixture can be resolubilized in OGP and redialyzed with a different buffer).

Following formation of SPLP, unencapsulated DNA is removed by DEAE-Sepharose chromatography. Up to 0.64 mg of DNA can be loaded per 1 ml volume of the DEAE-Sepharose CL-6B (Sigma) column. The column can be poured in a small plastic holder (such as a 3-ml disposable syringe stopped with a small plug of glass wool) and washed with 10 column volumes of 1$\times$ HBS. Once the column has settled, the formulation is slowly loaded onto the resin and eluted with HBS. The final volume equal to 1.5 times the sample volume is collected to completely elute all of the formulation from the column.

The final step in the purification of SPLP is removal of empty vesicles from those containing plasmid DNA. This is accomplished by stepwise sucrose density gradient ultracentrifugation in a Beckman ultracentrifuge using an SW41 rotor. Solutions containing 1.0, 2.5, and 10.0% w/v sucrose are prepared in 1$\times$ HBS, filter sterilized into a sterile container, and stored at 4°. Prior to pouring the gradients, a long glass pipette is pulled to a small point using forceps and a Bunsen burner and is broken at the tip to give a slow flow rate. The elongated pipette is placed in the ultracentrifuge tube (Beckman Ultra-clear polycarbonate centrifuge tubes, 14 $\times$ 89 mm,

#344059), and 3.6 ml of a 1.0% sucrose solution is poured into the pipette using a second (regular) pipette, avoiding air bubbles in the narrow part of the pipette. The sucrose layers are poured in order of increasing density; i.e., 3.6 ml of 1.0%, 3.6 ml of 2.5%, and 3.6 ml of 10.0%. One milliliter of SPLP is then gently layered on top of the gradients. All the tubes are then balanced with $1\times$ HBS to within 0.01 g and placed in the SW41 buckets. The tubes are centrifuged for 2 h at 36,000 rpm at 20°, ensuring that the brake is off during deceleration.

The SPLP will be visible as a pink (rhodamine) band layered at the interface of the 2.5 and 10% sucrose solutions (lower band). The empty vesicles will either be banded at the 1–2.5% interface or spread throughout the 1 and 2.5% solutions. To isolate the SPLP, the tube is punctured 2 mm below the SPLP band using an 18G1/2 needle with a 3-ml syringe, and the SPLP band is slowly aspirated. All of the SPLP bands are pooled and placed into a dialysis bag overnight in $1\times$ HBS to remove the sucrose. A picogreen assay is performed to determine the percent encapsulation and DNA concentration, and the size is measured by QELS. Ideally, the SPLP concentration should be in the range of 0.1 mg DNA/ml. If necessary, the sample can be concentrated using an Amicon filtration device or by using Aquazide. The final sample should be filter sterilized through a sterile Millipore 0.22-μm filter unit in a Biological Safety Cabinet. SPLP may be stored in sterile vials at 4° for up to 2 years. If desired, the lipid concentration can be determined from a phosphate assay, and agarose gel electrophoresis can be used to verify the integrity of the plasmid.

Stabilized Antisense–Lipid Particles (SALP): Vectors for Regulation of Gene Expression

Antisense oligonucleotides are a class of genetic drugs that inhibits gene expression by virtue of the ability to bind to specific mRNA sequences and interfere with protein synthesis (Stein and Cohen, 1988). These molecules may have clinical utility in the treatment of cancer. For example, the proliferation of human colon carcinoma cell lines expressing the protooncogene c-*myb* could be inhibited by antisense oligodeoxynucleotides directed against c-*myb* (Melani *et al.*, 1991). Similar effects have been observed for leukemia cell lines expressing either the c-*myc* protooncogene (Wickstrom *et al.*, 1988) or the proliferative protein BCR-ABL (Szczylik *et al.*, 1991) and for human glioma cell lines expressing basic fibroblast growth factor (Morrison, 1991). As with conventional drugs, the major problems hindering clinical applications of antisense therapy center on drug stability, transport, and uptake by cells *in vivo*. Recently, it has been shown that antisense molecules directed against the epidermal growth factor receptor (EGFR) of KB cells could be efficiently delivered to cells by encapsulation within folate-targeted liposomes (Wang *et al.*, 1995). Cells treated with the liposome-delivered antisense displayed growth inhibition, likely due to reduced EGFR expression (Wang *et al.*, 1995). This study, in which encapsulation of antisense was achieved by passive entrapment during extrusion of the liposomes, highlights the potential benefits of encapsulation within a liposomal carrier and the need for methods to achieve higher D/L ratios. Progress in this direction includes the development by Allen and co-workers of a novel system in which charge-neutralized cationic lipid–antisense particles are coated by a layer of neutral lipids. These coated cationic liposomes had an average diameter of 188 nm, entrapped 85–95% of the antisense particles, and exhibited an extended circulation half-life of >10 h compared with <1 h for free antisense (Stuart and Allen, 2000; Stuart *et al.*, 2000). A different approach has been reported by Madden and co-workers, who have developed a liposomal antisense formulation, involving the use of programmable fusogenic vesicles (PFV), that has been observed to cause down-regulation of bcl-2 mRNA levels in 518A2 melanoma cells (Hu *et al.*, 2001). Recently, we described a novel formulation process that utilizes an ionizable aminolipid [1,2-dioleoyl-3-dimethylammonium propane (DODAP)] and an ethanol-containing buffer for encapsulating large quantities of polyanionic antisense oligonucleotide in lipid vesicles (Maurer *et al.*, 2001; Semple *et al.*, 2000, 2001). The resulting particle is known as a "stabilized antisense–lipid particle" or SALP. Initially, an ethanolic liposome solution is formed by addition of lipids (DSPC/Chol/PEGCerC$_{14}$/DODAP) in ethanol to an aqueous buffer with subsequent extrusion (Semple *et al.*, 2001) or by addition of ethanol to preformed vesicles of the same composition (Maurer *et al.*, 2001). A citrate buffer is used to acidify the ethanol-containing buffer (pH 4) to ensure that the cationic lipid is protonated. The addition of oligonucleotide to the ethanolic liposome solution leads to the formation of multilamellar liposomes (as well as some unilamellar and bilamellar vesicles) that trap oligonucleotides between the bilayers. Subsequent dialysis against acidic and then neutral buffers removes the ethanol and causes release of any externally bound oligonucleotide from the uncharged liposome surface. Unencapsulated ODN is then removed by anion-exchange chromatography. The end result is a multilamellar vesicle with a small diameter (70–120 nm) and a maximum entrapment of 0.16 mg ODN/mg lipid, which corresponds to ~2200 oligonucleotide molecules per 100 nm liposome (Maurer *et al.*, 2001) (Fig. 7). The SALP exhibit extended circulation half-lives, ranging from 5–6 h for particles formed with PEGCerC$_{14}$ to 10–12 h for particles formed with PEGCerC$_{20}$ (Semple *et al.*, 2001). The combination of high entrapment efficiencies, small size, and extended

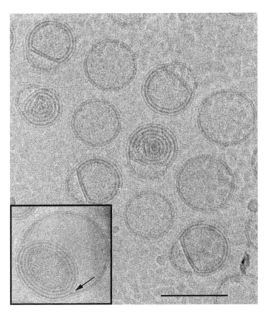

Fig. 7. Cryoelectron micrographs of DSPC/Chol/PEGCerC$_{14}$/DODAP (20:45:10: 25 mol%) liposomes entrapping oligonucleotides (SALP). The inset is an expanded view of a multilamellar liposome showing two initially separated membranes forced into close apposition by bound oligonucleotides (arrow). The entrapped antisense-to-lipid weight ratio was 0.125 mg/mg. The bar represents 100 nm. [Reprinted from Maurer, N., Wong, K. F., Stark, H., Louie, L., McIntosh, D., Wong, T., Scherrer, P., Semple, S. C., and Cullis, P. R. (2001). *Biophys. J.* **80,** 2310–2326, with permission.]

circulation lifetimes suggests that the SALP system should prove of utility for the liposomal delivery of antisense drugs. The following procedure describes the loading of antisense molecules via electrostatic attraction into preformed vesicles destabilized by ethanol.

Experimental Procedure: Spontaneous Entrapment of Antisense
    Oligonucleotides upon Electrostatic Interaction with
    Ethanol-Destabilized Cationic Liposomes

Stock solutions of DODAP (Inex Pharmaceuticals), DSPC, Chol, and PEGCerC$_{14}$ (Northern Lipids; Inex Pharmaceuticals) are prepared in chloroform or ethanol. Oligonucleotide (ODN) stock solutions are prepared at 20 mg/ml in distilled water or buffer [e.g., human c-*myc* (16-mer), 5'-TAACGTTGAGGGGCAT-3'] (Inex Pharmaceuticals).

A lipid mixture containing 25 mg total lipid is prepared, with a molar ratio of $DSPC/Chol/PEG-CerC_{14}/DODAP$ of 20:45:10:25. Solvent is removed under a stream of N_2 followed by high vacuum for 1 h. The lipid film is hydrated in 1 ml of 300 mM citrate buffer, pH 4, and subjected to five freeze/thaw cycles. LUVs are formed by extrusion at 60° as previously described.

Absolute ethanol ($\geq$99%) is slowly added to the LUVs with rapid vortexing up to a concentration of 40% (to avoid high local concentrations). The total lipid concentration should be 10 mg/ml. Slow addition of ethanol and rapid mixing are important as liposomes become unstable and coalesce into large lipid structures as soon as the ethanol concentration exceeds a certain upper limit, which depends on the lipid composition. The final liposomes should be around 100 nm in diameter. If necessary, the ethanolic liposome dispersions can be additionally extruded ($2\times$) to reduce size.

LUVs can also be prepared by slow addition of the lipids dissolved in ethanol (total volume of 0.4 ml) to citrate buffer at pH 4 (0.6 ml) followed by extrusion through two stacked 100-nm filters (two passes) at room temperature. Typical liposome diameters obtained from QELS should be 75–100 nm. The extrusion step can be omitted if ethanol is added very slowly under vigorous mixing to avoid high local concentrations of ethanol.

Entrapment of antisense is achieved by slow addition of the oligonucleotide solution (1–2 mg ODN) with vortexing to the ethanolic liposome dispersion (10 mg lipid/ml) to give an ODN-to-lipid ratio of 0.1–0.2 mg/mg.

The ethanolic ODN–lipid mixture is then incubated at 40° for 1 h, followed by dialysis for 2 h against 2 liters of citrate buffer (to remove most of the ethanol) and twice against 2 liters of HBS (20 mM HEPES, 145 mM NaCl, pH 7.5). At pH 7.5, DODAP becomes neutral, and oligonucleotides bound to the external membrane surface are released from their association with the cationic lipid.

Unencapsulated ODNs are removed by anion-exchange chromatography on DEAE-Sepharose CL-6B (Sigma-Aldrich) columns (3–4 ml of DEAE-Sepharose per 1 ml of ODN/liposome dispersion) equilibrated in HBS, pH 7.5.

To determine trapping efficiencies, the SALP is solubilized by addition of chloroform/methanol at a volume ratio of 1:2.1:1 chloroform/methanol/SALP. The oligonucleotide concentrations are determined from the absorbance at 260 nm on a Shimadzu UV160U spectrophotometer. If the solution is not completely clear after mixing, an additional 50–100 μl of methanol can be added. Alternatively, the absorbance can be read after solubilization of the samples in 100 mM OGP. The antisense concentrations are calculated according to c [μg/μl] $= A_{260} \times 1$ OD_{260} unit

$[\mu g/ml] \times$ dilution factor $[ml/\mu l]$, where the dilution factor is given by the total assay volume [ml] divided by the sample volume $[\mu l]$. OD_{260} units are calculated from pairwise extinction coefficients for individual deoxynucleotides, which take into account nearest-neighbor interactions. One OD corresponds to 30.97 $\mu g/ml$ anti-c-*myc*. Lipid concentrations are determined by the inorganic phosphate assay (Fiske and Subbarow, 1925) after separation of the lipids from the oligonucleotides by a Bligh and Dyer extraction (Bligh and Dyer, 1959) (initial aqueous volume = 250 μl).

Conclusions

The rapid development of liposome technology over the past 15 years has led to a wide variety of delivery systems for conventional drugs, some of which are in clinical trials or have been approved for use by the U.S. FDA. While liposomal systems for genetic drugs have not been developed to the same extent, several systems are available for both plasmid DNA and antisense oligonucleotides that have systemic potential. Further advances in this field are anticipated as tissue targeting and controlled drug release capabilities are included in future liposome designs.

References

Adlakha-Hutcheon, G., Bally, M. B., Shew, C. R., and Madden, T. D. (1999). Controlled destabilization of a liposomal drug delivery system enhances mitoxantrone antitumor activity. *Nat. Biotechnol.* **17,** 775–779.

Allen, T. M. (1994). Long-circulating (sterically stabilized) liposomes for targeted drug delivery. *Trends Pharmacol. Sci.* **15,** 215–220.

Allen, T. M. (1998). Liposomal drug formulations. Rationale for development and what we can expect for the future. *Drugs* **56,** 747–756.

Allen, T. M., Hansen, C., Martin, F., RedemAnn, C., and Yau-Young, A. (1991). Liposomes containing synthetic lipid derivatives of poly(ethylene glycol) show prolonged circulation half-lives *in vivo. Biochim. Biophys. Acta* **1066,** 29–36.

Bakker-Woudenberg, I. A., ten Kate, M. T., Guo, L., Working, P., and Mouton, J. W. (2001). Improved efficacy of ciprofloxacin administered in polyethylene glycol-coated liposomes for treatment of Klebsiella pneumoniae pneumonia in rats. *Antimicrob. Agents Chemother.* **45,** 1487–1492.

Bally, M. B., Mayer, L. D., Loughrey, H., Redelmeier, T., Madden, T. D., Wong, K., Harrigan, P. R., Hope, M. J., and Cullis, P. R. (1988). Dopamine accumulation in large unilamellar vesicle systems induced by transmembrane ion gradients. *Chem. Phys. Lipids* **47,** 97–107.

Bangham, A. D. (1968). Membrane models with phospholipids. *Prog. Biophys. Mol. Biol.* **18,** 29–95.

Batzri, S., and Korn, E. D. (1973). Single bilayer liposomes prepared without sonication. *Biochim. Biophys. Acta* **298,** 1015–1019.

Bligh, E. G., and Dyer, W. J. (1959). A rapid method of total lipid extraction and purification. *Can. J. Biochem. Physiol.* **37,** 911–917.

Boman, N. L., Mayer, L. D., and Cullis, P. R. (1993). Optimization of the retention properties of vincristine in liposomal systems. *Biochim. Biophys. Acta* **1152**, 253–258.

Boman, N. L., Masin, D., Mayer, L. D., Cullis, P. R., and Bally, M. B. (1994). Liposomal vincristine which exhibits increased drug retention and increased circulation longevity cuRes. mice bearing P388 tumors. *Cancer Res.* **54**, 2830–2833.

Brown, J. M., and Giaccia, A. J. (1998). The unique physiology of solid tumors: Opportunities (and problems) for cancer therapy. *Cancer Res.* **58**, 1408–1416.

Burstein, H. J., Ramirez, M. J., Petros, W. P., Clarke, K. D., Warmuth, M. A., Marcom, P. K., Matulonis, U. A., Parker, L. M., Harris, L. N., and Winer, E. P. (1999). Phase I study of Doxil and vinorelbine in metastatic breast cancer. *Ann. Oncol.* **10**, 1113–1116.

Campos, S. M., Penson, R. T., Mays, A. R., Berkowitz, R. S., Fuller, A. F., Goodman, A., Matulonis, U. A., Muzikansky, A., and Seiden, M. V. (2001). The clinical utility of liposomal doxorubicin in recurrent ovarian cancer. *Gynecol. Oncol.* **81**, 206–212.

Chang, C. W., Barber, L., Ouyang, C., Masin, D., Bally, M. B., and Madden, T. D. (1997). Plasma clearance, biodistribution and therapeutic properties of mitoxantrone encapsulated in conventional and sterically stabilized liposomes after intravenous administration in BDF1 mice. *Br. J. Cancer* **75**, 169–177.

Cheung, B. C. L., Sun, T. H. T., Leenhouts, J. M., and Cullis, P. R. (1998). Loading of doxorubicin into liposomes by forming Mn^{2+}-drug complexes. *Biochim. Biophys. Acta* **1414**, 205–216.

Cheung, T. W., Remick, S. C., Azarnia, N., Proper, J. A., Barrueco, J. R., and Dezube, B. J. (1999). AIDS-related Kaposi's sarcoma: A phase II study of liposomal doxorubicin. The TLC D-99 Study Group. *Clin. Cancer Res.* **5**, 3432–3437.

Chonn, A., and Cullis, P. R. (1995). Recent advances in liposomal drug-delivery systems. *Curr. Opin. Biotechnol.* **6**, 698–708.

Coukell, A. J., and Spencer, C. M. (1997). Polyethylene glycol-liposomal doxorubicin. A review of its pharmacodynamic and pharmacokinetic properties, and therapeutic efficacy in the management of AIDS-related Kaposi's sarcoma. *Drugs* **53**, 520–538.

Cullis, P. R. (2000). Liposomes by accident. *J. Liposome Res.* **10**, ix–xxiv.

Cullis, P. R., Hope, M. J., Bally, M. B., Madden, T. D., Mayer, L. D., and Fenske, D. B. (1997). Influence of pH gradients on the transbilayer transport of drugs, lipids, peptides and metal ions into large unilamellar vesicles. *Biochim. Biophys. Acta* **1331**, 187–211.

Dvorak, H. F., Nagy, J. A., Dvorak, J. T., and Dvorak, A. M. (1988). Identification and characterization of the blood vessels of solid tumors that are leaky to circulating macromolecules. *Am. J. Pathol.* **133**, 95–109.

Fenske, D. B., MacLachlan, I., and Cullis, P. R. (2001). Long-circulating vectors for the systemic delivery of genes. *Curr. Opin. Mol. Ther.* **3**, 153–158.

Fenske, D. B., MacLachlan, I., and Cullis, P. R. (2002). Stabilized plasmid-lipid particles: A systemic gene therapy vector. *Methods Enzymol.* **346**, 36–71.

Fenske, D. B., Maurer, N., and Cullis, P. R. (2003). Encapsulation of weakly-basic drugs, antisense oligonucleotides and plasmid DNA within large unilamellar vesicles for drug delivery applications. *In* "Liposomes: A Practical Approach" (V. P. Torchilin and V. Weissig, eds.), pp. 167–191. Oxford University Press, Oxford.

Fenske, D. B., Wong, K. F., Maurer, E., Maurer, N., Leenhouts, J. M., Boman, N., Amankwa, L., and Cullis, P. R. (1998). Ionophore-mediated uptake of ciprofloxacin and vincristine into large unilamellar vesicles exhibiting transmembrane ion gradients. *Biochim. Biophys. Acta* **1414**, 188–204.

Fiske, C. H., and Subbarow, Y. (1925). The colorimetric determination of phosphorus. *J. Biol. Chem.* **66**, 375–400.

Gabizon, A., and Papahadjopoulos, D. (1988). Liposome formulations with prolonged circulation time in blood and enhanced uptake by tumors. *Proc. Natl. Acad. Sci. USA* **85,** 6949–6953.

Gelmon, K. A., Tolcher, A., Diab, A. R., Bally, M. B., Embree, L., Hudon, N., Dedhar, C., Ayers, D., Eisen, A., Melosky, B., Burge, C., Logan, P., and Mayer, L. D. (1999). Phase I study of liposomal vincristine. *J. Clin. Oncol.* **17,** 697–705.

Gill, P. S., Wernz, J., Scadden, D. T., Cohen, P., Mukwaya, G. M., Von Roenn, J. H., Jacobs, M., Kempin, S., Silverberg, I., Gonzales, G., Rarick, M. U., Myers, A. M., Shepherd, F., Sawka, C., Pike, M. C., and Ross, M. E. (1996). Randomized phase III trial of liposomal daunorubicin versus doxorubicin, bleomycin, and vincristine in AIDS-related Kaposi's sarcoma. *J. Clin. Oncol.* **14,** 2353–2364.

Gokhale, P. C., Radhakrishnan, B., Husain, S. R., Abernethy, D. R., Sacher, R., Dritschilo, A., and Rahman, A. (1996). An improved method of encapsulation of doxorubicin in liposomes: Pharmacological, toxicological, and therapeutic evaluation. *Br. J. Cancer* **74,** 43–48.

Gordon, A. N., Granai, C. O., Rose, P. G., Hainsworth, J., Lopez, A., Weissman, C., Rosales, R., and Sharpington, T. (2000). Phase II study of liposomal doxorubicin in platinum- and paclitaxel-refractory epithelial ovarian cancer. *J. Clin. Oncol.* **18,** 3093–3100.

Grunaug, M., Bogner, J. R., Loch, O., and Goebel, F. D. (1998). Liposomal doxorubicin in pulmonary Kaposi's sarcoma: Improved survival as compared to patients without liposomal doxorubicin. *Eur. J. Med. Res.* **3,** 13–19.

Gutteridge, J. M. (1984). Lipid peroxidation and possible hydroxyl radical formation stimulated by the self-reduction of a doxorubicin-iron (III) complex. *Biochem. Pharmacol.* **33,** 1725–1728.

Gutteridge, J. M., and Quinlan, G. J. (1985). Free radical damage to deoxyribose by anthracycline, aureolic acid and aminoquinone antitumour antibiotics. An essential requirement for iron, semiquinones and hydrogen peroxide. *Biochem. Pharmacol.* **34,** 4099–4103.

Haran, G., Cohen, R., Bar, L. K., and Barenholz, Y. (1993). Transmembrane ammonium sulfate gradients in liposomes produce efficient and stable entrapment of amphipathic weak bases. *Biochim. Biophys. Acta* **1151,** 201–215.

Harrigan, P. R., Wong, K. F., Redelmeier, T. E., Wheeler, J. J., and Cullis, P. R. (1993). Accumulation of doxorubicin and other lipophilic amines into large unilamellar vesicles in response to transmembrane pH gradients. *Biochim. Biophys. Acta* **1149,** 329–338.

Hasinoff, B. B. (1989a). The interaction of the cardioprotective agent ICRF-187 [(+)-1,2-bis(3,5-dioxopiperazinyl-1-yL)propane); its hydrolysis product (ICRF-198); and other chelating agents with the Fe(III) and Cu(II) complexes of adriamycin. *Agents Actions* **26,** 378–385.

Hasinoff, B. B. (1989b). Self-reduction of the iron(III)-doxorubicin complex. *Free Radic. Biol. Med.* **7,** 583–593.

Hasinoff, B. B., Davey, J. P., and O'Brien, P. J. (1989). The Adriamycin (doxorubicin)-induced inactivation of cytochrome c oxidase depends on the presence of iron or copper. *Xenobiotica* **19,** 231–241.

Hope, M. J., Bally, M. B., Mayer, L. D., Janoff, A. S., and Cullis, P. R. (1986). Generation of multilamellar and unilamellar phospholipid vesicles. *Chem. Phys. Lipids* **40,** 89–107.

Hope, M. J., and Wong, K. F. (1995). Liposomal formulation of ciprofloxacin. *In* "Liposomes in Biomedical Applications" (P. N. Shek, ed.), pp. 121–134. Harwood Academic Publishers, London.

Hope, M. J., Bally, M. B., Webb, G., and Cullis, P. R. (1985). Production of large unilamellar vesicles by a rapid extrusion procedure. Characterization of size distribution, trapped volume and ability to maintain a membrane potential. *Biochim. Biophys. Acta* **812,** 55–65.

Hu, Q., Shew, C. R., Bally, M. B., and Madden, T. D. (2001). Programmable fusogenic vesicles for intracellular delivery of antisense oligodeoxynucleotides: Enhanced cellular uptake and biological effects. *Biochim. Biophys. Acta* **1514**, 1–13.

Huang, C. (1969). Studies on phosphatidylcholine vesicles. Formation and physical characteristics. *Biochemistry* **8**, 344–352.

Israel, V. P., Garcia, A. A., Roman, L., Muderspach, L., Burnett, A., Jeffers, S., and Muggia, F. M. (2000). Phase II study of liposomal doxorubicin in advanced gynecologic cancers. *Gynecol. Oncol.* **78**, 143–147.

Judson, I., Radford, J. A., Harris, M., Blay, J., van Hoesel, Q., le Cesne, A., van Oosterom, A. T., Clemons, M. J., Kamby, C., Hermans, C., Whittaker, J., Donato, d. P., Verweij, J., and Nielsen, S. (2001). Randomised phase II trial of pegylated liposomal doxorubicin (DOXIL(R)/CAELYX(R)) versus doxorubicin in the treatment of advanced or metastatic soft tissue sarcoma. A study by the EORTC Soft Tissue and Bone Sarcoma Group. *Eur. J. Cancer* **37**, 870–877.

Lasic, D. D., Ceh, B., Stuart, M. C., Guo, L., Frederik, P. M., and Barenholz, Y. (1995). Transmembrane gradient driven phase transitions within vesicles: Lessons for drug delivery. *Biochim. Biophys. Acta* **1239**, 145–156.

Lasic, D. D., Frederik, P. M., Stuart, M. C., Barenholz, Y., and McIntosh, T. J. (1992). Gelation of liposome interior. A novel method for drug encapsulation. *FEBS Lett.* **312**, 255–258.

Li, X., Cabral-Lilly, D., Janoff, A. S., and Perkins, W. R. (2000). Complexation of internalized doxorubicin into fiber bundles affects its release rate from liposomes. *J. Liposome Res.* **10**, 15–27.

Li, X., Hirsh, D. J., Cabral-Lilly, D., Zirkel, A., Gruner, S. M., Janoff, A. S., and Perkins, W. R. (1998). Doxorubicin physical state in solution and inside liposomes loaded via a pH gradient. *Biochim. Biophys. Acta* **1415**, 23–40.

Lichtenberg, D., and Barenholz, Y. (1988). Liposomes: Preparation, characterization, and preservation. *Methods Biochem. Anal.* **33**, 337–462.

Lim, H. J., Masin, D., Madden, T. D., and Bally, M. B. (1997). Influence of drug release characteristics on the therapeutic activity of liposomal mitoxantrone. *J. Pharmacol. Exp. Ther.* **281**, 566–573.

Lim, H. J., Masin, D., McIntosh, N. L., Madden, T. D., and Bally, M. B. (2000). Role of drug release and liposome-mediated drug delivery in governing the therapeutic activity of liposomal mitoxantrone used to treat human A431 and LS180 solid tumors. *J. Pharmacol. Exp. Ther.* **292**, 337–345.

Madden, T. D., Harrigan, P. R., Tai, L. C., Bally, M. B., Mayer, L. D., Redelmeier, T. E., Loughrey, H. C., Tilcock, C. P., Reinish, L. W., and Cullis, P. R. (1990). The accumulation of drugs within large unilamellar vesicles exhibiting a proton gradient: A survey. *Chem. Phys. Lipids* **53**, 37–46.

Maurer-Spurej, E., Wong, K. F., Maurer, N., Fenske, D. B., and Cullis, P. R. (1999). Factors influencing uptake and retention of amino-containing drugs in large unilamellar vesicles exhibiting transmembrane pH gradients. *Biochim. Biophys. Acta* **1416**, 1–10.

Maurer, N., Mori, A., Palmer, L., Monck, M. A., Mok, K. W., Mui, B., Akhong, Q. F., and Cullis, P. R. (1999). Lipid-based systems for the intracellular delivery of genetic drugs. *Mol. Membr. Biol.* **16**, 129–140.

Maurer, N., Wong, K. F., Hope, M. J., and Cullis, P. R. (1998). Anomalous solubility behavior of the antibiotic ciprofloxacin encapsulated in liposomes: A 1H-NMR study. *Biochim. Biophys. Acta* **1374**, 9–20.

Maurer, N., Wong, K. F., Stark, H., Louie, L., McIntosh, D., Wong, T., Scherrer, P., Semple, S. C., and Cullis, P. R. (2001). Spontaneous entrapment of polynucleotides upon electrostatic interaction with ethanol-destabilized cationic liposomes. *Biophys. J.* **80**, 2310–2326.

Mayer, L. D., Bally, M. B., and Cullis, P. R. (1986a). Uptake of Adriamycin into large unilamellar vesicles in response to a pH gradient. *Biochim. Biophys. Acta* **857**, 123–126.

Mayer, L. D., Hope, M. J., and Cullis, P. R. (1986b). Vesicles of variable sizes produced by a rapid extrusion procedure. *Biochim. Biophys. Acta* **858**, 161–168.

Mayer, L. D., Bally, M. B., and Cullis, P. R. (1990a). Strategies for optimizing liposomal doxorubicin. *J. Liposome Res.* **1**, 463–480.

Mayer, L. D., Bally, M. B., Cullis, P. R., Wilson, S. L., and Emerman, J. T. (1990b). Comparison of free and liposome encapsulated doxorubicin tumor drug uptake and antitumor efficacy in the SC115 murine mammary tumor. *Cancer Lett.* **53**, 183–190.

Mayer, L. D., Bally, M. B., Loughrey, H., Masin, D., and Cullis, P. R. (1990c). Liposomal vincristine preparations which exhibit decreased drug toxicity and increased activity against murine L1210 and P388 tumors. *Cancer Res.* **50**, 575–579.

Mayer, L. D., Nayar, R., Thies, R. L., Boman, N. L., Cullis, P. R., and Bally, M. B. (1993). Identification of vesicle properties that enhance the antitumour activity of liposomal vincristine against murine L1210 leukemia. *Cancer Chemother. Pharmacol.* **33**, 17–24.

Mayer, L. D., Tai, L. C., Ko, D. S., Masin, D., Ginsberg, R. S., Cullis, P. R., and Bally, M. B. (1989). Influence of vesicle size, lipid composition, and drug-to-lipid ratio on the biological activity of liposomal doxorubicin in mice. *Cancer Res.* **49**, 5922–5930.

Melani, C., Rivoltini, L., Parmiani, G., Calabretta, B., and Colombo, M. P. (1991). Inhibition of proliferation by c-myb antisense oligodeoxynucleotides in colon adenocarcinoma cell lines that expRes.s c-myb. *Cancer Res.* **51**, 2897–2901.

Millar, J. L., Millar, B. C., Powles, R. L., Steele, J. P., Clutterbuck, R. D., Mitchell, P. L., Cox, G., Forssen, E., and Catovsky, D. (1998). Liposomal vincristine for the treatment of human acute lymphoblastic leukaemia in severe combined immunodeficient (SCID) mice. *Br. J. Haematol.* **102**, 718–721.

Mimms, L. T., Zampighi, G., Nozaki, Y., Tanford, C., and Reynolds, J. A. (1981). Phospholipid vesicle formation and transmembrane protein incorporation using octyl glucoside. *Biochemistry* **20**, 833–840.

Mok, K. W., Lam, A. M., and Cullis, P. R. (1999). Stabilized plasmid-lipid particles: Factors influencing plasmid entrapment and transfection properties. *Biochim. Biophys. Acta* **1419**, 137–150.

Morrison, R. S. (1991). Suppression of basic fibroblast growth factor expression by antisense oligodeoxynucleotides inhibits the growth of transformed human astrocytes. *J. Biol. Chem.* **266**, 728–734.

Muggia, F. M. (2001). Liposomal encapsulated anthracyclines: New therapeutic horizons. *Curr. Oncol. Rep.* **3**, 156–162.

Muindi, J. R., Sinha, B. K., GiAnni, L., and Myers, C. E. (1984). Hydroxyl radical production and DNA damage induced by anthracycline-iron complex. *FEBS Lett.* **172**, 226–230.

Northfelt, D. W., Dezube, B. J., Thommes, J. A., Miller, B. J., Fischl, M. A., Friedman-Kien, A., Kaplan, L. D., Du, M. C., Mamelok, R. D., and Henry, D. H. (1998). Pegylated-liposomal doxorubicin versus doxorubicin, bleomycin, and vincristine in the treatment of AIDS-related Kaposi's sarcoma: Results of a randomized phase III clinical trial. *J. Clin. Oncol.* **16**, 2445–2451.

Olson, F., Hunt, C. A., Szoka, F. C., Vail, W. J., and Papahadjopoulos, D. (1979). Preparation of liposomes of defined size distribution by extrusion through polycarbonate membranes. *Biochim. Biophys. Acta* **557**, 9–23.

Pratt, G., Wiles, M. E., Rawstron, A. C., Davies, F. E., Fenton, J. A., Proffitt, J. A., Child, J. A., Smith, G. M., and Morgan, G. J. (1998). Liposomal daunorubicin: *In vitro* and *in vivo* efficacy in multiple myeloma. *Hematol. Oncol.* **16,** 47–55.

Rottenberg, H. (1979). The measurement of membrane potential and deltapH in cells, organelles, and vesicles. *Methods Enzymol.* **55,** 547–569.

Saravolac, E. G., Ludkovski, O., Skirrow, R., Ossanlou, M., Zhang, Y. P., Giesbrecht, C., Thompson, J., Thomas, S., Stark, H., Cullis, P. R., and Scherrer, P. (2000). Encapsulation of plasmid DNA in stabilized plasmid-lipid particles composed of different cationic lipid concentration for optimal transfection activity. *J. Drug Target.* **7,** 423–437.

Semple, S. C., Klimuk, S. K., Harasym, T. O., Dos, S. N., Ansell, S. M., Wong, K. F., Maurer, N., Stark, H., Cullis, P. R., Hope, M. J., and Scherrer, P. (2001). Efficient encapsulation of antisense oligonucleotides in lipid vesicles using ionizable aminolipids: Formation of novel small multilamellar vesicle structures. *Biochim. Biophys. Acta* **1510,** 152–166.

Semple, S. C., Klimuk, S. K., Harasym, T. O., and Hope, M. J. (2000). Lipid-based formulations of antisense oligonucleotides for systemic delivery applications. *Methods Enzymol.* **313,** 322–341.

Sessa, G., and Weissmann., G. (1968). Phospholipid spherules (liposomes) as a model for biological membranes. *J. Lipid Res.* **9,** 310–318.

Shields, A. F., Lange, L. M., and Zalupski, M. M. (2001). Phase II study of liposomal doxorubicin in patients with advanced colorectal cancer. *Am. J. Clin. Oncol.* **24,** 96–98.

Stein, C. A., and Cohen, J. S. (1988). Oligodeoxynucleotides as inhibitors of gene expression: A review. *Cancer Res.* **48,** 2659–2668.

Stuart, D. D., and Allen, T. M. (2000). A new liposomal formulation for antisense oligodeoxynucleotides with small size, high incorporation efficiency and good stability. *Biochim. Biophys. Acta* **1463,** 219–229.

Stuart, D. D., Kao, G. Y., and Allen, T. M. (2000). A novel, long-circulating, and functional liposomal formulation of antisense oligodeoxynucleotides targeted against MDR1. *Cancer Gene Ther.* **7,** 466–475.

Szczylik, C., Skorski, T., Nicolaides, N. C., Manzella, L., Malaguarnera, L., Venturelli, D., Gewirtz, A. M., and Calabretta, B. (1991). Selective inhibition of leukemia cell proliferation by BCR-ABL antisense oligodeoxynucleotides. *Science* **253,** 562–565.

Szoka, F. C., and Papahadjopoulos, D. (1978). Procedure for preparation of liposomes with large internal aqueous space and high capture by reverse-phase evaporation. *Proc. Natl. Acad. Sci. USA* **75,** 4194–4198.

Tam, P., Monck, M., Lee, D., Ludkovski, O., Leng, E. C., Clow, K., Stark, H., Scherrer, P., Graham, R. W., and Cullis, P. R. (2000). Stabilized plasmid-lipid particles for systemic gene therapy. *Gene Ther.* **7,** 1867–1874.

Tardi, P., Choice, E., Masin, D., Redelmeier, T., Bally, M., and Madden, T. D. (2000). Liposomal encapsulation of topotecan enhances anticancer efficacy in murine and human xenograft models. *Cancer Res.* **60,** 3389–3393.

Tokudome, Y., Oku, N., Doi, K., Namba, Y., and Okada, S. (1996). Antitumor activity of vincristine encapsulated in glucuronide-modified long-circulating liposomes in mice bearing Meth A sarcoma. *Biochim. Biophys. Acta* **1279,** 70–74.

Wang, S., Lee, R. J., Cauchon, G., Gorenstein, D. G., and Low, P. S. (1995). Delivery of antisense oligodeoxyribonucleotides against the human epidermal growth factor receptor into cultured KB cells with liposomes conjugated to folate via polyethylene glycol. *Proc. Natl. Acad. Sci. USA* **92,** 3318–3322.

Webb, M. S., Harasym, T. O., Masin, D., Bally, M. B., and Mayer, L. D. (1995). Sphingomyelin-cholesterol liposomes significantly enhance the pharmacokinetic and therapeutic properties of vincristine in murine and human tumour models. *Br. J. Cancer* **72**, 896–904.

Webb, M. S., Logan, P., Kanter, P. M., St-Onge, G., Gelmon, K., Harasym, T., Mayer, L. D., and Bally, M. B. (1998a). Preclinical pharmacology, toxicology and efficacy of sphingomyelin/cholesterol liposomal vincristine for therapeutic treatment of cancer. *Cancer Chemother. Pharmacol.* **42**, 461–470.

Webb, M. S., Boman, N. L., Wiseman, D. J., Saxon, D., Sutton, K., Wong, K. F., Logan, P., and Hope, M. J. (1998b). Antibacterial efficacy against an *in vivo Salmonella typhimurium* infection model and pharmacokinetics of a liposomal ciprofloxacin formulation. *Antimicrob. Agents Chemother.* **42**, 45–52.

Wheeler, J. J., Palmer, L., Ossanlou, M., MacLachlan, I., Graham, R. W., Zhang, Y. P., Hope, M. J., Scherrer, P., and Cullis, P. R. (1999). Stabilized plasmid-lipid particles: Construction and characterization. *Gene Ther.* **6**, 271–281.

Wheeler, J. J., Veiro, J. A., and Cullis, P. R. (1994). Ionophore-mediated loading of Ca2+ into large unilamellar vesicles in response to transmembrane pH gradients. *Mol. Membr. Biol.* **11**, 151–157.

Wickstrom, E. L., Bacon, T. A., Gonzalez, A., Freeman, D. L., Lyman, G. H., and Wickstrom, E. (1988). Human promyelocytic leukemia HL-60 cell proliferation and c-myc protein expression are inhibited by an antisense pentadecadeoxynucleotide targeted against c-myc mRNA. *Proc. Natl. Acad. Sci. USA* **85**, 1028–1032.

Woodle, M. C., Newman, M. S., and Cohen, J. A. (1994). Sterically stabilized liposomes: Physical and biological properties. *J. Drug Target.* **2**, 397–403.

Zhang, Y. P., Sekirov, L., Saravolac, E. G., Wheeler, J. J., Tardi, P., Clow, K., Leng, E., Sun, R., Cullis, P. R., and Scherrer, P. (1999). Stabilized plasmid–lipid particles for regional gene therapy: Formulation and transfection properties. *Gene Ther.* **6**, 1438–1447.

[2] Preparation, Characterization, and Biological Analysis of Liposomal Formulations of Vincristine

By Dawn N. Waterhouse, Thomas D. Madden, Pieter R. Cullis, Marcel B. Bally, Lawrence D. Mayer, and Murray S. Webb

Abstract

Vincristine is a dimeric Catharanthus alkaloid derived from the Madagascan periwinkle that acts by binding to tubulin and blocking metaphase in actively dividing cells. While vincristine is widely used in the treatment of a number of human carcinomas, its use is associated with dose-limiting neurotoxicity, manifested mainly as peripheral neuropathy. It is known that the therapeutic activity of vincristine can be significantly enhanced after its encapsulation in appropriately designed liposomal systems. Enhanced efficacy is also associated with a slight decrease in drug toxicity. Thus, the therapeutic index of vincristine can be enhanced significantly through the use of a liposomal delivery system. Vincristine may be encapsulated into liposomes of varying lipid composition by several techniques, including passive loading, pH gradient loading, and ionophore-assisted loading. However, most research has focused on the encapsulation of vincristine in response to a transbilayer pH gradient, which actively concentrates the drug within the aqueous interior of the liposome. This chapter details the preparation and evaluation of liposomal vincristine. Specifically, we elaborate on the components (choice of lipids, molar proportions, etc.), methods (preparation of liposomes, drug loading methods, etc.), critical design features (size, surface charge, etc.), and key biological endpoints (circulation lifetime, bioavailability, efficacy measurements) important to the development of a formulation of vincristine with enhanced therapeutic properties.

Introduction

Vincristine is a bisindole alkaloid that was initially purified from the periwinkle *Catharanthus roseus* (*Vinca rosea*) in the late 1950s and early 1960s (Svoboda, 1961). The resulting agent (Fig. 1) is a lipophilic amine, a weak base, with pK_as at 5.0–5.5 and 7.4 and a partition coefficient (*P*) between octanol and water of log $P = 2.82$ (Leo *et al.*, 1971). In its pure form, vincristine is a solid white to off-white powder, with a melting temperature between 218 and 220° and a molecular weight of 824.94 (Budavari *et al.*, 1989).

The activity of vincristine is cell-cycle specific, manifested in metaphase by the inhibition of tubulin polymerization (Rowinsky and Donehower, 1997). Based on this mechanism of action, it is not surprising that vincristine cytotoxicity is observed against a broad spectrum of tumor cell lines including leukemias, lung (small cell and non-small cell), colon, central

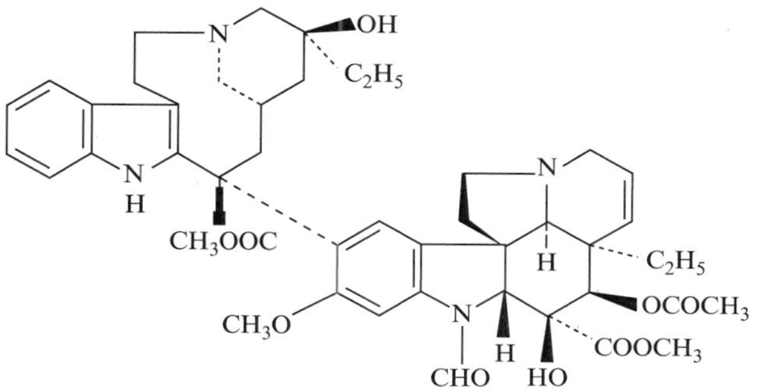

FIG. 1. Chemical structure of vincristine.

nervous system (CNS), melanoma, ovarian, renal, prostate, and breast carcinomas [NCI Cancer Screen Data for NSC 67574 (vincristine)]. Moreover, it might also be anticipated that increased duration of exposure of cells to vincristine would substantially improve the cytotoxicity of the drug. This expectation is supported by the *in vitro* observation that increasing the duration of exposure of L1210 leukemia cells to vincristine from 1 to 72 h was associated with a 10^5-fold increase in cytotoxicity compared with only a 40-fold increase in the cytotoxicity of doxorubicin, whose activity is believed to be less cell-cycle specific (Boman *et al.*, 1995; Mayer *et al.*, 1995).

Vincristine received its first approval by the U.S. Food and Drug Administration (FDA) in 1963 for the treatment of acute leukemia in children. Since then, vincristine has become one of the most commonly used anticancer drugs. Current approvals for vincristine include the treatment of a variety of adult and pediatric cancers such as acute leukemia, Hodgkin's disease, non-Hodgkin's lymphomas, rhabdomyosarcoma, neuroblastoma, and Wilms' tumor and is also used in the treatment of breast cancer and small cell lung cancer (Rowinsky and Donehower, 1997). Vincristine is rarely used as a single agent; instead, vincristine is almost exclusively used as a component of combination chemotherapy protocols. Currently, approximately 50% of the use of vincristine is in the treatment of lymphomas, for example, as part of the CHOP (cyclophosphamide, doxorubicin, vincristine, prednisone) combination for the first-line treatment of non-Hodgkin's lymphomas (Shipp *et al.*, 1997). Based on the cell-cycle–dependent activity of vincristine described above, Jackson *et al.* (1981, 1984) attempted to increase its clinical activity by increased duration of exposure to the drug achieved by continuous infusion. Substantial clinical activity was observed in some patients but was also associated with significant to severe toxicities, particularly neurotoxicities (Jackson *et al.*, 1984).

An approach to the treatment of cancers that would facilitate a significant increase in the duration of exposure to vincristine would be to encapsulate the drug in a liposomal delivery system. The general principles and benefits of encapsulating drugs in liposomes have been described in detail elsewhere (Bally *et al.*, 1998; Tardi *et al.*, 1996). Briefly, encapsulation of therapeutic drugs in appropriately designed liposomal carriers can accomplish some or all of the following: (1) significant increases in the plasma concentration of the drug and extension of its circulation lifetime; (2) increased drug accumulation and duration of exposure at disease sites that are characterized by increased vascular permeability, including tumor sites and sites of inflammation and/or infection; (3) significant increases in the efficacy of the drug, coupled with (4) a decrease, or no increase, in the toxicity of the encapsulated drug compared with the unencapsulated drug. The specific benefits associated with the encapsulation of vincristine have been reviewed previously (Boman *et al.*, 1997; Webb *et al.*, 1995a) and are all described by these general characteristics. The success of liposomal vincristine in preclinical testing has led to the advanced clinical evaluation of an optimized formulation in advanced refractory non-Hodgkin's lymphomas (Sarris *et al.*, 1999, 2000; Webb *et al.*, 1998) and in other indications. A pivotal Phase IIb trial on liposomal vincristine (vincristine sulfate liposome injection; VSLI) in treating non-Hodgkin's lymphoma at second or greater relapse has been completed, and an application for registration of this drug by the U.S. FDA has been submitted.

Preparation and Characterization of Liposomal Vincristine

Methods For Encapsulation of Vincristine in Liposomes

There have been numerous published reports of successful entrapment of vincristine into liposomal carriers. This chapter will discuss several of these, with a focus on the most biologically relevant formulations, and those that are most clinically advanced. Initial consideration will be given to the lipid components and liposome formation, followed by specific methodology for vincristine uptake into preformed liposomes.

Choice of Lipids

Inclusion of Cholesterol. Liposomes traditionally contain a significant proportion of cholesterol (Chol) as a stabilizing lipid. The inclusion of cholesterol in liposome formulations has been shown to prevent lipoprotein-induced vesicle destabilization and concomitant release of the encapsulated drug (Kirby *et al.*, 1980; Scherphof *et al.*, 1978). This leads to increased circulation longevity of drug and can enhance drug accumulation at tumor sites. The cholesterol molecule inserts into a phospholipid bilayer with its hydroxyl group oriented toward the aqueous surface and the planar steroid ring systems parallel to the phospholipid acyl chain orientation. Below the phase transition temperature of the primary phospholipid component of a liposomal formulation, the inclusion of cholesterol has the effect of disrupting acyl chain packing and increasing the fluidity of the gel phase, as well as increasing the membrane permeability. These effects are reversed above the phase transition temperature of the lipids, with reduced chain fluidity and membrane permeability. Cholesterol is typically incorporated at a molar ratio of up to 45% if there is only one other lipid component in the liposome; for example, a distearoylphosphatidylcholine (DSPC)/Chol formulation would have a molar ratio of 55:45. When other lipids are included in the membrane, it may be either the amount of the primary lipid or the cholesterol that is adjusted accordingly, such as the inclusion of 5 mol% poly(ethylene glycol) (PEG)-conjugated lipid in the above example, giving DSPC/Chol/DSPE-PEG, at a molar ratio of 50:45:5 or 55:40:5.

Choice of Membrane Lipid Components. Liposomes may be prepared with a wide range of lipids, including DSPC, egg phosphatidylcholine (EPC), dipalmitoylphosphatidylcholine (DPPC), distearoylphosphatidyl-glycerol (DSPG), sphingomyelin (SM), and others (lipids are obtained from specialty suppliers such as Avanti Polar Lipids Inc., Alabaster, AL or Northern Lipids Inc., Vancouver, BC, Canada). EPC was one of the earliest lipids used in the formation of liposomes but, with the exception of a liposomal doxorubicin formulation approved in Europe and Canada (Myocet), is not used as commonly today as a result of its relatively high membrane permeability coefficients. Consideration must be given to the charge of the lipids used, as well as the degree of saturation and length of the acyl chains. In general, lipids with longer, saturated acyl chains (i.e., DPPC and DSPC) produce liposomes with lower solute per-meability and increased stability and blood residence times (Senior and Gregoriadis, 1982). The inclusion of acidic lipids such as PG or phospha-tidylserine (PS) lowers the tendency of liposomes to aggregate during formation and enhances *in vitro* uptake by cells, however, it will also lead to a shorter blood circulation half-life following intravenous (iv) injection (Senior *et al.*, 1983).

Liposomes often include stabilizing lipids such as the ganglioside G_{M1} (Allen *et al.*, 1989; Gabizon and Papahadjopoulos, 1988) or the more commonly used PEG-conjugated lipids (Papahadjopoulos *et al.*, 1991). PEG-conjugated lipids serve to provide a steric barrier around the lipo-some, most likely protecting it from opsonization and clearance by the mononuclear phagocyte system (MPS). Although they can have varying polymer lengths, PEG chains of average molecular weight 2000 are typi-cally employed. Liposomes possessing PEG lipids are usually eliminated less rapidly by the MPS due to the exclusion of macromolecules such as opsonins from the periliposomal space. This increased circulation lifetime may also result in altered biodistribution compared with liposomes without PEG lipids. It has more recently been suggested that the primary effect of liposome steric stabilization is due to elimination of surface–surface inter-actions that can lead to liposome aggregation or liposome–cell interactions (Johnstone *et al.*, 2001).

The decreased interaction with serum proteins observed for liposomes possessing PEG-conjugated lipids or G_{M1} is also seen in liposomes prepared with the naturally occurring phospholipid sphingomyelin (Webb *et al.*, 1995a). This lipid has the additional advantage of excellent chemical stability. The lability of the ester linkages can limit the "shelf life" of liposomes composed of PC.

Preparation of Liposomes

Preparation of Lipid Films by Solvent Evaporation or Freeze Drying. The amount of material to prepare must depend upon the experimental requirements, and, for simplicity, this chapter will describe formulations with a starting amount of 100 mg total lipid. This may be scaled up or down as required, remembering that if liposomal formulations are to be extruded (see below), a minimum extrusion volume of 1 ml is recommended to minimize sample loss during the procedure. Unless great care is taken, extrusion through a 10-ml extruder can result in a loss of as much as 200 μl (representing 20% of a 1-ml sample).

For a 1-ml preparation containing 100 mg of lipid, the required amounts of lipids, based on the desired molar ratio as described above (e.g., DSPC/Chol 55:45, mol:mol; 71.4:28.6, wt:wt), must be dissolved in organic solvent to ensure homogeneous lipid mixing. This is typically done in chloroform ($CHCl_3$) or $CHCl_3$:methanol mixtures, with a lipid concentration of 10–100 mg/ml, depending on solubility of the individual lipids. We have found that with most lipids, a final solvent volume of 500–1000 μl is sufficient to completely solubilize the lipids. If sphingomyelin is being used, it may be necessary to add several drops of methanol to the $CHCl_3$ to completely dissolve the lipid. The dissolved lipids are combined in one glass test tube, mixing thoroughly to ensure even lipid distribution in the solvent. If liposomes are being used for *in vitro* applications or for pharmacokinetic experiments, a radiolabeled lipid marker can be added to the lipids in chloroform at this stage. We typically use the nonexchangeable, nonmetabolizable marker cholesterylhexadecylether (CHE) labeled with either ^{3}H or [14]C. For specific activity determination, small aliquots are taken from the lipid/$CHCl_3$ solution, with care being taken to prewet the pipette tip to prevent drips during dispensing of small volumes of the organic solvent. It should also be remembered that $CHCl_3$ is a potent quenching agent and must be removed prior to addition of a scintillation cocktail. The dpm values obtained by liquid scintillation counting may then be correlated to the precise amount of lipid known to be in the aliquot taken. If a fluorescent lipid is required as a tracer, we recommend DiI (1,1'-dioctadecyl-3,3,3',3'-tetramethylindocarbocyanine perchlorate; Molecular Probes Eugene, OR), a nonexchangeable lipid, the metabolism of which has not yet been established.

Excess solvent is removed from the lipid solution by evaporation under a gentle stream of nitrogen gas. At this point, the lipids should be left in a small volume of $CHCl_3$, in a slurry-like consistency. If too much solvent is removed at this point, it will be difficult to subsequently obtain a completely dried film. The remaining solvent is removed under a high vacuum (approximately 75 cm Hg) for a minimum of 3 h, or until no solvent remains in the tube. If the starting amount of lipid is above approximately 20 mg, the application of a vacuum should result in a lipid "puff" rising to halfway up the test tube. This puff is desirable, as it maximizes the lipid surface during the drying procedure. Alternately, excess solvent may be removed by rotary evaporation, yielding a thin lipid film on the sides of the flask. Lipid surface area may be maximized in rotary evaporation by the use of glass beads in the flask. This step is also followed by removal of residual solvent on a vacuum pump. When mixtures of chloroform and methanol are used initially to solubilize the lipids, care must be taken to ensure that selective precipitation of one lipid component, for example, cholesterol, does not occur during solvent removal when the methanol content in the mixture increases due to preferential evaporation of chloroform.

Following complete removal of solvent, the lipid film is hydrated with 1 ml of aqueous buffer. For employment of the pH gradient loading technique (Mayer *et al.*, 1993), this buffer is typically 300 mM citric acid, pH 4.0. Other methods to produce a pH gradient include ammonium sulfate (300 mM) gradients or manganese sulfate (300 mM) gradients followed by the use of an ionophore such as A23187 (A.G. Scientific, Inc., San Diego, CA), which acts as a divalent cation shuttle or pump and results in an acidified liposomal interior. For other applications, the hydration solution may be saline, distilled water, HEPES-buffered saline (HBS), or sugar solutions, taking care to keep the osmolality of solutions within a physiological range (270–290 mOsm/kg). For optimal lipid hydration, the film and buffer should be preheated to above the phase transition temperature of the lipid with the highest phase transition temperature (T_c) of the formulation (see Table I) and maintained at this temperature during the hydration procedure. Vigorous vortex mixing or agitation aids in optimization of the hydration. This results in the formation of multilamellar vesicles (MLVs), which are large (>1 μm) and have a heterogeneous size distribution. The suspension of MLVs is then transferred to cryovials and subjected to five freeze–thaw cycles, alternating between liquid nitrogen and a water bath set above the phase transition temperature of the lipid employed. This step allows for equilibration of solute across the bilayers of the vesicles, full hydration of the head groups of the lipids, as well as dissociation of large lipid aggregates (Mayer *et al.*, 1985).

TABLE I
AVERAGE TEMPERATURES OF THE MAIN GEL-TO-LIQUID CRYSTALLINE PHASE TRANSITION (T_c)
FOR LIPIDS COMMONLY USED IN LIPOSOMAL VINCRISTINE FORMULATIONS[a]

Lipid name	Number of acyl carbons: number of cis-unsaturated bonds	T_c (°C)
Dilauroylphosphatidylcholine (DLPC)	12:0	−1.1
Dimyristoylphosphatidylcholine (DMPC)	14:0	23.5
Dipalmitoylphosphatidylcholine (DPPC)	16:0	41.4
Distearoylphosphatidylcholine (DSPC)	18:0	55.1
Dioleoylphosphatidylcholine (DOPC)	18:1	−18.4
Dimyristoylphosphatidylethanolamine (DMPE)	14:0	49.6
Dipalmitoylphosphatidylethanolamine (DPPE)	16:0	64.0
Dioleoylphosphatidylethanolamine (DOPE)	18:1	−16
Sphingomyelin (SM), N-palmitoyl-sphingosyl	16:0	40.9
Dipalmitoylphosphatidylserine (DPPS), Na salt	16:0	54
Dipalmitoylphosphatidylglycerol (DPPG), Na salt	16:0	40.9

[a] Data obtained from Marsh, D. (1990). "CRC Handbook of Lipid Bilayers." CRC Press, Boca Raton, FL.

As an alternative to the preceding method, MLVs may also be generated without forming lipid films by direct combination of lipids dissolved in EtOH with aqueous buffer. Briefly, lipids in the appropriate molar amounts (i.e., DSPC/Chol or SM/Chol, 55:45, mol:mol) dissolved in 95% EtOH (1 ml/100 mg lipid) are heated at 60° for 30 min. This ethanol solution of lipids is then added dropwise to preheated 300 mM citric acid (3 ml/100 mg lipid)(pH 4.0 or 2.0) with constant vortexing. The solution is heated an additional 30 min, resulting in MLVs that may be taken directly to the size reduction step (without using a freeze–thaw procedure) as discussed below. Following extrusion (below), liposomes must be dialyzed against 300 mM citric acid (pH 4.0 or 2.0) to remove ethanol, then further dialyzed against the desired external aqueous buffer, such as HBS, pH 7.5, hence establishing a pH gradient across the liposomal membrane (Boman et al., 1994).

For in vivo applications of liposomes, it is important to further reduce the size of these liposomes. It has been experimentally determined that vesicles between 100 and 200 nm in diameter are optimal for intravenous injection (Allen and Everest, 1983). These vesicles are small enough to gain access to tissues in areas of inflammation and at tumor sites due to the fenestrated/leaky vasculature in these areas. Smaller vesicles are also eliminated less rapidly from the circulation, hence increasing the blood residence time for the encapsulated drug. To obtain smaller vesicle sizes, MLVs may be either sonicated or extruded. We use the Lipex Extruder (Northern Lipids, Inc., Vancouver, BC, Canada) for this purpose, as described below. For formation of approximately 120-nm-diameter vesicles, or large unilamellar vesicles (LUVs), with a homogeneous size distribution, the MLVs are forced through polycarbonate membranes with 100-nm-diameter pores, under high pressure (300–600 psi) and with the thermobarrel of the extruder equilibrated to a temperature 5° above the T_c of the lipids. It is recommended that the filters be prewet with the buffer in which the liposomes were prepared to minimize volume loss during the extrusion procedure. A minimum of 10 passes through the extruder will typically result in homogeneous size distributions of liposomes (Hope *et al.*, 1985) (see discussion of size analysis below).

The extrusion procedure, if the hydration buffer is 300 mM citrate, pH 4.0, results in a suspension of liposomes in which both the interior aqueous volume and the external buffer are at pH 4.0. To establish a pH gradient for drug loading, the external buffer must either be raised to a pH of 7.0–7.5 by the addition of, for example, 0.5 M Na$_2$HPO$_4$, or by passing the liposomes over a gel filtration column such as G-50 Sephadex that has been preequilibrated in the desired buffer, such as 150 mM NaCl, 20 mM HEPES, pH 7.5 (HBS).

Loading Preformed LUVs with Vincristine. To the formed LUVs with an established pH gradient, sufficient vincristine sulfate solution (e.g., Oncovin, Eli Lilly & Co., Indianapolis, IN) is added for a final drug to lipid ratio (wt/wt) of between 0.05 and 0.2. For a starting solution of 100 mg total lipid, this will be between 5 and 20 ml of a 1 mg/ml vincristine sulfate solution. The drug and liposome mixture is incubated above about 45–65° (the actual temperature required depends upon lipid composition) for a minimum of 10 min to effect drug uptake into the liposomes in response to the pH gradient. Vincristine is highly permeable to lipid membranes in its neutral form. Upon entering the liposome interior, it becomes protonated and is no longer as readily able to cross the lipid bilayer, effectively trapping it within the aqueous core of the liposome. It is important to note that the protonation of vincristine in the liposome interior reduces the hydrogen ion pool; therefore, it is important that the starting pH gradient is sufficient to ensure complete uptake of vincristine (typically pH 4.0 or 2.0 within the liposomes, pH 7.2 to 7.5 in the external buffer). Redistribution of vincristine between the intravesicular medium and the external solution is in accordance with the Henderson–Hasselbach equation. In a simplified form, this predicts that the drug concentration gradient across the liposomal membrane will be equal to the proton concentration gradient. Also, it should be noted that the lower the internal pH the lower the efflux rate of vincristine from the liposome (Boman *et al.*, 1993). Interior pH values less than 4 are, of course, difficult to employ experimentally due to the potential for lipid and drug degradation in the loaded liposome.

Ionophore (A23187 or Nigericin)-Mediated Loading of Vincristine into Liposomes. It is possible to use DSPC/Chol or SM/Chol liposomes with either the channel ionophore A23187 or nigericin to encapsulate several drugs, including vincristine. DSPC/Chol or SM/Chol (55:45, mol:mol) liposomes are prepared as outlined above, with the lipid film being hydrated in either 300 mM MnSO$_4$ or MgSO$_4$ (for A23187-mediated loading) or 300 mM K$_2$SO$_4$ (for nigericin-mediated loading). Once extruded, liposomes are dialyzed against two changes of 100 volumes of 300 mM sucrose to exchange the external buffer, thus creating a salt gradient across the liposomal membrane.

For A23187-mediated uptake of vincristine, the ionophore is added to the liposomes (0.1 μg A23187/μmol lipid) and incubated at 65° for 5 min before the addition of vincristine (0.05:1.0–0.2:1.0 drug-to-lipid ratio, wt/wt) and 200 mM ethylenediaminetetraacetic acid (EDTA), pH 7.0 (final concentration of 30 mM). The presence of EDTA in this system is important to chelate manganese or magnesium ions effluxing from the liposomes to drive drug uptake to completion. Following a further incubation of 10–60 min at 65°, the vincristine is encapsulated within the liposomes, with entrapment approaching 100% by virtue of the pH gradient created with the coupling of external transport of one manganese/magnesium ion to the internal transport of two protons. Removal of ionophore may be easily accomplished by either dialysis or column chromatography. Vincristine encapsulation using nigericin (1 ng/μmol lipid) is performed in a similar manner, except that a potassium ion gradient is employed and EDTA is not required, with nigericin catalyzing a one-for-one exchange of K$^+$ for H$^+$ (Fenske *et al.*, 1998).

Formation of Vincristine Precipitates Using Suramin, Heparin Sulfate, or Dextran Sulfate. It has been determined for some formulations that leakage of vincristine from the liposomal carrier occurs during the plasma distribution phase, thereby diminishing the amount of liposomal drug that will accumulate within the tumor site, as evidenced by faster elimination of vincristine than the lipid carrier. It is possible to extend the plasma distribution phase of vincristine to more closely match that of the liposomal carrier, by causing a polyanion/vincristine precipitate to form within the liposomes. We will not provide extensive detail in this area, since experiments were unable to demonstrate any enhanced efficacy, although precipitates form within the liposomes and the leakage rate is reduced (Zhu *et al.*, 1996). These researchers formed pegylated liposomes using an ethanol injection method, where lipids dissolved in ethanol were injected into 125 mM ammonium sulfate containing heparin sulfate, dextran sulfate, or suramin. Polyanion to vincristine ratios were estimated to be 8.7:1, 1.8:1, and 2.8:1 for vincristine-loaded liposomes containing dextran sulfate, heparin, and suramin, respectively.

Characterization of Liposomal Vincristine

Determination of Trapping Efficiency. To determine the trapping efficiency of vincristine in liposomes, a comparison of the drug-to-lipid ratio is made before and after a column chromatography step to remove unencapsulated drug. Specifically, aliquots are taken both before and after running a portion of the sample down a size exclusion chromatographic column such as G-50 Sephadex, equilibrated in the appropriate buffer, and measured for both lipid and vincristine as outlined above. The drug-to-lipid ratio (wt/wt) in the column eluate is compared with the precolumn drug-to-lipid ratio for determination of percent encapsulation of the drug within the liposomes. Methods for quantifying the drug and lipid are described below.

Determination of both lipid and vincristine concentrations may be either by addition of a radiolabeled marker in the initial formulation or by spectrophotometric assay. We use [^{3}H]- or [^{14}C]CHE as a lipid label as discussed above, bearing in mind that the concentration of CHE must not be so high as to prevent adequate incorporation and [^{3}H]vincristine (Amersham Canada Inc.) for determination of vincristine concentrations. The [^{3}H]vincristine should be pipetted into a clean, dry test tube and the solvent evaporated off prior to thorough mixing with the stock vincristine sulfate solution. Aliquots for determination of specific activity by liquid scintillation counting are taken prior to addition of desired amount to liposomes. This activity is then correlated to the concentration of vincristine in the aliquot and a specific activity (dpm/μmol or dpm/μg) determined. Caution should be exercised when working with [^{3}H]vincristine, as over time the marker becomes less accurate, requiring calibration with a spectrophotometric determination of vincristine concentration. If spectrophotometric assays are preferred for vincristine, the following procedures may be employed. Sample aliquots and standards (0–100 μg/ml Oncovin in H$_2$O) are brought to a volume of 200 μl with distilled, deionized water. Liposomes are solubilized by addition of 800 μl EtOH. Absorbance is measured at 297 nm and compared with the standard curve for determination of concentration.

Phosphate assays follow the method of Fiske and Subbarow (1925). Standards are prepared from 2 mM Na$_2$HPO$_4$, from 0–200 nmol total phosphate. Standards and samples are pipetted into thick-walled test tubes, to which 750 μl perchloric acid is added and heated at 180° for a minimum of 2 h or until solutions become colorless. To prevent sample loss, tubes are topped with glass marbles, allowing venting of excess steam during the heating steps. After cooling to room temperature, 750 μl Fiske solution (150 g NaHSO$_3$ and 5 g Na$_2$SO$_3$ dissolved in 1 liter distilled H$_2$O, plus 2.5 g 1-amino-2-naphthol-4-sulfonic acid; covered and stirred at 40° to dissolve crystals, stored overnight in the dark, filtered, and stored in the dark) and 7 ml ammonium molybdate solution (800 ml distilled H$_2$O plus 2.2 g ammonium molybdate and 20 ml H$_2$SO$_4$, made up to 1 liter with distilled H$_2$O) are added to the tubes, and the contents are gently mixed. Tubes are further heated at 100° for 20 min for color development, cooled to room temperature, and the absorbance measured at 815 nm.

Determination of Liposome Size. The elimination rate of liposomes from the circulation is very sensitive to liposome size, with larger liposomes eliminated more quickly than small ones. It is therefore critical to measure this parameter prior to administration of a liposomal drug. Several techniques are available, but we have found that quasielastic light scattering (QELS) is the most reliable and efficient when dealing with relatively homogeneous samples. This method provides analysis of the mean size, as well as giving information with respect to the distribution of diameters within the sample. We use the NICOMP model Nicomp 270 submicron particle sizer (Pacific Scientific, Santa Barbara, CA), with an argon laser operating at 632.8 nm according to the manufacturer's instructions. Liposomes prepared by extrusion typically exhibit size distributions in good agreement with a Gaussian fit. Mean diameters can then be expressed on a number, volume, or light-scattering intensity basis. By convention, mean diameter on a volume basis is generally used.

pH Gradient Determination. When utilizing the pH gradient method for loading vincristine into liposomes, there is an anticipated drug loading of close to 100%. Should the experimenter achieve less than this, a possible explanation is the lack of a sufficient transmembrane pH gradient. This may be easily verified by the use of [^{14}C]methylamine (Harrigan *et al.*, 1992). The uncharged form of methylamine (MeNH$_3$) is highly lipid permeable and will rapidly equilibrate across a membrane (Rottenberg *et al.*, 1972) according to the following equation:

$$\frac{[\text{MeNH}]^+_{\text{in}}}{[\text{MeNH}_3^+]_{\text{out}}} = \frac{[\text{H}^+]_{\text{in}}}{[\text{H}^+]_{\text{out}}} \tag{1}$$

Upon entry to the acidified aqueous interior of the liposome, the methylamine becomes protonated (MeNH$_3^+$) and is hence unable to readily cross the membrane. The subsequent measurement of trapped versus untrapped probe allows the proton gradient (Δ pH) to be determined.

To measure ΔpH, aliquots of liposomes are diluted into the same buffer used to initially create the gradient (i.e., HBS, pH 7.5), with the addition of 1 μCi [^{14}C]methylamine, to a final lipid concentration of approximately 2 mM. Following a 10-min incubation at a temperature just over the phase transition temperature of the lipid, an aliquot is passed over a G-50 Sephadex minispin column (1 ml bed volume, column spun at 2000 g for 2 min) then ^{14}C and phosphorus determined in the prespin and eluate samples. The transmembrane ΔpH is then simply determined using the following formula:

$$\Delta\text{pH} = \log \frac{[\text{MeNH}_3]_{\text{in}}}{[\text{MeNH}_3]_{\text{out}}} \qquad (2)$$

This method of determining a transmembrane proton gradient has been found to be accurate to a difference of up to 3 pH units given a sufficient interior buffering capacity (i.e., citrate concentration of 20 mM or higher) and absence of significant transmembrane osmotic gradient and requires measurement of the interior (trapped) volume of the liposomes, as described previously (Harrigan *et al.*, 1992).

Biological Analysis of Liposomal Vincristine

It has been our experience that the most informative and rapid means of evaluating and characterizing liposomal formulations of cytotoxic agents, including vincristine, is by *in vivo* experiments. For example, it is possible to compare vincristine release rates from different liposomal formulations using *in vitro* methods (Webb *et al.*, 1995a,b). While drug release rates determined *in vitro* may show a good qualitative correlation, for different liposomes, with the *in vivo* drug release rates, the absolute drug release kinetics are significantly faster *in vivo* (Webb *et al.*, 1995a). Similarly, *in vitro* evaluation of the efficacy of liposomal cytotoxic drugs such as vincristine is complicated by the slow or negligible release of drug from liposomes under *in vitro* conditions. For this and other reasons, we focus on the use of *in vivo* assessments for the characterization and optimization of liposomal formulations of agents such as vincristine.

Pharmacokinetics

Pharmacokinetic characterization is the most useful tool for the rapid screening and optimization of formulations of liposomal vincristine, as well as formulations of other liposomal agents. For liposomal vincristine, intravenous administration is the route of choice and is typically achieved via the tail vein with volumes not exceeding 200 μl per 20-g mouse. Care needs to be taken administering formulations of drug loaded using the citrate-pH gradient method, particularly at high doses and/or low drug-to-lipid ratios, that adverse toxicities are not caused by citrate-mediated chelation of cations in the blood. Formulations in this category may require the exchange of external citrate for saline or dextrose solutions using dialysis, column chromatography, or tangential flow methods prior to administration of the formulation. The choice of mouse strain for appropriate pharmacokinetic studies should reflect the ultimate use of the formulation for planned efficacy studies. While pharmacokinetic parameters are not markedly different in different strains, the dose for the pharmacokinetic study needs to be at or near the maximum tolerated dose (MTD). On single administration of liposomal vincristine, MTDs can vary from about 4–5 mg/kg for outbred mice (i.e., CD1) to 2 mg/kg for immunocompromised mice (i.e., SCID). As a general comment, all *in vivo* studies should be done in accordance with local animal care guidelines, such as those defined in Canada by the Canadian Council of Animal Care (http://www.ccac.ca). As lethal dose assessments are not permitted by these guidelines, we determine MTD values using small dose range–finding studies (involving less than 10 mice in total) in which the MTD is defined as the dose that achieves a nadir weight loss of 15%.

Acceptable pharmacokinetic criteria for liposomal formulations of vincristine are based on all major components of the preparation. Specifically, (1) liposome elimination half-life of at least 8 h, based on the elimination of a nonexchangeable lipid marker such as CHE; (2) half-life for the release of vincristine from the liposomal carrier, based on the change of the vincristine/lipid ratio in the plasma of greater than 15 h; and (3) vincristine elimination half-life of at least 4 h. The best available formulations of vincristine have lipid elimination half-lives greater than 50 h, half-lives for the vincristine release from the liposome in excess of 25 h, and vincristine elimination half-lives of at least 12 h (Boman *et al.*, 1997). To date, no formulations of liposomal vincristine have been described in which the rate of drug release is reduced sufficiently to result in decreased antitumor efficacy.

Since free vincristine is rapidly eliminated from the circulation, it is reasonable to assume that all vincristine observed in the plasma after intravenous administration of liposomal drug represents only that drug that is encapsulated in the liposomes. However, in some instances it may be necessary to quantify the contributions of both unencapsulated and liposomally encapsulated vincristine to the total plasma concentration. To separate unencapsulated vincristine from encapsulated vincristine in plasma after intravenous administration, aliquots of plasma are placed into

Microcon-30 ultracentrifugation devices (Millipore, Bedford, MA) and centrifuged at 10,000 rpm at 4° for 15 min (Mayer and St.-Onge, 1995; Krishna, et al., 2001). The resultant ultrafiltrate contains free vincristine only, but does not quantitatively account for protein-bound vincristine. However, as protein-bound vincristine has been experimentally determined to represent 40% of the total nonliposomal vincristine plasma content (Mayer and St.-Onge, 1995), the values of vincristine obtained from the ultrafiltrate represent 60% of the total nonliposomal vincristine concentration in the plasma.

Antitumor Efficacy Endpoints

Therapeutic activity of vincristine encapsulated in liposomal delivery vehicles may be measured *in vivo* in a range of tumor model types and assessed by several different means. Methods of determining the therapeutic activity in animal tumor models are described below.

Solid (Subcutaneous) Tumor Model Evaluation

1. Tumor volume (mass): Mean tumor volumes are determined from vernier caliper measurements of perpendicular length and width measurements (height measurements can often be obtained as well). Tumor volume (mass; units of ml^3 or mg) is calculated from

$$\text{Volume} = (\text{length} \times \text{width}^2)/2 \qquad (3)$$

   or

$$\text{Volume} = \pi/6 \times (\text{length} \times \text{width} \times \text{height}) \qquad (4)$$

   Data are plotted with respect to time.
2. Tumor weight inhibition (TWI%): At a defined time point the mean tumor weight of a treated group divided by the mean tumor weight of the control group, minus 1. This value is then multiplied by 100 to define a percent change.
3. Tumor growth delay (T–C): Median time in days for the treated (T) groups to reach an arbitrarily determined tumor size (i.e., 300 mg) minus median time in days for the control group (C) to reach the same tumor size.
4. Tumor regression: Treatment results in reductions in tumor size (mass) often with disappearance of the tumor.

Intravenous and/or Intraperitoneal Model Evaluation

1. Increase in life span (ILS%): Percentage increase in life span (days) of treated groups versus control or untreated groups.

2. Tumor growth delay (T–C): Median time in days for treated (T) group survival minus median time in days for control (C) group survival.
3. Long-term survivors (Cures): Treatment results in long-term survival where treatment groups survive up to and beyond three times the survival times of untreated or control groups.

Conclusions

The preparation of a liposomal drug formulation having enhanced therapeutic value is a challenging process that requires attention to a variety of important physicochemical characteristics (liposome size, trapped volume, transmembrane pH gradient, etc.), biochemical properties of both lipid and drug components (charge, pK_as, solubilities in polar and nonpolar environments, drug stability at acidic pH, etc.), biological performance (drug retention, liposome circulation longevity, efficacy, etc.), as well as methodological issues. In general terms, the processes described in this chapter for liposomal encapsulation of vincristine are similar to those that would be encountered in the development of many liposomal cytotoxic drugs (see Chapter 4). More specifically, the dramatic increases in therapeutic value occurring as a consequence of liposomal encapsulation of vincristine would also be expected to occur for other cytotoxic drugs whose activity is primarily cell-cycle dependent. Agents with this mechanism of action include the additional vinca alkaloids, as well as other agents, such as the taxanes, with activity against microtubules.

References

Allen, T. M., and Everest, J. M. (1983). Effect of liposome size and drug release properties on pharmacokinetics of encapsulated drug in rats. *J. Pharmacol. Exp. Ther.* **226,** 539–544.

Allen, T. M., Hansen, C., and Rutledge, J. (1989). Liposomes with prolonged circulation times: Factors affecting uptake by reticuloendothelial and other tissues. *Biochim. Biophys. Acta* **981,** 27–35.

Bally, M. B., Lim, H., Cullis, P. R., and Mayer, L. D. (1998). Controlling the drug delivery attributes of lipid-based drug formulations. *J. Liposome Res.* **8,** 299–335.

Boman, N. L., Mayer, L. D., and Cullis, P. R. (1993). Optimization of the retention properties of vincristine in liposomal systems. *Biochim. Biophys. Acta* **1152,** 253–258.

Boman, N. L., Masin, D., Mayer, L. D., Cullis, P. R., and Bally, M. B. (1994). Liposomal vincristine which exhibits increased drug retention and increased circulation longevity cures mice bearing P388 tumors. *Cancer Res.* **54,** 2830–2833.

Boman, N. L., Bally, M. B., Cullis, P. R., Mayer, L. D., and Webb, M. S. (1995). Encapsulation of vincristine in liposomes reduces its toxicity and improves its antitumor efficacy. *J. Liposome Res.* **5,** 523–541.

Boman, N. L., Cullis, P. R., Mayer, L. D., Bally, M. B., and Webb, M. S. (1997). Liposomal vincristine: The central role of drug retention in defining therapeutically optimized anticancer formulations. *In* "Long Circulating Liposomes: Old Drugs, New Therapeutics" (M. C. Woodle and G. Storm, eds.), p. 29. Landes Bioscience, Georgetown, TX.

Budavari, S., O'Neil, M. J., and Smith, A. (1989). "The Merck Index." Merck & Co., Inc., Rahway, NJ..

Fenske, D. B., Wong, K. F., Maurer, E., Maurer, N., Leenhouts, J. M., Boman, N., Amankwa, L., and Cullis, P. R. (1998). Ionophore-mediated uptake of ciprofloxacin and vincristine into large unilamellar vesicles exhibiting transmembrane ion gradients. *Biochim. Biophys. Acta* **1414,** 188–204.

Fiske, C. H., and Subbarow, Y. (1925). The colorimetric determination of phosphorus. *J. Biol. Chem.* **66,** 375–400.

Gabizon, A., and Papahadjopoulos, D. (1988). Liposome formulations with prolonged circulation time in blood and enhanced uptake by tumors. *Proc. Natl. Acad. Sci. USA* **85,** 6949–6953.

Harrigan, P. R., Hope, M. J., Redelmeier, T. E., and Cullis, P. R. (1992). Determination of transmembrane pH gradients and membrane potentials in liposomes. *Biophys. J.* **63,** 1336–1345.

Hope, M. J., Bally, M. B., Webb, G., and Cullis, P. R. (1985). Production of large unilamellar vesicles by a rapid extrusion procedure: Characterization of size, trapped volume, and ability to maintain a membrane potential. *Biochim. Biophys. Acta* **812,** 55–65.

Jackson, D. V., Jr., Paschold, E. H., Spurr, C. L., Muss, H. B., Richards, F., II, Cooper, M. R., White, D. R., Stuart, J. J., Hopkins, J. O., Rich, R., Jr., and Wells, H. B. (1984). Treatment of advanced non-Hodgkin's lymphoma with vincristine infusion. *Cancer* **53,** 2601–2606.

Jackson, D. V., Jr., Sethi, V. S., Pharm, M., Spurr, C. L., Willard, V., White, D. R., Richards, F., Stuart, J. J., Muss, H. B., Cooper, M. R., Homesley, H. D., Jobson, V. W., and Castle, M. C. (1981). Intravenous vincristine infusion: Phase I trial. *Cancer* **48,** 2559–2564.

Johnstone, S. A., Masin, D., Mayer, L., and Bally, M. B. (2001). Surface-associated serum proteins inhibit the uptake of phosphatidylserine and poly(ethylene glycol) liposomes by mouse macrophages. *Biochim. Biophys. Acta* **1513,** 25–37.

Kirby, C., Clarke, J., and Gregoriadis, G. (1980). Cholesterol content of small unilamellar liposomes controls phospholipid loss to high density lipoproteins in the presence of serum. *FEBS Lett.* **111,** 324–327.

Krishna, R., Webb, M. S., St.-Onge, G., and Mayer, L. D. (2001). Liposomal and non-liposomal drug pharmacokinetics after administration of liposome-encapsulated vincristine and their contribution to drug tissue distribution properties. *J. Pharmacol. Exp. Ther.* **298,** 1206–1212.

Leo, A., Hansch, C., and Elkins, D. (1971). Partition coefficients and their uses. *Chem. Rev.* **71,** 525–616.

Mayer, L. D., and St.-Onge, G. (1995). Determination of free and liposome-associated doxorubicin and vincristine levels in plasma under equilibrium conditions employing ultrafiltration techniques. *Anal. Biochem.* **232,** 149–157.

Mayer, L. D., Gelmon, K., Cullis, P. R., Boman, N., Webb, M. S., Embree, L., Tolcher, T., and Bally, M. B. (1995). Preclinical and clinical studies with liposomal vincristine. *In* "Progress in Drug Delivery Systems IV" (S. Hirota, ed.), p. 151. Biomedical Research Foundation, Tokyo.

Mayer, L. D., Hope, M. J., Cullis, P. R., and Janoff, A. S. (1985). Solute distributions and trapping efficiencies observed in freeze-thawed multilamellar vesicles. *Biochim. Biophys. Acta* **817,** 193–196.

Mayer, L. D., Madden, T. D., Bally, M. B., and Cullis, P. R. (1993). Preparation of streptavidin liposomes for use in ligand specific targeting applications. *In* "Liposome Technology, Volume II. Entrapment of Drugs and Other Materials" (G. Gregoriadis, ed.), p. 27. CRC Press, Boca Raton, FL.

Papahadjopoulos, D., Allen, T. M., Gabizon, A., Mayhew, E., Matthay, K., Huang, S. K., Lee, K. D., Woodle, M. C., Lasic, D. D., Redemann, C., and Martin, F. J. (1991). Sterically stabilized liposomes: improvements in pharmacokinetics and antitumor therapeutic efficacy. *Proc. Natl. Acad. Sci. USA* **88,** 11460–11464.

Rottenberg, H., Grunwald, T., and Avron, M. (1972). Determination of pH in chloroplasts. I. Distribution of (14 C) methylamine. *Eur. J. Biochem.* **25,** 54–63.

Rowinsky, E. K., and Donehower, R. C. (1997). Non-small cell lung cancer. *In* "Cancer. Principles and Practices of Oncology" (V. T. DeVita, Jr., S. Hellman, and S. A. Rosenberg, eds.), p. 467. Lippincott-Raven, Philadelphia.

Sarris, A. H., Hagemeister, F., Romaguera, J., Rodriguez, M. A., McLaughlin, P., Tsimberidou, A. M., Medeiros, L. J., Samuels, B., Pate, O., Oholendt, M., Kantarjian, H., Burge, C., and Cabanillas, F. (2000). Liposomal vincristine in relapsed non-Hodgkin's lymphomas: Early results of an ongoing phase II trial. *Ann. Oncol.* **11,** 69–72.

Sarris, A. H., Romaguera, J., Hagemeister, F., Rodriguez, M. A., McLaughlin, P., Dang, N., Tsimberidou, A. M., Medeiros, L. J., Samuels, B., Oholendt, M., Pate, O., Burge, C., and Cabanillas, F. (1999). Liposomal vincristine: A phase II trial in relapsed or refractory non-Hodgkin's lymphoma (NHL). *American Society of Hematology 41st Annual Meeting* Abstract #412.

Scherphof, G., Roerdink, F., Waite, M., and Parks, J. (1978). Disintegration of phosphatidyl-choline liposomes in plasma as a result of interaction with high-density lipoproteins. *Biochim. Biophys. Acta* **542,** 296–307.

Senior, J., and Gregoriadis, G. (1982). Is half-life of circulating liposomes determined by changes in their permeability? *FEBS Lett.* **145,** 109–114.

Senior, J., Gregoriadis, G., and Mitropoulos, K. A. (1983). Stability and clearance of small unilamellar liposomes. Studies with normal and lipoprotein-deficient mice. *Biochim. Biophys. Acta* **760,** 111–118.

Shipp, M. A., Mauch, P. M., and Harris, N. L. (1997). Non-Hodgkin's lymphomas. *In* "Cancer. Principles and Practices of Oncology" (V. T. DeVita, Jr., S. Hellman, and S. A. Rosenberg, eds.), p. 2165. Lippincott-Raven, Philadelphia.

Svoboda, G. H. (1961). Alkaloids of Vinca rosea. IX. Extraction and characterization of leurosidine and leurocristine. *Lloydia* **24,** 173–178.

Tardi, P. G., Boman, N. L., and Cullis, P. R. (1996). Liposomal doxorubicin. *J. Drug Target* **4,** 129–140.

Webb, M. S., Harasym, T. O., Masin, D., Bally, M. B., and Mayer, L. D. (1995a). Sphingomyelin-cholesterol liposomes significantly enhance the pharmacokinetic and therapeutic properties of vincristine in murine and human tumour models. *Br. J. Cancer* **72,** 896–904.

Webb, M. S., Wheeler, J. J., Bally, M. B., and Mayer, L. D. (1995b). The cationic lipid stearylamine reduces the permeability of the cationic drugs verapamil and prochlorper-azine to lipid bilayers: Implications for drug delivery. *Biochim. Biophys. Acta* **1238,** 147–155.

Webb, M. S., Logan, P., Kanter, P. M., St-Onge, G., Gelmon, K., Harasym, T., Mayer, L. D., and Bally, M. B. (1998). Preclinical pharmacology, toxicology and efficacy of sphingomyelin/cholesterol liposomal vincristine for therapeutic treatment of cancer. *Cancer Chemother. Pharmacol.* **42,** 461–470.

Zhu, G., Oto, E., Vaage, J., Quinn, Y., Newman, M., Engbers, C., and Uster, P. (1996). The effect of vincristine-polyanion complexes in STEALTH liposomes on pharmacokinetics, toxicity and anti tumor activity. *Cancer Chemother. Pharmacol.* **39,** 138–142.

[3] Lipophilic Arabinofuranosyl Cytosine Derivatives in Liposomes

By Reto Schwendener and Herbert Schott

Abstract

Highly lipophilic drugs can be used therapeutically only by the addition of possibly toxic solubilizing agents or by development of complex pharmaceutical formulations. One way of overcoming these disadvantages is the incorporation of such drugs into the bilayer matrix of phospholipid liposomes. To this end, we chose the approach of chemical transformation of water-soluble nucleosides of known cytotoxic properties into lipophilic drugs or prodrugs. Due to their insolubility, we developed formulations that can be used for intravenous applications in which the lipophilic molecules are incorporated into lipid bilayer membranes of small liposomes. We chose 1-β-D-arabinofuranosylcytosine (ara-C) as a cytotoxic nucleoside, and we demonstrated that N^4-acyl derivatives of ara-C were active *in vivo* in various murine tumor models as liposomal formulations. However, the protection against enzymatic deamination was only partially achieved and was insufficient for significant improvement of cytotoxic properties. Thus, we synthesized a new class of N^4-alkyl-ara-C derivatives. The most effective derivative, N^4-octadecyl-ara-C (NOAC), is highly lipophilic and extremely resistant toward deamination. NOAC exerts excellent antitumor activity after oral and parenteral therapy. The activity of NOAC against freshly explanted clonogenic cells from human tumors was determined and compared with conventional antitumor agents. NOAC was used in two liposomal preparations, a stable lyophilized and a freshly prepared liquid formulation. Both formulations inhibited tumor colony formation equally in a concentration-dependent fashion. At optimal conditions, liposomal NOAC had significantly better activity compared with the clinically used drugs cisplatin, doxorubicin, 5-fluorouracil, gemcitabine, mitomycin C, and etoposide. Furthermore, in a hematopoietic stem cell assay, NOAC was less toxic than ara-C and doxorubicin by factors ranging from 2.5 to 200, indicating that this drug is well tolerated at high doses.

Introduction

In the field of drug delivery, liposomes are predominantly used as carriers for hydrophilic molecules that are entrapped within their aqueous inner volume. Long circulating poly(ethylene glycol)–modified liposomes with the cytotoxic drug doxorubicin are an example of such a clinically used formulation (Gabizon and Martin, 1997). Highly lipophilic drugs can be used therapeutically only by the addition of possibly toxic solubilizing agents or by development of complex pharmaceutical formulations. One practical way of overcoming these disadvantages is the incorporation of such drugs into the bilayer matrix of phospholipid liposomes.

We chose the approach of the chemical transformation of water-soluble nucleosides of known cytotoxic properties into lipophilic drugs or pro-drugs. Due to the insolubility of these compounds, we developed formulations that can be used for intravenous applications in which the lipophilic molecules are incorporated into lipid bilayer membranes of small uni-lamellar liposomes. Initially, we chose 1-β-D-arabinofuranosylcytosine (ara-C) as a cytotoxic nucleoside because the major untoward properties of the drug are the short plasma half-life and rapid degradation by dea-mination to the inactive metabolite 1-β-D-arabinofuranosyluracil (ara-U), a disadvantage that also impedes the oral application of ara-C.

To overcome these shortcomings, a large number of 5'- and N^4-sub-stituted ara-C derivatives have been synthesized and characterized in the past. We demonstrated that N^4-acyl derivatives of ara-C were active *in vivo* as liposomal formulations at concentrations two to four times lower than ara-C (Rubas *et al.*, 1986). However, the protection of the N^4-acyl-ara-C derivatives against enzymatic deamination was only partially achieved and was not sufficient for significant improvement of cytotoxic properties. Based on these findings, we synthesized a new class of N^4-alkyl-ara-C derivatives (Schott *et al.*, 1994; Schwendener and Schott, 1992). These compounds show a typical structure–activity correlation between the length of the alkyl side chain and their antitumor activity (Schwendener and Schott, 1992). The most effective derivative, N^4-octadecyl-ara-C (NOAC), is highly lipophilic and extremely resistant toward deamination. NOAC exerts excellent antitumor activity after oral and parenteral therapy (Horber *et al.*, 1995a,b; Schwendener *et al.*, 1995a,b). In a recent study, the activity of NOAC against freshly explanted clonogenic cells from human tumors was determined and compared with conventional antitumor agents. NOAC was used in two liposomal preparations, a stable lyophilized and a freshly prepared liquid formulation. Both formulations inhibited tumor colony formation equally in a concentration-dependent fashion. At optimal conditions, liposomal NOAC had significantly better activity compared with the clinically used drugs cisplatin, doxorubicin, 5-fluorouracil, gemci-tabine, mitomycin C, and etoposide. Furthermore, in a hematopoietic stem cell assay, NOAC was less toxic than ara-C and doxorubicin by factors ranging from 2.5 to 200, indicating that this drug is well tolerated at high doses (Schwendener *et al.*, 2001). In contrast to the parent drug ara-C, the cellular uptake of NOAC is nucleoside-transporter independent, and only insignificant amounts are phosphorylated to ara-C triphosphate (ara-CTP). Furthermore, NOAC is cytotoxic in ara-C-resistant HL-60 cells, and treatment of multidrug resistant tumor cells did not induce P-170 glycoprotein expression, suggesting that the N^4-alkyl-ara-C derivatives are able to circumvent MDR1 multidrug resistance (Horber et al., 1995c; Schwendener and Schott, 1996; Schwendener et al., 1996). We conclude therefore that the mechanisms of action of the N^4-alkyl-ara-C derivatives are different from those of ara-C.

Recently, we further modified NOAC by the synthesis of new duplex drugs that combine the clinically used drugs ara-C and 5-fluorodeoxyuridine (5-FdU) with NOAC, yielding the heterodinucleoside phosphates arabinocytidylyl-(5' → 5')-N^4-octadecyl-1-β-D-arabinocytidine (ara-C-NOAC) and 5-fluoro-2'-deoxyuridylyl-(5' → 5')-N^4-octadecyl-1-β-D-arabinocytidine (5-FdU-NOAC). Due to the combination of the effects of both active molecules that can be released in the cells as monomers or as the corresponding monophosphates, the cytotoxic activity of the duplex drugs is expected to be more effective than the monomers. Furthermore, we anticipate that monophosphorylated ara-C (ara-CMP) or 5-FdU (5-FdUMP), respectively, are directly formed in the cell after enzymatic cleavage of the duplex drugs. Thus, monophosphorylated molecules would not have to pass the first phosphorylation step, which is known to be rate limiting.

The chemical modification of ara-C and other nucleosides and their formulation in liposomes render these new cytotoxic drugs interesting candidates for further development.

Materials and Methods

Chemicals

Soy phosphatidylcholine (SPC) was obtained from L. Meyer (Hamburg, Germany). Cholesterol (Fluka AG, Buchs, Switzerland) was recrystallized from methanol. Poly(ethylene glycol)-dipalmitoylphosphatidylethanolamine (PEG-DPPE, M_r 2750) was synthesized as described (Allen et al., 1991). DL-α-Tocopherol, all buffer salts, and other chemicals used are of analytical grade and were obtained from Merck (Darmstadt, Germany) or Fluka.

Synthesis of Lipophilic Arabinofuranosylcytosine Derivatives and Heterodinucleoside Phosphate Duplex Drugs

NOAC and all other N^4-alkyl derivatives of ara-C are synthesized by starting from uridine that is converted to an intermediary 1,2,4-triazolyl compound and reacted with octadecylamine. After cleavage of the O-acetyl protection groups on the arabinose, N^4-octadecyl-ara-C is obtained as crystalline solid at yields of 80–90% (Schwendener and Schott, 1992). The heterodinucleoside duplex drugs ara-C-NOAC and 5-FdU-NOAC are synthesized by condensation of NOAC via a $5' \rightarrow 5'$ phosphodiester linkage to ara-C or 5-FdU, respectively (Schott $et\ al.$, 1994, 1997). Figure 1 shows the chemical structures of NOAC and the lipophilic nucleoside duplex drugs.

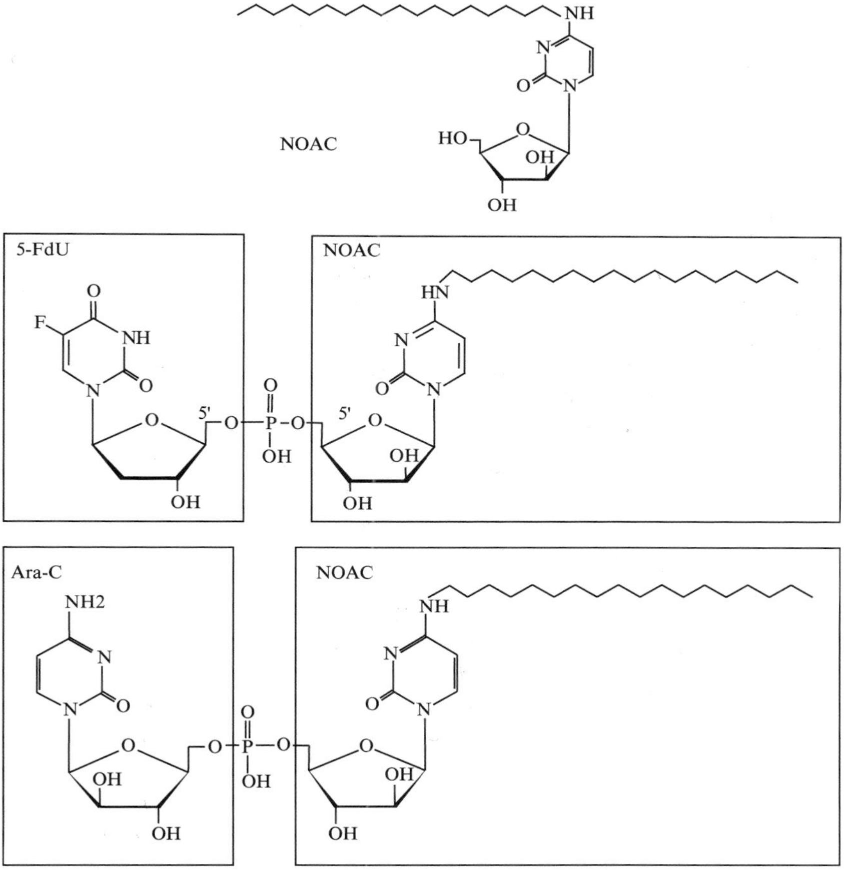

FIG. 1. Chemical structures of NOAC (N^4-octadecyl-1-β-D-arabinofuranosylcytosine, M_r 495.7) and the duplex drugs 5-FdU-NOAC (5-fluoro-2'-deoxyuridylyl-($5' \rightarrow 5'$)-N^4-octadecyl-1-β-D-arabinocytidine, M_r 803.9) and ara-C-NOAC (arabinocytidylyl-($5' \rightarrow 5'$)-N^4-octadecyl-1-β-D-arabino cytidine, M_r 800.9).

Liposome Preparation

A stable lyophilized liposome preparation of NOAC is prepared as follows. The lipids SPC (40 mg/ml), cholesterol (4 mg/ml), DL-α-tocopherol (0.2 mg/ml), and NOAC (5 mg/ml) are dissolved in methanol/methylene chloride (1:1, v/v) in a round-bottom flask. PEG-modified liposomes are obtained by addition of PEG(2000)-DPPE (14 mg/ml) to the basic lipids. The organic solvent is removed on a rotatory evaporator (Büchi, Flawil, Switzerland) at 40° for 30–60 min. The lipids and NOAC are solubilized by the addition of an appropriate volume of phosphate buffer (10 mM, pH 7.4) containing 230 mM mannitol (Fluka, PB-mannitol) that serves as a cryoprotectant, followed by careful shaking of the flask until all lipids are suspended in the aqueous medium. Some glass beads can be added to facilitate the detachment of the lipids from the glass of the flask. Small unilamellar liposomes can be prepared either by high-pressure filter extrusion or by detergent dialysis. In the case of dialysis, the detergent sodium cholate (Merck) is added initially to the organic lipid mixture at a molar ratio of 0.6–0.7 (detergent to total lipid including NOAC) (Rubas *et al.*, 1986). Alternatively, the detergents octylglucoside or octanoyl-*N*-methylglucamide (MEGA-8) can be used at a molar ratio of 0.2 (Schwendener *et al.*, 1981). After solubilization with PB-mannitol, the lipid/drug/detergent mixed micelles are dialyzed against PB-mannitol using a Lipoprep instrument (Diachema, Munich, Germany). Larger volumes (50–500 ml) of liposomes are prepared using a capillary dialysis instrument (Schwendener, 1993). Resulting liposomes are sterile filtered (0.2 μm, Gelman Sciences, Ann Arbor, MI), filled at appropriate volumes (e.g., 10 ml liposomes with 5 mg NOAC/ml) into sterile vials, frozen in liquid nitrogen, and lyophilized for 28–48 h (Dura-Dry lyophilizer, FTS Systems, New York).

The amount of incorporated NOAC can be determined by the addition of a trace amount of tritium-labeled NOAC ([^{3}H]NOAC, Amersham Int., Amersham, UK), followed by liquid scintillation counting. Alternatively, NOAC can be quantified by high-performance liquid chromatography (HPLC) (Rentsch *et al.*, 1995, 1997).

The lyophilized liposomes are reconstituted shortly before use with sterile water or 0.9% sodium chloride. The size and homogeneity of the liposomes are monitored by laser light scattering (Submicron Particle Sizer Model 370, Nicomp, Santa Barbara, CA). Liposome formulations containing the duplex drugs ara-C-NOAC or 5-FdU-NOAC are prepared in a similar manner.

The concentration of NOAC and the duplex drugs ara-C-NOAC or 5-FdU-NOAC in the liposomes can be varied from 1 mg/ml to about 10 mg/ml, depending on the concentration of phospholipids, the lipid composition, and the method of preparation. Reconstituted lyophilized liposomes retain their size and homogeneity longer than 72 h after reconstitution. The drugs remain chemically stable during the freeze drying process and the liposome preparations are ready for parenteral use within 1–2 h of reconstitution.

Properties of NOAC and the Duplex Drugs

The cytotoxic activity of the nucleoside drugs was tested *in vitro* on human tumor cell lines using a dye reduction assay (Cattaneo-Pangrazzi *et al.*, 2000a,b). The IC$_{50}$ values for NOAC and the duplex drugs are summarized in Table I. The antitumor activity of NOAC and the duplex drugs ara-C-NOAC and 5-FdU-NOAC was further confirmed by the National Cancer Institute (NCI) *in vitro* drug screening program where they were found to be active against several types of human tumors (National Cancer Institute, Drug Screening Program, 1999, unpublished observations).

In an experiment using the L1210 mouse leukemia model, we compared the cytotoxic activity of NOAC formulated in small unilamellar plain and in PEG liposomes of 80–160 nm mean diameter. As summarized in Table II, the antitumor activity of NOAC was excellent, irrespective of the liposome lipid composition. The PEG coating of the liposomes did not improve the antitumor effect in the L1210 leukemia model. However, it is conceivable that NOAC incorporated in long circulating PEG liposomes might have improved activity in solid tumors, despite of the transfer of the drug into plasma proteins (see below).

The inhibition of tumor colony formation from freshly explanted human tumors treated with liposomal NOAC is shown in Fig. 2, which

TABLE I
IN VITRO CYTOTOXICITY (IC$_{50}$) AND INHIBITION OF THYMIDYLATE SYNTHASE (TS)^a

Drug	IC$_{50}$ (μM) DU-145	50% TS inhibition (μM) DU-145	IC$_{50}$ (μM) PC-3	50% TS inhibition (μM) PC-3
5-FdU	3.4	0.005	3.4	0.006
5-FdU-NOAC	4.2	0.66	8.2	0.64
Ara-C	5.4	na^b	7.3	na
Ara-C-NOAC	12	na	123	na
NOAC	134	na	110	na

^{a} On DU-145 and PC-3 prostate tumor cells by the duplex drugs in comparison to 5-FdU, ara-C, and NOAC.
^{b} na, not applicable.

TABLE II

Cytostatic Activity of NOAC in Plain and in PEG-Liposomes in the
L1210 Mouse Leukemia Model[a]

Preparation[b]	Total dose (μmol/kg)	Survival time (days) Mean $\pm$ SD	T/C[c] (%)	Survivors 60 days
NOAC in PEG liposomes	100	60	857	6/6
	50	52 $\pm$ 18	750	5/6
NOAC in plain liposomes	100	60	857	6/6
	50	52 $\pm$ 19	744	5/6
Ara-C in PB[d]	200	20 $\pm$ 19	282	1/6
	400	23 $\pm$ 20	297	1/5
Controls	—	7	100	0/6

[a] After intravenous therapy on days 2 and 6 after tumor cell inoculation.

[b] On day 0, 10^5 L1210 cells were injected iv into BDF1 mice; iv treatment was on days 2 and 6.

[c] Increase of life span T/C% calculated including the 60-day survivors. SD, standard deviation.

[d] PB, phosphate buffer (67 mM, pH 7.4).

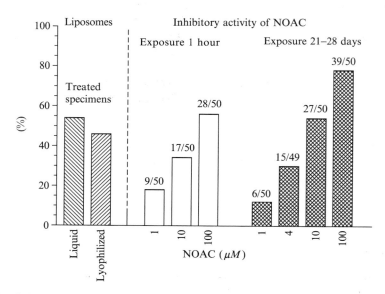

Fig. 2. Inhibition of tumor colony formation by NOAC from freshly explanted human tumors.

summarizes the growth inhibitory activity of NOAC after short- and long-term exposure with increasing drug concentrations obtained with lyophilized and freshly prepared liposomes. The lipophilic ara-C derivative had a strong concentration-dependent effect on colony formation on a panel of 50 tumor samples of various tumor types. Maximal inhibition of tumor growth was found in 28 of 50 (56%) samples with short-term incubation (1 h) and in 39 of 50 (78%) samples with long-term incubation (21–28 days). At 100 μM and long-term exposure NOAC proved to be active against tumor cells that were resistant to cisplatin, doxorubicin, 5-fluorouracil, gemcitabine, mitomycin C, and etoposide. There was no difference in NOAC activity between the application of lyophilized and freshly prepared liquid liposomes (Schwendener *et al.*, 2001).

In an *in vivo* experiment a human acute lymphatic leukemia (ALL) xenotransplanted to NOD/SCID-3 mice was used to evaluate the antileukemic effect of NOAC compared with standard agents. As summarized in Table III, liposomal NOAC induced the prevention of leukemia development without causing severe side effects at lower doses. The prevention of peritoneal tumors and ascites in the ALL-SCID-3 model was achieved with 8-fold lower drug concentrations of liposomal NOAC compared with ara-C (Horber *et al.*, 2000).

TABLE III
ANTILEUKEMIC ACTIVITY OF LIPOSOMAL NOAC IN THE ALL-SCID-3 LEUKEMIA MODEL

Substance[a]	Route	Dose/inj. mg/kg	Dose/inj. μmol/kg	Peritoreal tumors	Average tumor burden (ml or g/mouse)[b]
Experiment A					
Saline	ip		—	6/7	1.8 ± 1.16
Ara-C	ip	100	411	0/8	0
NOAC	ip	100	200	0/8	0
Vincristine	ip	1	1.2	0/7	0
Experiment B					
Saline	ip		—	8/8	2.92 ± 1.69
NOAC	ip	50	100	0/8	0
NOAC	ip	25	50	0/8	0
NOAC	Oral	100	200	0/8	0
NOAC	iv	25	50	0/3[c]	0

[a] Leukemia cells (10^6/mouse) were inoculated ip on day 0 (six to eight animals/group). Treatment started on day 1. Treatment schedule was on days 1, 4, 7, and 11 after tumor inoculation.

[b] Average tumor burden is given as the sum of peritoneal ascites fluid and solid tumor nodules in milliliters or gram per mouse.

[c] Three of six mice died of toxicity.

In several aspects, the lipophilic ara-C derivative NOAC was found to have different pharmacological effects compared with ara-C. Total uptake after incubation of liposomal NOAC with tumor cells was generally 5- to 10-fold higher than that of ara-C. More than 90% of NOAC distributes into the membrane fraction of a cell extract, whereas more than 95% of ara-C is found in the cytoplasmic fraction. Furthermore, the uptake mechanism of NOAC was shown to be independent of the nucleoside transport mechanism involved for ara-C and was known to be sensitive to the blocking agents dipyridamole and deoxycytidine. NOAC uptake and cytotoxicity were not decreased in the presence of these agents. Additionally, NOAC is not a substrate for deoxycytidine kinase, the key enzyme of ara-C phosphorylation, suggesting other mechanisms of action (Horber *et al.*, 1995a–c).

NOAC has several advantages compared with other lipophilic ara-C derivatives such as ara-C ocfosfate (ara-CMP-stearyl ester, YNK-01) (Ueda *et al.*, 1994) and thioether-linked lipid conjugates (Hong *et al.*, 1991), which are orally active prodrugs of ara-C. NOAC is both orally and parenterally active, whereas ara-C ocfosfate and the thioether-linked lipid conjugates cannot be applied parenterally due to their hemolytic toxicity. Furthermore, NOAC is active against solid and drug-resistant tumors, whereas cytarabine ocfosfate and BH-AC are classic prodrugs of ara-C with an activity spectrum that is not significantly different from that of the parent drug. In an experimental Phase I study with liposomal NOAC, the plasma elimination half-life in cancer patients ranged between 11 and 16 h with peak plasma levels of 180–240 μM at 600 mg NOAC/m^2 without untoward toxic effects, warranting further clinical development of this new drug (B. Pestalozzi and R. A. Schwendener, 1999, unpublished observations)

The duplex drugs were found to be strong inhibitors of the cell cycle. They mainly arrest tumor cells in the early S phase (Cattaneo-Pangrazzi *et al.*, 2000a,b). The cell uptake, intracellular distribution and fate, and metabolism of the duplex drugs have not yet been investigated. However, due to their similarity to NOAC, it can be assumed that they have comparable properties. 5-FdU-NOAC was found to overcome 5-FdU resistance in p53-mutated and androgen-independent DU-145 and PC-3 cells (Cattaneo-Pangrazzi *et al.*, 2000b). It is likely that the heteronucleoside dimer is cleaved to 5-FdUMP and NOAC, resulting in sustained intracellular drug concentrations over an extended period, consequently increasing the duration and magnitude of the cytotoxic effect of the cleaved molecules. This hypothesis is supported by the fact that the duplex drug specifically inhibits thymidylate synthase (TS) activity (see Table I) and that it exerts a cell cycle phase-dependent cytotoxicity, two mechanisms characteristic of 5-FdU (Cattaneo-Pangrazzi *et al.*, 2000b). The higher concentrations and longer incubation periods that are required *in vitro* for the cytotoxic effect of 5-FdU-NOAC can be explained by its prodrug nature, requiring enzymatic cleavage resulting in persisting intracellular concentrations of the active metabolites. The slow hydrolysis of the duplex drug by phosphodiesterase-I and human serum further indicate the pro-drug nature of 5-FdU-NOAC. Thus, due to expected changes of the phar-macokinetic properties and the prodrug character of 5-FdU-NOAC, the lipophilic duplex drug may have more favorable *in vivo* properties than the individual compounds 5-FdU and NOAC. In previous studies performed with similar heterodinucleoside phosphate dimers composed of the anti-viral nucleosides azidothymidine (AZT) and dideoxycytidine (ddC) and formulated in liposomes, we found significantly different pharmacokinetic properties and superior antiviral effects compared with the parent hydrophilic nucleosides in the murine Rauscher leukemia virus model (Schwendener *et al.*, 1994).

Distribution of NOAC to Blood Cells and Plasma Proteins

An important finding was that NOAC is not tightly anchored to the liposomal lipid bilayer and thus is readily distributed in the blood mainly into lipoproteins and erythrocyte membranes (Koller-Lucae *et al.*, 1997). It is known that unilamellar liposomes aggregate with low-density lipoprotein (LDL) and allow the transfer of incorporated drugs from the liposomes to LDL. Similar to the observation of others for other lipophilic drugs (Matsushita *et al.*, 1981), we found that after intravenous injection of liposomal NOAC, the drug is transferred within a short time from the liposome bilayers to plasma proteins (<67%), erythrocyte membranes (30%), and leukocytes (<2%) (Koller-Lucae *et al.*, 1999a).

Thus, liposomes provide an ideal pharmaceutical formulation for NOAC to ensure the transfer of the drug to lipoproteins, preferentially to LDL. The natural affinity of NOAC for LDL provides an interesting rationale for the specific delivery of the drug to tumors. Growing and dividing cells require cholesterol for membrane synthesis that is delivered by LDL. This also accounts for cancer cells in which an increased LDL uptake in tumors with high metastatic potential and aggressive or undiffer-entiated character is known. The LDL-mediated uptake and cytotoxic effects of NOAC were also studied *in vitro* on Daudi lymphoma cells. NOAC was either incorporated into LDL or liposomes. Specific binding of NOAC-LDL to Daudi cells was five times higher than to human lymphocytes. LDL receptor binding of NOAC-LDL could be blocked and up- or downregulated with free LDL. In a cytotoxicity test the IC_{50} of NOAC-LDL was about 160 μM and after blocking of the receptors with

LDL cytotoxicity was reduced by 50%. The natural affinity of NOAC for LDL provides an interesting rationale for the specific delivery of the drug to tumors that express elevated numbers of LDL receptors (Koller-Lucae *et al.*, 1999b).

Conclusions and Prospects

With the chemical transformation of water-soluble nucleosides into lipophilic compounds and their incorporation into liposomes as their carriers, we developed a new class of cytotoxic and antiviral drugs that can be applied for treatment by parenteral routes. Lipophilic ara-C derivatives, particularly NOAC and the duplex drugs composed of NOAC and ara-C or 5-FdU, represent very promising new anticancer drugs with high cytotoxic activity and strong apoptosis-inducing capability.

Based on our findings, we conclude that due to the chemical modification of ara-C and 5-FdU to molecules that have new physicochemical properties such as high lipophilicity and stability against enzymatic degradation, together with the possibility of preparing stable lyophilized liposome formulations of these compounds, more potent drugs are made available. The mechanisms of action of the new drugs seem to be different from those of the parent water-soluble nucleosides, and as an important finding they seem to be able to overcome drug resistance mechanisms. The composition of the liposomes does not seem to represent a crucial factor in the pharmacological properties of the new derivatives. As shown with the L1210 leukemia model, the addition of PEG-lipids did not contribute to an enhancement of the cytotoxic effect (see Table II). This property can be explained with the *in vivo* distribution experiments and the *in vitro* incubation studies with plasma, where we observed a fast transfer of NOAC from the liposome membranes to plasma proteins and erythrocytes (Koller-Lucae *et al.*, 1997, 1999a). Thus, the liposomes serve mainly as a pharmaceutical formulation to permit solubilization and parenteral administration of the new derivatives.

In conclusion, the chemical modification of water-soluble molecules by attachment of long alkyl chains and their stable incorporation into the bilayer membranes of small unilamellar liposomes represent a very promising method for the development of new drugs not only for the treatment of tumors or infections but also for many other diseases.

Acknowledgments

The authors thank Daniel Horber, Sibylle Koller-Lucae, Rosanna Cattaneo-Pangrazzi, Iduna Fichtner, and Kathrin Friedl for their valuable contributions.

References

Allen, T. M., Hansen, C., Martin, F., Redemann, C., and Yau Young, A. (1991). Liposomes containing synthetic lipid derivatives of poly(ethylene glycol) show prolonged circulation half-lives *in vivo*. *Biochim. Biophys. Acta* **1066**, 29–36.

Cattaneo-Pangrazzi, R. M. C., Schott, H., Wunderli-Allenspach, H., Derighetti, M. I., and Schwendener, R. A. (2000a). New amphiphilic heterodinucleoside phosphate dimers of 5-fluorodeoxyuridine (5FdUrd): Cell cycle dependent cytotoxicity and induction of apoptosis in PC-3 prostate tumor cells. *Biochem. Pharmacol.* **60**, 1887–1896.

Cattaneo-Pangrazzi, R. M. C., Schott, H., and Schwendener, R. A. (2000b). The novel heterodinucleoside dimer 5-FdU-NOAC is a potent cytotoxic drug and a p53-independent inducer of apoptosis in the androgen-independent human prostate cancer cell lines PC-3 and DU-145. *The Prostate* **45**, 8–18.

Gabizon, A., and Martin, F. (1997). Polyethylene glycol-coated (pegylated) liposomal doxorubicin. Rationale for use in solid tumours. *Drugs* **54**(Suppl. 4), 15–21.

Hong, C. I., West, C. R., Bernacki, R. J., Tebbi, C. K., and Berdel, W. E. (1991). 1-β-D-arabinofuranosylcytosine conjugates of ether and thioether phospholipids. A new class of ara-C prodrugs with improved antitumor activity. *Lipids* **26**, 1437–1444.

Horber, D. H., Ottiger, C., Schott, H., and Schwendener, R. A. (1995a). Pharmacokinetic properties and interactions with blood components of N^4-hexadecyl-1-β-D-arabinofuranosylcytosine (NHAC) incorporated into liposomes. *J. Pharm. Pharmacol.* **47**, 282–288.

Horber, D. H., Ottiger, C., Schott, H., and Schwendener, R. A. (1995b). Cellular pharmacology of a liposomal preparation of N^4-hexadecyl-1-β-D-arabinofuranosylcytosine, a lipophilic derivative of 1-β-D-arabinofuranosylcytosine. *Br. J. Cancer* **71**, 957–962.

Horber, D. H., Schott, H., and Schwendener, R. A. (1995c). Cellular pharmacology of N^4-hexadecyl-1-β-D-arabinofuranosylcytosine (NHAC) in the human leukemic cell lines K-562 and U-937. *Cancer Chemother. Pharmacol.* **36**, 483–492.

Horber, D. H., Cattaneo-Pangrazzi, R. M. C., von Ballmoos, P., Schott, H., Ludwig, P. S., Eriksson, S., Fichtner, I., and Schwendener, R. A. (2000). Cytotoxicity, cell cycle perturbations and apoptosis in human tumor cells by lipophilic N^4-alkyl-1-β-D-arabinofuranosylcytosine derivatives and the new heteronucleoside phosphate dimer arabinocytidylyl-(5′→5′)-N^4-octadecyl-1-β-D-ara-C. *J. Cancer Res. Clin. Oncol.* **126**, 311–319.

Koller-Lucae, S. K. M., Schott, H., and Schwendener, R. A. (1997). Pharmacokinetic properties in mice and interactions with human blood *in vitro* of liposomal N^4-octadecyl-1-β-D-arabinofuranosylcytosine (NOAC), a new anticancer drug. *J. Pharmacol. Exp. Ther.* **282**, 1572–1580.

Koller-Lucae, S. K. M., Suter, M. J.-F., Rentsch, K. M., Schott, H., and Schwendener, R. A. (1999a). Metabolism of the new liposomal anticancer drug N^4-octadecyl-1-β-D-arabinofuranosylcytosine (NOAC) in mice. *Drug Metab. Disposition* **27**, 342–350.

Koller-Lucae, S. K. M., Schott, H., and Schwendener, R. A. (1999b). Low density lipoprotein and liposome mediated uptake and cytotoxic effect of N^4-octadecyl-1-β-D-arabinofuranosylcytosine (NOAC) in Daudi lymphoma cells. *Br. J. Cancer* **80**, 1542–1549.

Matsushita, T., Ryu, E. K., Hong, C. I., and MacCoss, M. (1981). Phospholipid derivatives of nucleoside analogs as prodrugs with enhanced catabolic stability. *Cancer Res.* **41**, 2707–2713.

Pestalozzi, B., and Schwendener, R. A. (1999). Unpublished data.

Rentsch, K. M., Schwendener, R. A., Schott, H., and Hänseler, E. (1995). A sensitive high-performance liquid chromatographic method for the determination of N^4-hexadecyl- and

N^4-octadecyl-1-β-D-arabinofuranosylcytosine in plasma and erythrocytes. *J. Chromatogr. B.* **673**, 259–266.

Rentsch, K. M., Schwendener, R. A., Schott, H., and Hänseler, E. (1997). Pharmacokinetics of N^4-octadecyl-1-β-D-arabinofuranosylcytosine (NOAC) in plasma and whole blood after intravenous and oral application in mice. *J. Pharm. Pharmacol.* **49**, 1076–1081.

Rubas, W., Supersaxo, A., Weder, H. G., Hartmann, H. R., Hengartner, H., Schott, H., and Schwendener, R. A. (1986). Treatment of murine L1210 leukemia and melanoma B16 with lipophilic cytosine arabinoside prodrugs incorporated into unilamellar liposomes. *Int. J. Cancer* **37**, 149–154.

Schott, H., Häussler, M. P., and Schwendener, R. A. (1994). Synthese und Eigenschaften von N^4-hexadecyl-2'-desoxycytidylyl-(3'-5')-5-ethyl-2' desoxyuridin und 2'-Desoxythymidylyl-(3'-5')-N^4-hexadecyl-1-β-D-arabinofuranosylcytosin, zwei Vertreter einer neuen Prodrug-Gruppe. *Liebigs Annalen der Chemie.* 277–282.

Schott, H., Schwendener, R. A., and Guerin, A. (1997). Amphiphilic nucleosidephosphate analogues. U.S. Patent 5,679,652.

Schwendener, R. A. (1993). The preparation of large volumes of sterile liposomes for clinical applications. *In* "Liposome Technology" (G. Gregoriadis, ed.), 2nd ed., pp. 487–500. CRC Press, Boca Raton, FL.

Schwendener, R. A., and Schott, H. (1992). Treatment of L1210 murine leukemia with liposome—incorporated N^4-hexadecyl-1-β-D-arabinofuranosylcytosine. *Int. J. Cancer* **51**, 466–469.

Schwendener, R. A., and Schott, H. (1996). Lipophilic 1-β-D-arabinofuranosylcytosine derivatives in liposomal formulations for oral and parenteral antileukemic therapy in the murine L1210 leukemia model. *J. Cancer Res. Clin. Oncol.* **122**, 723–726.

Schwendener, R. A., Asanger, M., and Weder, H. G. (1981). The preparation of large bilayer liposomes: controlled removal of n-alkyl-glucoside detergents from lipid/detergent micelles. *Biochem. Biophys. Res. Commun.* **100**, 1055–1062.

Schwendener, R. A., Horber, D. H., Gowland, P., Zahner, R., Schertler, A., and Schott, H. (1994). New lipophilic acyl/alkyl dinucleoside phosphates as derivatives of 3'-azido-3'-deoxythymidine: Inhibition of HIV-1 replication *in vitro* and antiviral activity against Rauscher leukemia virus infected mice with delayed treatment regimens. *Antiviral Res.* **24**, 79–93.

Schwendener, R. A., Horber, D. H., Ottiger, C., Rentsch, K. M., Fiebig, H. H., and Schott, H. (1995a). Monograph: Alkasar-18, 1-(β-D-arabinofuranosyl)-4-octadecylamino-2(1H)-pyrimidin-one, N^4-octadecyl-ara-C, NOAC. *Drugs Future* **20**, 11–15.

Schwendener, R. A., Horber, D. H., Ottiger, C., and Schott, H. (1995b). Preclinical properties of N^4-hexadecyl- and N^4-octadecyl-1-β-D-arabinofuranosylcytosine in liposomal preparations. *J. Liposome Res.* **5**, 27–47.

Schwendener, R. A., Horber, D. H., Odermatt, B., and Schott, H. (1996). Antitumor activity and pharmacological properties of orally administered lipophilic N^4alkyl derivatives of 1-β-D-arabinofuranosylcytosine (arc-C). *J. Cancer Res. Clin. Oncol.* **122**, 102–108.

Schwendener, R. A., Friedl, K., Depenbrock, H., Schott, H., and Hanauske, A.-R. (2001). *In vitro* activity of liposomal N^4-octadecyl-1-β-D-arabinofuranosylcytosine (NOAC), a new lipophilic derivative of 1-β-D-arabinofuranocylcytosine on biopsized clonogenic human tumour cells and haematopoietic precursor cells. *Invest. New Drugs* **19**, 203–210.

Ueda, T., Kamiya, K., Urasaki, Y., Wataya, S., Kawai, Y., Tsutani, H., Sugiyama, M., and Nakamura, T. (1994). Clinical pharmacology of 1-beta-D-arabinofuranosylcytosine-5'-stearylphosphate, an orally administered long-acting derivative of low-dose 1-beta-D-arabinofuranosylcytosine. *Cancer Res.* **54**, 109–113.

[4] The Liposomal Formulation of Doxorubicin

By SHEELA A. ABRAHAM, DAWN N. WATERHOUSE, LAWRENCE D. MAYER,
PIETER R. CULLIS, THOMAS D. MADDEN, and MARCEL B. BALLY

Abstract

Doxorubicin is the best known and most widely used member of the anthracycline antibiotic group of anticancer agents. It was first introduced in the 1970s, and since that time has become one of the most commonly used drugs for the treatment of both hematological and solid tumors. The therapy-limiting toxicity for this drug is cardiomyopathy, which may lead to congestive heart failure and death. Approximately 2% of patients who have received a cumulative (lifetime) doxorubicin dose of 450–500 mg/m^2 will experience this condition. An approach to ameliorating doxorubicin-related toxicity is to use drug carriers, which engender a change in the pharmacological distribution of the drug, resulting in reduced drug levels in the heart. Examples of these carrier systems include lipid-based (liposome) formulations that effect a beneficial change in doxorubicin biodistribution, with two formulations approved for clinical use. Drug approval was based, in part, on data suggesting that beneficial changes in doxorubicin occurred in the absence of decreased therapeutic activity. Preclinical (animal) and clinical (human) studies showing that liposomes can preferentially accumulate in tumors have provided a rationale for improved activity. Liposomes represent ideal drug delivery systems, as the microvasculature in tumors is typically discontinuous, having pore sizes (100–780 nm) large enough for liposomes to move from the blood compartment into the extravascular space surrounding the tumor cells (Hobbs *et al.*, 1998). Liposomes, in the size range of 100–200 nm readily extravasate within the site of tumor growth to provide locally concentrated drug delivery, a primary role of liposomal formulation. Although other liposomal drugs have been prepared and characterized due to the potential for liposomes to improve antitumor potency of the encapsulated drug, the studies on liposomal doxorubicin have been developed primarily to address issues of acute and chronic toxicity that occur as a consequence of using this drug. It is important to recognize that research programs directed toward the development of liposomal doxorubicin occurred concurrently with synthetic chemistry programs attempting to introduce safer and more effective anthracycline analogues. Although many of these drugs are approved for use, and preliminary liposomal formulations of these analogues have been prepared, doxorubicin continues to be a mainstay of drug cocktails used

METHODS IN ENZYMOLOGY, VOL. 391

in the management of most solid tumors. It will be of great interest to observe how the approved formulations of liposomal doxorubicin are integrated into combination regimes for treatment of cancer. In the meantime, we have learned a great deal about liposomes as drug carriers from over 20 years of research on different liposomal doxorubicin formulations, the very first of which were identified in the late 1970s (Forssen *et al.*, 1979; Rahman *et al.*, 1980). This chapter will discuss the various methods for encapsulation of doxorubicin into liposomes, as well as some of the important interactions between the formulation components of the drug and how this may impact the biological activity of the associated drug. This review of methodology, in turn, will highlight research activities that are being pursued to achieve better performance parameters for liposomal formulations of doxorubicin, as well as other anticancer agents being considered for use with lipid-based carriers.

Introduction

When considering the biological activity of liposomal formulations of doxorubicin, it is important to remember that (1) doxorubicin is released from the liposomes following intravenous administration, (2) the distribution of doxorubicin following administration of a liposomal formulation is dependent, in part, on the biodistribution characteristics of the liposomes, and (3) the biological activity of liposomal doxorubicin is dependent on when, where, and at what rate the drug is released. In consideration of the above, we can make two broad statements about all the different liposomal formulations described for doxorubicin. First, differences in doxorubicin-mediated toxicity and efficacy are dependent on liposomal lipid composition, which in turn influences drug release attributes and liposome biodistribution patterns. Second, the biologically active component of all formulations described is doxorubicin, and a general understanding of the chemical and physical properties, as well as the mechanism(s) of action of this drug, is necessary before considering liposomal formulations.

Doxorubicin

Chemistry

Doxorubicin is an anthracycline antibiotic originally isolated from *Streptomyces peucetius* var. *caesius*. This amphipathic molecule possesses a water-insoluble aglycone (adriamycinone: $C_{21}H_{18}O_9$) and a water-soluble, basic, reducing amino-sugar moiety (daunosamine: $C_6H_{13}NO_3$) (Fig. 1).

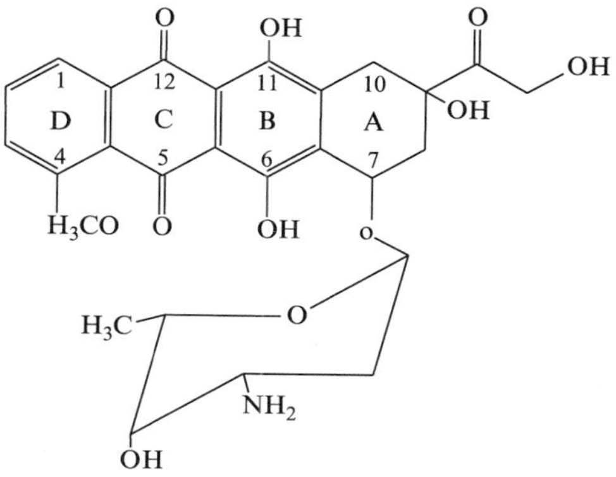

FIG. 1. The chemical structure of doxorubicin.

Of note, doxorubicin has three significant prototropic functions with associated pK_as: (1) the amino group in the sugar moiety ($pK_1 = 8.15$), (2) the phenolic group at C_{11} ($pK_2 = 10.16$), and (3) the phenolic group at C_6 ($pK_3 = 13.2$) (Bouma et al., 1986; Fiallo et al., 1998).

Doxorubicin hydrochloride typically exists as a hygroscopic crystalline powder composed of orange-red thin needles. It has a melting point of 229–231° and absorption maximums (in methanol) of 233, 252, 288, 479, 496, and 529 nm due to the dihydroxyanthraquinone chromophore (Budavari et al., 2000). Any variation of groups on the chromophore ultimately leads to changes in the absorption spectrum, therefore the spectrum depends on pH, binding ions and their concentration, drug concentration, solvent type, and ionic strength. Deprotonation of the chromophore results in a modification of the UV, visible, and circular dichroic (CD) spectra, causing a red shift due to the deprotonation of the phenolic functions of the chromophore. Doxorubicin is documented to appear orange at pH 7, violet at pH 11, and blue at pH 13 (Fiallo et al., 1999).

It has also been well documented that doxorubicin has a propensity to self-associate (Chaires et al., 1982; Menozzi et al., 1984). Self-association constants in the literature vary dramatically, most likely due to differences in methods, experimental conditions, pH, and buffer compositions. This phenomenon can be detected in aqueous media at concentrations as low as 1 μM and can be monitored by measuring the ratio of absorbance at 470/550 nm. Interactions between the π-electron systems have a measured thermodynamic value of $\Delta H = -33$ kJ/mol (Bouma et al., 1986), making self-association a favorable interaction, with the mechanism of self-association being attributed to interactions between planar aromatic rings of individual molecules. The nonprotonated neutral anthracycline species is thought to be the dominant species involved in self-association. Increasing the ionic strength of the buffer solution increases the constant (K) of association of doxorubicin, and increasing methanol concentration decreases K (Menozzi et al., 1984).

Mechanism of Activity

The precise mechanism of action of doxorubicin is not understood; however, it is known to intercalate between DNA base pairs, resulting in DNA and DNA-dependent RNA synthesis inhibition due to template disordering and steric obstruction. Intercalation leads to single and double strand breaks, as well as exchange of sister chromatids. Scission of DNA is believed to be mediated through the action of topoisomerase II or by the iron-catalyzed generation of free radicals, both hydrogen peroxide and hydroxyl, which are highly destructive to cells. Doxorubicin also induces the formation of covalent topoisomerase–DNA complexes, resulting in inhibition of the religation portion of the ligation–religation reaction in replicating DNA. Although it is active throughout the cell cycle, the maximal toxicity occurs during the DNA synthesis (S) phase. At low concentrations of drug, cells will continue through the S phase, and die in G_2 (Hortobagyi, 1997).

Toxicity

The therapy-limiting toxicity for this drug is cardiomyopathy, which may lead to congestive heart failure (CHF) and death. Approximately 2% of patients who have received a cumulative (lifetime) doxorubicin dose of 450–500 mg/m^2 will experience this condition. Recently, this toxicity has been highlighted in clinical studies evaluating the use of doxorubicin with a humanized monoclonal antibody targeting the oncoprotein HER-2/neu, referred to as trastuzumab (Pegram et al., 1999, 2000). Combinations of trastuzumab and doxorubicin provide particularly promising thera-peutic effects, but the use of the combination exacerbated cardiac-related toxicities.

It is important to remember that doxorubicin, administered in free form, exhibits many side effects other than cardiotoxicity. Similar to other anticancer drugs targeting proliferating cell populations, doxorubicin causes significant gastrointestinal toxicity, with nausea, vomiting, and diarrhea being common soon after therapy, and stomatitis that occurs within 7–10 days after administration. The typical acute dose-limiting toxicity for doxorubicin is that of a high incidence of myelosuppression, typically leukopenia and thrombocytopenia. In severe cases, this may lead to neutropenic fever and sepsis, requiring hospitalization (Gabizon *et al.*, 1994).

Liposomal Formulations of Doxorubicin: General Considerations

The primary aim of doxorubicin encapsulation in liposomes has been to decrease nonspecific organ toxicity. Liposomes are able to direct the doxorubicin away from sites with tight capillary junctions such as the heart muscle. Instead, they distribute in areas where fenestrations or gaps exist in the vasculature (liver, spleen, and bone marrow, areas of inflammation, and neoplasms). Phagocytic cells that comprise the mononuclear phagocyte system (MPS)[1] can recognize these particulate carrier systems as "foreign". Although distribution to these phagocytic cells is dependent on the physical (size) and chemical (charge) attributes of the liposomes used, it should be noted that when liposomes contain doxorubicin, the cells of the MPS are adversely affected. More specifically, following uptake into a phagocytic cell, release of doxorubicin causes these cells to die, thus reducing the capacity of the MPS to accumulate the injected liposomes. This is reflected, in turn, by significantly increased liposome circulation lifetimes (Bally *et al.*, 1990). Liposomal formulations containing poly(ethylene glycol) (PEG)-modified lipids are known to exhibit increased circulation time because of a reduced tendency to aggregate following iv administration (Allen *et al.*, 2002); however, these formulations also exhibit MPS toxicities (Parr *et al.*, 1993). Whether the enhanced circulation lifetime of a liposomal formulation of doxorubicin is due to subtle changes in lipid composition, inclusion of doxorubicin, which affects MPS cell function, or the use of PEG-modified lipids, increased circulation longevity of the liposomes allows enhanced extravasation across the leaky endothelium of solid tumors (Papahadjopoulos *et al.*, 1991).

Two liposomal doxorubicin formulations that have received clinical approval are Doxil (United States) or Caelyx (Canada and Europe) and Myocet, which was given community marketing authorization from the European Commission in August 2000 for the treatment of metastatic breast cancer. The Doxil/Caelyx liposomal formulation is composed of

---

[1]Abbreviations: A, absorbance; Chol, cholesterol; [^{3}H]CHE, [^{3}H]cholesteryl hexadecyl ether; cTEM, cryotransmission electron microscopy; DMPC, 1,2-dimyristoyl-*sn*-glycero-3-phosphocholine; DOX, doxorubicin; $\Delta\Psi$, electrochemical gradient; EDTA, ethylenediaminetetraacetic acid; HEPES, N-[2-hydroxyethyl]piperazine-N'-[2-ethanesulfonic acid]; HBS, HEPES-buffered saline; iv, intravenous; LUV, large unilamellar vesicle; MLV, multilamellar vesicle; MPS, mononuclear phagocyte system; SHE, sucrose-HEPES-EDTA.

hydrogenated soya phosphatidylcholine, cholesterol (Chol) and PEG-modified phosphatidylethanolamine (55:40:5 molar ratio) whereas Myocet is composed of egg phosphatidylcholine (EPC) and Chol (55:45 molar ratio). The EPC–Chol formulation releases drug fairly rapidly and has a relatively short circulation lifetime (approximately three times that of free doxorubicin) but has been shown clinically to reduce doxorubicin-induced cardiotoxicity and gastrotoxicity. In contrast, Doxil/Caelyx has a much longer circulation lifetime due to the steric barrier provided by the surface-grafted PEG, which leads to large changes in biodistribution and particularly increased amounts of drug being delivered to the skin. This has advantages for the treatment of skin localized cancers such as Kaposi's sarcoma but disadvantages in the observation of new dose-limiting toxicities such as the hand and foot syndrome (reviewed in Waterhouse *et al.*, 2001).

General Description of Materials and Techniques

When generating a pharmaceutically viable liposomal doxorubicin formulation, several important factors must be considered. The methodology must be straightforward and conceptually easy, using the most economically available materials possible. Liposomes must be uniformly generated with reproducible size distributions; an optimal loading procedure would approach 100% trapping efficiency at the desired drug-to-lipid ratio, thereby negating the need to remove unencapsulated doxorubicin from the sample. The drug-loaded sample must also exhibit drug release rates that are conducive to improvements in drug activity through decreases in toxicity or increases in efficacy (Mayer *et al.*, 1994). In the extreme, rapid (instantaneous) doxorubicin release from liposomal formulations following iv administration will result in a drug that is substantially no different from the free, unencapsulated, drug. Such a formulation would typically not be classified as a drug carrier. In the opposite extreme, complete drug retention (no drug release) should, theoretically, result in a formulation that is neither toxic nor efficacious. Such a formulation may exhibit substantial improvements in drug delivery to sites of tumor growth and substantial reductions in drug delivery to cardiac tissue, but such a formulation would be of little therapeutic value or interest. Thus *in vivo* drug release parameters between these extremes has become the guiding light through which liposomal formulations of doxorubicin have been developed.

The lipid composition selected, when preparing liposomal formulations of doxorubicin, largely dictates rates of doxorubicin binding, partitioning, and retention in the liposomes, as well as the elimination rate of the liposomal formulation. In general, the chemical attributes of the specific lipid molecules chosen affect bilayer permeability in a manner that is somewhat predictable on the phase transition temperature of the bulk lipid component (Table I). Liposomes prepared of phospholipids with short acyl chains (e.g., C14:0, DMPC) release doxorubicin more rapidly than phospholipids with longer acyl chains (e.g., C18:0, DSPC). Cholesterol is a common constituent of liposomal doxorubicin formulations due to its ability to modulate membrane permeability and biological stability. Membrane permeability is thought to be dependent on the amount of cholesterol being incorporated and the transition temperature of the bulk phospholipid being used. For example, cholesterol at concentrations above 30 mol% lowers the amount of energy required to melt the acyl chains, eliminates the gel-to-liquid-crystalline phase transition (Cullis *et al.*, 1987), and causes an increase in acyl chain disorder at temperatures below the phase transition temperature of the bulk phospholipid component. In addition to the lipid components, the chemical interactions of doxorubicin with the carrier should be considered, particularly when charge–charge interactions with the phospholipid head groups or partitioning into the lipid hydrocarbon chains may affect drug loading and release attributes.

Further, although this chapter describes what are believed to be the most efficient and versatile methods for preparing liposomal formulations of doxorubicin, other methods could be used. In general, methods of encapsulation are categorized into two procedures, passive and active. Passive encapsulation involves the hydration of a dried lipid film with an

TABLE I
EXAMPLES OF SOLUTIONS AND THEIR CONCENTRATIONS IN ACTIVE LOADING

Active encapsulation procedures	Concentration of hydrating buffer	pH	Hydrating buffer osmolarity (mOsm/liter)	Concentration of external buffer	pH	Osmolarity of external buffer
Citrate	$C_6H_8O_7$ 300 mM	4.0	550	(HEPES buffered saline) 150 mM NaCl 25 mM HEPES	7.5	326
Ammonium sulfate	$(NH_4)_2SO_4$ 120 mM	5.5	276	NaCl 145 mM	5.5	268
Manganese sulfate	$MnSO_4$ 300 mM	3.5	319	300 mM sucrose 30 mM HEPES	7.5	380
Manganese sulfate +A23187	$MnSO_4$ 300 mM	3.5	319	300 mM sucrose 20 mM HEPES 15 mM EDTA	7.5	517

aqueous solution of doxorubicin. This method takes advantage of the fact that during the preparation of liposomes, a certain aqueous volume is obtained within each liposome. Drug and liposomes are codispersed with a certain fraction of the drug entrapped directly, resulting from the combination of the hydrophilic, hydrophobic, and ionic interactions. As an amphipathic drug, doxorubicin, depending on the pH, has the potential to reside in the aqueous core, as well as partition into the lipid bilayer. Typically, passive trapping methods are not efficient (maximum efficiencies of 80%; Cullis *et al.*, 1989), and the maximum drug-to-lipid ratio achievable is low and dependent on the maximum solubility of doxorubicin (<10 mM). Passive encapsulation methods also require removal of the unencapsulated drug. Although this can be accomplished easily in a laboratory setting, the methods for passive encapsulation are not well suited to scaled production of pharmaceutical batches for use in clinical trials.

Active trapping procedures involve adding doxorubicin to preformed liposomes that possess a trans-bilayer ion gradient. Under appropriate conditions, when the drug is added to liposomes possessing such a gradient, a redistribution of the drug occurs such that the drug crosses the liposomal bilayer and is subsequently trapped within the core of the liposome. Trapping occurs for reasons that involve internal aqueous pH and induced drug precipitation along with associated chemical reactions that affect the nature of the precipitate formed. The active trapping procedure results in drug-to-lipid ratios as high as 0.3:1 (wt:wt), which corresponds to approximately 48,000 doxorubicin molecules per 100-nm-diameter liposome. Active trapping methods can be used with any lipid composition that is able to form a bilayer and maintain a transmembrane ion gradient. These methods have been used to prepare liposomal doxorubicin formulations approved for clinical use, and, thus, the techniques for preparing these formulations are the focus of this chapter.

Encapsulation of Doxorubicin Using Active Loading Methods

Until recently (Abraham *et al.*, 2002; Cheung *et al.*, 1998), active loading methods have been dependent on an established and/or a created transmembrane pH gradient, as well as the accumulation of doxorubicin to levels within the liposome that exceed the solubility of the drug (Madden *et al.*, 1990). Transmembrane pH gradients across liposomes result in doxorubicin precipitation through drug self-association or through interaction with salts present in the aqueous core of the liposome. There are two principal ways to achieve a pH gradient across the liposome bilayer: directly or indirectly. To directly establish the trans-bilayer pH gradient, liposomes are prepared in the presence of an acidic buffer, and the exterior buffer of the liposomes is then adjusted to a desired pH using either added bases that increase the pH or, alternatively, exchanging the outside buffer using column chromatography or dialysis. Obviously, it is critical that the pH gradient established across the bilayer is stable. Interestingly, protons (H^+), relative to other cations, are highly permeable. As H^+ diffuse outward, they create an electrochemical gradient ($\Delta\Psi$ inside negative) (Harrigan et al., 1992), and this prevents/limits further H^+ efflux. If, for example, there is a ΔpH of three units, this translates to $\Delta\Psi$ of -177 mV. Cullis et al. (1991) suggested that only 150 protons must diffuse outwardly to maintain this potential and allow maintenance of a stable ΔpH and $\Delta\Psi$. A commonly used procedure to prepare liposomes with a transmembrane pH gradient relies on the entrapment of a 300 mM citrate buffer, pH 4.0 (Fig. 2A). Citrate is a triprotic buffer and possesses a large buffering capacity in the range of pH 3–6.5 ($pK_{a1} = 3.13$, $pK_{a2} = 4.76$, $pK_{a3} = 5.41$ or 6.4) (Budavari et al., 2000).

There are several ways to indirectly establish a pH gradient, including the use of electrochemical gradients and ionophores. For example, liposomes can be synthesized in a potassium (K^+)-based buffer and the external buffer replaced with a sodium-containing buffer by column chromatography or dialysis. In the presence of the created K^+ gradient, the addition of an ionophore such as valinomycin (an H^+/K^+ exchanger) will shuttle K^+ to the liposomal exterior, thus creating an electrochemical gradient ($\Delta\Psi$), with the interior of the liposomes being negative. Stable electrical potentials in excess of 150 mV can be generated in large unilamellar vesicles (LUVs) using this technique (Redelmeier et al., 1989). H^+ readily cross the lipid bilayer into the interior in response to this $\Delta\Psi$ gradient, thus establishing a pH gradient (Cullis et al., 1991). It has been shown that this pH gradient is smaller than theoretically predicted for electrochemical equilibrium and is sensitive to the ionic composition of the external buffer.

By virtue of being able to exchange cations, other ionophores can also be used to form pH gradients from chemical gradients. Several researchers have documented that both nigericin (K^+ ionophore) and A23187 (divalent cation ionophore) exchange specific cations for either one or two protons, respectively (Fenske et al., 1998) (Fig. 2B). Liposomes possessing a transmembrane salt gradient (the liposome interior containing either K_2SO_4 or $MnSO_4$) can be incubated with the specific ionophore and drug. The ionophores are able to transport the outward movement of cations for the inward movement of protons, thus creating a pH gradient, which in turn promotes drug loading.

Another method capable of causing formation of a pH gradient is one that relies on encapsulation of ammonium sulfate (Lasic et al., 1992)

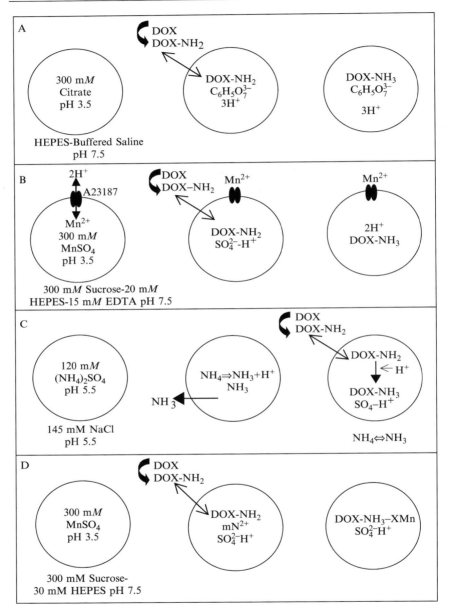

FIG. 2. Methods of doxorubicin encapsulation into liposomes exhibiting the indicated gradients. (A) The citrate loading procedure. Liposomes are prepared in 300 mM citrate buffer, pH 3.5, and outside buffer exchanged to HEPES-buffered saline at pH 7.5. (B) The

(Fig. 2C). Following encapsulation of $(NH_4)_2SO_4$, the external buffer is exchanged to a solution such as 145 mM NaCl to establish an $(NH_4)_2SO_4$ gradient. Due to the high permeability of NH_3 (1.3×10^4 cm/s), it readily crosses the liposome bilayer, leaving behind one proton for every molecule of NH_3 lost (Bolotin et al., 1994). This creates a pH gradient, the magnitude of which is determined by the $[NH_4^+]_{in}/[NH_4^+]_{out}$ gradient. This approach differs from other chemical methods in that the liposomes are not prepared in an acidic buffer nor is there a need to alkalinize the exterior liposome solution (Haran et al., 1993).

The choice of aqueous buffer in which liposomes are formed has direct effects on the subsequent loading of doxorubicin. The osmolarity, ionic strength, pH, counterions (e.g., citrate, sulfate, or glutamate), and their concentrations interplay to generate different precipitated structures, but it is unclear whether this affects the biological behavior of the resulting product. As indicated earlier, doxorubicin (Menozzi et al., 1984) self-associates with stacking of the planar aromatic rings under specific conditions (Chaires et al., 1982). It has been suggested that doxorubicin forms a precipitate when encapsulated in liposomes in response to a pH gradient, an observation that has been confirmed by several research groups (Lasic, 1996; Lasic et al., 1992; Li et al., 1998). Cryotransmission electron microscopy (cTEM) reveals doxorubicin precipitates as fibrous-bundle aggregates in both citrate- and sulfate-containing liposomes (Fig. 3). The planar aromatic anthracycline rings are thought to stack longitudinally to form linear fibers. These fibers are aligned in a hexagonal arrangement to form bundles, with approximately 12–60 fibers per bundle. Doxorubicin aggregates in the presence of sulfate typically have rigid linear fiber bundles (interfiber spacing is approximately 27 Å) compared with the doxorubicin–citrate aggregates in the presence of citrate, which appear mostly linear or curved (interfiber spacing is approximately 30–35 Å) (Li et al., 1998). These results suggest that the sulfate anion, being smaller than the citrate anion, may allow a tighter packing arrangement, resulting in a decreased flexibility of fiber bundles.

---

manganese sulfate loading procedure with the A23187 ionophore. Liposomes are prepared in 300 mM manganese sulfate, pH 3.5, and outside buffer is exchanged to 300 mM sucrose/20 mM HEPES/15 mM EDTA at pH 7.5 with the A23187 added to the liposomes prior to doxorubicin addition. (C) The ammonium sulfate loading procedure. Liposomes are prepared in 120 mM ammonium sulfate, pH 5.5, and outside buffer is exchanged to 145 mM sodium chloride, pH 5.5. (D) The manganese sulfate loading procedure. Liposomes are prepared in 300 mM manganese sulfate, pH 3.5, and outside buffer is exchanged to 300 mM sucrose/20 mM HEPES/15 mM EDTA at pH 7.5.

Cheung *et al.* (1998) described a novel encapsulation procedure that takes advantage of the ability of doxorubicin to complex or coordinate with manganese (Fig. 2D). Work from our laboratory and others suggests that manganese can theoretically bind doxorubicin at three sites: (1) the anthraquinone (hydroxyketone) moiety at the $C_5–C_6$ position, (2) the anthraquinone (hydroxyketone) moiety at the $C_{11}–C_{12}$ position, and (3) the

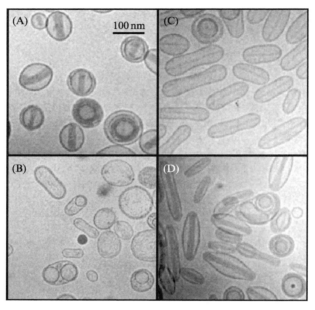

FIG. 3. Cryotransmission electron microscopy (cTEM) images of either DMPC/Chol (55/45 mol ratio) or DSPC/Chol (55/45 mol ratio) liposomes after drug loading achieving a final drug-to-lipid ratio of 0.1/0.2:1 (wt:wt). The liposomes were prepared and loaded with doxorubicin as illustrated in Fig. 2. Briefly, the samples were incubated at 60° for 30 min to facilitate >95% doxorubicin encapsulation. All images are representative of the entire sample. (A) DSPC/Chol liposomes prepared in 300 mM citrate, pH 3.5, and with the outside buffer changed to HEPES-buffered saline, pH 7.5, as the exterior buffer followed by the encapsulation of doxorubicin. (B) DMPC/Chol liposomes prepared in 300 mM MnSO$_4$, pH 3.5, and with the outside buffer changed to 300 mM sucrose–30 mM HEPES, pH 7.5, followed by the encapsulation of doxorubicin. (C) DMPC/Chol liposomes prepared in 120 mM (NH$_4$)$_2$SO$_4$, pH 5.5; with the outside buffer changed to 145 mM NaCl, pH 5.5, and followed by the addition of doxorubicin. (D) DMPC/Chol liposomes prepared in 300 mM MnSO$_4$, pH 3.5, with the outside buffer changed to 300 mM sucrose–20 mM HEPES–15 mM EDTA, pH 7.5, with the addition of the ionophore A23187, followed by the encapsulation of doxorubicin. The bar in (A) is equivalent to 100 nm, and all micrographs are shown at the same magnification. (Electron micrographs courtesy of Katarina Edwards and Göran Karlsson, Uppsala University, Uppsala, Sweden.)

side chain at C_9, possibly in combination with the C_9 hydroxyl position and the amino-sugar group at the C_3–C_4 position (Abraham *et al.*, 2002; Bouma *et al.*, 1986;). Several studies using spectroscopic methods, circular dichroism (CD), electron paramagnetic resonance (EPR), proton nuclear magnetic resonance (^{1}H-NMR), and infrared spectroscopy (IR) have concluded that the deprotonated hydroxy-anthraquinone moieties are the predominant metal-binding sites. Complexation is favored between pH 2 and 8; at pH < 2, no complexation occurs, and at pH > 8, metal hydroxide formation is highly favored. Protonation of the amine group inhibits complex formation, even though this group is not directly involved in metal ion binding.

Specific Description of Techniques: Preparation of Liposomal Doxorubicin through the Use of Ion Gradients

There are four principal stages involved in the preparation of liposomal doxorubicin: (1) preparation of the multilamellar vesicles (MLVs), (2) formation of LUVs from MLVs, (3) establishment of an ion gradient, and (4) encapsulation of doxorubicin. Over the course of this process, a number of critical parameters should be routinely checked, including liposomal lipid concentrations, drug concentrations, drug-to-lipid ratio, encapsulation efficiency, liposomal size, and pH gradient, both before and after loading. The following methods should serve as a guide through this process.

Stage I: Preparation of Multilamellar Vesicles

Weighing and Dissolving Lipids. In general, lipids (purchased from Avanti Polar Lipids, Alabaster, AL, or Northern Lipids Inc., Vancouver, BC, Canada, or other companies specializing in lipids) should be stored at −80° for optimal stability. Prior to weighing, lipids should be brought to room temperature in a closed container containing a desiccant [e.g., calcium chloride (anhydrous) desiccant 20-mesh, catalog no. C77-500/500 g, purchased from Fisher Scientific Company, Pittsburgh, PA]. This decreases the likelihood of contamination by condensation of the lipid stock and ensures accuracy when dispensing the lipid. Lipids are weighed using an analytical balance, e.g., 71.4 mg of DSPC and 28.6 mg of cholesterol would be weighed into a 16 × 100-mm glass test tube and mixed together to prepare 100 mg of DSPC/cholesterol liposomes at a molar ratio of 55:45. All lipids are initially dissolved in organic solvent, e.g., chloroform, methylene chloride, or a chloroform–methanol solution (95:5 v/v). High-performance liquid chromatograph (HPLC) grade solvents should be used to avoid contamination by impurities. Approximately 50–200 mg of total lipid dissolves well in 1 ml of solvent. When using lipids such as sphingomyelin and lysolipid, the addition of several drops of methanol aids in solubilization.

Quantitation of Lipids. At this point it must be decided which method will be used to subsequently quantitate the liposomal lipid concentration. Lipid can be quantitated by measuring the total concentration of phosphorus and calculating the phospholipid concentration. This can be done at any stage during the processing of the liposomes or lipid mixtures; however, if the concentration is to be determined at the stage when solvents are present, exhaustive steps must be taken to remove all traces of solvents prior to performance of the phosphate assay. The phosphate assay is described below in detail (see Stage IV: Doxorubicin Encapsulation).

An alternative method of lipid quantitation involves the addition of a radioactive marker to the dissolved lipid. The ideal marker must be nonexchangeable and nonmetabolizable if it is to be used as a tracer of liposomal lipid *in vivo*. In this regard, we utilize cholesteryl hexadecyl ether (CHE), a lipid that can be bought in a ^{3}H or ^{14}C form from PerkinElmer Life Sciences Inc. (Boston, MA). The specific activity of ^{3}H is typically 40–60 Ci/mmol (which we prefer), and the specific activity of ^{14}C is typically 40–60 mCi/mmol. For *in vitro* experiments, typically 1 μCi of [^{3}H]CHE is added per 100 mg of lipid, and for *in vivo* experiments, approximately 10 μCi of lipid marker is added to 100 mg of lipid. For accuracy, we recommend determining the ^{3}H or ^{14}C disintegrations per minute (dpm) to bulk lipid ratio at the stage when the weighed lipids are dissolved in solvents. The lipids and [^{3}H]- or [^{14}C]CHE are transferred to a 1-ml volumetric flask and topped with organic solvent to the 1-ml mark. Three 10-μl samples are taken using a 50-μl Hamilton syringe and transferred to three separate scintillation vials. The solvent is then evaporated completely using a steady stream of nitrogen gas, as even the smallest amounts of chloroform can cause significant quenching of radioactivity. Five milliliters of scintillation cocktail is added to each vial, and the dpm are obtained using a scintillation counter such as the Packard 1900TR Liquid Scintillation Analyzer. Knowing the amount of lipid weighed and dissolved per 1 ml of solvent allows calculation of the average dpm per mg or μmol lipid, which will serve as the specific activity to be subsequently used to calculate all lipid concentrations.

Solvent Removal by Evaporation. Once the lipid has been fully dissolved with the specific activity determined, the solvent must be removed completely to create a lipid film. It is advantageous to slowly evaporate the bulk of the solvent with a stream of nitrogen gas at a constant temperature of 37–45° until a thick viscous lipid slurry remains. This is followed by a quick transfer to a vacuum pump, where the sample remains under a pressure of $\sim$76 cm Hg for several hours to ensure the removal of residual solvent. Solvent must be uniformly removed to prevent certain lipid components from crystallizing, leading to a less homogeneous lipid mixture. In particular, care must be taken in removing chloroform–methanol solvent mixtures from cholesterol-containing formulations. As the solvent is being evaporated, the cholesterol can precipitate in alcohols, leading to cholesterol microcrystals that will not incorporate well into the lipid bilayer after hydration and can block the polycarbonate filters during the extrusion process (described in Stage II, under Extrusion). The lipid sample should eventually assume a dried white fluffy appearance to obtain maximal surface area for hydration. All of the organic solvent must be removed from the lipid sample, as any residual solvent has the potential to intercalate within the lipid bilayer and disrupt the ability of the liposomes to retain the ion gradient or the doxorubicin itself.

Solvents can be removed by other methods, including rotary evaporation (Waterhouse *et al.*, this volume), lyophilization (benzene–methanol solvents) (Mayer *et al.*, 1984), or even reverse-phase evaporation to directly prepare LUVs (Szoka and Papahadjopoulos, 1978, 1980).

Lipid Hydration. The dried lipid, prepared as described above, is then hydrated with the desired buffer to form MLVs. If using an extruder as a secondary processing technique to generate unilamellar vesicles, and in consideration of the concentration of the final lipid sample, a good rule is to aim for a minimum final volume of 1 ml if 100 mg total lipid is used. Table I outlines the hydrating buffers, their concentrations, and pHs for the various loading procedures. For saturated acyl chain phospholipids, the hydration buffer should be initially heated for 5 min at a temperature approximately 5° above the phase transition temperature of the highest melting lipid component (Table II). The heated buffer is added to the lipid to achieve the desired lipid concentration (50–200 mg/ml final lipid concentration is recommended), and the lipid sample is intermittently vortexed during the 5-min hydration process to ensure maximal hydration. The resulting MLVs are very heterogeneous and range from 500 nm to several micrometers in size.

Freeze and Thaw Cycles. Mayer *et al.* (1985) documented that upon hydration of the lipid film, the buffer solute does not equally distribute between the aqueous core of the liposome and the outside buffer. Cyclic dehydration and rehydration of the MLV lipid head groups are thought to assist in increasing the aqueous trapped volume without changing the overall vesicle size distribution. Dehydration can be accomplished by evaporation, lyophilization, or freezing (the formation of ice crystals in this case also physically disrupts the multilamellar structure). Rehydration

TABLE II
LIPIDS COMMONLY USED IN LIPOSOMAL DOXORUBICIN FORMULATIONS

Phospholipids	Carbon chain	Phase transition temperatures	Molecular weight	Charge
1,2-Dilauroyl-*sn*-glycerol-3-phosphocholine	12:0	−1	621.89	Neutral
1,2-Dimyristoyl-*sn*-glycerol-3-phosphocholine	14:0	23	677.95	Neutral
1,2-Dipalmitoyl-*sn*-glycerol-3-phosphocholine	16:0	41	734.05	Neutral
1,2-Distearoyl-*sn*-glycerol-3-phophocholine	18:0	55	790.15	Neutral
1,2-Distearoyl-*sn*-glycero-3-[phospho-rac-(1-glycerol)]	18:0	55	801.06	Negative
1,1′,2,2′-Tetramyristoyl cardiolipin	14:0	59	1284.97	Negative
1,2-Dioleoyl-*sn*-glycero-3-[phospho-L-serine]	18:1	−11	810.03	Negative

is achieved by hydrating (for the evaporated and the lyophilized sample) or thawing (for freezing), respectively. The freeze and thaw method, in general, allows efficient distribution of solute. In this method, the hydrated lipid sample is subjected to five repeated freezing and thawing cycles. The lipid sample is transferred to a 5-ml cryogenic vial (e.g., Nalgene cryoware, Nalge Company, Rochester, NY) and is immersed for 5 min into a Dewar jar containing liquid nitrogen. The sample is then immersed in a water bath kept at approximately 5° above the transition temperature of the highest melting lipid component for an additional 5 min. This freeze and thaw cycle is repeated four more times. It is important to wear safety goggles during the freeze and thaw cycles due to the potential hazards of working with liquid nitrogen. At the completion of this stage, the frozen MLVs can be either stored in a −20° freezer or can be carried to the next step of the formation of LUVs.

Stage II: Formation of Unilamellar Vesicles

Extrusion. After producing the MLVs, several processing methods are available to produce LUVs. The repetitious physical extrusion of MLVs, under moderate pressures of 400–600 psi through polycarbonate filters of a defined pore size, results in a homogeneous population of single bilayer vesicles of decreased particle size. The extrusion of liposomes is a procedure based on the work of Olson *et al.* (1979), who described a technique involving extruding MLVs under low pressure (<50 psi) through filters with sequentially decreasing pore size in order to obtain small liposomes. This methodology has been improved and developed further through the efforts of our research group (Hope *et al.*, 1985; Mayer *et al.*, 1986), and commercial devices are available for extrusion of lipids under high pressures, such as the Thermobarrel Extruder System, which can be purchased from Northern Lipids Inc. (Vancouver, BC, Canada).

The extrusion process is started by making sure the extruder is scrupulously clean. While wearing gloves, a typical cleaning cycle consists of removing all grease and dirt from each extruder component with a brush and phosphate-free soap followed by rinsing with hot tap water, isopropanol, and finally distilled water. All components without rubber or plastic parts may be subsequently sonicated in chloroform in a chemical hood for 30–60 s; each component is sonicated separately to avoid scratching or rubbing. The chloroform is allowed to evaporate completely before assembling the unit. Prior to assembling the extrusion unit, a new support drain disk is placed on the metal mesh filter, followed by polycarbonate filters with pore sizes selected to produce a defined size liposome. To prepare liposomes with mean diameters between 100 and 120 nm, two stacked 0.1-μm pore-sized filters are typically used; however, some investigators use a combination of a 0.08-μm pore-size filter and a 0.1-μm pore-size filter. Care must be taken to ensure that the drain disk and the filters adhere well to the metal mesh. This can be accomplished by the use of one or two drops of the hydrating buffer applied directly to the positioned filters.

Once assembled, the extruder should be positioned in a fume hood (a radioactive aerosol may be produced during the extrusion process) and allowed to equilibrate to the desired extrusion temperature (5° higher than the phase transition temperature of the lipid possessing the highest melting temperature). The MLV sample is added through the extruder inlet port with a $5\frac{3}{4}$-inch glass Pasteur pipette, taking care that the sample is not added close to the extruder cap but more toward the bottom of the barrel without piercing the filters. Once the cap is securely fastened and the pressure release valve is closed, the nitrogen gas (at 300–600 psi) is slowly released into the extruder chamber. This will force the lipid sample out through the rubber tubing and into a glass collection tube making one extrusion pass. This is repeated nine more times. Ideally, the collection tube is kept at approximately the same temperature as the extruder using a heat block. This extrusion process consistently yields unilamellar vesicles exhibiting a narrow size dispersity of 100–120 nm. The resulting lipid sample should have an increased opalescence and a more transparent appearance compared with the opaque chalky white color of the MLVs.

Unilamellar vesicles can be stored at 4° for 1–2 weeks depending on the susceptibility of the specific lipid composition to the pH of the buffer. For liposomes to be used *in vivo*, it is recommend that the LUVs are used within 24 h after preparation.

Determination of Liposome Size. Size serves as an important determinant in the characteristics of the liposomal doxorubicin formulation. Particle size is an important indicator of batch-to-batch reproducibility, and changes may indicate liposome instability. The most accurate technique available to determine individual particle size is the electron microscope (Woodle and Papahadjopoulos, 1989). Because equipment associated with this technique may not be readily available to most laboratories, a less arduous and simpler way of determining the average particle size and the relative distribution for sample particles involves using a particle sizer, e.g., NICOMP 380/DLS Submicron Particle Sizer (Santa Barbara, CA). This instrument employs a laser-based technique using quasielastic light scattering to determine the hydrodynamic equivalent diameter of particles. Samples are simply prepared by diluting the liposomes, for example 20–25 μl of a 100 mg/ml lipid solution is added to 400 μl filtered saline in a borosilicate tube of the correct dimensions for the particular particle sizer. Size is determined using the software program accompanying the equipment. If the particle size distribution is greater than anticipated, the sample may be reextruded to obtain the desired size range.

Stage III: Establishing an Ion Gradient

The buffer employed during lipid hydration establishes the chemical conditions of the interior aqueous compartment of the liposomal preparation. For the creation of an ion gradient, the exterior buffer of the liposomes must be exchanged. The external buffer can be exchanged by several techniques. Creating a transmembrane pH gradient with an acidic interior requires an increase in pH of the exterior liposomal environment. One method involves the addition of an alkalinizing reagent [e.g., a concentrated base or an alkaline buffer such as dibasic phosphate (0.1–1 M) or sodium carbonate] to neutralize the extravesicular pH. Creating an ion (or again a pH) gradient can also be accomplished by exchanging the exterior media using size-exclusion column chromatography. A gel media such as Sephadex G-50 (medium), which has a dry bead size of 50–150 μm and a globular protein fractionation range of 1500–30,000 MW, is suitable for this purpose. Sephadex G-50 is hydrated in an excess amount of the desired exterior buffer (75% of settled gel) for approximately 3 h at 20°. The suspension is fully degassed for approximately 20 min prior to pouring the column. The volume of the column is dependent on the sample size; a sample volume of 0.5–5% of the bed volume is recommended. To stabilize the column bed and equilibrate with the exterior buffer, two or three volumes of eluent should be passed through the column. The lipid sample should be applied carefully and evenly to the column with the least amount of dilution. The fractions containing the lipid can be clearly identified and will appear opaque white compared with the eluent buffer-containing fractions. The liposomes are always collected in the void volume of the column. This technique allows the freedom of choice for a specific external buffer. The extravesicular buffer can also be exchanged using dialysis techniques. The lipid sample is placed within dialysis bags of a molecular cut-off range of 12–14 kDa and placed in a large volume of the extravesicular buffer. The buffer is replaced two or three times over the course of 24–48 h in a 4° cold room. This technique also provides flexibility when choosing an external buffer.

Stage IV: Doxorubicin Encapsulation

Following the establishment of an ion gradient, doxorubicin must be encapsulated within a time frame that does not compromise the gradient; this is dependent on both the lipid composition and the buffer. Drug uptake requires incubation at specific elevated temperatures. The time of incubation and the temperature are dependent on the lipid composition and the desired drug-to-lipid-ratio and, therefore, must be determined experimentally (see Fig. 4 for an example).

The liposome sample concentrations are calculated prior to doxorubicin encapsulation. If using a radiolabeled lipid marker, liposome lipid concentrations are determined by adding a small aliquot to 5 ml of scintillation cocktail, where the radioactivity of the sample is subsequently determined by scintillation counting with the liquid scintillation analyzer. If using the phosphate concentration to calculate the total liposome concentration, the phosphate assay initially described by Fiske and Subbarow (1925) can be followed. This procedure requires all equipment in contact with the lipid sample to be phosphate free, which necessitates that all glassware be acid washed. The glassware can be soaked overnight in Fisher's chromic/sulfuric acid mix (Fisher Scientific Company, Pittsburgh, PA). Briefly, 700 μl of 70% perchloric acid is added to lipid samples and standards in 16 × 125-mm-thick walled borosilicate heat-resistant tubes (e.g., Pyrex or Kimax Brand, Fisher Scientific Company, Pittsburgh, PA). Standards can be prepared in the 0–200 nmol total phosphate range using a 2 mM sodium phosphate solution. Tubes are capped with acid-washed marbles to minimize evaporation and allow the venting of pressure. Samples and standards are heated to approximately 180–200° using a block

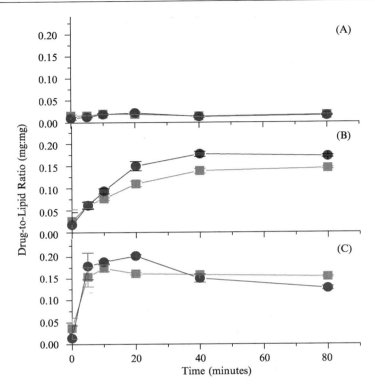

FIG. 4. Doxorubicin encapsulation in DMPC/Chol (55/45 mol ratio) liposomes using either the manganese chloride–loading procedure (■) or the manganese chloride–loading procedure with the A23187 ionophore (●). Liposomes were prepared in the presence of 300 mM MnCl$_2$ at pH 3.5, and the outer buffers were exchanged using column chromatography to create a pH or an Mn^{2+} gradient. For the MnCl$_2$ loading procedure with the A23187 ionophore, the A23187 was added and incubated 5 min prior to the addition of drug. Doxorubicin was added to the liposomes to achieve a 0.2:1 drug-to-lipid ratio (w:w) and incubated at 20° (A), 40° (B), or 60° (C). At the indicated time points, aliquots were fractionated on 1-ml spin columns to separate the encapsulated drug (collected in the void volume) from the unencapsulated drug. Lipid concentrations were determined using [^{3}H]CHE, and doxorubicin was quantitated by reading the A_{480} of a detergent solubilized sample as described in Section IV. Data points represent the mean drug-to-lipid ratios of at least three replicate experiments, and the error bars indicate the standard deviation.

heater for 2 h or until the samples are colorless. The block heater is typically set up in a fume hood with constant hydration (the water faucet in the fume hood is left on during the entire procedure). The tubes must not be allowed to completely dry, as this is an explosion hazard. Great care must be taken when using perchloric acid; refer to appropriate Material

Safety Data Sheets prior to use. Samples are cooled, and 700 μl of Fiske reagent (as described below) is added to each tube. Seven milliliters of a 5% ammonium molybdate solution is also added, and samples are gently vortexed and subsequently reheated again at 100° for 20 min until the samples are at a full boil. Samples are then cooled to room temperature, and the absorbance is read at 820 nm. Phospholipid concentrations may then be calculated based on standard curve results. The Fiske and Subbarow reagent can be prepared by weighing 150 g of $NaHSO_3$, 5 g Na_2SO_3, and 2.5 g 1-amino-2-naphthol-4-sulfonic acid and dissolving in 1 liter of distilled water with slight warming to assist in dissolution. The solution should be stored in an amber bottle at 4° and discarded within 6 months. The solution should be filtered using a 0.45-μm filter before use.

Doxorubicin can be purchased in a pure powdered form (e.g., Handetech USA, Inc., Houston, TX) or in a powder injectable form of 10, 50, and 150 mg of doxorubicin-hydrochloride with 52.6, 263.1, and 789.4 mg lactose, respectively, in each vial [e.g., Faulding (Canada) Inc., Kirkland, QC, Canada, or Faulding Pharmaceuticals, South Paramus, NJ). As with the handling of any cytotoxic agent, extreme care must be taken when weighing doxorubicin; gloves and masks must be used if handling doxorubicin powder directly. Direct handling of doxorubicin in powdered form is not recommended, and the injectable powder, for example, can be dissolved in injectable water, as specified by the manufacturer, prior to any dispensing of the drug.

The desired quantities of doxorubicin and the liposomes are individually incubated at the determined temperature for optimal loading; this temperature is typically selected to give greater than 95% loading within 30 min following mixing of drug with the liposomes. Once lipid and drug have reached the specific temperature, the doxorubicin is rapidly added to the prepared liposomes and vortexed intermittently during the incubation period. An example of typical loading conditions is provided in Table III.

The accumulation of doxorubicin into liposomes can be determined by measuring the drug-to-lipid ratio as a function of time. This is determined

TABLE III
DOXORUBICIN LOADING

Component	Amount	Volume (total = 2.0 ml)
Lipid (20 mg/ml)	10 mg	500 μl
Doxorubicin (10 mg/ml)	2 mg	200 μl
External buffer	N/A	1300 μl [2.0−(0.5+0.2)]

by removing small (e.g., 100 μl) aliquots from a loading sample and separating unencapsulated drug from encapsulated drug on 1-ml Sephadex G-50 (medium) spin columns equilibrated with the appropriate buffer. One-milliliter spin columns can be purchased commercially or can be easily made for a fraction of the price. Sephadex G-50 (medium) beads are hydrated in the external buffer used for the liposomes. The solution is degassed as in a typical column preparation. One-milliliter tuberculin syringes without needles are used, with the plunger removed. A small portion of glass wool is rolled into a small ball (the size of a pea) and firmly inserted into the empty syringe using the plastic plunger (with the sealing rubber tip removed), thus creating a plug for the column. This should be done in a fume hood to avoid inhalation of the glass particles that can arise when working with glass wool. Columns are placed into empty 16 × 100-mm glass tubes, and each syringe is filled with the Sephadex, making sure no air bubbles are present. Columns are gently shaken to compact the solution, and more Sephadex is added until the tops of the syringes are almost overflowing. Excess buffer is removed from the glass tube, and columns are centrifuged at 760g for 1 min. Columns should typically pack to the 1 ml mark. The columns are then put into clean test tubes and are ready to be used within 1 h of preparation. Before adding the liposome sample, the columns are always "prewet" with 50–100 μl of external buffer, and the sample is immediately added. Columns are recentrifuged at 760g for 3 min, and the liposomes and associated drug are collected in the voided fraction.

The concentration of doxorubicin in the excluded fraction is easily determined using either spectrophotometric or fluorescent assays. Doxorubicin absorbance is typically read at 480 nm from a solution consisting of the sample or standard. Standards are prepared using diluted doxorubicin in the 10–60 μg/ml range. Samples or standards are adjusted to 100 μl with extravesicular buffer, to which 900 μl of 1% Triton X-100 is added. Prior to assessing absorbance at 480 nm, the sample is placed in a water bath at >90° until the cloud point of the detergent is observed. The amount of doxorubicin-associated fluorescence can be determined by first solubilizing the liposomes using a suitable detergent devoid of contaminating fluorescent chemicals that may interfere with doxorubicin fluorescence. Typically, our laboratory uses n-octyl β-D-glucopyranoside, a high-quality nonionic detergent. Doxorubicin fluorescence can be read using a luminescence spectrometer with excitation/emission wavelengths depending on an initial prescan of doxorubicin to obtain optimal values. The fluorescence readings are compared with a standard curve of doxorubicin prepared using known amounts of doxorubicin in the 0–500 ng range in a solution similar to that used for the unknowns.

Conclusions

This chapter has described active loading methods for the encapsulation of doxorubicin into unilamellar liposomes for use *in vitro* and *in vivo*. It is hoped that sufficient insight has been provided into the physicochemical properties of doxorubicin and how this drug interacts with lipids and with the specific buffers used to promote encapsulation of the drug. This chapter has not, however, defined how the physicochemical properties of the formulation influence biological activity. It is believed that the *in vivo* pharmacokinetics and biodistribution of iv-injected liposomal doxorubicin are dependent on a fine balance between the rate of doxorubicin release from the lipid carrier and the elimination rate of the carrier itself. The rate of doxorubicin release from the lipid carrier is dictated by several factors. The internal pH of the liposome dictates the proportion of drug in the neutral or charged form, and as the proportion shifts to the neutral form (e.g., as the interior pH increases), the rate of drug dissociation increases. Some investigators believe that the initial stage of drug release involves dissolution of the drug itself from the precipitated form adopted within the liposome; however, results from our laboratory suggest that this is not a rate-limiting step (Abraham *et al.*, 2002).

For doxorubicin to move from the inside of the liposome to the outside, it must pass through the lipid bilayer, a process that involves interactions with the membrane interface on the inside of the liposome, the lipid headgroups, the lipid acyl chains, and the membrane interface on the outside of the liposome. Obviously, the pH within the aqueous core of the liposome and the pH at the interface dictate the proportion of drug in the neutral and in the charged form. It is believed that both neutral and charged drugs can permeate the bilayer, but it is acknowledged that the rate of permeation of the neutral form of the drug is far faster than the charged form. Evidence in support of membrane partitioning of the protonated form of doxorubicin arises from studies evaluating the effects of charged lipid species, which increase hydration at the interface and, in turn, will increase drug release. More specifically, the presence of anionic lipids likely increases the concentration of charged doxorubicin at the interface, and this results in an increased rate of drug release even under conditions where the internal pH is kept below 5. It is also established that inclusion of PEG-modified phosphatidylethanolamine (an anionic lipid) increases doxorubicin release rates, an effect that could be attributed to surface-grafted PEGs increasing hydration at the interface or to the fact that the PEG lipid is anionic.

Assuming that the lipid composition consists primarily of the zwiterionic lipid phosphatidylcholine and cholesterol (55:45 mol%), drug release is primarily a consequence of drug partitioning through the bilayer. This statement is not intended to minimize the important parameters such as trapping efficiency, drug-to-lipid ratios, and their link to maintenance of a large pH gradient (Mayer *et al.*, 1989). Rather it acknowledges that under conditions in which the loading parameters are identical and the interior pHs are comparable, the most important parameter controlling doxorubicin release is the liposomal lipid composition. The importance of this was best illustrated in a review published in 1994 (Mayer *et al.*, 1994), where the drug release rates and acute toxicity parameters of five different liposomal formulations of doxorubicin were compared. All formulations were comparable on the basis of drug-to-lipid ratio, liposome size, and liposome elimination rates following iv administration, yet there was more than a 10-fold range in the dose that could be tolerated in mice. Using the maximum tolerated dose (MTD) of free drug (20 mg/kg) as a comparator, the MTD of doxorubicin encapsulated in DMPC/Chol liposomes was less than 5 mg/kg, while the MTD of doxorubicin encapsulated in DSPC/Chol liposomes was greater than 80 mg/kg. The rate of drug release from liposomes in the plasma compartment was shown to be linear over a 24-h time course for both liposomal formulations, but the rate of drug release from the DMPC/Chol liposomes was significantly faster.

As a final note, we strongly believe that the lessons learned while developing liposomal formulations of doxorubicin are valuable when considering the development of other liposomal drug formulations. There are several therapeutic applications being considered that will benefit from continued research on liposomal formulations of doxorubicin. In particular, we and other investigators are considering the use of liposomal anticancer drug formulations that are designed to retain a drug well in the plasma compartment but to release it rapidly following accumulation at sites of disease. The rationale for these formulations is simple. An ideal liposomal anticancer drug would exhibit little or no drug release while in the plasma compartment, thus ensuring limited exposure of the drug to healthy tissue. This feature would also maximize drug delivery to disease sites, as mediated by the movement of the drug-loaded liposomes from the plasma compartment to the extravascular space at disease sites, such as a region of tumor growth. Following localization, however, the drug-loaded liposome must transform itself from a stable carrier to an unstable carrier. This would ensure that the drug, which has localized in the diseased site, is bioavailable. In this regard, our research team has placed specific emphasis on the development of liposomal anthracycline formulations that meet these design attributes, including the use of programmable fusogenic liposomes (Adlakha-Hutcheon *et al.*, 1999) and thermosensitive liposomes combined with use of hyperthermia (Needham *et al.*, 2000).

References

Abraham, S. A., Edwards, K., Karlsson, G., MacIntosh, S., Mayer, L. D., McKenzie, C., and Bally, M. B. (2002). Formation of transition metal-doxorubicin complexes inside liposomes. *Biochim. Biphys. Acta* **1565**, 41–45.

Adlakha-Hutcheon, G., Bally, M. B., Shew, C. R., and Madden, T. D. (1999). Controlled destabilization of a liposomal drug delivery system enhances mitoxantrone antitumor activity. *Nat. Biotechnol.* **17**, 775–779.

Allen, C., Dos Santos, N., Gallagher, R., Chiu, G. N. C., Shu, Y., Li, W. M., Johnstone, S. A., Mayer, L. D., Webb, M. S., and Bally, M. B. (2002). Controlling the physical behavior and biological performance of liposome formulations through use of surface grafted poly(ethylene glycol). *Biosci. Rep.* **22**, 225–250.

Bally, M. B., Nayar, R., Masin, D., Hope, M. J., Cullis, P. R., and Mayer, L. D. (1990). Liposomes with entrapped doxorubicin exhibit extended blood residence times. *Biochim. Biophys. Acta* **1023**, 133–139.

Bolotin, E. M., Cohen, R., Bar, L. K., Emanuel, N., Ninio, S., Lasic, D. D., and Barenholz, Y. (1994). Ammonium sulfate gradients for efficient and stable remote loading of amphipathic weak bases into liposomes and ligandsomes. *J. Liposome Res.* **4**, 455–479.

Bouma, J., Beijnen, J. H., Bult, A., and Underberg, W. J. (1986). Anthracycline antitumour agents. A review of physicochemical, analytical and stability properties. *Pharm. Weekbl. Sci.* **8**, 109.

Budavari, S., O'Neil, M. J., and Smith, A. (2000). *In* "The Merck Index," (S. Budavari, ed.), p. 541. Merck and Co. Inc., Rahway, NJ.

Chaires, J. B., Dattagupta, N., and Crothers, D. M. (1982). Self-association of daunomycin. *Biochemistry* **21**, 3927–3932.

Cheung, B. C., Sun, T. H., Leenhouts, J. M., and Cullis, P. R. (1998). Loading of doxorubicin into liposomes by forming Mn^{2+} drug complexes. *Biochim. Biophys. Acta* **1414**, 205–216.

Cullis, P. R., Hope, M. J., Bally, M. B., Madden, T. D., Mayer, L. D., and Janoff, A. S. (1987). *In* "Liposomes: From Biophysics to Therapeutics" (M. J. Ostro, ed.), p. 39. Marcel Dekker Inc, New York.

Cullis, P. R., Mayer, L. D., Bally, M. B., Madden, T. D., and Hope, M. J. (1989). Generating and loading of liposomal systems for drug-delivery applications. *Adv. Drug Deliv. Rev.* **3**, 267–282.

Cullis, P. R., Bally, M. B., Madden, T. D., Mayer, L. D., and Hope, M. J. (1991). pH gradients and membrane transport in liposomal systems. *Trends Biotechnol.* **9**, 268–272.

Fenske, D. B., Wong, K. F., Maurer, E., Maurer, N., Leenhouts, J. M., Boman, N., Amankwa, L., and Cullis, P. R. (1998). Ionophore-mediated uptake of ciprofloxacin and vincristine into large unilamellar vesicles exhibiting transmembrane ion gradients. *Biochim. Biophys. Acta* **1414**, 188.

Fiallo, M. M., Tayeb, H., Suarato, A., and Garnier-Suillerot, A. (1998). Circular dichroism studies on anthracycline antitumor compounds. Relationship between the molecular structure and the spectroscopic data. *J. Pharm. Sci.* **87**, 967–975.

Fiallo, M. M., Garnier-Suillerot, A., Matzanke, B., and Kozlowski, H. (1999). How Fe^{3+} binds anthracycline antitumor compounds. The myth and the reality of a chemical sphinx. *J. Inorg. Biochem.* **75**, 105–115.

Fiske, C. H., and Subbarow, Y. (1925). The colorimetric determination of phosphorus. *J. Biol. Chem.* **2**, 375.

Forssen, E. A., and Tokes, Z. A. (1979). *In vitro* and *in vivo* studies with adriamycin liposomes. *Biochem. Biophys. Res. Commun.* **91**, 1295–1301.

Gabizon, A., Catane, R., Uziely, B., Kaufman, B., Safra, T., Cohen, R., Martin, F., Huang, A., and Barenholz, Y. (1994). Prolonged circulation time and enhanced accumulation in malignant exudates of doxorubicin encapsulated in polyethylene-glycol coated liposomes. *Cancer Res.* **54,** 987–992.

Haran, G., Cohen, R., Bar, L. K., and Barenholz, Y. (1993). Transmembrane ammonium sulfate gradients in liposomes produce efficient and stable entrapment of amphipathic weak bases. *Biochim. Biophys. Acta* **1151,** 201–215.

Harrigan, P. R., Hope, M. J., Redelmeier, T. E., and Cullis, P. R. (1992). Determination of transmembrane pH gradients and membrane potentials in liposomes. *Biophys. J.* **63,** 1336–1345.

Hobbs, S. K., Monsky, W. L., Yuan, F., Roberts, W. G., Griffith, L., Torchilin, V. P., and Jain, R. K. (1998). Regulation of transport pathways in tumor vessels: Role of tumor type and microenvironment. *Proc. Natl. Acad. Sci. USA* **95,** 4607–4612.

Hope, M. J., Bally, M. B., Webb, G., and Cullis, P. R. (1985). Production of large unilamellar vesicles by a rapid extension procedure: Characterization of size, trapped volume, and ability to maintain a membrane potential. *Biochim. Biophys. Acta* **812,** 55–65.

Hortobagyi, G. N. (1997). Anthracyclines in the treatment of cancer. An overview. *Drugs* **54**(Suppl 4), 1–7.

Lasic, D. D. (1996). Doxorubicin in sterically stabilized liposomes. *Nature* **380,** 561–562.

Lasic, D. D., Frederik, P. M., Stuart, M. C., Barenholz, Y., and McIntosh, T. J. (1992). Gelation of liposome interior. A novel method for drug encapsulation. *FEBS Lett.* **312,** 255–258.

Li, X., Hirsh, D. J., Cabral-Lilly, D., Zirkel, A., Gruner, S. M., Janoff, A. S., and Perkins, W. R. (1998). Doxorubicin physical state in solution and inside liposomes loaded via a pH gradient. *Biochim. Biophys. Acta* **1415,** 23–40.

Madden, T. D., Harrigan, P. R., Tai, L. C., Bally, M. B., Mayer, L. D., Redelmeier, T. E., Loughrey, H. C., Tilcock, C. P., Reinish, L. W., and Cullis, P. R. (1990). The accumulation of drugs within large unilamellar vesicles exhibiting a proton gradient: A survey. *Chem. Phys. Lipids* **53,** 37–46.

Mayer, L. D., Madden, T. D., Bally, M. B., and Cullis, P. R. (1984). *In* "Liposome Technology" (G. Gregoriadis, ed.), p. 27. CRC Press, Boca Raton, FL.

Mayer, L. D., Hope, M. J., Cullis, P. R., and Janoff, A. S. (1985). Solute distributions and trapping efficiencies observed in freeze-thawed multilamellar vesicles. *Biochim. Biophys. Acta* **817,** 193–196.

Mayer, L. D., Bally, M. B., Hope, M. J., and Cullis, P. R. (1986). Techniques for encapsulating bioactive agents into liposomes. *Chem. Phys. Lipids* **40,** 333–345.

Mayer, L. D., Tai, L. C., Ko, D. S., Masin, D., Ginsberg, R. S., Cullis, P. R., and Bally, M. B. (1989). Influence of vesicle size, lipid composition, and drug-to-lipid ratio on the biological activity of liposomal doxorubicin in mice. *Cancer Res.* **49,** 5922–5930.

Mayer, L. D., Cullis, P. R., and Bally, M. B. (1994). The use of transmembrane pH gradient-driven drug encapsulation in the pharmacodynamic evaluation of liposomal doxorubicin. *J. Liposome Res.* **4,** 529–553.

Menozzi, M., Valentini, L., Vannini, E., and Arcamone, F. (1984). Self-association of doxorubicin and related compounds in aqueous solution. *J. Pharm. Sci.* **73,** 766–770.

Needham, D., Anyarambhatla, G., Kong, G., and Dewhirst, M. W. (2000). A new temperature-sensitive liposome for use with mild hyperthermia: Characterization and testing in a human tumor xenograft model. *Cancer Res.* **60,** 1197–1210.

Olson, F., Hunt, C. A., Szoka, F. C., Vail, W. J., and Papahadjopoulos, D. (1979). Preparation of liposomes of defined size distribution by extrusion through polycarbonate membranes. *Biochim. Biophys. Acta* **557,** 9–23.

Papahadjopoulos, D., Allen, T. M., Gabizon, A., Mayhew, E., Matthay, K., Huang, S. K., Lee, K. D., Woodle, M. C., Lasic, D. D., Redemann, C., and Martin, F. J. (1991). Sterically stabilized liposomes: Improvements in pharmacokinetics and antitumor therapeutic efficacy. *Proc. Natl. Acad. Sci. USA* **88,** 11460–11464.

Parr, M. J., Bally, M. B., and Cullis, P. R. (1993). The presence of GM1 in liposomes with entrapped doxorubicin does not prevent RES blockade. *Biochim. Biophys. Acta* **1168,** 249–252.

Pegram, M., Hsu, S., Lewis, G., Pietras, R., Beryt, M., Sliwkowski, M., Coombs, D., Baly, D., Kabbinavar, F., and Slamon, D. (1999). Inhibitory effects of combinations of HER-2/neu antibody and chemotherapeutic agents used for treatment of human breast cancers. *Oncogene* **18,** 2241–2251.

Pegram, M. D., Lopez, A., Konecny, G., and Slamon, D. J. (2000). Trastuzumab and chemotherapeutics: Drug interactions and synergies. *Semin. Oncol.* **27,** 21–25.

Rahman, A., Kessler, A., and More, N. (1980). Liposomal protection of adriamycin-induced cardiotoxicity in mice. *Cancer Res.* **40,** 1532–1537.

Redelmeier, M. J., Mayer, L. D., Wong, K. F., Bally, M. B., and Cullis, P. R. (1989). Proton flux in large unilamellar vesicles in response to membrane potentials and pH gradients. *Biophys. J.* **56,** 385–393.

Szoka, F., Jr., and Papahadjopoulos, D. (1978). Procedure for preparation of liposomes with large internal aqueous space and high capture by reverse-phase evaporation. *Proc. Natl. Acad. Sci. USA* **75,** 4194–4198.

Szoka, F., Jr., and Papahadjopoulos, D. (1980). Comparative properties and methods of preparation of lipid vesicles (liposomes). *Annu. Rev. Biophys. Bioeng.* **9,** 467–508.

Waterhouse, D. N., Tardi, P. G., Mayer, L. D., and Bally, M. B. (2001). A comparison of liposomal formulations of doxorubicin with drug administered in free form: Changing toxicity profiles. *Drug Safety* **24,** 903–920.

Woodle, M. C., and Papahadjopoulos, D. (1989). Liposome preparation and size characterization. *Methods Enzymol.* **171,** 193–217.

[5] Preparation and Characterization of Taxane-Containing Liposomes

By Robert M. Straubinger and
Sathyamangalam V. Balasubramanian

Abstract

Drug carriers such as liposomes provide a means to alter the biodisposition of drugs and to achieve concentration–time exposure profiles in tissue or tumor that are not readily accomplished with free drug. These changes in biodisposition can improve treatment efficacy. For hydrophobic drugs, incorporation in liposome carriers can increase drug solubility markedly. The taxanes paclitaxel (taxol) and docetaxel (Taxoteré) are members of one of the most important new classes of oncology drugs. However, their poor solubility presents pharmaceutical challenges, and emerging data suggest that specific tissue exposure profiles, such as low drug concentrations for extended times, can enhance beneficial antitumor mechanisms. Incorporation of the taxanes into liposomes eliminates not only the toxic effects of cosolvents required to administer these drugs clinically but also increases drug efficacy in animal tumor models, usually through a reduction in dose-limiting tissue toxicities. Although the taxanes are poorly water soluble, the preparation of physically stabile taxane/lipo-liposome formulations requires the balancing of three factors: (1) the drug:lipid ratio, (2) the liposome composition, and (3) the duration of storage in aqueous media. Biophysical evaluation of formulation characteristics, principally using circular dichroism (CD) and differential scanning calorimetry (DSC), can provide the information necessary to develop stable taxane–liposome formulations. These techniques provide information on drug–drug and drug–lipid interactions that underlie the events that lead to taxane formulation instability. Owing to the unusually low solubility of the taxanes, special consideration is necessary to devise methods for resolving drug-containing liposomes from released or precipitated drug to obtain reliable estimates of drug incorporation and retention in liposomes.

Introduction

The taxanes represent one of the most important new classes of oncology drugs approved in the past two decades. An unprecedented drug development effort resulted from the observation that taxol (paclitaxel), the prototype of this class, showed activity in recurrent, platinum-resistant ovarian cancer (Suffness, 1993). Taxoteré (docetaxel), a semisynthetic derivative, followed paclitaxel into the clinic, and both are now FDA approved. Through widespread clinical experience, the taxanes have progressed from drug of last resort to first-line therapy for a variety of cancers, such as refractory ovarian, breast, and non-small cell lung cancer (Adler et al., 1994; Cortes et al., 2003; Guastalla et al., 1994; Kubota et al., 1997; Murphy et al., 1993; Sledge et al., 2003).

The main mechanism of taxane action results from interaction with cellular microtubules (Schiff et al., 1979), which promotes their assembly and stabilization. However, the mechanisms of tumor growth control may vary with drug concentration (Derry et al., 1995; Gan et al., 1996; Jordan et al., 1993; Milross et al., 1996; Yen et al., 1996) or the concentration versus time pharmacokinetic profile (Bocci et al., 2002). At the highest (μM) concentrations, the taxanes induce cytoplasmic microtubule bundling and aster formation (De Brabander et al., 1981; Manfredi et al., 1982; Rowinsky et al., 1988; Schiff and Horwitz, 1981; Schiff et al., 1979). In the low nM range, regarded as more relevant clinically, paclitaxel blocks the cell cycle at the G_2/M interface by kinetic stabilization of microtubule dynamics. Cell death through the apoptosis pathway has been observed, with activation of a variety of signaling cascades (Fan, 1999; Huang *et al.*, 2000; Jordan and Wilson, 1995; Jordan *et al.*, 1993; Kawasaki *et al.*, 2000; Lee *et al.*, 1998; Ling *et al.*, 1998; Manfredi *et al.*, 1982; Ojima *et al.*, 1999; Roy and Horwitz, 1985; Schiff *et al.*, 1979; Tudor *et al.*, 2000; Wang *et al.*, 1998, 1999; Yeung *et al.*, 1999; Yvon *et al.*, 1999). Most recently, it has been observed that the taxanes possess significant antiangiogenic activity at low concentrations (mid pM range) or synergize with antiangiogenic agents (Bocci *et al.*, 2002; Farinelle *et al.*, 2000; Grant *et al.*, 2003; Satoh *et al.*, 1998; Wang *et al.*, 2003). For paclitaxel, inhibition of vascular endothelial cell proliferation and migration has been observed at concentrations 10- to 100-fold lower than those inducing mitotic arrest in nonendothelial cells. Docetaxel appears even more active, with an estimated 10-fold greater potency than paclitaxel in *in vitro* antiangiogenic assays (Grant *et al.*, 2003). This spectrum of concentration- or exposure (concentration × time)-dependent effects suggests that drug delivery approaches that control the exposure profile can alter therapy in beneficial ways.

The taxanes are complex diterpenoid natural products or semisynthetic derivatives (Fig. 1). They consist of a bulky, fused ring system and an extended side chain (at C13) that is required for activity. Paclitaxel, the prototype, is a natural product (Wani *et al.*, 1971), and docetaxel is a semisynthetic analogue that differs from paclitaxel by substitutions at C10 and C13 (Bissery *et al.*, 1991; Guéritte-Voegelein *et al.*, 1991; Ringel and Horwitz, 1991).

One common characteristic of the clinically used taxanes is poor aqueous solubility. Although the taxane molecular structure has several relatively hydrophilic domains (in the vicinity of C7–C10 and C1′–C2′), hydrophobic domains of the fused ring system and side chain (Balasubramanian *et al.*, 1994; Guénard *et al.*, 1993) contribute to the overall poor aqueous solubility. Estimates of paclitaxel solubility vary

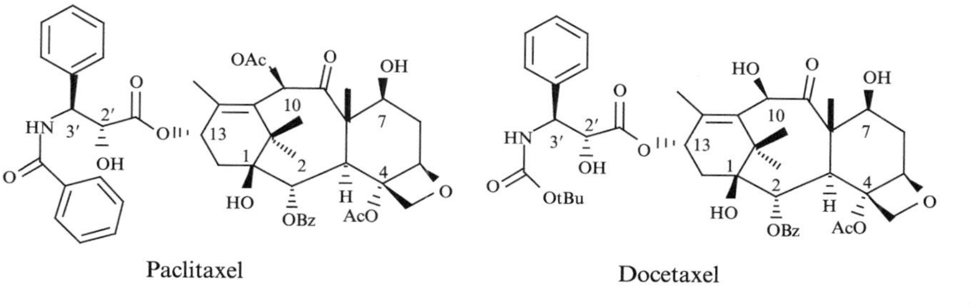

Paclitaxel   Docetaxel

FIG. 1. Chemical structure of paclitaxel and docetaxel. Bu, butyl; Ac, acetyl; Bz, benzoyl.

widely, including ~35 μM (Ringel and Horwitz, 1991; Swindell *et al.*, 1991), ~7 μM (Tarr and Yalkowsky, 1987), and $\leq 0.77\ \mu M$ (~0.7 $\mu g/ml$) (Mathew *et al.*, 1992). The discrepancy may reflect the measurement of solubility under equilibrium versus nonequilibrium conditions (Adams *et al.*, 1993). A time-dependent decline in solubility was reported for paclitaxel (Sharma *et al.*, 1995b); over 24 h, the measured solubility fell 10-fold from initial values to a final value of 0.4 μM. Detailed studies of paclitaxel solid state (Liggins *et al.*, 1997) confirmed the low equilibrium solubility of paclitaxel, and demonstrated that polymorphic crystalline forms exist (anhydrous and a dihydrate solvate), each with distinct dissolution properties.

The taxanes are poorly soluble, not only in water but also in the oils and surfactants commonly used in preparing emulsions or other formulations (Adams *et al.*, 1993; Rose, 1992; Straubinger, 1995; Suffness, 1993). Therefore, paclitaxel is formulated for clinical use (Taxol, Bristol Myers Squibb, Inc., Princeton, NJ) at a concentration of 6 mg/ml in a mixture containing 50% (v/v) of the organic solvent ethanol plus the surfactant polyethoxylated castor oil (Cremophor EL) (USPDI, 2003a). Prior to administration, the solution is diluted with saline or dextrose to a concentration of 0.3–1.2 mg/ml, producing a microemulsion. The solution is administered through an in-line filter (U.S. Public Health Service, 1990). These procedures and guidelines are necessary to avoid precipitation upon dilution.

Formulation considerations for docetaxel have been described previously (Bissery *et al.*, 1991; Bisset *et al.*, 1993; Extra *et al.*, 1993; Rhone Poulenc Rorer S.A.). For clinical use (Taxoteré, Aventis, Inc., Bridgewater, NJ), it is also formulated and administered in a cosolvent system. The drug is packaged at 40 mg/ml in polysorbate-80 (USPDI, 2003b). Prior to use, it is diluted to 10 mg/ml with a solution containing 13% (v/v) ethanol in water. Before administration, the drug is further diluted in 250 ml saline or dextrose, achieving a final concentration of 0.3–0.9 mg/liter. The solution is used within 4 h.

The ethanol:Cremophor vehicle required to solubilize paclitaxel is toxic and has been observed to cause life-threatening anaphylactoid reactions (Donehower *et al.*, 1987; Dye and Watkins, 1980; Friedland *et al.*, 1992; Grem *et al.*, 1987; Lorenz *et al.*, 1977; Rowinsky *et al.*, 1992; Weiss *et al.*, 1990). Furthermore, Cremophor appears to modify the pharmacological activity of paclitaxel (Webster *et al.*, 1993) and to contribute to nonlinear pharmacokinetics (Sparreboom *et al.*, 1996). Toxicities associated with docetaxel administration partially overlap those of paclitaxel. However, some apparently unique adverse effects are associated with docetaxel administration. Delayed-onset pleural effusions and edema (Aapro *et al.*, 1993; Behar *et al.*, 1997; Burris *et al.*, 1993; Fumoleau *et al.*, 1993; Irvin *et al.*,

1993) have led in some cases to the discontinuation of treatment. The polysorbate-80 vehicle has been suspected of contributing to this unique spectrum of side effects (Irvin *et al.*, 1993). Most adverse effects of taxane administration are managed by premedication of patients with corticosteroids and antihistamines (Arbuck *et al.*, 1993; Behar *et al.*, 1997; Onetto *et al.*, 1993; Rowinsky *et al.*, 1991; Runowicz *et al.*, 1993; Schrijvers *et al.*, 1993; Weiss *et al.*, 1990).

Because of the toxicities associated with the cosolvents required for taxane administration, a variety of alternative formulation strategies have been investigated. Here we discuss methods for the preparation and characterization of liposomes containing active taxanes, focused principally on paclitaxel.

Formulation Considerations

A growing body of literature reports the preparation and characterization of liposomes containing taxanes. The majority of studies describe paclitaxel-containing formulations; less information is available for docetaxel-containing liposomes (Immordino *et al.*, 2003; Sharma *et al.*, 1995a). One common conclusion supported by much of the literature is that physical stability of taxane–liposome formulations is determined by interaction among three factors: (1) the drug/lipid molar ratio, (2) the liposome composition, and (3) the duration of storage in aqueous media. Drug–drug interactions determine the propensity for drug intermolecular aggregation that may lead to precipitation. Drug–lipid interactions, affected by liposome membrane factors such as lipid miscibility, phase separation, and the membrane phase state, determine the degree to which the drug is accommodated in the bilayer. Chemical stability is also a concern, particularly deesterification of the C13 side chain. However, several approaches reduce or avoid chemical instability. Physical stability remains one of the most important issues in taxane formulation.

Drug/Lipid Ratio

Maximizing the ratio of drug to lipid reduces the amount of lipid that must be administered for a given drug dose. High lipid doses may raise concerns of toxicity and reduce the economic feasibility of pharmaceutical-scale production. In the absence of other considerations, optimal taxane liposomes contain the highest achievable drug/lipid molar ratio. However, because drug–lipid interactions are a major determinant of formulation physical stability, and these are concentration dependent, a tradeoff exists between taxane content and duration of stability in aqueous media (Sharma and Straubinger, 1994b). In many studies, liposomes containing a maximum of ~3–4 mol% drug (with respect to phospholipid) possess stability of sufficient duration as to be clinically usable. In several notable exceptions, liposomes appear to accommodate much higher drug/lipid ratios. However, physical stability is often not characterized rigorously. Little information is available on the stability characteristics of docetaxel formulations, but liposomes containing 3 mol% (drug–lipid) or more of docetaxel have been prepared and are stable (Immordino *et al.*, 2003; Sharma *et al.*, 1997).

Compositional Dependence of Physical Stability

Most studies utilize liposomes in which the majority phospholipid is a zwitterionic neutral component such as phosphatidylcholine (PC), either derived from natural sources or produced synthetically to define acyl chain composition. However, liposomes of 100% PC aggregate (Sharma and Straubinger, 1994b; Straubinger *et al.*, 1993) and therefore anionic (Sharma and Straubinger, 1994b) or cationic (Campbell *et al.*, 2001) phospholipids (or other amphiphiles) are included to inhibit aggregation. The effect of cholesterol, which is included commonly in liposome formulations to increase stability, has not been determined definitively in terms of paclitaxel liposome stability. Phospholipids modified covalently on the headgroup with simple sugars, or extended-chain poly(ethylene glycols) (PEG) show prolonged circulation time *in vivo* (Allen and Chonn, 1987; Gabizon and Papahadjopoulos, 1988; Klibanov *et al.*, 1991). However, some reports indicate that the inclusion of PEG-modified lipids decreases the physical stability of paclitaxel liposomes (Crosasso *et al.*, 2000).

Duration of Aqueous Stability

One basic composition of taxane liposomes that is physically stable in aqueous solution on the time scale of weeks or months contains PC, a small fraction of anionic lipid to reduce particle aggregation, and 3–3.5 mol% paclitaxel (with respect to phospholipid) (Sharma and Straubinger, 1994b). Liposomes of this composition but containing 4–5 mol% paclitaxel may be stable on the scale of hours to a day, and 8 mol% paclitaxel liposomes may be physically stable for 15 min or less.

Preparation and Characterization of Liposomes

Preparation of Liposomes

For lipophilic drugs such as the taxanes, a simple method for incorporation into liposomes is to dissolve the drug in a volatile organic solvent and mix it with a solution containing the desired phospholipids, also in organic solvent. The solvents used must be miscible. The organic solvent is then removed using a rotary evaporator, producing a thin lipid film. Hydration of the dried drug–lipid film results in spontaneous formation of multilamellar liposomes (MLVs) (Bangham *et al.*, 1965), and the hydrophobic drug partitions into the bilayer membrane.

Taxanes can be incorporated at small scale using this simple MLV method (Bartoli *et al.*, 1990; Riondel *et al.*, 1992; Sharma *et al.*, 1993), but our unpublished experiments suggested problems in scaling up to the larger preparations required for preclinical or animal studies. One hypothesis for the observed problems was that the drug concentrates in the organic solvent during rotary evaporation and undergoes self-aggregation, as was observed in detailed physical studies (Balasubramanian *et al.*, 1994). This self-aggregation of drug arises from specific intermolecular interactions among taxane molecules, in which one face of the molecule interacts with the opposite face of another, leading to a "stacked" structure capable of infinite propagation (Balasubramanian *et al.*, 1994). The conformation of this intermolecular aggregate in organic solvent (chloroform) is shown in Fig. 2. The conformation of the aggregate in organic and aqueous solvents differs somewhat, in that head-to-tail interactions, reminiscent of interactions observed in the taxane crystal structure (Mastropaolo *et al.*, 1995),

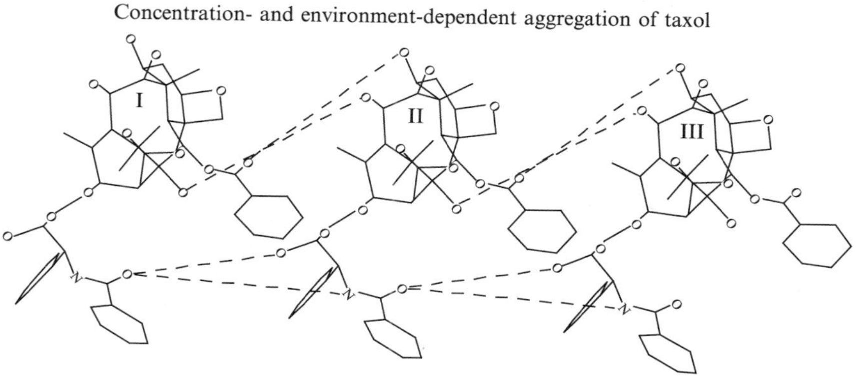

Concentration- and environment-dependent aggregation of taxol

Fig. 2. Paclitaxel–paclitaxel interactions in nonpolar media. NMR and CD spectroscopy indicate that paclitaxel intermolecular interactions can occur at sub-mM concentrations in organic solvents such as chloroform (Balasubramanian *et al.*, 1994). Dashed lines indicate intermolecular hydrogen bonds inferred from NMR spectroscopy. The stacked structure may propagate from both faces and thus grow indefinitely. Conformation of the stack in aqueous environments can include head/tail interactions as well (our unpublished observations). (Reprinted with permission from Balasubramanian *et al.*, 1994.)

are observed in water. We hypothesize that this "stacking" can nucleate destabilization of formulations and precipitation of the drug, and procedures or conditions that reduce this type of drug–drug interaction enhance formulation stability.

To avoid achieving drug concentrations that could permit intermolecular stacking when the drug–lipid mixture is initially dried from organic solvents, a freeze-drying method (Perez-Soler et al., 1990) was adapted for the preparation of taxane-containing liposomes (Sharma and Straubinger, 1994b; Sharma et al., 1993). Because the method results in a dried drug–lipid mixture, greater chemical or physical stability during long-term storage is an additional benefit.

The freeze-drying method includes several of the steps described above for the preparation of MLV. Typically, taxane is dissolved in methanol at 10–30 mM and mixed with phospholipids dissolved in chloroform at 10–150 mM. The critical nature of the drug/lipid ratio was discussed above. The drug–lipid mixture is dried on a rotary evaporator, and the thin film is redissolved at 50 mM in tert-butanol. The solution is flash frozen in liquid nitrogen, freeze dried, and stored until use.

Immediately prior to use, the freeze-dried drug–lipid powder is reconstituted with an isotonic aqueous solution such as normal saline (0.9% w/v) or nonionic solutions such as glucose, sucrose, or mannitol. The aqueous reconstitution medium is added to produce a final phospholipid concentration of 50–150 mM, depending on the application. Higher concentrations may become too viscous to administer through a small-gauge needle. Final drug concentrations are typically 1.6–38 mM. Following addition of the reconstitution medium, the solution is mixed vigorously by vortexing to ensure the suspension and hydration of the dried lipid.

Size Definition of Liposomes

After aqueous reconstitution, the reduction of liposome size may be desirable for reasons of subsequent sterilization by filtration or extending the circulation time in vivo. Size reduction may also simplify the analysis of physical stability during the development of taxane formulations (Sharma and Straubinger, 1994b) (below). Two methods for reducing the size distribution of liposomes include (1) sonication or (2) extrusion through polycarbonate filters having well-defined pores (Olson et al., 1979; Szoka et al., 1980).

Sonication. Following hydration of the dried drug–lipid mixture, the resulting liposomes are sonicated under nitrogen for 30 min at room temperature, using a high-energy bath-type sonicator (Laboratory Supplies, Inc., Hicksville, NY). The milky MLV suspension clears as particle size decreases, and extended sonication produces a nearly clear suspension of liposomes of ~25–35 nm diameter.

Extrusion. Extrusion of liposomes through successively smaller defined-pore polycarbonate filters (Olson *et al.*, 1979; Szoka *et al.*, 1980) can be used to reduce liposome diameter, as can passage through a restricted orifice at high pressure using an emulsifier. These techniques are reproducible and reduce free radical formation that can damage lipids. Under certain conditions, the operation can be performed while maintaining sterility. Devices from Avestin, Inc (Ottawa, Canada) and Northern Lipids Inc. (Vancouver, Canada) permit both small-scale (0.5 ml) and larger scale extrusion or emulsification.

Characterization of Liposomes

Although useful liposome-based taxane formulations have been achieved, the development of optimal formulations remains the subject of active investigation. Formulations must be characterized for drug content and for physical and chemical stability. Routinely, the concentration of both drug and lipid should be quantified and the drug/lipid ratio determined. This information should be obtained before and after any series of processing steps to detect changes in chemical and physical stability; instability in either would be manifested as an alteration of the drug/lipid ratio. The taxanes are easily quantified by high-performance liquid chromatography (HPLC) assay (Sharma *et al.*, 1994a) or by spectrophotometry. For the latter, concentrated liposome solutions are dissolved by serial dilution in methanol. Extinction coefficients at specific wavelengths in standard solvents are published; for paclitaxel, the molar extinction coefficient in methanol is 28,500 at 227 nm (U.S. Public Health Service, 1990). Phospholipid can be determined by a variety of methods; one simple method is inorganic phosphorus analysis following acid hydrolysis (Bartlett, 1959; Düzgüneş, 2003).

Determination of Physical Stability by Measurement of Drug Retention

Optical Microscopy. The destabilization of paclitaxel-containing formulations results in the formation of crystalline drug or drug-rich–lipid-poor complexes, often leaving drug-poor–lipid-rich components that resemble liposomes (Sharma and Straubinger, 1994b; Straubinger *et al.*, 1993). This type of destabilization can be observed readily by optical microscopy using phase contrast or differential interference contrast (DIC) optics (Sharma and Straubinger, 1994b; Straubinger *et al.*, 1993). Samples are obtained at timed intervals from hydrated taxane–lipid formulations, and multiple fields are scanned for the presence of characteristic needlelike or crystalline precipitates. The time at which the first such structures are observed is recorded, and the precipitate-free interval is regarded as a relative, qualitative indication of stability.

Procedures that can separate precipitated drug from liposomes, in conjunction with methods to determine drug and phospholipid concentration, can be used to quantify drug retention in liposomes (Sharma and Straubinger, 1994b).

Dialysis. Dialysis of drug-containing liposomes has been used frequently to investigate stability of liposome formulations; detection of drug in the dialysate, or a reduction in the drug/phospholipid ratio of the dialyzed material, indicates leakage or efflux of the drug from the carrier. In some studies, dialysis has been used to analyze the stability of paclitaxel-containing liposomes. However, the aqueous solubility of the drug is $\sim$0.4 μM (Sharma *et al.*, 1995b); if the paclitaxel concentration of a typical taxane liposome formulation were in the range of 10–150 mM, equilibration of 1 ml formulation with 1 liter of dialysate could remove only 4.0–0.27% (respectively) of the drug in each cycle of dialysis. If precipitated drug is not removed from the formulation within the period of dialysis, then the stability of formulations would be overestimated. Thus, dialysis may represent a poor method for determining the physical stability of taxane liposome formulations.

Centrifugation/Filtration. Centrifugation methods are capable of separating dense drug precipitates from liposomes, which are less dense. Similarly, filtration methods are capable of separating large drug precipitates from liposomes that are smaller. A combination of these simple methods has been used as a rapid screen of large numbers of drug–lipid compositions (Sharma and Straubinger, 1994b). For both of these methods, the size of liposomes is reduced after the preparation of MLV, as described above.

Centrifugation. MLVs sediment when centrifuged for 15 min at approximately 15,000g to 20,000g, but small unilamellar vesicles (SUVs) require much higher forces for longer times (Düzgüneş *et al.*, 1983). Therefore, SUVs are centrifuged for 15 min at 20,000g (20°) immediately after preparation in order to remove any residual MLVs (Düzgüneş, 2003). At intervals thereafter, the formulation is recentrifuged for 15 min at 15,000g (20°). Both drug and phospholipid are quantified. A reduction in the drug/lipid ratio, without a reduction in the phospholipid concentration in the supernatant, is interpreted as destabilization of the formulation, accompanied by the formation of dense taxane-rich crystals. A reduction in the drug/lipid ratio, accompanied by a reduction in the phospholipid concentration, may result from either loss of the taxane from the liposome or the formation of dense liposome aggregates. Optical microscopy (above) can be used to discriminate these possibilities.

Filtration. Various types of filters permit the passage of small liposomes and retain larger aggregates or precipitates. The polycarbonate filters used for extrusion (above) are one example; a solvent-resistant polycarbonate filter having pores larger than the liposomes but smaller than drug precipitates can be incorporated into a syringe-compatible holder (Avestin, Inc., Ottawa, Canada), which permits the analysis of small-volume samples. Filtration of the SUV formulation is performed at intervals, and both drug and phospholipid concentrations in the filtrate are determined (above). A reduction in the drug/lipid ratio without a reduction in the phospholipid concentration in the filtrate is interpreted as destabilization accompanied by drug precipitation. By optical microscopy, the observation of taxane-rich crystals would be expected. If both the drug concentration and the phospholipid concentration of the filtrate decrease, dense liposome aggregates may be forming. In parallel, drug may be precipitating. Optical microscopy (above) can be used to discriminate between these possibilities.

Combined Methods. In practice, filtration or centrifugation alone may be unsatisfactory. We have observed that filtration alone can be difficult to perform if excessive drug precipitation or liposome aggregation occurs. Also, immediately after sonication, crystals may be reduced to a size range that is able to pass through filter pores and thus lead to an underestimate of the amount of precipitated drug. Centrifugation may be inadequate for removing very small drug precipitates, particularly when destabilization is proceeding rapidly. As a result, we developed a combined procedure, in which the preparation is centrifuged and filtered repeatedly (Sharma and Straubinger, 1994). This process continues until no visible precipitate can be seen after centrifugation. Each successive filtrate or supernatant is collected after the separation step; this fraction should contain stable taxane-containing liposomes. The drug and phospholipid concentrations are assayed for each filtrate and supernatant; if the process has completely removed any drug precipitates and liposome aggregates, then the drug/lipid ratio will reach a constant value and indicate that a stable drug/lipid ratio has been achieved.

Determination of Physical Stability by Spectroscopic Techniques

Circular Dichroism Spectropolarimetry (CD). The taxanes possess several chromophores linked through chiral centers and therefore are CD active (Balasubramanian and Straubinger, 1994; Balasubramanian *et al.*, 1994). The conformation of the taxanes, and therefore the taxane

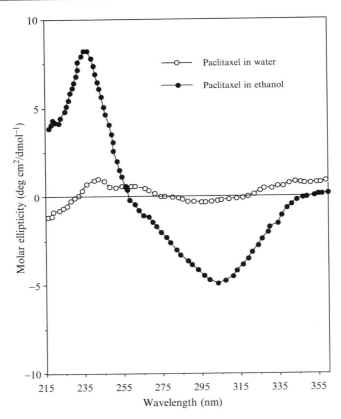

FIG. 3. Circular dichroism spectra of paclitaxel. Paclitaxel was dissolved at 0.1 mM in ethanol (filled circles) or suspended in water (open circles). CD spectra were acquired at 25° over the range of 360–215 nm, using a Jasco J500 spectropolarimeter calibrated with d_{10} camphor sulfonic acid and a 1-mm pathlength cuvette. (Reprinted with permission from Campbell *et al.*, 2001.)

CD spectrum, is sensitive to the polarity of the molecular environment (Balasubramanian *et al.*, 1994; Williams *et al.*, 1993; Vander Velde *et al.*, 1993) (Fig. 3). This phenomenon can be used to monitor both the conformation and stability of taxanes incorporated into liposomes (Balasubramanian and Straubinger, 1994; Campbell *et al.*, 2001) or other carriers (Alcaro *et al.*, 2002; Sharma *et al.*, 1995b). Because CD measurements are rapid and nondestructive, this technique provides a useful adjunct to the centrifugation/filtration method of investigating taxane formulation stability (Balasubramanian and Straubinger, 1994; Campbell *et al.*, 2001).

Interpretation of Paclitaxel CD Spectra. Figure 3 shows the CD spectrum of 100 μM paclitaxel dissolved in methanol or suspended in water. In methanol, the soluble drug shows a deep, symmetrical negative band centered at 295 nm, which arises from a $\pi-\pi^*$ transition of the aromatic rings in the C13- and C2–O-benzoyl side chains (Fig. 1); this band shifts to shorter wavelengths in solvents of lower polarity (Balasubramanian *et al.*, 1994). A shoulder at ~265 nm may be more or less prominent, depending on solvent polarity, and this band corresponds to the conformation of the small O-benzoyl side chain at C2 (Fig. 1). Both the C2–O-benzoyl ring and the C13 side chain are markedly sensitive to solvent polarity (Balasubramanian and Straubinger, 1994; Balasubramanian *et al.*, 1994; Vander Velde *et al.*, 1993). A positive band at approx. 230 nm exists in methanol and arises from n–π^* transitions involving C=O groups at C3′ and C9′ and several chiral monoolefins (Balasubramanian and Straubinger, 1994; Balasubramanian *et al.*, 1994).

The CD spectrum of paclitaxel in water (Fig. 3) represents aggregated, precipitating drug. As the drug aggregates, the negative 295-nm band decreases drastically in intensity and shifts to higher wavelengths, indicating a change in the conformation and environment of the C13 side chain. The formation of exciton split bands at >305 nm marks the formation of dimeric drug aggregates. Changes in the intensity of the 265-nm band implicate changes in the C2–O-benzoyl side chain as well.

The CD spectrum of paclitaxel incorporated into liposomes at low drug/lipid ratios resembles the CD spectrum of fully soluble drug in 70–100% methanol (Balasubramanian and Straubinger, 1994). As the drug/lipid ratio increases to proportions that destabilize the liposome formulations, the 265-nm band may become more prominent, and the 295-nm band both shifts to longer wavelengths and decreases in intensity. Both changes signify drug aggregation in the liposome membrane. As destabilization and precipitation of drug occur, the CD spectrum resembles that of paclitaxel in water (Fig. 3).

Acquisition of CD Spectra for Taxane Liposomes. To acquire CD spectra of taxanes incorporated in liposomes, the liposomes must be prepared in solutions that do not contribute to the CD spectrum in the range of 350–200 nm; water or physiological saline (0.9% w/v) is acceptable. Typically, liposomes are prepared by the MLV method described above. Large liposomes cause an intense scattering signal, and two steps may be taken to reduce interference in the CD spectra. First, the concentration of liposomes can be reduced, and a short pathlength (0.1 mm) cuvette may be used. Second, the liposome diameter may be reduced, usually by brief sonication, to limit the delay between liposome preparation and the initiation of CD observations. Spectra are acquired repetitively over the range of 350–200 nm to capture time-dependent changes that may signify liposome destabilization. In performing experiments to investigate the role of membrane composition on formulation stability, liposome preparations containing a wide range of drug/lipid ratios are prepared, and spectra are acquired for each to determine the time until the spectral signature of drug aggregation/precipitation occurs. Because formulation destabilization may be time delayed, measurements include not only the immediate (<2 h) temporal changes following hydration of the dried drug–lipid mixture but also include more detailed measurements at times that reflect the desired duration of stability in the hydrated state. Typically, 24-h stability in the hydrated state is required for formulations to have a clinical potential.

Quantitative Interpretation of Data. The complex CD spectral changes that occur during drug aggregation and precipitation may be simplified by analysis to produce a simple means to represent stability data (Balasubramanian and Straubinger, 1994; Campbell *et al.*, 2001). Conversion from ellipticity to molar ellipticity normalizes for differences in drug concentration in the formulation. Therefore, comparison of molar ellipticity at wavelengths that are particularly sensitive to destabilization (e.g., the negative band at ~295 nm) provides a relative measure of formulation stability. Typically, molar ellipticity of the ~295 nm band as a function of time enables comparison of the temporal stability of various formulations.

A second manipulation of the CD spectral data permits estimation of the fraction of taxane incorporated in the liposome membrane (Balasubramanian and Straubinger, 1994; Campbell *et al.*, 2001). For this method, CD spectra are acquired for 100 μM taxane in water and in 70% (v/v) methanol. Molar ellipticity is calculated, and the value of the negative-going peak at ~295 nm in methanol is taken as the molar ellipticity of taxane that is 100% incorporated in the membrane. The molar ellipticity at the same wavelength for drug suspended in water is taken to represent 0% incorporation. Time-dependent changes in the percentage incorporated in membranes is taken as an indication of destabilization. The benefit of this second method is that if separation techniques are used to remove precipitated material, calculation of the molar ellipticity reflects the fraction of original drug retained in the membrane and thus provides a means to compare formulations.

Determination of Physical Stability by Other Techniques

Differential Scanning Calorimetry (DSC). The formation of packing defects in the membrane bilayer lipids, mediated by taxane aggregation, is one mechanism hypothesized for formulation destabilization. DSC is a relatively simple technique that can reveal drug partitioning into the bilayer and the presence of packing defects in the bilayer. Liposomes composed of phospholipids that possess defined acyl chains exhibit a sharp, cooperative transition at characteristic temperatures as the bilayer changes from the gel to the liquid crystalline phase. Drug interaction with the membrane lipids thus can be observed easily with liposomes composed of phospholipids that contain defined hydrocarbon moieties such as dimyristoyl-, dipalmitoyl-, or distearoyl-acyl chains. The incorporation of the drug into the liposome membrane can be observed readily, as the drug may alter the phase transition temperature or broaden the thermotropic peak (Pedroso de Lima *et al.*, 1990). At lower drug/lipid ratios, taxane incorporation induces transition peak broadening (Balasubramanian and Straubinger, 1994), indicating that the drug partitions in the upper portion of the bilayer (i.e., the domain bordering the aqueous interface). Such partitioning is expected, based on the hydrophobic nature of the taxanes. At higher taxane/lipid ratios, a shift in the peak position can be observed, indicating the appearance of packing defects in the bilayer that can induce instability. At high drug/lipid ratios that exceed the capacity of the membrane to incorporate the drug, the DSC thermogram reverts to a sharp peak similar to that of membranes in the absence of drug. This observation indicates that destabilization of the formulation results in the loss of most drug from the membrane bilayer upon drug precipitation.

Typically, thermograms are recorded at heating rates of 2.5–5°K/min. For each thermogram, 15 μl of a 40 mM liposomal solution is loaded into the sample pan, and the samples are held at the initial temperature for 15 min. The DSC instrument is calibrated with standard samples covering a wide range of temperatures.

The peak position and peak width at half height are generally indicative of liposome–drug interaction. A broadening of the main transition without any change in the peak melting is classified as an A type change. A shift in transition temperature in addition to peak broadening suggests the formation of packing defects. The appearance of new peaks on either side of the main transition indicates the formation of discrete domains—drug-rich or drug-poor—and may indicate phase separation and immiscibility.

Concluding Remarks

The unusual molecular structure of the active taxanes appears to contribute to both poor solubility and to the desired pharmacological activity. Although a wide variety of new, more water-soluble analogues are under investigation, drug carrier approaches that address the solubility problems of the currently used agents, which have been studied extensively in large numbers of clinical trials, provide a rational and attractive alternative. Novel drug delivery approaches may provide a useful formulation tool that can eliminate vehicle toxicity and potentially improve the antitumor efficacy of these clinically important agents.

References

Aapro, M., Pujade-Lauraine, E., Lhomme, C., Lentz, M., Le Bail, N., Fumoleau, P., and Chevallier, B. (1993). Phase II study of Taxotere in ovarian cancer. EORTC Clinical Screening Group (CSG). *Proc. Am. Soc. Clin. Oncol.* **12,** A809.

Adams, J. D., Flora, K. P., Goldspiel, B. R., Wilson, J. W., Finley, R., Arbuck, S. G., and Finley, R. (1993). Taxol: A history of pharmaceutical development and current pharmaceutical concerns. *J. Natl. Cancer Inst. Monogr.* **15,** 141–147.

Adler, L. M., Herzog, T. J., Williams, S., Rader, J. S., and Mutch, D. G. (1994). Analysis of exposure times and dose escalation of paclitaxel in ovarian cancer cell lines. *Cancer* **74,** 1891–1898.

Alcaro, S., Ventura, C. A., Paolino, D., Battaglia, D., Ortuso, F., Cattel, L., Puglisi, G., and Fresta, M. (2002). Preparation, characterization, molecular modeling, and *in vitro* activity of paclitaxel-cyclodextrin complexes. *Bioorg. Med. Chem. Lett.* **12,** 1637–1641.

Allen, T. M., and Chonn, A. (1987). Large unilamellar liposomes with low uptake into the reticuloendothelial system. *FEBS Lett.* **223,** 42–46.

Arbuck, S. G., Canetta, R., Onetto, N., and Christian, M. C. (1993). Current dosage and schedule issues in the development of paclitaxel (TAXOL). *Semin. Oncol.* **20,** 31–39.

Balasubramanian, S. V., Alderfer, J. L., and Straubinger, R. M. (1994). Solvent- and concentration-dependent molecular interactions of taxol (paclitaxel). *J. Pharm. Sci.* **83,** 1470–1476.

Balasubramanian, S. V., and Straubinger, R. M. (1994). Taxol-lipid interactions: taxol-dependent effects on the physical properties of model membranes. *Biochemistry* **33,** 8941–8947.

Bangham, A., Standish, M., and Watkins, J. (1965). Diffusion of univalent ions across the lamellae of swollen phospholipids. *J. Mol. Biol.* **13,** 238–252.

Bartlett, G. R. (1959). Phosphorus assay in column chromatography. *J. Biol. Chem.* **234,** 466–468.

Bartoli, M.-H., Boitard, M., Fessi, H., Beriel, H., Devissaguet, J.-H., Picot, F., and Puisieux, F. (1990). *In vitro* and *in vivo* antitumor activity of free and encapsulated taxol. *J. Microencapsulation* **7,** 191–197.

Patent. New compositions containing taxane derivatives. France, 9108527 (*Assigned to* Rhone Poulenc Rorer SA, Antony, France).

Behar, A., Pujade-Lauraine, E., Maurel, A., Brun, M., Chauvin, F., Feuilhade de Chauvin, F., D, O.-A., and Hille, D. (1997). The pathophysiological mechanism of fluid retention in advanced cancer patients treated with docetaxel, but not receiving corticosteroid comedication. *Br. J. Clin. Pharmacol.* **43,** 653–658.

Bissery, M., Guénard, D., Guéritte-Voegelein, F., and Lavelle, F. (1991). Experimental antitumor activity of taxotere (RP 56976, NSC 628503), a taxol analogue. *Cancer Res.* **51,** 4845–4852.

Bisset, D., Setanoians, A., Cassidy, J., Graham, M. A., Chadwick, G. A., Wilson, P., Auzannet, V., Le Bail, N., Kaye, S. B., and Kerr, D. J. (1993). Phase I and pharmacokinetic study of taxotere (RP 56976) administered as a 24-hour infusion. *Cancer Res.* **53,** 523–527.

Bocci, G., Nicolaou, K. C., and Kerbel, R. S. (2002). Protracted low-dose effects on human endothelial cell proliferation and survival *in vitro* reveal a selective antiangiogenic window for various chemotherapeutic drugs. *Cancer Res.* **62,** 6938–6943.

Burris, H., Eckardt, J., Fields, S., Rodriguez, G., Smith, L., Thurman, A., Peacock, N., Kuhn, J., Hodges, S., Bellet, R., Bayssas, M., Le Bail, N., and Von Hoff, D. (1993). Phase II trials of Taxotere in patients with non-small cell lung cancer. *Proc. Am. Soc. Clin. Oncol.* **12,** A1116.

Campbell, R. B., Balasubramanian, S. V., and Straubinger, R. M. (2001). Influence of cationic lipids on the stability and membrane properties of paclitaxel-containing liposomes. *J. Pharm. Sci.* **90,** 1091–1105.

Cortes, J., Rodriguez, J., Aramendia, J. M., Salgado, E., Gurpide, A., Garcia-Foncillas, J., Aristu, J. J., Claver, A., Bosch, A., Lopez-Picazo, J. M., Martin-Algarra, S., Brugarolas, A., and Calvo, E. (2003). Front-line paclitaxel/cisplatin-based chemotherapy in brain metastases from non-small-cell lung cancer. *Oncology* **64,** 28–35.

Crosasso, P., Ceruti, M., Brusa, P., Arpicco, S., Dosio, F., and Cattel, L. (2000). Preparation, characterization and properties of sterically stabilized paclitaxel-containing liposomes. *J. Contr. Release* **63,** 19–30.

De Brabander, M., Geuens, G., Nuydens, R., Willebrords, R., and De Mey, J. (1981). Taxol induces the assembly of free microtubules in living cells and blocks the organizing capacity of centrosomes and kinetochores. *Proc. Natl. Acad. Sci. USA* **78,** 5608–5612.

Derry, W. B., Wilson, L., and Jordan, M. A. (1995). Substoichiometric binding of taxol suppresses microtubule dynamics. *Biochemistry* **34,** 2203–2211.

Donehower, R. C., Rowinsky, E. K., Grochow, L. B., Longnecker, S. M., and Ettinger, D. S. (1987). Phase I trial of taxol in patients with advanced cancer. *Cancer Treat. Rep.* **71,** 1171–1177.

Düzgüneş, N. (2003). *Methods Enzymol.* **367,** 23.

Düzgüneş, N., Wilschut, J., Hong, K., Fraley, R., Perry, C., Friend, D., James, T., and Papahadjopoulos, D. (1983). Physicochemical characterization of large unilamellar phospholipid vesicles prepared by reverse-phase evaporation. *Biochim. Biophys. Acta.* **732,** 289–299.

Dye, D., and Watkins, J. (1980). Suspected anaphylactic reaction to Cremophor EL. *Br. Med. J.* **280,** 1353.

Extra, J.-M., Rousseau, F., Bruno, R., Clavel, M., Le Bail, N., and Marty, M. (1993). Phase I and pharmacokinetic study of taxotere (RP 56976; NSC 628503) given as a short intravenous infusion. *Cancer Res.* **53,** 1037–1042.

Fan, W. (1999). Possible mechanisms of paclitaxel-induced apoptosis. *Biochem. Pharmacol.* **57,** 1215–1221.

Farinelle, S., Malonne, H., Chaboteaux, C., Decaestecker, C., Dedecker, R., Gras, T., Darro, F., Fontaine, J., Atassi, G., and Kiss, R. (2000). Characterization of TNP-470-induced modifications to cell functions in HUVEC and cancer cells. *J. Pharm. Tox. Meth.* **43,** 15–24.

Friedland, D., Gorman, G., and Treat, J. (1992). Hypersensitivity reactions from taxol and etoposide. *J. Natl. Cancer Inst.* **85,** 2036.

Fumoleau, P., Chevallier, B., Kerbrat, P., Dieras, V., Le Bail, N., Bayssas, M., and van Glabbeke, M. (1993). First line chemotherapy with Taxotere in advanced breast cancer: A Phase II study of the EORTC clinical screening group. *Proc. Am. Soc. Clin. Oncol.* **12,** A27.

Gabizon, A., and Papahadjopoulos, D. (1988). Liposome formulations with prolonged circulation time in blood and enhanced uptake by tumors. *Proc. Natl. Acad. Sci. USA* **85,** 6949–6953.

Gan, Y., Wientjes, M. G., Schuller, D. E., and Au, J. L.-S. (1996). Pharmacodynamics of taxol in human head and neck tumors. *Cancer Res.* **56**, 2086–2093.

Grant, D. S., Williams, T. L., Zahaczewsky, M., and Dicker, A. P. (2003). Comparison of antiangiogenic activities using paclitaxel (taxol) and docetaxel (taxotere). *Int. J. Cancer* **104**, 121–129.

Grem, J. L., Tutsch, K. D., Simon, K. J., Alberti, D. B., Willson, J. V. K., Tormey, D. C., Swaminathan, S., and Trump, D. L. (1987). Phase I study of taxol administered as a short iv infusion daily for 5 days. *Cancer Treat. Rep.* **71**, 1179–1184.

Guastalla, J. P., Lhomme, C., Dauplat, J., Namer, M., Bonneterre, J., Oberling, F., Pouillart, P., Fumoleau, P., Kerbrat, P., and Tubiana, N. (1994). Taxol (paclitaxel) safety in patients with platinum pretreated ovarian carcinoma: An interim analysis of a phase II multicenter study. *Ann. Oncol.* **5**, S33–38.

Guénard, D., Guéritte-Voegelein, F., Dubois, J., and Potier, P. (1993). Structure-activity relationships of taxol and taxotere analogues. *J. Natl. Cancer Inst. Monogr.* **15**, 79–82.

Guéritte-Voegelein, F., Guénard, D., Lavelle, F., Le Goff, M.-T., Mangatal, L., and Potier, P. (1991). Relationships between the structure of taxol analogues and their antimitotic activity. *J. Med. Chem.* **34**, 992–998.

Huang, Y., Johnson, K., Norris, J., and Fan, W. (2000). Nuclear Factor-kB/IkB signaling pathway may contribute to the mediation of paclitaxel-induced apoptosis in solid tumor cells. *Cancer Res.* **60**, 4426–4432.

Immordino, M. L., Brusa, P., Arpicco, S., Stella, B., Dosio, F., and Cattel, L. (2003). Preparation, characterization, cytotoxicity and pharmacokinetics of liposomes containing docetaxel. *J. Control Release* **91**, 417–429.

Irvin, R., Burris, H., Kuhn, J., Kalter, S., Smith, L., Rodriguez, G., Weiss, G., Eckardt, J., Cook, G., Wall, J., Le Bail, N., Bayssas, M., and VonHoff, D. (1993). Phase I trial of a 2- and 3-hr infusion of Taxotere. *Proc. Am. Soc. Clin. Oncol.* **12**, A347.

Jordan, M. A., Toso, R. J., Thrower, D., and Wilson, L. (1993). Mechanism of mitotic block and inhibition of cell proliferation by taxol at low concentrations. *Proc. Natl. Acad. Sci. USA* **90**, 9552–9556.

Jordan, M. A., and Wilson, L. (1995). Microtubule polymerization dynamics, mitotic block, and cell death by paclitaxel at low concentrations. *In* "Taxane Anticancer Agents: Basic Science and Current Status." (G. Georg, T. T. Chen, I. Ojima, and D. Vyas, eds.), pp. 138–153. ACS Symposium Series, Washington, DC.

Kawasaki, K., Akashi, S., Shimazu, R., Yoshida, T., Miyake, K., and Nishijima, M. (2000). Mouse Toll-like receptor 4MD-2 complex mediates lipopolysaccharide-mimetic signal transduction by taxol. *J. Biol. Chem.* **275**, 2251–2254.

Klibanov, A., Maruyama, K., Beckerleg, A., Torchilin, V., and Huang, L. (1991). Activity of amphipathic poly(ethylene glycol) 5000 to prolong the circulation time of liposomes depends on the liposome size and is unfavorable for immunoliposome binding to target. *Biochim. Biophys. Acta.* **1062**, 142–148.

Kubota, T., Matsuzaki, S. W., Hoshiya, Y., Watanabe, M., Kitajima, M., Asanuma, F., Yamada, Y., and Koh, J. I. (1997). Antitumor activity of paclitaxel against human breast carcinoma xenografts serially transplanted into nude mice. *J. Surg. Oncol.* **64**, 115–121.

Lee, L. F., Li, G., Templeton, D. J., and Ting, J. P. (1998). Paclitaxel (Taxol)-induced gene expression and cell death are both mediated by the activation of c-Jun NH2-terminal Kinase (JNK/SAPK). *J. Biol. Chem.* **273**, 28253–28260.

Liggins, R., Hunter, W., and Burt, H. (1997). Solid-state characterization of paclitaxel. *J. Pharm. Sci.* **86**, 1458–1463.

Ling, Y. H., Yang, Y., Tornos, C., Singh, B., and Perez-Soler, R. (1998). Paclitaxel-induced apoptosis is associated with expression and activation of c-Mos gene product in human ovarian carcinoma SKOV3. *Cancer Res.* **58,** 3633–3640.

Lorenz, W., Riemann, H. J., Schmal, A., Schult, H., Lang, S., Ohmann, C., Weber, D., Kapp, B., Luben, L., and Doenicke, A. (1977). Histamine release in dogs by Cremophor EL and its derivatives: Oxyethylated oleic acid is the most effective constituent. *Agents Actions* **7,** 63–67.

Manfredi, J. J., Parness, J., and Horwitz, S. B. (1982). Taxol binds to cellular microtubules. *J. Cell Biol.* **94,** 688–696.

Mastropaolo, D., Camerman, A., Luo, Y. G., Brayer, G. D., and Camerman, N. (1995). Crystal and molecular structure of paclitaxel (taxol). *Proc. Natl. Acad. Sci. USA* **92,** 6920–6924.

Mathew, A. E., Mejillano, M. R., Nath, J. P., Himes, R. H., and Stella, V. J. (1992). Synthesis and evaluation of some water-soluble prodrugs and derivatives of taxol with antitumor activity. *J. Med. Chem.* **35,** 145–151.

Milross, C. G., Mason, K. A., Hunter, N. R., Chung, W.-K., Peters, L. J., and Milas, L. (1996). Relationship of mitotic arrest and apoptosis to antitumor effect of paclitaxel. *J. Natl. Cancer Inst.* **88,** 1308–1314.

Murphy, W. K., Fossella, F. V., Winn, R. J., Shin, D. M., Hynes, H. E., Gross, H. M., Davilla, E., Leimert, J., Dhingra, H., and Raber, M. N. (1993). Phase II study of taxol in patients with untreated advanced non-small-cell lung cancer. *J. Natl. Cancer Inst.* **85,** 384–388.

Ojima, I., Chakravarty, S., Inoue, T., Lin, S., He, L., Horwitz, S. B., Kuduk, S. D., and Danishefsky, S. J. (1999). A common pharmacophore for cytotoxic natural products that stabilize microtubules. *Proc. Natl. Acad. Sci. USA* **96,** 4256–4261.

Olson, F., Hunt, C. A., Szoka, F. C., Jr., Vail, W. J., and Papahadjopoulos, D. (1979). Preparation of liposomes of defined size by extrusion through polycarbonate membranes. *Biochim. Biophys. Acta* **557,** 9–23.

Onetto, N., Canetta, R., Winograd, B., Catane, R., Dougan, M., Grechko, J., Burroughs, J., and Rozencweig, M. (1993). Overview of taxol safety. *J. Natl. Cancer Inst. Monogr.* **15,** 131–139.

Pedroso de Lima, M. C., Chiche, B. H., Debs, R. J., and Düzgüneş, N. (1990). Interaction of antimycobacterial and anti-pneumocystis drugs with phospholipid membranes. *Chem. Phys. Lipids* **53,** 361–371.

Perez-Soler, R., Lopez-Berestein, G., Lautersztain, J., al-Baker, S., Francis, K., Macias-Kiger, D., Raber, M. N., and Khokhar, A. R. (1990). Phase I clinical and pharmacological study of liposome-entrapped cis-bis-neodecanoato-trans-R,R-1,2-diaminocyclohexane platinum(II). *Cancer Res.* **50,** 4254–4259.

Rhone Poulenc Rorer, S. A. French Patent #N.65109, Antony, France.

Ringel, I., and Horwitz, S. B. (1991). Studies with RP 56976 (taxotere): A semisynthetic analogue of taxol. *J. Natl. Cancer Inst.* **83,** 288–291.

Riondel, J., Jacrot, M., Fessi, H., Puisieux, F., and Poiter, P. (1992). Effects of free and liposome-encapsulated taxol on two brain tumors xenografted into nude mice. *In Vivo* **6,** 23–28.

Rose, W. C. (1992). Taxol: A review of its preclinical *in vivo* antitumor activity. *Anti-Cancer Drugs* **3,** 311–321.

Rowinsky, E., McGuire, W., Guarnieri, T., Fisherman, J., Christian, M., and Donehower, R. (1991). Cardiac disturbances during the administration of taxol. *J. Clin. Oncol.* **9,** 1704–1712.

Rowinsky, E. K., Donehower, R. C., Jones, R. J., and Tucker, R. W. (1988). Microtubule changes and cytotoxicity in leukemic cell lines treated with taxol. *Cancer Res.* **48,** 4093–4100.

Rowinsky, E. K., Onetto, N., Canetta, R. M., and Arbuck, S. G. (1992). Taxol: The first of the taxanes, and important new class of antitumor agents. *Semin. Oncol.* **19**, 646–662.

Roy, S. N., and Horwitz, S. B. (1985). A phosphoglycoprotein associated with taxol resistance in J774.2 cells. *Cancer Res.* **45**, 3856–3863.

Runowicz, C. D., Wiernik, P. H., Einzig, A. I., Goldberg, G. L., and Horwitz, S. B. (1993). Taxol in Ovarian Cancer. *Cancer* **71**, 1591–1596.

Satoh, H., Ishikawa, H., Fujimoto, M., Fujiwara, M., Yamashita, Y. T., Yazawa, T., Ohtsuka, M., Hasegawa, S., and Kamma, H. (1998). Combined effects of TNP-470 and taxol in human non-small cell lung cancer cell lines. *Anticancer Res.* **18**, 1027–1030.

Schiff, P. B., Fant, J., and Horwitz, S. B. (1979). Promotion of microtubule assembly *in vitro* by taxol. *Nature* **277**, 665–667.

Schiff, P. B., and Horwitz, S. B. (1981). Taxol assembles tubulin in the absence of exogenous guanosine 5' triphosphate or microtubule associated proteins. *Biochemistry* **20**, 3247–3252.

Schrijvers, D., Wanders, J., Dirix, L., Prove, A., Vonck, I., van Oosterom, A., and Kaye, S. (1993). Coping with toxicities of docetaxel (Taxotere). *Ann. Oncol.* **4**, 610–611.

Sharma, A., Mayhew, E., and Straubinger, R. M. (1993). Antitumor effect of taxol-containing liposomes in a taxol-resistant murine tumor model. *Cancer Res.* **53**, 5877–5881.

Sharma, A., Conway, W. D., and Straubinger, R. M. (1994a). Reversed-phase high-performance liquid chromatographic determination of taxol in mouse plasma. *J. Chromatog. B, Anl. Tech. Biomed. Life Sci.* **655**, 315–319.

Sharma, A., and Straubinger, R. M. (1994b). Novel taxol formulations: Preparation and characterization of taxol-containing liposomes. *Pharm. Res.* **11**, 889–896.

Sharma, A., Straubinger, R. M., Ojima, I., and Bernacki, R. J. (1995a). Antitumor efficacy of taxane liposomes on a human ovarian tumor xenograft in nude athymic mice. *J. Pharm. Sci.* **84**, 1400–1404.

Sharma, U. S., Balasubramanian, S. V., and Straubinger, R. M. (1995b). Pharmaceutical and physical properties of paclitaxel (taxol) complexes with cyclodextrins. *J. Pharm. Sci.* **84**, 1223–1230.

Sharma, A., Mayhew, E., Bolcsak, L., Cavanaugh, C., Harmon, P., Janoff, A., and Bernacki, R. J. (1997). Activity of paclitaxel liposome formulations against human ovarian tumor xenografts. *Int. J. Cancer* **71**, 103–107.

Sledge, G. W., Neuberg, D., Bernardo, P., Ingle, J. N., Martino, S., Rowinsky, E. K., and Wood, W. C. (2003). Phase III trial of doxorubicin, paclitaxel, and the combination of doxorubicin and paclitaxel as front-line chemotherapy for metastatic breast cancer: An intergroup trial (E1193). *J. Clin. Oncol.* **21**, 588–592.

Sparreboom, A., van Tellingen, O., Nooijen, W. J., and Beijnen, J. H. (1996). Nonlinear pharmacokinetics of paclitaxel in mice results from the pharmaceutical vehicle Cremophor EL. *Cancer Res.* **56**, 2112–2115.

Straubinger, R. M., Sharma, A., Murray, M., and Mayhew, E. (1993). Novel taxol formulations: Taxol-containing liposomes. *J. Natl. Cancer Inst. Monogr.* **15**, 69–78.

Straubinger, R. M. (1995). Biopharmaceutics of paclitaxel (taxol): Formulation, activity, and pharmacokinetics. *In* "Taxol. Science and Applications," (M. Suffness, ed.), pp. 237–258. CRC Press, Boca Raton, FL.

Suffness, M. (1993). Taxol: From discovery to therapeutic use. *Ann. Rep. Med. Chem.* **28**, 305–314.

Swindell, C. S., Krauss, N. E., Horwitz, S. B., and Ringel, I. (1991). Biologically active taxol analogues with deleted A-ring side chain substituents and variable C-2' configurations. *J. Med. Chem.* **34**, 1176–1184.

Szoka, F. C., Jr., Olson, F., Heath, T., Vail, W., Mayhew, E., and Papahadjopoulos, D. (1980). Preparation of unilamellar liposomes of intermediate size (0.1–0.2 μm) by a combination of reverse phase evaporation and extrusion through polycarbonate membranes. *Biochim. Biophys. Acta* **601,** 559–571.

Tarr, B., and Yalkowsky, S. (1987). A new parenteral vehicle for the administration of some poorly water soluble anti-cancer drugs. *J. Parenteral Sci. Technol.* **41,** 31–33.

Tudor, G., Aguilera, A., Halverson, D. O., Laing, N. D., and Sausville, E. A. (2000). Susceptibility to drug-induced apoptosis correlates with differential modulation of Bad, Bcl-2 and Bcl-xL protein levels. *Cell Death Differ.* **7,** 574–586.

USPDI (2003a). Paclitaxel (systemic). *In* "USP Dispensing Information: Drug Information for the Health Professional." (U. S. P. Convention, ed.) Micromedex, Inc., Englewood, CO.

USPDI (2003b). Docetaxel systemic. *In* "USP Dispensing Information: Drug Information for the Health Professional." (U. S. P. Convention, ed.) Micromedex, Inc., Englewood, CO.

USPHS (1990). Taxol. *In* "NCI Investigational Drugs, Pharmaceutical Data," pp. 151–153. National Cancer Institute, National Institutes of Health, Department of Health and Human Services, U.S Public Health Service, Bethesda, MD.

Vander Velde, D. G., Georg, G. I., Gruewald, G. L., Gunn, C. W., and Mitscher, L. A. (1993). Hydrophobic collapse of taxol and taxotere solution conformations in mixtures of water and organic solvent. *J. Am. Chem. Soc.* **115,** 11650–11651.

Wang, J., Lou, P., Lesniewski, R., and Henkin, J. (2003). Paclitaxel at ultra low concentrations inhibits angiogenesis without affecting cellular microtubule assembly. *Anticancer Drugs* **14,** 13–19.

Wang, T. H., Wang, H. S., Ichijo, H., Giannakakou, P., Foster, J. S., Fojo, T., and Wimalasena, J. (1998). Microtubule-interfering agents activate c- Jun N-terminal kinase/Stress-activated protein kinase through both Ras and apoptosis signal-regulating kinase pathways. *J. Biol. Chem.* **273,** 4928–4936.

Wang, T. H., Popp, D. M., Wang, H. S., Saitoh, M., Mural, J. G., Henley, D. C., Ichijo, H., and Wimalasena, J. (1999). Microtubule dysfunction induced by paclitaxel initiates apoptosis through both c-Jun N-terminal Kinase (JNK)-dependent and – independent pathways in ovarian cancer cells. *J. Biol. Chem.* **274,** 8208–8216.

Wani, M. C., Taylor, H. L., Wall, M. E., Coggon, P., and McPhail, A. T. (1971). Plant antitumor agents. VI. The isolation and structure of taxol, a novel antileukemic and antitumor agent from *Taxus brevifolia. J. Am. Chem. Soc.* **93,** 2325–2327.

Webster, L., Linsenmeyer, M., Millward, M., Morton, C., Bishop, J., and Woodcock, D. (1993). Measurement of Cremophor EL following taxol: Plasma levels sufficient to reverse drug exclusion mediated by the multidrug-resistant phenotype. *J. Natl. Cancer Inst.* **85,** 1685–1690.

Weiss, R. B., Donehower, R. C., Wiernik, P. H., Ohnuma, T., Gralla, R. J., Trump, D. L., Baker, J. R., VanEcho, D. A., VonHoff, D. D., and Leyland-Jones, B. (1990). Hypersensitivity reactions from taxol. *J. Clin. Oncol.* **8,** 1263–1268.

Williams, H. J., Scott, A. I., Dieden, R. A., Swindell, C. S., Chirlian, L. E., Francl, M. M., Heerding, J. M., and Krauss, N. E. (1993). NMR and Molecular modeling study of the conformations of taxol and its side chain methyl ester in aqueous and non-aqueous solution. *Tetrahedron* **49,** 6545–6560.

Yen, W.-C., Wientjes, M. G., and Au, J. L.-S. (1996). Differential effect of taxol in rat primary and metastatic prostate tumors: Site-dependent pharmacodynamics. *Pharm. Res.* **13,** 1305–1312.

Yeung, T. K., Germond, C., Chen, X., and Wang, Z. (1999). The mode of action of taxol: Apoptosis at low concentration and necrosis at high concentration. *Biochem. Biophys. Res. Comm.* **263,** 398–404.

Yvon, A. M., Wadsworth, P., and Jordan, M. A. (1999). Taxol suppresses dynamics of individual microtubules in living human tumor cells. *Mol. Biol. Cell* **10,** 947–959.

[6] Cisplatin Nanocapsules

By Anton I. P. M. de Kroon, Rutger W. H. M. Staffhorst,
Ben de Kruijff, and Koert N. J. Burger

Abstract

Cisplatin nanocapsules represent a novel lipid formulation of the anticancer drug cis-diamminedichloroplatinum(II), in which nanoprecipitates of cisplatin are covered by a phospholipid bilayer coat consisting of an equimolar mixture of phosphatidylcholine and phosphatidylserine. Cisplatin nanocapsules are characterized by an unprecedented cisplatin-to-lipid molar ratio and exhibit strongly improved cytotoxicity against tumor cells in vitro compared with the free drug. Here, methods for preparing and characterizing cisplatin nanocapsules are reported.

Introduction

Cisplatin, cis-diamminedichloroplatinum(II), is an anticancer drug that is commonly used in the treatment of a variety of solid tumors, including genitourinary, head and neck, and lung (Lippert, 1999). The main cellular target of the drug appears to be the nuclear DNA, with which it forms stable adducts that interfere with transcription and replication, and trigger apoptosis, eventually resulting in the death of the cancer cell (Pinto and Lippard, 1985). The clinical use of cisplatin faces a number of serious problems. Due to its reactive nature, most of the drug is rapidly inactivated by binding to proteins upon entry in the blood by intravenous administration and never reaches the tumor in an active form (Howe-Grant and Lippard, 1980). Moreover, binding to proteins is considered a major cause of the many dose-limiting toxicities exhibited by cisplatin such as nephro-, oto-, and neurotoxicity (Calvert et al., 1993; Hacker, 1991). Finally, the clinical utility of cisplatin is limited by the emergence of resistance in many tumor types (Perez, 1998).

One approach to try and circumvent these drawbacks of cisplatin is to encapsulate the drug in liposomes. Several liposomal formulations of cisplatin have been developed, the most recent being SPI-077, in which cisplatin is encapsulated in pegylated ("stealth") liposomes consisting of hydrogenated soy phosphatidylcholine, cholesterol, and poly(ethyleneglycol) derivatized distearoylphosphatidylethanolamine (Newman et al., 1999). Preclinical studies showed that compared with the free drug, SPI-077 exhibited improved stability, prolonged circulation time, increased antitumor effect, and reduced toxicity (Newman *et al.*, 1999; Vaage *et al.*, 1999). However, in Phase I/II studies, essentially no antitumor activity of SPI-077 was observed in patients (Harrington *et al.*, 2001; Kim *et al.*, 2001; Veal *et al.*, 2001).

A major problem of the conventional liposomal formulations of cisplatin, such as SPI-077, appears to be the limited bioavailability of the drug in the tumor (Bandak *et al.*, 1999; Meerum Terwogt *et al.*, 2002). A key factor is likely to be the low water solubility (7 mM at 37°) and low lipophilicity of cisplatin, which lead to liposomal formulations with low drug-to-lipid molar ratios (in the order of 0.02). Serendipitously, we discovered an alternative method to encapsulate cisplatin in a lipid formulation with superior efficiency (Burger *et al.*, 2002). Our method takes advantage of the limited solubility of the drug in water and results in cisplatin nanocapsules. Cisplatin nanocapsules are aggregates of cisplatin surrounded by a single lipid bilayer and have an unprecedented drug-to-lipid ratio and an unprecedented *in vitro* cytotoxicity.

Principle

Nanocapsules are prepared by repeatedly freezing and thawing a concentrated aqueous solution of cisplatin in the presence of negatively charged phospholipids. The preparation of cisplatin nanocapsules was shown to depend critically on the presence of negatively charged phospholipids and on the presence of positively charged aquo species of cisplatin, pointing to an essential role for electrostatic interaction in the mechanism of formation. A solution of cisplatin in water, in the absence of added chloride, contains a mixture of the neutral dichloride and dihydroxo species of cisplatin with low solubility in water and positively charged aqua species of cisplatin with a much higher solubility (Lippert, 1999). In our current model (Fig. 1) for the mechanism of nanocapsule formation (Burger *et al.*, 2002), cisplatin is concentrated in the residual fluid during freezing, and it forms small aggregates when the solubility limit of the dichloro species is exceeded. Subsequently, the aggregates of the dichloro species of cisplatin are covered by the positively charged aqua species, which have a higher solubility limit. The negatively charged membranes interact with the positively charged cisplatin aggregates and then reorganize to wrap the aggregates in a phospholipid bilayer coat. The resulting nanocapsules do not redissolve upon thawing.

Method

Cisplatin (Sigma, St. Louis, MO) is dissolved in MilliQ water to a concentration of 5 mM, which is facilitated by incubating at 55° for

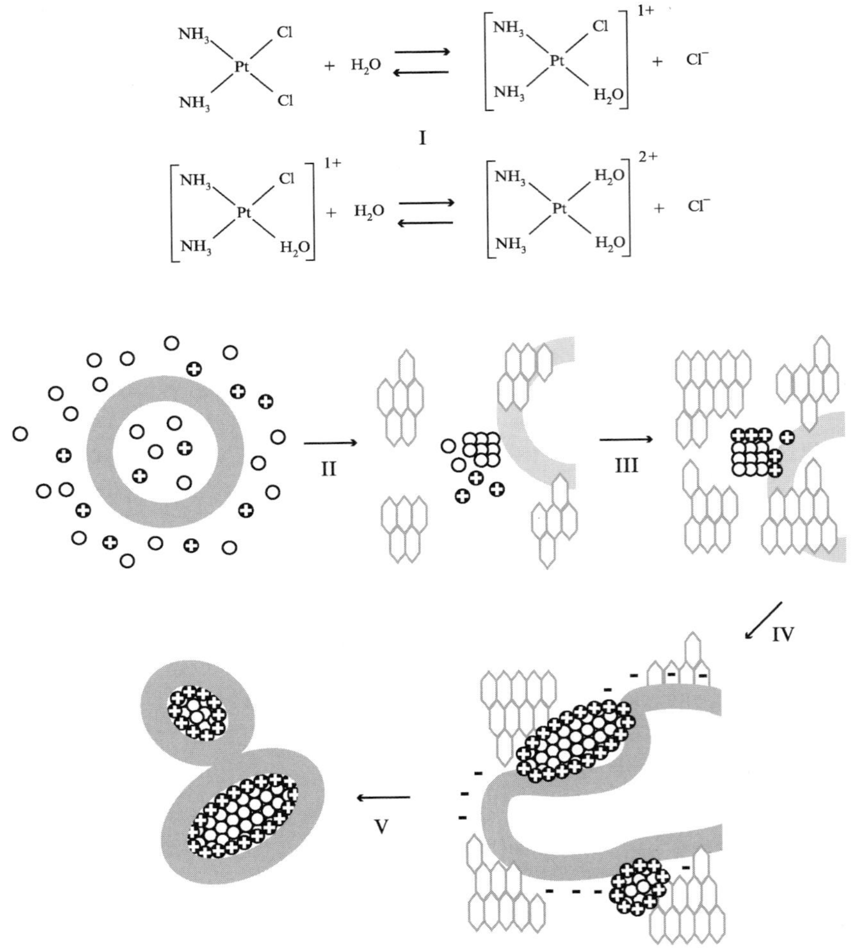

FIG. 1. Model of cisplatin nanocapsule formation. Partial hydrolysis of cisplatin in water yields positively charged aqua species (I). When a suspension of negatively charged liposomes is frozen in the presence of neutral (white spheres) and positively charged species (+ marked black spheres) of cisplatin, the formation of ice crystals drives the aggregation of the neutral species (II). As the ice crystals grow and the fluid phase is further concentrated, the positively charged species coaggregate (III), eventually covering the neutral aggregates, which then interact with the negatively charged liposomes (IV). The membranes reorganize to surround the cisplatin aggregates, and upon thawing, nanocapsules have been formed (V).

30 min. The solution is incubated overnight in the dark at 37° to ensure full equilibration.

Stock solutions of 1,2-dioleoyl-*sn*-glycer-3-phosphocholine (DOPC) and 1,2-dioleoyl-*sn*-glycero-3-phosphoserine (DOPS), obtained from Avanti Polar Lipids, Inc. (Alabaster, AL), are prepared in chloroform (concentration ~5 mM). The exact concentrations are determined by phosphate analysis (Rouser *et al.*, 1970). Aliquots corresponding to 0.6 μmol of each phospholipid are mixed, the solvent is removed by rotary evaporation, and the lipid film is further dried under vacuum overnight.

The dry lipid film is hydrated by adding 1.2 ml of the 5 mM cisplatin solution and incubation for 15 min at 37°. After brief homogenization in a vortex mixer, the dispersion is transferred to a glass tube and subjected to 10 freeze–thaw cycles using ethanol/dry-ice (−70°) and a waterbath (37°).

The resulting colloidal solution is transferred to microfuge tubes and centrifuged for 4 min at 470g (2100 rpm) in an Eppendorf centrifuge to collect the nanocapsules. After removal of the supernatant, the fluffy white layer on top of the yellow pellet, corresponding to large liposomes, is removed by a micropipette. The yellow pellet containing the cisplatin nanocapsules is resuspended in 1 ml water and centrifuged as above to wash away nonencapsulated cisplatin. Upon resuspending the final pellet in 0.5 ml water, the nanocapsules are stored at 4° until use.

Alternatively, the nanocapsules are purified from contaminating liposomes by density gradient centrifugation (Burger *et al.*, 2002). Briefly, the dispersion obtained after the freeze–thaw cycles is loaded on top of a step gradient consisting of 1 ml of each 1.8 M, 0.6 M, and 0.2 M sucrose in 10 mM PIPES-NaOH, 1 mM EGTA, pH 7.4. After centrifugation at 4° for 30 min at 400,000g (SW60 rotor, Beckman), the pellet fraction corresponding to the nanocapsules is collected and washed as above.

Remarks

Instead of DOPS, the negatively charged phospholipids dioleoyl-phoshatidylglycerol (DOPG) and dioleoylphosphatidic acid (DOPA) can also be used to prepare cisplatin nanocapsules. Likewise, DOPC can be replaced by dioleoylphosphatidylethanolamine (DOPE) or sphingomyelin (Burger *et al.*, 2002).

The method is sensitive to high chloride concentrations and alkaline pH, because these conditions prevent the formation of the positively charged aqua species (Lippert, 1999).

Instead of hydrating a lipid film with 5 mM cisplatin, it is also possible to add the cisplatin solution to preformed DOPC/DOPS liposomes and then start the freeze–thaw cycles.

The cisplatin nanocapsules can be stored after lyophilization and retain cisplatin upon rehydration.

Characterization

Encapsulation Efficiency

The phospholipid content of the nanocapsules is determined by phosphate analysis (Rouser *et al.*, 1970). The cisplatin content of the nanocapsules is assessed by flameless atomic absorption spectrometry (NFAAS) using K_2PtCl_2 (Sigma) as a standard (Burger *et al.*, 1999; van Warmerdam *et al.*, 1995). For this purpose, an aliquot of the cisplatin nanocapsule suspension is dissolved and diluted in 0.05% (w/v) Triton X-100, 0.05% (w/v) HNO_3 to a Pt concentration in the calibration range of the spectrometer (e.g., 2–10 ng K_2PtCl_2 per 30 μl sample volume for a Varian SpectrAA-400 Zeeman spectrometer).

Analysis of the cisplatin nanocapsules prepared as above typically yields a Pt/phosphate molar ratio of 11 ± 1. Based on the size of the nanocapsules (see below), this number is estimated to correspond to an internal cisplatin concentration exceeding 5 *M*, which is far beyond the solubility limit of cisplatin and consistent with the quasicrystalline structure of the encapsulated cisplatin (Burger *et al.*, 2002). The method encapsulates cisplatin with an efficiency of approximately 30%.

Shape and Size

Analysis by negative stain electron microscopy reveals bean-shaped particles consisting of an electron-dense core surrounded by a bright layer (excluding stain), corresponding to the bilayer coat (Fig. 2) (Burger *et al.*, 2002). The nanocapsules have a heterogeneous size distribution, with 75% of the population having a length between 50 and 250 nm and a width of around 50 nm (Fig. 2). Size analysis by dynamic light scattering yields consistent results (Burger *et al.*, 2002). Smaller size nanocapsules with a narrower size distribution have been obtained by high-pressure extrusion through polycarbonate filters with a 200-nm pore size (Burger *et al.*, 2002; Hope *et al.*, 1986).

Cytotoxicity

The cytotoxicity of the cisplatin nanocapsules toward the human ovarian carcinoma IGROV-1 cell line has been compared with that of free cisplatin. The IC_{50} value (the drug concentration at which cell growth is inhibited by 50%) of cisplatin administered as nanocapsules is two orders

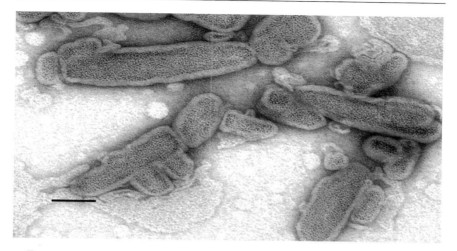

FIG. 2. Electron micrograph of cisplatin nanocapsules visualized by negative staining. A dilute suspension of nanocapsules was transferred to a carbon-Formvar–coated grid, dried, and stained with 4% (w/v) uranyl acetate for 45 s. Scale bar, 50 nm. Figure reproduced from Velinova *et al.* (2004) with permission.

of magnitude smaller than that of the free drug (Burger *et al.*, 2002). The higher cytotoxicity is explained by the reduced inactivation of the drug, due to the lipid coat sequestering it from reaction with substrates in the extracellular environment. Upon binding to the cell surface or endocytotic uptake of the nanocapsules, the coat is destabilized, and after membrane passage, cisplatin exerts its cytotoxic effect.

Conclusion

The cisplatin nanocapsules represent a new lipid formulation of cisplatin, distinct from conventional liposomal formulations. The distinctive feature is that the drug is present as aggregates surrounded by a bilayer. This results in a drug-to-lipid ratio that is two to three orders of magnitude higher than that of liposomal formulations (Newman *et al.*, 1999; Peleg-Shulman *et al.*, 2001) and probably accounts for the characteristic bean shape of the cisplatin nanocapsules. The increased encapsulation efficiency of cisplatin in nanocapsules is expected to increase the bioavailability of the drug and thus improve the therapeutic index compared with liposomal formulations of cisplatin. The method for preparing cisplatin nanocapsules may be applicable to other compounds that, due to their limited solubility in water and low lipophilicity, are not amenable to encapsulation in conventional liposomes.

Acknowledgments

The authors thank W. Mulder and M. J. Velinova for experimental support. The financial support by the Dutch Cancer Society (project UU2001-2493) is gratefully acknowledged.

Note added in proof: The molecular architecture of the cisplatin nanocapsules was recently solved (Chupin, V., de Kroon, A. I. P. M., and de Kruÿff, B. (2004). Molecular architecture of nanocapsules, bilayer-enclosed solid particles of cisplatin. *J. Am. Chem. Soc.* **126,** 13816–13821).

References

Bandak, S., Goren, D., Horowitz, A., Tzemach, D., and Gabizon, A. (1999). Pharmacological studies of cisplatin encapsulated in long-circulating liposomes in mouse tumor models. *Anticancer Drugs* **10,** 911–920.

Burger, K. N. J., Staffhorst, R. W. H. M., and de Kruijff, B. (1999). Interaction of the anticancer drug cisplatin with phosphatidylserine in intact and semi-intact cells. *Biochim. Biophys. Acta* **1419,** 43–54.

Burger, K. N. J., Staffhorst, R. W. H. M., de Vijlder, H. C., Velinova, M. J., Bomans, P. H., Frederik, P. M., and de Kruijff, B. (2002). Nanocapsules: Lipid-coated aggregates of cisplatin with high cytotoxicity. *Nat. Med.* **8,** 81–84.

Calvert, H., Judson, I., and van der Vijgh, W. J. (1993). Platinum complexes in cancer medicine: Pharmacokinetics and pharmacodynamics in relation to toxicity and therapeutic activity. *Cancer Surv.* **17,** 189–217.

Hacker, M. P. (1991). *In* "The Toxicity of Anticancer Drugs" (G. Powis and M. P. Hacker, eds.), p. 82. McGraw-Hill, New York.

Harrington, K. J., Lewanski, C. R., Northcote, A. D., Whittaker, J., Welbank, H., Vile, R. G., Peters, A. M., and Stewart, J. S. (2001). Phase I-II study of pegylated liposomal cisplatin (SPI-077) in patients with inoperable head and neck cancer. *Ann. Oncol.* **12,** 493–496.

Hope, M. J., Bally, M. B., Mayer, L. D., Janoff, A. S., and Cullis, P. R. (1986). Generation of multilamellar and unilamellar phospholipid vesicles. *Chem. Phys. Lipids* **40,** 89–107.

Howe-Grant, M. E., and Lippard, S. J. (1980). Aqueous platinum(II) chemistry; binding to biological molecules. *In* "Metal Ions in Biological Systems" (H. Sigel, ed.), Vol. XI, pp. 63–125. Marcel Dekker, New York.

Kim, E. S., Lu, C., Khuri, F. R., Tonda, M., Glisson, B. S., Liu, D., Jung, M., Hong, W. K., and Herbst, R. S. (2001). A phase II study of STEALTH cisplatin (SPI-77) in patients with advanced non-small cell lung cancer. *Lung Cancer* **34,** 427–432.

Lippert, B. (1999). "Cisplatin: Chemistry and Biochemistry of a Leading Anticancer Drug." Wiley-VCH, New York.

Meerum Terwogt, J. M., Groenewegen, G., Pluim, D., Maliepaard, M., Tibben, M. M., Huisman, A., ten Bokkel Huinink, W. W., Schot, M., Welbank, H., Voest, E. E., Beijnen, J. H., and Schellens, J. H. M. (2002). Phase I and pharmacokinetic study of SPI-77, a liposomal encapsulated dosage form of cisplatin. *Cancer Chemother. Pharmacol.* **49,** 201–210.

Newman, M. S., Colbern, G. T., Working, P. K., Engbers, C., and Amantea, M. A. (1999). Comparative pharmacokinetics, tissue distribution, and therapeutic effectiveness of cisplatin encapsulated in long-circulating, pegylated liposomes (SPI-077) in tumor-bearing mice. *Cancer Chemother. Pharmacol.* **43,** 1–7.

Peleg-Shulman, T., Gibson, D., Cohen, R., Abra, R., and Barenholz, Y. (2001). Characterization of sterically stabilized cisplatin liposomes by nuclear magnetic resonance. *Biochim. Biophys. Acta* **1510,** 278–291.

Perez, R. P. (1998). Cellular and molecular determinants of cisplatin resistance. *Eur. J. Cancer* **34,** 1535–1542.

Pinto, A. L., and Lippard, S. J. (1985). Binding of the antitumor drug cis-diamminedichloroplatinum(II) (cisplatin) to DNA. *Biochim. Biophys. Acta* **780,** 167–180.

Rouser, G., Fleischer, S., and Yamamoto, A. (1970). Two-dimensional thin-layer chromatographic separation of polar lipids and determination of phospholipids by phosphorus analysis of spots. *Lipids* **5,** 494–496.

Vaage, J., Donovan, D., Wipff, E., Abra, R., Colbern, G., Uster, P., and Working, P. (1999). Therapy of a xenografted human colonic carcinoma using cisplatin or doxorubicin encapsulated in long-circulating pegylated stealth liposomes. *Int. J. Cancer* **80,** 134–137.

van Warmerdam, L. J. C., van Tellingen, O., Maes, R. A. A., and Beijnen, J. H. (1995). Validated method for the determination of carboplatin in biological-fluids by Zeeman atomic-absorption spectrometry. *Fresenius J. Anal. Chem.* **351,** 777–781.

Veal, G. J., Griffin, M. J., Price, E., Parry, A., Dick, G. S., Little, M. A., Yule, S. M., Morland, B., Estlin, E. J., Hale, J. P., Pearson, A. D., Welbank, H., and Boddy, A. V. (2001). A phase I study in paediatric patients to evaluate the safety and pharmacokinetics of SPI-77, a liposome encapsulated formulation of cisplatin. *Br. J. Cancer* **84,** 1029–1035.

Velinova, M. J., Steffhorst, R. W. H. M., Mulder, W. J. M., Dries, A. J., Jensen, B. A. J., de Kruÿff, B., and de Kroon, A. I. P. M. (2004). Preparation and stability of lipid-coated nanocapsules of cisplatin: Anionic phospholipid specificity. *Biochim. Biophys. Acta* **1663,** 135–142.

[7] Liposomal Cytokines in the Treatment of Infectious Diseases and Cancer

By Timo L. M. ten Hagen

Abstract

Despite of the demonstrated activity of cytokines *in vitro*, their use in the clinical setting is often disappointing. Cytokine-related toxicity seriously limits optimal use *in vivo*. In addition, rapid degradation and excretion, neutralization and binding to receptors, or metabolization of the molecule results in a short half-life in serum when injected intravenously. As the dose–response curve of cytokines is relatively steep, outcome greatly benefits from improved delivery and bioavailability. One way to improve the pharmacokinetics of cytokines after systemic application is encapsulation in liposomes. An advantage of liposomes is that the encapsulated drug is protected from (rapid) degradation and excretion, and it eliminates the binding to neutralizing antibodies or (soluble) receptors. Moreover, liposomes can be tailored in such a way that they exhibit favorable pharmacokinetics, i.e., increased serum half-life and improved targeting to tissues or cells of interest. In this chapter, the use of liposomal cytokines in the treatment of cancer and infectious disease is discussed.

Introduction

The use of cytokines for the treatment of diseases, such as cancer, (chronic) inflammation, sepsis, or immunodeficiencies, is seriously limited by the unfavorable characteristics of these agents. Cytokines are generally rapidly degraded and excreted, neutralized, bound by receptors, or otherwise metabolized, resulting in a short half-life in serum when injected intravenously. These effects negatively influence the pharmacokinetics of the cytokines, limiting the potential of the drugs to reach the site of interest. Moreover, most cytokines inflict strong side effects, which limits the dose used, and, therefore, the effective concentration is not easily reached. For instance, tumor necrosis factor-α (TNF-α), which will be discussed in more detail below, has a half-life in serum of 25 min after systemic injection. Moreover, application of high dosages of this cytokine is limited, as it inflicts life-threatening side effects at relatively low dosages. One way to improve the pharmacokinetics of cytokines after systemic application is encapsulation in liposomes. An advantage of liposomes is that the encapsulated drug is protected from (rapid) degradation and excretion, and it eliminates the binding to neutralizing antibodies or (soluble) receptors. Moreover, liposomes can be tailored in such a way that they exhibit favorable pharmacokinetics, i.e., increased serum half-life and improved targeting to tissues or cells of interest. In this chapter, two possible liposomal preparations will be discussed. First, the natural fate of liposomes is utilized to target cytokines to macrophages. After systemic injection, liposomes are generally rapidly taken up by macrophages, which is normally an unwanted side effect of the use of liposomes. Here the use of liposomal interferon-γ (IFN-γ) for the treatment of sepsis will be discussed. Second, successful attempts have been made to abrogate rapid clearance of liposomes by macrophages to improve circulation time. This can be achieved to some extent by coating liposomes with poly(ethylene glycol) (PEG), also called Stealth liposomes. The PEG coating not only diminishes recognition of the liposomes by macrophages but also helps the liposomes to accumulate better in tumors or inflamed tissues. The use of Stealth liposomal TNF for the treatment of solid tumors will be discussed below.

Treatment of Sepsis by Activation of the Host Defense

Background

Severe infections represent a continuing threat to immunocompromised patients. The availability of a broad spectrum of antibiotics and the large number of studies on optimal dosing and administration protocols have not negated this threat. Gram-negative septicemia, a relatively rare clinical diagnosis a few decades ago, is perhaps the most serious infectious disease problem in hospitals today. Even the recent shift in infection type from gram-negative to gram-positive has not diminished the mortality due to gram-negative infections. Among patients who have complications of shock and organ failure develop, mortality can reach 90%. Septicemia thus represents a leading cause of death in the developed countries, and its incidence has increased significantly over the past decade (Bone, 1993).

Cure rates from infection is poorest among those patients who have persistent neutropenia. Various strategies are available to improve treatment of immunodeficient patients with severe infections. One is to intensify antibiotic treatment by applying more drugs at the same time; this is, however, not always effective. Poor results may at least partly be due to the occurrence of antimicrobial resistance among microorganisms causing septicemia. The inability to eliminate all microorganisms from the focus of infection results in the emergence of resistant microorganisms and the recurrence of infection. Failure of the host defense to support antibiotic treatment permits this to happen. A possible way to improve therapeutic results might thus be to stimulate the host defense, either a nonspecific defense (granulocytes and macrophages) or a specific humoral and cellular defense.

Host Defense Activation by Immunomodulation

Activation of a nonspecific host defense can be effective in different types of infection and does not induce tolerance of the microorganisms to treatment. Cells of the mononuclear phagocyte system (MPS) play a key role in nonspecific host defense. Activation of these cells will result first in enhanced killing of intracellular microorganisms infecting the MPS. However, activation of the MPS may enhance resistance to more systemic (extracellular) infections as well. Increased blood clearance and the microbial killing capacity of the MPS, as well as an enhanced granulocyte blood count, may result from treatment with so-called immunomodulators. Activation of a nonspecific host defense can be achieved with biological or synthetic immunomodulators that influence or modify (parts of) the innate resistance in a direct or indirect way, independently of challenge. Many different agents have been tested for their immunomodulatory capacity.

Immunomodulation with Liposomal Immunomodulators

Interferon-γ

The introduction of recombinant DNA technology has paved the way for the use of cytokines as immunomodulatory agents. Large quantities of relatively pure cytokine can be produced under controlled circumstances. IFN is a cytokine primarily produced by stimulated Th1 cells, CD8[+] T cells, and natural killer (NK) cells. It influences all major macrophage functions, including major histocompatibility complex (MHC) II expression, antigen presentation, $Fc_\gamma R1$ receptor expression, uptake and intracellular killing of microorganisms, tumor cell cytotoxicity, and the production of monokines (Baron *et al.*, 1991; Williams *et al.*, 1993). Thus far, promising studies with IFN in patients with chronic granulomatous disease (CGD) have led to the approval of IFN for prophylaxis against opportunistic infections in these patients (The International Chronic Granulomatous Disease Cooperative Study Group, 1991). Moreover, as will be discussed in more detail, data have been obtained supporting the possible usefulness of IFN for prophylaxis in other groups of patients.

Although significant macrophage activation occurs after exposure to immunomodulators *in vitro, in vivo* results are often quite disappointing. Actually this should not be surprising. *In vivo* experiments are complicated by several factors, which include pharmacokinetically dictated effects (short half-life, dilution, lack of significant localization at the site of interest, serum protein binding, and degradation). In addition, the lack of synergy and concomitant toxicity including sensitization are important. The use of IFN for immunotherapy is limited by rapid clearance of this cytokine from circulation and by toxicity seen with high dosage regimens (Bennett *et al.*, 1986; Gutterman *et al.*, 1984; Kurzrock *et al.*, 1985). Free IFN has a serum half-life in humans of approximately 20 min and is degraded and secreted from the body. IFN induces undesirable side effects such as fever, weight loss, and liver and kidney dysfunction.

Application of Liposomes to Improve the Therapeutic Window

To overcome some of these disadvantages, liposomes may be used as carriers. Liposomes are microscopic vesicles consisting of one or more lipid bilayers surrounding an internal aqueous compartment. By encapsulating agents in liposomes, lifetime in the body is prolonged, high concentrations at specific sites can be reached, and by coencapsulation of agents synergy *in vivo* can be ensured. Toxicity of the encapsulated compound may be reduced, and an immunological reaction may be prevented. In addition, liposomes avoid serum protein binding and dilution (Pak and Fidler, 1991). Liposomes are biodegradable and nonimmunogenic if composed of natural phospholipids. A variety of agents can be entrapped in liposomes: hydrophobic agents with high efficiency in the lipid bilayers and hydrophilic agents in the inner aqueous space. Liposome encapsulation of IFN (LE-IFN) increases its half-life time and, as will be discussed below, also increases its ability to stimulate the host defense.

The increased activity LE-IFN over free IFN was shown in an *in vivo* infection model using *Listeria monocytogenes* (Melissen *et al.*, 1993).

Encapsulation of IFN increased efficacy 66-fold in mice infected with *L. monocytogenes*. Reduction of toxicity of IFN by liposomal encapsulation has also been demonstrated (Hockertz *et al.*, 1989, 1991).

As macrophages are believed to be the most important target cells for immunomodulation, the use of classic liposomes, which tend to localize in large numbers in these cells, is quite obvious. Classic liposomes are relatively large (ranging from 1 to 20 μm in diameter) and do not have features that diminish uptake by the MPS, in contrast to the newer generation of liposomes such as the MPS-avoiding Stealth liposomes (see the section on TNF Encapsulated in Sterically Stabilized Liposomes).

The tendency of classic liposomes to rapidly localize in MPS cells, primarily macrophages residing in the liver and spleen (Melissen *et al.*, 1994a), results in an augmented accumulation of encapsulated agent in macrophages. Phospholipid composition, charge, and size have a strong influence on uptake and degradation of liposomes by macrophages. Specifically, inclusion of negatively charged phospholipids such as phosphatidylserine (PS) in liposomes consisting of phosphatidylcholine (PC) enhances binding to and phagocytosis by macrophages, whereas uncharged liposomes composed exclusively of PC are less efficiently taken up by macrophages (Pak and Fidler, 1991). Pharmacokinetics in rats showed that 30 min after intravenous injection, liposomes (PC:PS, molar ratio 7:3) mainly localized in the liver (55% of injected dose), lung (19%), and spleen (15%) (Schumann *et al.*, 1989). Similarly, in mice liposomes, were preferentially taken up by the liver and spleen (32% and 17%, respectively, after 60 min) (Melissen *et al.*, 1994a).

Together, these results demonstrate that liposome encapsulation will reduce the toxicity of IFN, probably by reducing undesirable localization to sensitive sites (site avoiding delivery) but also will increase the localization at the site of interest, the macrophage (site-specific delivery). This means that lower dosages can be applied, and better results may be obtained *in vivo* when these immunomodulators are encapsulated into liposomes. It might be expected that administration of high dosages of liposomes results in saturation of the MPS cells, and by doing so reduces the host defense. The phagocytic capacity of the MPS for bacteria might consequently be reduced. However, 5-fold repeated injection for 7 days of a high dose of placebo liposomes (125 mg lipid per kg per dose) in mice did not negatively affect the clearance capacity of the MPS (ten Hagen *et al.*, 1995).

Liposome-Encapsulated IFN in Klebsiella pneumoniae *Septicemia*

The effect of LE-IFN was studied in the *K. pneumoniae* septicemia model. Intravenous injection of a single dose of LE-IFN 24 h before intraperitoneal infection resulted in 15% survival (ten Hagen *et al.*, 1995).

In parallel with results obtained with LE-muramyl-tripeptide phosphatidylethanolamine (MTPPE), multiple dosages of LE-IFN could further increase survival of mice to 65%. Moreover, the combination of MTPPE and IFN by coencapsulation in liposomes further enhanced host resistance: 100% survival was obtained (ten Hagen *et al.*, 1995). The effect of coencapsulated MTPPE and IFN on host defense was additive, suggesting that these two immunomodulators produce their effect on different levels in the same chain of activity. Importantly, it was observed that in the *K. pneumoniae* infection models, antimicrobial effects were maximal if immunomodulators were administered 24 h or more before infection (Melissen *et al.*, 1991, 1994b; Parant and Chedid, 1985).

Coencapsulation: Platform for Synergistic Macrophage Activation In Vivo

The possibility of coencapsulation of agents into liposomes also provides an important tool for drug delivery *in vivo*. Synergy between immunomodulating agents can be obtained *in vitro*. However, *in vivo* synergy is questionable, since simultaneous exposure of macrophages to multiple immunomodulators after intravenous administration is expected to be minimal due to wide differences in their pharmacokinetic behavior. With agents coencapsulated in liposomes, simultaneous delivery of agents to macrophages is guaranteed. Synergy between MTPPE and IFN in the free form was shown *in vitro* using *L. monocytogenes*–infected peritoneal macrophages (Melissen *et al.*, 1993). Striking synergy between muramyl-peptides and IFN was also shown in several tumor models (Dukor and Schumann, 1987) *in vitro* and *in vivo*. As stated above, coencapsulation of MTPPE and IFN improved survival of mice suffering from a *K. pneumoniae* septicemia compared with these agents in the free form (ten Hagen *et al.*, 1995).

Mechanisms of Action

It has been observed that exposure of macrophages to IFN or other immunomodulators results in increased phagocytic activity, enhanced production of oxygen or nitrogen radicals, and excretion of elevated levels of proinflammatory cytokines (Billiau and Dijkmans, 1990). Next to that, treatment of mice with LE-IFN (with or without LE-MTPPE) resulted in an increased macrophage number, as well as elevated levels of monocytes, granulocytes, and lymphocytes (Wiltrout *et al.*, 1989; personal observations). These immunomodulators most likely enhance cell numbers by initiating an increased number of progenitor cells (Wiltrout *et al.*, 1989; unpublished observations) (Fig. 1).

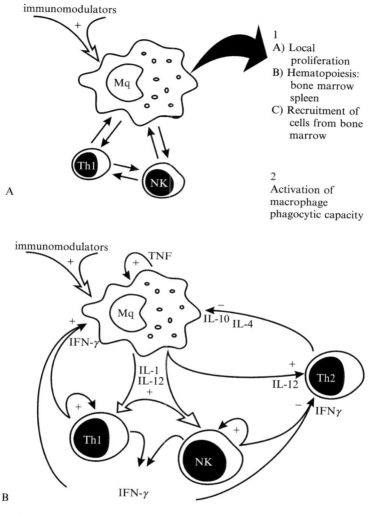

FIG. 1. Proposed mechanism of host defense activation by immunomodulators such as muramyl-peptides or interferon, which is thought to be controlled by macrophages. (A) As described in the text, immunomodulation *in vivo* results in activation of macrophages, which also involves other cells of the immune system such as T and NK cells. This activation results in local proliferation of macrophages, increased hematopoiesis, and recruitment of monocytes and granulocytes from bone marrow to the periphery. These activities are thought to be of major importance for host defense activation to infection. Direct antimicrobial activation of macrophages is thought to be of only minor importance. (B) Stimulation and subsequent down-modulation of macrophage activity are controlled by a number of cytokines. The major cytokines involved in these processes are shown. Upon triggering of macrophages by

Method of IFN-Liposome Preparation

The liposomes used for macrophage targeting are made from a dry lyophilisate composed of phosphatidylcholine and phosphatidylserine in a molar ratio of 7:3. IFN (1 mg/ml) is diluted in HEPES buffer, and 2.5 ml is mixed with 250 mg of lipids. The dispersion is vigorously shaken. Free IFN is removed by ultracentrifugation at 44,000g for 1 h at 4° after addition of phosphate-buffered saline (PBS). The final liposomes contained 48% of the IFN and had an average diameter between 2.0 and 3.5 μm.

The encapsulation of IFN is determined by labeling IFN with ^{125}I using the iodobead method. Briefly, IFN (400 μm/ml) diluted in iodination buffer is mixed with iodobeads and 15 MBq ^{125}I. After 15 min of incubation at room temperature, the beads are discarded, and the free ^{125}I is removed by gel filtration over a PD10 column. A trace amount of labeled IFN is added to unlabeled IFN, after which the liposomes are prepared. The percentage of encapsulated IFN is calculated by dividing the liposome-associated label by the total activity added. For this purpose, the liposomes are washed three times by ultracentrifugation as described above. The activity of the supernatants is also measured to check the efficacy of the wash steps. At most, 1% activity is generally found in the supernatant of the last wash procedure.

Application of Liposomal Cytokine Formulation in the Treatment of Solid Tumors

Tumor Necrosis Factor-α: Applicability and Limitations

Spontaneous regression of cancer was observed in patients with concurrent bacterial infection (Coley, 1896). Many years passed before a factor causing the hemorrhagic necrosis in experimental tumors, later named tumor necrosis factor-α (TNF), was discovered (Carswell *et al.*, 1975). TNF is a homotrimeric complex of 52 kDa, which is produced by many cell types but mainly secreted by activated monocytes/macrophages. Its expression and regulation are affected by a variety of other cytokines, including IFN-γ, interleukins (IL-1, IL-2, IL-12), granulocyte-macrophage

---

immunomodulators, these cells produce cytokines such as TNF, which inflicts autostimulation, and IL-1 and IL-12, which act on T cells and NK cells. A major cytokine produced by Th1 cells and NK cells in this process is interferon-γ (IFN-γ), which stimulates both the Th1 and NK cells, as well as macrophages, the latter especially in conjunction with TNF. On Th2 cells, IFN-γ has an inhibitory effect, diminishing production of cytokines such as IL-4 and IL-10, which have an inhibitory effect on macrophages. Also, macrophages produce IL-10, which is thought to provide negative feedback on macrophage activation and prevent inflammatory overshoot.

colony-stimulating factor (GM-CSF), platelet aggregating factor (PAF), as well as TNF itself (Sidhu and Bollon, 1993). TNF has pleiotropic effects, which depend on concentration. It has been shown to have vasculotoxic effects at high concentrations, while at low concentrations it promotes DNA synthesis and angiogenesis (Fajardo *et al.*, 1992). The effects of TNF are exerted by binding to two types of receptor, with molecular weights of 55 kDa (TNF-R1) and 75 kDa (TNF-R2), respectively, which are present on nearly all mammalian cells and in soluble form in the circulation.

It was not until the large-scale production of TNF, using recombinant DNA techniques, that clinical application became feasible (Pennica *et al.*, 1984). Phase I/II trials proved to be disappointing, mainly caused by the deleterious side effects after systemic administration of TNF, hypotension being the dose-limiting toxicity (Hieber and Heim, 1994). The maximal tolerated dose (MTD) varied between 200 and 400 $\mu g/m^2$, depending on treatment schedule, which is only 1/50 of the effective dose in murine tumor models (Brouckaert *et al.*, 1986). Phase II studies revealed a meager 1–2% complete remission rate after intravenous administration of TNF as a single agent or combination therapy with chemotherapeutic drugs (Hieber and Heim, 1994). Locoregional administration of TNF seemed the only option for successful application of TNF. Intratumoral and intraperitoneal administration or hepatic artery infusions were reported to increase response rates, however, only marginally (Bartsch *et al.*, 1989). This all changed when TNF-based isolated extremity perfusion in melanoma and in soft tissue sarcoma was reported to yield impressive response rates (Eggermont *et al.*, 1997). This also led to the exploration of its use in an isolated organ perfusion setting, which was attempted in lungs, livers, and kidneys. In general, the isolation procedures in organs proved to be technically difficult, and organ toxicity limited the dose of TNF that could be used. Moreover, disseminated disease still remained an untreatable entity.

Clinical Use of TNF: Isolated Perfusion

The first consistent antitumor results with TNF were obtained after isolated limb perfusion (ILP). With ILP, 15–20 times higher local drug concentrations can be achieved than those reached after systemic administration (Benckhuijsen *et al.*, 1988). ILP facilitates leakage-free treatment of the limb and an effective washout thereafter. In combination with hyperthermia, isolated limb perfusion with melphalan yielded a 54% complete response rate for patients with melanoma in transit metastases confined to the limb (Eggermont, 1996). ILP with high-dose TNF and IFN was undertaken for multiple melanoma in transit metastases and irresectable high-grade soft tissue sarcomas of the limb resulting in high complete response rates, with limb salvage rates of more than 80% (Eggermont *et al.*, 1997). Also other cytotxic drugs were investigated in the TNF-based ILP such as doxorubicin, an anthracycline. Anthracyclines are among the most active agents against solid tumors, and doxorubicin is the most widely used agent of this class. Moreover, doxorubicin is the agent of choice for the treatment of sarcoma and has shown good antitumor activity in clinical and experimental perfusion settings for the treatment of lung metastases (Abolhoda *et al.*, 1997).

The success of ILP with cytotoxic drugs in combination with TNF encouraged translation of this therapy to other settings such as isolated organ perfusion. Various organs, which are amenable to isolated perfusion, have been described in which the applicability and antitumor efficacy of TNF were evaluated. Isolated lung perfusion proved to be relatively safe; Phase I studies for isolated renal perfusion are underway, and recent reports described the efficacy of isolated liver perfusion with TNF (Pass *et al.*, 1996; van der Veen *et al.*, 1999). Experiments in pigs showed that drug levels comparable to those achieved in isolated limb perfusion can be obtained in a leakage-free isolated hepatic perfusion setting (Borel Rinkes *et al.*, 1997). In a Phase I study in patients with unresectable colorectal cancer liver metastases, isolated hepatic perfusion with TNF was performed and resulted in high mortality and morbidity (de Vries *et al.*, 1998). Toxicity was less dramatic in the series reported by Alexander *et al.* (1998). In patients, the liver was perfused via the portal vein and the hepatic artery simultaneously. This may, in part, account for the unexpected high toxicity. Introduction of a relatively new balloon-catheter technique allows isolated perfusion of the liver without massive surgery as shown in pigs (Van Ijken *et al.*, 1998). These results initiated clinical trials for perfusion of the liver, pelvic area, and abdomen with a combination of melphalan and mitomycin C using balloon catheters. TNF will be included at a later time point in some of these settings.

Preclinical Studies with TNF: In Vitro *and* In Vivo *Evaluations*

For a small minority of cancer cell lines, TNF is directly cytostatic or cytotoxic. On other cell types, TNF shows growth-inhibitory or even growth-stimulatory effects. The mechanism by which TNF exerts its cytotoxic effects is not yet fully understood. The number of receptors on the tumor cell is probably of less importance than the role of oxygen free radicals and activation of lysosomal enzymes (Sidhu and Bollon, 1993). Several cell lines were tested for their susceptibility to TNF and certain cytotoxic drugs (melphalan, doxorubicin, cisplatin). In contrast to what was expected, no direct cytotoxic effects of TNF and no synergism with melphalan, doxorubicin, or cisplatin could be observed in all tumor cell lines thus far investigated (Manusama *et al.*, 1996a,b; van der Veen *et al.*, 2002). Additive antitumor effects, however, were noted. Cytotoxic effects of TNF can be enhanced by a number of other biological response modifiers such as IFN or IL-1, hyperthermia, and irradiation.

Many animal studies demonstrated antitumor effects of TNF. Making use of the rapidly growing soft tissue sarcoma BN 175 and the osteosarcoma ROS-1, we developed two isolated limb perfusion models that resemble the clinical situation closely (Manusama *et al.*, 1996a,b). In both tumor systems, highly synergistic antitumor effects were observed when TNF was combined with melphalan, resulting in a complete remission rate of 70–80% (de Wilt *et al.*, 1999; Manusama *et al.*, 1996a,b). Histopathologically extensive hemorrhagic necrosis was observed after ILP with the combination and not after ILP with melphalan alone (Nooijen *et al.*, 1996a). Early endothelial damage and platelet aggregation in the tumor vessels are observed after ILP with TNF and melphalan, and this is believed to lead to ischemic (coagulative) necrosis, which is in line with observations in patients (Fraker and Alexander, 1994). ILP with TNF and doxorubicin in solid tumor-bearing rats demonstrated a comparable outcome (van der Veen *et al.*, 2000).

Application of Liposomes in Anticancer Therapy

Cytotoxic drugs are characterized by a low therapeutic index (TI) due to (1) the inability to achieve therapeutic concentrations at the site of the tumor, (2) nonspecific cytotoxicity to normal tissues, or (3) problems arising with the formulation of the drug. In general, the purpose of encapsulation of cytotoxic drugs will be a reduction of toxicity, and at the same time, enhancement of selective delivery of the encapsulated drug to tumor tissue. For cytotoxic drugs to reach their target site, the carrier system must leave the circulation. Solid tumors show a disorganized microvasculature, with increased capillary permeability due to wider endothelial gaps and depleted endothelial layers and basal lamina (Dvorak *et al.*, 1988). In addition, physiological parameters such as the increased interstitial fluid pressure and the absence of an adequate lymphatic barrier play a role in the deposition of liposomes in the tumor (Jain, 1989). It was proven that sterically stabilized liposomes, which show a higher vascular permeability than conventional liposomes, preferentially are taken up in tumor tissue as opposed to nontumor tissue with relatively impermeable capillaries (Wu *et al.*, 1993; Yuan *et al.*, 1995). A high accumulation of Stealth liposomes in tumor tissue was demonstrated, as well as uptake in liver and spleen, confirming previous results (van der Veen *et al.*, 1998). It has been suggested that tumor-bearing animals have an increased splenic clearance, possibly related to an altered opsonic activity of serum in these animals. Depletion of macrophages by liposomal clodronate completely abolished splenic clearance, while hepatic clearance was relatively unaffected (unpublished observations), making phagocytic mechanisms of clearance likely.

Liposomal Formulation of Doxorubicin and Cisplatin

The altered pharmacokinetics and efficacy of anthracyclines (e.g., doxorubicin) encapsulated in sterically stabilized liposomes have been described for a variety of animal tumor models. All showed an increased accumulation of the encapsulated drug versus the nonencapsulated drug, with peak levels in tumor tissue reached at a later time. The encapsulation of anthracyclines, especially doxorubicin, epirubicin, and daunorubicin, in liposomes has been reviewed (Gabizon and Martin, 1997).

In human studies, the alteration of pharmacokinetics of doxorubicin after encapsulation in sterically stabilized liposomes was confirmed. The plasma elimination of Doxil was even slower than in animal studies (Gabizon, 1993). In a subsequent Phase I study, patients with a variety of malignant tumors were treated with escalating doses between 25 and 60 mg/m^2 every 3–4 weeks. The efficacy of PEG-liposomal doxorubicin is most obvious in the management of AIDS-related Kaposi's sarcoma, where augmented tumor concentrations of doxorubicin were measured (Coukell and Spencer, 1997). Higher response rates were achieved with Doxil (20 mg/m^2) than with the standard regimen containing doxorubicin, bleomycin, and vincristine given systemically (Northfelt et al., 1998). The drug formulation is generally well tolerated with dose-limiting toxicity being skin toxicity (Alberts and Garcia, 1997). These studies led to the approval of Doxil (CAELYX) for clinical use. Several hundred patients with HIV-related refractory Kaposi's sarcoma have been treated. At present, multiple Phase I/II and III studies are being performed for a variety of tumors, like (metastatic or recurrent) breast cancer, refractory ovarian cancer, and prostate cancer, as well as non-small cell lung cancer and soft tissue sarcomas (Muggia, 1997). The deposition of Doxil was followed not only for solid tumors but also in a liver metastasis model. A homogeneous distribution to metastatic liver tumors after intraarterial and intraportal administration was shown, and a sustained release of encapsulated drug was observed (Cay et al., 1997).

Liposomal Formulation of Cytokines

Incorporation of cytokines (IL-1, IL-2, IL-6, TNF, GM-CSF, and IFN) in conventional liposomes was carried out by several groups. To address the main adverse effect of TNF, its severe toxicity, liposomal encapsulation has been studied by several authors (Debs *et al.*, 1989; Kedar *et al.*, 1997; Nii *et al.*, 1991; Utsumi *et al.*, 1991). Encapsulated TNF retained immuno-modulatory and antitumor activity *in vivo* or *in vitro*, even in the presence of antibodies to TNF or toward normally TNF-resistant cells, at the same time reducing toxicity compared with free TNF (Debs *et al.*, 1989; Nii *et al.*, 1991). It was soon recognized that encapsulation of TNF in liposomes was hampered by the low hydrophobicity of the cytokine, and the preparation of a more lipophilic variety was therefore attempted (Klostergaard *et al.*, 1992; Utsumi *et al.*, 1991). Encapsulation of TNF in liposomes was also enhanced by varying the calcium concentration in the membrane (Saito *et al.*, 1995).

The application of another cytokine in cancer therapy, IL-2, is similarly hampered by its severe toxicity. IL-2 controls the proliferation and activation of cytotoxic T-lymphocytes (CTL) and NK cells, as well as the proliferation of B-lymphocytes. Similar to TNF, its pharmacokinetics are characterized by a short blood-residence time (Schwartzentruber, 1995). Liposomal encapsu-lation of IL-2 was demonstrated to enhance the therapeutic effect of IL-2 *in vitro* as well as *in vivo* (Kedar *et al.*, 1997). On the other hand, in several preclinical studies, an antitumor effect of liposomal IL-2 was observed when given locoregionally, but the effect was considerably less when administered systemically (Anderson *et al.*, 1990; Konno *et al.*, 1991).

TNF Encapsulated in Sterically Stabilized Liposomes

The major drawback of conventional liposomes remains the rapid clearance by the MPS. Therefore, encapsulation of cytokines in Stealth liposomes was attempted. Incorporation of IL-2 in sterically stabilized lipo-somes (SSL) enhanced immunomodulatory activity. SSL-IL-2 was shown to have superior therapeutic efficacy in mice with advanced metastatic tumors, either alone or in combination with cyclophosphamide pretreatment (Kedar *et al.*, 1994b).

We demonstrated that TNF could be incorporated in sterically stabi-lized liposomes of 95 nm diameter with an encapsulation efficiency of 24% without loss of activity (TNF-SL) (van der Veen *et al.*, 1998). The blood circulation of TNF-SL was comparable to the circulation times mentioned in the literature for other long circulating liposome formulations. Initial release of entrapped drugs from the circulating liposomes is more rapid, resulting in a faster rate of clearance in the first few hours. This phenome-non is also known for IL-2 in sterically stabilized liposomes (Kedar *et al.*, 1994a). TNF-SL showed a localization of almost 8% of the injected dose 24 h after injection in soft tissue sarcoma (van der Veen *et al.*, 1998).

Subsequently, in soft tissue sarcoma–bearing rats, the antitumor efficacy of different combinations of TNF-SL and Doxil was tested. An improved tumor response after administration of Doxil plus TNF-SL was observed compared with administration of Doxil alone (ten Hagen *et al.*, 2002). Recently, we observed that application of sterically stabilized liposomal TNF in combination with radiotherapy resulted in an improved tumor response (Kim *et al.*, 2001).

Preparation of Sterically Stabilized Long Circulating Liposomes

A mixture of PHEPC, cholesterol, and PEG-DSPE, in a molar ratio of 1.85:1:0.15 suspended in chloroform/methanol, is evaporated to dryness in a rotary evaporator. The PEG-DSPE was kindly provided by Dr. P. Working [SEQUUS Pharmaceuticals Inc. (now ALZA, Inc.) Menlo Park, CA]. The dried lipid film is resuspended with HEPES buffer. The hydrated lipid film is vortexed, and liposomes are brought to a specific size by sonication for 10–15 min at amplitude 9 using an ultrasonic disintegrator supplied with an exponential microprobe (diameter 3.5 mm) (Soniprep 150, Sanyo, UK). The liposomes are sterilized through a 0.2-μm filter. Liposome size is determined by dynamic light scattering (DLS, Malvern 4700 system, Malvern, UK). The liposomes are stored at 4° until further use after determination of total lipid concentration.

The concentration of encapsulated TNF is determined by two methods. In the first, encapsulation of TNF is determined by labeling TNF with ^{125}I using the iodobead method. See the method for the production of liposomal IFN described above. Second, the TNF-sensitive tumor cell line WEHI-164 may be used to detect TNF. As TNF is easily destroyed during manipulation, it is very important to check bioactivity of the formulation. We noticed that sonication, or labeling, had a marginal effect on the cytotoxicity of TNF. However, encapsulation of TNF in liposomes resulted in a 30-fold reduced cytotoxicity (ten Hagen *et al.*, 2002).

Use of Low-Dose TNF in Combination with Liposomal Antitumor Formulations

It seems probable that to achieve an antitumor response in the combination treatment protocol described above, only low local concentrations of TNF are needed. We therefore explored the possibility of combining low-dose TNF, systemically injected in combination with Doxil (ten Hagen *et al.*, 2000). In soft tissue sarcoma–bearing rats, addition of low-dose (15 μg/kg) TNF to Doxil resulted in an increased antitumor response. Enhanced interstitial accumulation of doxorubicin in tumor tissue when the liposomes were combined with TNF appeared to be crucial.

Effect of TNF in Antitumor Therapy: Dual Targeting as the Mechanism of Action

The tumor vasculature consists of preexistent vessels in tissues in which the tumor has invaded, as well as microvessels as a result of angiogenesis. These vessels are characterized by high permeability caused by incomplete endothelial linings and basement membranes, and the vasculature is not well organized. The vessels run a tortuous and convoluted path, with venous lakes, reversed flow, and stasis (Dewhirst *et al.*, 1989; Dvorak *et al.*, 1988). In addition, tumor stroma is characterized by the absence of lymphatics and high interstitial pressure. Drug delivery to solid tumors is therefore influenced by the microvasculature and tumor stroma.

The effects on the tumor-associated vasculature after ILP with TNF were described as early endothelium activation, up-regulation of adhesion molecules, and invasion of polymorphonuclear cells, leading to coagulative necrosis with or without hemorrhagic necrosis (Nooijen *et al.*, 1996a; Renard *et al.*, 1994, 1995). Thrombus formation and negative influence on tumor blood flow also seemed to play a pivotal role. However, examination of melanomas and sarcomas of patients treated with TNF-based ILP did not show differences in expression of adhesion molecules such as ICAM-1, E-selectin (ECAM-1), VCAM-1, or PECAM-1 in tumors compared with healthy tissue (Nooijen *et al.*, 1996b).

In patients treated with ILP with TNF, IFN, and melphalan, detachment and apoptosis of endothelial cells were demonstrated *in vivo* in melanoma metastases resulting from suppression of integrin $\alpha_V\beta_3$, an adhesion receptor expressed by angiogenic endothelial cells that binds to extracellular matrix components (Ruegg *et al.*, 1998). Subsequent injury to the endothelial cells may be the beginning of a cascade of events. This was further indicated by angiographic studies that clearly showed the disappearance of tumor-associated vessels after TNF-ILP (Eggermont *et al.*, 1994). The endothelial damage, proven by a change in distribution of von Willebrand factor, already occurred 3 h after the onset of TNF-based perfusion (Renard *et al.*, 1995). However, a delayed type of hyperpermeability may also be present, explaining the fact that complete tumor regression frequently requires longer periods after TNF-based isolated perfusion (Nooijen *et al.*, 1998). Vasculotoxic effects of TNF lead to a significant drop in tumor interstitial pressure and permeability changes as well, which leads to better penetration of the cytotoxic drug into tumor tissue (Kristensen *et al.*, 1996; Suzuki *et al.*, 1990). We showed a strongly augmented accumulation of doxorubicin or melphalan in tumor tissue when TNF was added to ILP (de Wilt *et al.*, 2000; Suzuki *et al.*, 1990). Synergism between TNF and cytotoxic drugs in a high-dose perfusion setting may be explained by dual targeting, whereas immune factors also play an important role. Moreover, as stated above, low-dose TNF improved the tumor response when combined with Doxil by augmenting intratumoral accumulation of doxorubicin.

Acknowledgments

The studies with LE-IFN were performed at the Department of Medical Microbiology and Infectious Diseases of the Erasmus University Medical Center, Rotterdam.

References

Abolhoda, A., Brooks, A., Nawata, S., Kaneda, Y., Cheng, H., and Burt, M. E. (1997). Isolated lung perfusion with doxorubicin prolongs survival in a rodent model of pulmonary metastases. *Ann. Thorac. Surg.* **64,** 181–184.

Alberts, D. S., and Garcia, D. J. (1997). Safety aspects of pegylated liposomal doxorubicin in patients with cancer. *Drugs* **54**(Suppl. 4), 30–35.

Alexander, H. R., Jr., Bartlett, D. L., Libutti, S. K., Fraker, D. L., Moser, T. X., and Rosenberg, S. A. (1998). Isolated hepatic perfusion with tumor necrosis factor and melphalan for unresectable cancers confined to the liver. *J. Clin. Oncol.* **16,** 1479–1489.

Anderson, P. M., Katsanis, E., Leonard, A. S., Schow, D., Loeffler, C. M., Goldstein, M. B., and Ochoa, A. C. (1990). Increased local antitumor effects of interleukin 2 liposomes in mice with MCA-106 sarcoma pulmonary metastases. *Cancer Res.* **50,** 1853–1856.

Baron, S., Tyring, S. K., Fleischmann, W. R., *et al.* (1991). The interferons: Mechanisms of action and clinical applications. *JAMA* **266,** 1375–1383.

Bartsch, H. H., Pfizenmaier, K., Schroeder, M., and Nagel, G. A. (1989). Intralesional application of recombinant human tumor necrosis factor alpha induces local tumor regression in patients with advanced malignancies. *Eur. J. Cancer Clin. Oncol.* **25,** 287–291.

Benckhuijsen, C., Kroon, B. B., Van Geel, A. N., and Wieberdink, J. (1988). Regional perfusion treatment with melphalan for melanoma in a limb: An evaluation of drug kinetics. *Eur. J. Surg. Oncol.* **14,** 157–163.

Bennett, C. L., Vogelzang, N. J., Ratain, M. J., and Reich, S. D. (1986). Hyponatremia and other toxic effects during a phase I trial of recombinant human gamma interferon and vinblastine. *Cancer Treat. Rep.* **70,** 1081–1084.

Billiau, A., and Dijkmans, R. (1990). Interferon-γ: Mechanism of action and therapeutic potential. *Biochem. Pharmacol.* **40**(7), 1433–1439.

Bone, R. C. (1993). Gram-negative sepsis: A dilemma of modern medicine. *Clin. Microbiol. Rev.* **6,** 57–68.

Borel Rinkes, I. H., de Vries, M. R., Jonker, A. M., Swaak, T. J., Hack, C. E., Nooyen, P. T., Wiggers, T., and Eggermont, A. M. (1997). Isolated hepatic perfusion in the pig with TNF-alpha with and without melphalan. *Br. J. Cancer* **75,** 1447–1453.

Brouckaert, P. G., Leroux-Roels, G. G., Guisez, Y., Tavernier, J., and Fiers, W. (1986). *In vivo* anti-tumour activity of recombinant human and murine TNF, alone and in combination with murine IFN-gamma, on a syngeneic murine melanoma. *Int. J. Cancer* **38,** 763–769.

Carswell, E. A., Old, L. J., Kassel, R. L., Green, S., Fiore, N., and Williamson, B. (1975). An endotoxin-induced serum factor that causes necrosis of tumors. *Proc. Natl. Acad. Sci. USA* **72,** 3666–3670.

Cay, O., Kruskal, J. B., Nasser, I., Thomas, P., and Clouse, M. E. (1997). Liver metastases from colorectal cancer: Drug delivery with liposome-encapsulated doxorubicin. *Radiology* **205,** 95–101.

Coley, W. B. (1896). The therapeutic value of the mixed toxins of the streptococcus of erysipelas and bacillus prodigious in the treatment of inoperable malignant tumors. *Am. J. Med. Sci.* **112**, 251–281.

Coukell, A. J., and Spencer, C. M. (1997). Polyethylene glycol-liposomal doxorubicin. A review of its pharmacodynamic and pharmacokinetic properties, and therapeutic efficacy in the management of AIDS-related Kaposi's sarcoma. *Drugs* **53**, 520–538.

Debs, R. J., Düzgüneş, N., Brunette, E. N., Fendly, B., Patton, J., and Philip, R. (1989). Liposome-associated tumor necrosis factor retains bioactivity in the presence of neutralizing anti-tumor necrosis factor antibodies. *J. Immunol.* **143**, 1192–1197.

de Vries, M. R., Rinkes, I. H., van de Velde, C. J., Wiggers, T., Tollenaar, R. A. X., Kuppen, P. J., Vahrmeijer, A. L., and Eggermont, A. M. (1998). Isolated hepatic perfusion with tumor necrosis factor alpha and melphalan: Experimental studies in pigs and phase I data from humans. *Recent Results Cancer Res.* **147**, 107–119.

de Wilt, J. H. W., Manusama, E. R., van Tiel, S. T., Van Ijken, M. G., ten Hagen, T. L. M., and Eggermont, A. M. M. (1999). Prerequisites for effective isolated limb perfusion using tumor necrosis alpha and melphalan in rats. *Br. J. Cancer* **80**, 161–166.

de Wilt, J. H. W., ten Hagen, T. L. M., de Boeck, G., van Tiel, S. T., de Bruijn, E. A., and Eggermont, A. M. M. (2000). Tumour necrosis factor alpha increases melphalan concentration in tumour tissue after isolated limb perfusion. *Br. J. Cancer* **82**, 1000–1003.

Dewhirst, M. W., Tso, C. Y., Oliver, R., Gustafson, C. S., Secomb, T. W., and Gross, J. F. (1989). Morphologic and hemodynamic comparison of tumor and healing normal tissue microvasculature. *Int. J. Radiat. Oncol. Biol. Phys.* **17**, 91–99.

Dukor, P., and Schumann, G. (1987). Modulation of non-specific resistance by MTP-PE. *In* "Immunopharmacology of Infectious Diseases. Vaccine Adjuvants and Modulators of Non-specific Resistance of Infectious Diseases" (J. A. Majde, ed.), pp. 255–265. Alan R. Liss Inc., New York.

Dvorak, H. F., Nagy, J. A., Dvorak, J. T., and Dvorak, A. M. (1988). Identification and characterization of the blood vessels of solid tumors that are leaky to circulating macromolecules. *Am. J. Pathol.* **133**, 95–109.

Eggermont, A. M. M. (1996). Treatment of melanoma in-transit metastases confined to the limb. *Cancer Surv.* **26**, 335–349.

Eggermont, A. M. M., Schraffordt Koops, H., Lienard, D., Lejeune, F. J., and Oudkerk, M. (1994). Angiographic observations before and after high dose TNF isolated limb perfusion in patients with extremity soft tissue sarcomas. *Eur. J. Surg. Oncol.* **20**, 323–329.

Eggermont, A. M. M., Schraffordt Koops, H., Klausner, J. M., Lienard, D., Kroon, B. B. R., Schlag, P. M., Ben-Ari, G., and Lejeune, F. J. (1997). Isolation limb perfusion with tumor necrosis factor alpha and chemotherapy for advanced extremity soft tissue sarcomas. *Semin. Oncol.* **24**, 547–555.

Fajardo, L. F., Kwan, H. H., Kowalski, J., Prionas, S. D., and Allison, A. C. (1992). Dual role of tumor necrosis factor-alpha in angiogenesis. *Am. J. Pathol.* **140**, 539–544.

Fraker, D. L., and Alexander, H. R. (1994). Isolated limb perfusion with high-dose tumor necrosis factor for extremity melanoma and sarcoma. *Adv. Oncol.* 179–192.

Gabizon, A. (1993). Tailoring liposomes for cancer drug delivery: From the bench to the clinic. *Ann. Biol. Clin. (Paris)* **51**, 811–813.

Gabizon, A., and Martin, F. (1997). Polyethylene glycol-coated (pegylated) liposomal doxorubicin. Rationale for use in solid tumours. *Drugs* **54**(Suppl. 4), 15–21.

Gutterman, J. U., Rosenblum, M. G., Rios, A., Fritsche, H. A., and Quesada, J. R. (1984). Pharmacokinetic study of partially pure gamma-interferon in cancer patients. *Cancer Res.* **44**, 4164–4171.

Hieber, U., and Heim, M. E. (1994). Tumor necrosis factor for the treatment of malignancies. *Oncology* **51,** 142–153.

Hockertz, S., Franke, G., Kniep, E., and Lohmann-Matthes, M. L. (1989). Mouse interferon-gamma in liposomes; pharmacokinetics, organ-distribution, and activation of spleen and liver macrophages *in vivo. J. Interferon Res.* **9,** 591–602.

Hockertz, S., Franke, G., Paulini, I., and Lohmann-Matthes, M. L. (1991). Immunotherapy of murine visceral leishmaniasis with murine recombinant interferon-gamma and MTP-PE encapsulated in liposomes. *J. Interferon Res.* **11,** 177–185.

The International Chronic Granulomatous Disease Cooperative Study Group. (1991). A controlled trial of interferon gamma to prevent infection in chronic granulomatous disease. *N. Engl. J. Med.* **324,** 509–516.

Jain, R. K. (1989). Delivery of novel therapeutic agents in tumors: Physiological barriers and strategies. *J. Natl. Cancer Inst.* **81,** 570–576.

Kedar, E., Rutkowski, Y., Braun, E., Emanuel, N., and Barenholz, Y. (1994a). Delivery of cytokines by liposomes. I. Preparation and characterization of interleukin-2 encapsulated in long-circulating sterically stabilized liposomes. *J. Immunother. Emphasis Tumor Immunol.* **16,** 47–59.

Kedar, E., Braun, E., Rutkowski, Y., Emanuel, N., and Barenholz, Y. (1994b). Delivery of cytokines by liposomes. II. Interleukin-2 encapsulated in long-circulating sterically stabilized liposomes: Immunomodulatory and anti-tumor activity in mice. *J. Immunother. Emphasis Tumor Immunol.* **16,** 115–124.

Kedar, E., Palgi, O., Golod, G., Babai, I., and Barenholz, Y. (1997). Delivery of cytokines by liposomes. III. Liposome-encapsulated GM-CSF and TNF-alpha show improved pharmacokinetics and biological activity and reduced toxicity in mice. *J. Immunother.* **20,** 180–193.

Kim, D. W., Andres, M. L., Li, J., Kajioka, E. H., Miller, G. M., Seynhaeve, A. L. B., ten Hagen, T. L. M., and Gridley, D. S. (2001). Liposome-encapsulated tumor necrosis factor-alpha enhances the effects of radiation against human colon tumor xenografts. *J. Interferon Cytokine Res.* **21,** 885–897.

Klostergaard, J., Utsumi, T., Macatee, S., Suen, T. C., Leroux, E., Levitan, A., and Hung, M. C. (1992). Characterization and modeling of monocyte/macrophage presentation of membrane TNF. *In* "Tumor Necrosis Factor: Structure-Function Relationship and Clinical Application" (T. Osawa and B. Bonavida, eds.), pp. 101–110. Karger, Basel.

Konno, H., Yamashita, A., Tadakuma, T., and Sakaguchi, S. (1991). Inhibition of growth of rat hepatoma by local injection of liposomes containing recombinant interleukin-2. Antitumor effect of IL-2 liposome. *Biotherapy* **3,** 211–218.

Kristensen, C. A., Nozue, M., Boucher, Y., and Jain, R. K. (1996). Reduction of interstitial fluid pressure after TNF-alpha treatment of three human melanoma xenografts. *Br. J. Cancer* **74,** 533–536.

Kurzrock, R., Rosenblum, M. G., Sherwin, S. A., Rios, A., Talpaz, M., Quesada, J. R., and Gutterman, J. U. (1985). Pharmacokinetics, single-dose tolerance, and biological activity of recombinant gamma-interferon in cancer patients. *Oncology* **42,** 41–50.

Manusama, E. R., Nooijen, P. T. G. A., Stavast, J., Durante, N. M. C., Marquet, R. L., and Eggermont, A. M. M. (1996a). Synergistic antitumour effect of recombinant human tumour necrosis factor alpha with melphalan in isolated limb perfusion in the rat. *Br. J. Surg.* **83,** 551–555.

Manusama, E. R., Stavast, J., Durante, N. M. C., Marquet, R. L., and Eggermont, A. M. M. (1996b). Isolated limb perfusion with TNF alpha and melphalan in a rat osteosarcoma model: A new anti-tumour approach. *Eur. J. Surg. Oncol.* **22,** 152–157.

Melissen, P. M. B., van Vianen, W., Rijsbergen, Y., and Bakker-Woudenberg, I. A. J. M. (1991). Free versus liposome-encapsulated muramyl tripeptide phosphatidylethanolamine in treatment of experimental Klebsiella pneumoniae infection. *Infect. Immun.* **60,** 95–101.

Melissen, P. M. B., van Vianen, W., Bidjai, O., van Marion, M., and Bakker-Woudenberg, I. A. J. M. (1993). Free versus liposome-encapsulated muramyl tripeptide phosphatidylethanolamide (MTPPE) and interferon-γ (IFN-γ) in experimental infetion with Listeria monocytogenes. *Biotherapy* **6,** 113–124.

Melissen, P. M. B., van Vianen, W., Leenen, P. J. M., and Bakker-Woudenberg, I. A. J. M. (1994a). Tissue distribution and cellular distribution of liposomes encapsulating muramyltripeptide phosphatidyl ethanolamide. *Biotherapy* **7,** 71–78.

Melissen, P. M. B., van Vianen, W., and Bakker-Woudenberg, I. A. J. M. (1994b). Treatment of Klebsiella pneumoniae septicemia in normal and leukopenic mice by liposome-encapsulated muramyl tripeptide phosphatidylethanolamide. *Antimicrob. Agents Chemother.* **38,** 147–150.

Muggia, F. M. (1997). Clinical efficacy and prospects for use of pegylated liposomal doxorubicin in the treatment of ovarian and breast cancers. *Drugs* **54**(Suppl. 4), 22–29.

Nii, A., Fan, D., and Fidler, I. J. (1991). Cytotoxic potential of liposomes containing tumor necrosis factor-alpha against sensitive and resistant target cells. *J. Immunother.* **10,** 13–19.

Nooijen, P. T. G. A., Manusama, E. R., Eggermont, A. M. M., Schalkwijk, L., Stavast, J., Marquet, R. L., de Waal, R. M., and Ruiter, D. J. (1996a). Synergistic effects of TNF-alpha and melphalan in an isolated limb perfusion model of rat sarcoma: A histopathological, immunohistochemical and electron microscopical study. *Br. J. Cancer* **74,** 1908–1915.

Nooijen, P. T. G. A., Eggermont, A. M. M., Verbeek, M. M., Schalkwijk, L., Buurman, W. A., de Waal, R. M., and Ruiter, D. J. (1996b). Transient induction of E-selectin expression following TNF alpha-based isolated limb perfusion in melanoma and sarcoma patients is not tumor specific. *J. Immunother. Emphasis Tumor Immunol.* **19,** 33–44.

Nooijen, P. T., Eggermont, A. M., Schalkwijk, L., Henzen-Logmans, S., de Waal, R. M. X., and Ruiter, D. J. (1998). Complete response of melanoma-in-transit metastasis after isolated limb perfusion with tumor necrosis factor alpha and melphalan without massive tumor necrosis: A clinical and histopathological study of the delayed-type reaction pattern. *Cancer Res.* **58,** 4880–4887.

Northfelt, D. W., Dezube, B. J., Thommes, J. A., Miller, B. J., Fischl, M. A. X., Friedman-Kien, A., Kaplan, L. D., Du Mond, C., Mamelok, R. D., and Henry, D. H. (1998). Pegylated-liposomal doxorubicin versus doxorubicin, bleomycin, and vincristine in the treatment of AIDS-related Kaposi's sarcoma: Results of a randomized phase III clinical trial. *J. Clin. Oncol.* **16,** 2445–2451.

Pak, C. C., and Fidler, I. J. (1991). Liposomal delivery of biological response modifiers to macrophages. *Biotherapy* **3,** 55–64.

Parant, M., and Chedid, L. (1985). Stimulation of non-specific resistance to infections by synthetic immunoregulatory agents. *Infection* **13,** S251–255.

Pass, H. I., Mew, D. J., Kranda, K. C., Temeck, B. K., Donington, J. S., and Rosenberg, S. A. (1996). Isolated lung perfusion with tumor necrosis factor for pulmonary metastases. *Ann. Thorac. Surg.* **61,** 1609–1617.

Pennica, D., Nedwin, G. E., Hayflick, J. S., Seeburg, P. H., Derynck, R., Palladino, M. A., Kohr, W. J., Aggarwal, B. B., and Goeddel, D. V. (1984). Human tumour necrosis factor: Precursor structure, expression and homology to lymphotoxin. *Nature* **312,** 724–729.

Renard, N., Lienard, D., Lespagnard, L., Eggermont, A., Heimann, R., and Lejeune, F. (1994). Early endothelium activation and polymorphonuclear cell invasion precede specific necrosis of human melanoma and sarcoma treated by intravascular high-dose tumour necrosis factor alpha (rTNF alpha). *Int. J. Cancer* **57,** 656–663.

Renard, N., Nooijen, P. T., Schalkwijk, L., de Waal, R. M., Eggermont, A. M., Lienard, D, Kroon, B. B., Lejeune, F. J., and Ruiter, D. J. (1995). VWF release and platelet aggregation in human melanoma after perfusion with TNF alpha. *J. Pathol.* **176,** 279–287.

Ruegg, C., Yilmaz, A., Bieler, G., Bamat, J., Chaubert, P., and Lejeune, F. J. (1998). Evidence for the involvement of endothelial cell integrin $\alpha_v\beta_3$ in the disruption of the tumor vasculature induced by TNF and IFN-gamma. *Nat. Med.* **4,** 408–414.

Saito, M., Fan, D., and Lachman, L. B. (1995). Antitumor effects of liposomal IL1 alpha and TNF alpha against the pulmonary metastases of the B16F10 murine melanoma in syngeneic mice. *Clin. Exp. Metastasis* **13,** 249–259.

Schumann, G., van Hoogevest, P., Frankhauser, P., *et al.* (1989). Comparison of free and liposomal MTPPE: Pharmacological, toxicological and pharmacokinetic aspects. *In* "Liposomes in the Therapy of Infectious Disease and Cancer" (G. Lopez-Berestein and I. J. Fidler, eds.), "Symposium on Molecular and Cellular Biology, New Series," Vol. 89, pp. 191–203. UCLA, Alan R. Liss. Inc., New York.

Schwartzentruber, D. J. (1995). Biological therapy with interleukin-2: Clinical applications. Principles of administration and management of side effects. *In* "Biologic Therapy of Cancer" (V. T. J. DeVita, S. Hellman, and S. A. Rosenberg, eds.), pp. 235–249. Lippincott, Philadelphia.

Sidhu, R. S., and Bollon, A. P. (1993). Tumor necrosis factor activities and cancer therapy—a perspective. *Pharmacol. Ther.* **57,** 79–128.

Suzuki, S., Ohta, S., Takashio, K., Nitanai, H., and Hashimoto, Y. (1990). Augmentation for intratumoral accumulation and anti-tumor activity of liposome-encapsulated adriamycin by tumor necrosis factor-alpha in mice. *Int. J. Cancer* **46,** 1095–1100.

ten Hagen, T. L. M., van Vianen, W., and Bakker-Woudenberg, I. A. J. M. (1995). Modulation of nonspecific antimicrobial resistance of mice to Klebsiella pneumoniae septicemia by liposome-encapsulated muramyl tripeptide phosphatidylethanolamine and interferon-gamma alone or combined. *J. Infect. Dis.* **171,** 385–392.

ten Hagen, T. L. M., van der Veen, A. H., Nooijen, P. T. G. A., van Tiel, S. T., Seynhaeve, A. L. B., and Eggermont, A. M. M. (2000). Low-dose tumor necrosis factor-alpha augments antitumor activity of stealth liposomal doxorubicin (DOXIL) in soft tissue sarcoma-bearing rats. *Int. J. Cancer* **87,** 829–837.

ten Hagen, T. L. M., Seynhaeve, A. L. B., van Tiel, S. T., Ruiter, D. J., and Eggermont, A. M. M. (2002). Pegylated liposomal tumor necrosis factor-alpha results in reduced toxicity and synergistic antitumor activity after systemic administration in combination with liposomal doxorubicin (Doxil) in soft tissue sarcoma-bearing rats. *Int. J. Cancer* **97,** 115–120.

Utsumi, T., Hung, M. C., and Klostergaard, J. (1991). Preparation and characterization of liposomal-lipophilic tumor necrosis factor. *Cancer Res.* **51,** 3362–3366.

van der Veen, A. H., Eggermont, A. M., Seynhaeve, A. L., van Tiel, S. T., and ten Hagen, T. L. (1998). Biodistribution and tumor localization of stealth liposomal tumor necrosis factor-alpha in soft tissue sarcoma bearing rats. *Int. J. Cancer* **77,** 901–906.

van der Veen, A. H., Seynhaeve, A. L. B., Breurs, J., Nooijen, P. T. G. A., Marquet, R. L., and Eggermont, A. M. M. (1999). *In vivo* isolated kidney perfusion with TNF-alpha in tumour bearing rats. *Br. J. Cancer* **79,** 433–439.

van der Veen, A. H., de Wilt, J. H. W., Eggermont, A. M. M., van Tiel, S. T., Seynhaeve, A. L. B., and ten Hagen, T. L. M. (2000). TNF-alpha augments intratumoural concentrations of doxorubicin in TNF-alpha-based isolated limb perfusion in rat sarcoma models and enhances anti-tumour effects. *Br. J. Cancer* **82,** 973–980.

van der Veen, A. H., ten Hagen, T. L., Seynhaeve, A. L., and Eggermont, A. M. (2002). Lack of cell-cycle specific effects of tumor necrosis factor-alpha on tumor cells *in vitro*: Implications for combination tumor therapy with doxorubicin. *Cancer Invest.* **20,** 499–508.

Van Ijken, M. G., de Bruijn, E. A., de Boeck, G., ten Hagen, T. L. M., van der Sijp, J. R. M., and Eggermont, A. M. M. (1998). Isolated hypoxic hepatic perfusion with tumor necrosis factor-alpha, melphalan and mitomycin C using balloon catheter techniques: A pharmacokinetic study in pigs. *Ann. Surg.* **228,** 763–770.

Williams, J. G., Jurkovich, G. J., and Maier, R. V. (1993). Interferon-γ: A key immuno-regulatory lymphokine. *J. Surg. Res.* **54,** 79–93.

Wiltrout, R. H., Pilaro, A. M., Gruys, M. E., Taldmadge, J. E., Longo, D. L., Ortaldo, J. R., and Reynolds, C. W. (1989). Augmentation of mouse liver-associated natural killer activity by biologic response modifiers occurs largely via rapid recruitment of large granular lymphocytes from the bone marrow. *J. Immunol.* **143,** 372–378.

Wu, N. Z., Da, D., Rudoll, T. L., Needham, D., Whorton, A. R., and Dewhirst, M. W. (1993). Increased microvascular permeability contributes to preferential accumulation of Stealth liposomes in tumor tissue. *Cancer Res.* **53,** 3765–3770.

Yuan, F., Dellian, M., Fukumura, D., Leunig, M., Berk, D. A., Torchilin, V. P., and Jain, R. K. (1995). Vascular permeability in a human tumor xenograft: Molecular size dependence and cutoff size. *Cancer Res.* **55,** 3752–3756.

[8]   Glucuronate-Modified, Long-Circulating Liposomes for the Delivery of Anticancer Agents

By Naoto Oku and Yukihiro Namba

Abstract

Liposomes are useful as drug carriers in drug delivery systems, especially for drugs with severe side effects such as antitumor agents. The conventional formulations of liposomes are opsonized by plasma proteins in the bloodstream and trapped in the reticuloendothelial system (RES). Therefore, liposomes with reduced opsonization are expected to have prolonged circulation and to accumulate in tumor tissue due to the leaky endothelium of the tissue. To avoid RES trapping of liposomes, two approaches have been considered. Liposomes may mimic cells circulating in the blood to escape host recognition as foreign substances, or liposomes may be covered with a hydrophilic barrier to escape recognition. For the latter purpose, poly(ethylene glycol) is widely used. For the former purpose, here we focus on the characteristics, *in vivo* trafficking, and usage in cancer therapy of glucuronate-modified liposomes. Glucuronate-modified liposomes bind to a lower extent to macrophage-like cells *in vitro* and passively accumulate in tumor tissue evaluated by a technique using positron emission tomography. Glucuronate-modified liposomes with extended circulation are useful for delivering anticancer agents to tumors and reducing the toxic side effects of the agents.

Introduction

If a drug could be delivered to only the right site in the right concentration at the right time, we could expect more drastic therapeutic effects of the drug without side effects. Thus, drug delivery systems have been widely investigated and utilized in medical treatment. Liposomes are employed as drug carriers, since they can encapsulate both hydrophilic and hydrophobic materials, are biodegradable, and essentially are nontoxic. Liposomes are the assembly of amphiphilic molecules such as polar lipids and cholesterol, and, therefore, their size, charge, components, and modifications with various molecules are easily controlled.

In general, anticancer agents cause severe side effects. Therefore, many anticancer agents or their derivatives such as doxorubicin (Gabizon et al., 1998; Krishna and Mayer, 1997; Sadzuka and Hirota, 1998; Symon et al., 1999; Vaage et al., 1999; Vail et al., 1998; Valero et al., 1999), daunorubicin (Verdonck et al., 1998; Zucchetti et al., 1999), vincristine (Embree et al., 1998; Webb et al., 1998), paclitaxel (Cabanes et al., 1998; Scialli et al., 1997), annamycin (Zou et al., 1995), cisplatin (Working et al., 1998), camptothecin (Colbern et al., 1998; Sadzuka et al., 1998), and 5-fluorouracil (Doi et al., 1994; van Borssum et al., 1998) have been encapsulated in liposomes to reduce side effects and enhance therapeutic efficacy. Furthermore, anthracyclines and some other agents can be encapsulated with about 100% trapping efficiency in preformed liposomes by the remote-loading method (Mayer et al., 1990). Therefore, such anticancer agents are quite favorable for liposomal formulations.

Delivery of entrapped agents is well achieved by liposomal formulations when the targets are elements of the reticuloendothelial system (RES) such as the liver and spleen, since the RES readily traps most conventional liposomes. In contrast, for a broader utilization of liposomes, formulations that avoid RES would be essentially required. Such liposomes are called long-circulating liposomes, since RES trapping is responsible for the clearance of liposomes. To avoid or reduce RES trapping of liposomes, two approaches have been considered (Oku, 1999; Oku et al., 2000). Liposomes may mimic cells circulating in the blood to escape host recognition as foreign substances, or liposomes may be covered with a hydrophilic barrier to escape recognition. Long-circulating liposomes thus obtained have characteristics favorable for usage in cancer therapy: They passively accumulate in tumor tissue due to the leaky endothelium of the tissue. Furthermore, since the lymphatic system is not developed in tumor tissues, extravasated liposomes tend to reside and accumulate in the interstitial space of tumor tissues.

Here we focus on the characteristics, in vivo trafficking, and usage in cancer therapy of glucuronate-modified liposomes that have prolonged circulation. Glucuronate-modified liposomes bind to a lower extent to macrophage-like cells *in vitro*, have long-circulating and tumor-accumulating characteristics *in vivo*, and show enhanced efficacy and reduced toxicity of anticancer agents encapsulated in them.

Preparation of Glucuronate-Modified Liposomes

Glycoproteins and glycolipids, especially sialylglycoconjugates, on the surface of blood cells are thought to play an important role in recognition as self. Therefore, liposomal modification with natural glycolipids or synthetic glycoderivatives has been attempted. Allen and co-workers first observed that the modification of liposomes with monosialoganglioside G_{M1} achieved prolonged circulation in mice (Allen and Chonn, 1987). We modified liposomes with a glucuronate derivative, since glucuronic acid resembles sialic acid to some extent: Both sugars contain a carboxyl group and are biocompatible. After the development of such glycomodified liposomes, another approach for preparing long-circulating liposomes was employed using poly(ethylene glycol) (PEG), which endowed liposomes with high hydrophilicity (Allen *et al.*, 1991; Blume and Cevc, 1990; Klibanov *et al.*, 1990).

Synthesis of Palmitylglucuronide

Palmityl-D-glucuronide (PGlcUA) is prepared as follows (Oku *et al.*, 1992). Hexadecyl alcohol (4.1 g) is dissolved in 100 ml $CHCl_3$. After addition of 6.4 g of silver oxide, 2.3 g of iodine, and 60 g of activated $CaSO_4$, 2 g of methyl(2,3,4-tri-*O*-acetyl-α-D-glucopyranosyl bromide)uronate dissolved in 50 ml $CHCl_3$ is added dropwise. The mixture is stirred for 18 h, monitoring the progress of the reaction by thin-layer chromatography (TLC). After the reaction is over, the mixture is filtered, and the solvent is evaporated. The product is purified by silica gel chromatography, and the acetyl group is removed by the addition of sodium methoxide. After deacetylation, 10 ml of potassium hydroxide solution is added to hydrolyze the methyl ester. The final product, 1.5 g of PGlcUA, was identified by TLC, nuclear magnetic resonance (NMR), and IR spectrum.

Preparation of Glucuronate-Modified Liposomes

PGlcUA liposomes are prepared as follows: Dipalmitoylphosphatidylcholine (DPPC), cholesterol, and PGlcUA (4:4:1 as molar ratio) dissolved in $CHCl_3$/MeOH are dried under reduced pressure and stored *in vacuo* for at least 1 h. The lipid film is then hydrated with 0.3 M glucose or appropriate buffered solution. The resulting multilamellar vesicles are frozen and thawed three times and extruded through a polycarbonate membrane with 100-nm pore size.

Characteristics of Glucuronate-Modified Liposomes

Size Distribution and ζ-Potential

The size and charge of liposomes, as well as the rigidity of liposomal membrane, are important factors for *in vivo* behavior. Small-sized liposomes with less fluid membrane are taken up to a lower extent by macrophages (Harashima *et al.*, 1994) and have longer half-lives in the bloodstream (Oku and Namba, 1994). Liposomal size is usually determined by dynamic light scattering.

The mean size of prepared PGlcUA-liposomes is in the range of 135–145 nm, with a standard deviation of about 35 nm. The ζ-potential of PGlcUA liposomes (10 mol% PGlcUA) is −53.1 mV, and 76.8% of the liposomes have a ζ-potential between −36.0 and −69.7 mV (Namba *et al.*, 1992).

Hemolytic Activity

Although liposomes are essentially nontoxic, hemolytic activity is sometimes observed by a component of liposomes alone. Hemolytic activity is determined as follows: Heparinized blood is drawn from a healthy donor and washed three times with saline. Then, 0.5% (v/v) erythrocytes in 1 ml of saline or 50% serum are incubated for 30 min at 37° with various amounts of liposomes or liposomal components. After centrifugation, released hemoglobin is determined at 541 nm. One hundred percent hemolysis is achieved with hypotonic treatment.

PGlcUA liposomes do not show any hemolytic activity up to 10 mM as lipids. However, PGlcUA alone causes 100% hemolysis at 0.05 mM. This result indicates that PGlcUA in liposomes does not come out from the lipid bilayer (Namba *et al.*, 1992).

Stability in the Presence of Serum

Aggregation of liposomes enhances the RES trapping of liposomes. Liposomal agglutinability in the presence of serum is evaluated by turbidity change or light-scattering change. For example, liposomes prepared in 0.3 M mannitol are incubated in 50% serum for 30 min at 37° (the final concentration of liposomes is 1 mM as lipids). The turbidity of the liposomal solution is determined at 450 nm, and the relative turbidity compared with that in mannitol is calculated.

Although liposomes resemble biomembrane, they are still foreign to the host. Therefore, liposomes are recognized by the mononuclear phagocytic system (MPS) after interaction with plasma proteins. The binding of plasma proteins, such as complement components, immunoglobulins, and fibronectin, enhances the trapping of liposomes by macrophages (Bradley *et al.*, 1999; Derksen *et al.*, 1988). Chonn and co-workers developed a spin column procedure that enables the separation of proteins bound to liposomes from unbound proteins (Chonn *et al.*, 1991). They showed that the clearance of liposomes depends on the amount of proteins bound to liposomes and that the opsonization of liposomes promotes their clearance from the circulation. We used the following spin column procedure: One-milliliter tuberculin syringes plugged with 30 mg of glass wool are filled with BioGel A-15 m equilibrated with phosphate-buffered saline (PBS) and packed by centrifugation at 800 rpm for 30 s. Liposomes (0.1 ml) prepared in 0.3 *M* mannitol are incubated with 0.1 ml of serum at 37° for 30 min. Then a 0.1-ml aliquot of the reaction mixture is applied to a spin column. The column is eluted with 0.1 ml volumes of PBS by centrifugation at 500 rpm for 30 s. The liposomal fraction, i.e., the void volume, is collected, and the amounts of protein and phospholipid are determined.

Agglutinability of various kinds of liposomes including PGlcUA liposomes and PEG liposomes (both liposomes having long-circulating characteristics) in the presence of serum and serum protein binding to these liposomes are shown in Fig. 1 (Oku *et al.*, 1996a). Liposomes composed of dipalmitoylphosphatidylcholine and cholesterol with or without charged lipid are prepared in the presence of mannitol, and the turbidity change in the presence of serum is determined. Increase in turbidity is not observed for long-circulating liposomes, i.e., PGlcUA or PEG-modified liposomes. Negatively charged liposomes containing dicetyl phosphate (DCP), phosphatidylglycerol, or phosphatidylserine are also less aggregated. However, a significant turbidity increase is observed in positively charged liposomes. These liposomes bind a high amount of serum proteins as determined by separation of unbound serum proteins by use of a spin column.

Binding to Macrophage-Like Cells In Vitro

Long-circulating liposomes may be less recognized by MPS. This characteristic is examined *in vitro* by use of macrophage-like cells. J774 cells derived from mouse macrophages are seeded on 24-well plates at 1×10^6 cells/1 ml and incubated for 24 h. Liposomes (500 nmol as phospholipids) that are fluorescence labeled with N,N'-dioctadecyloxacarbocyanine *p*-toluene sulfonate (DOCS) are added to each well, and the plates are incubated for selected times at 37°. After incubation, the medium is

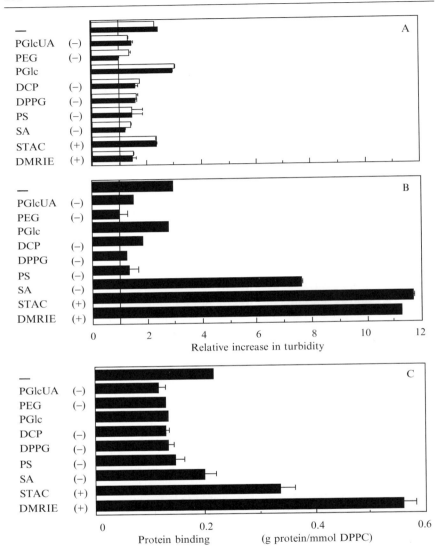

Fig. 1. Agglutinability of liposomes and serum protein binding to liposomes of various compositions. (A) Liposomes composed of dipalmitoylphosphatidylcholine (DPPC), cholesterol, and modifying lipids (10:10:1 as a molar ratio) as indicated in the figure were incubated at $4°$ for 1 week (open bars) or at $37°$ for 30 min (closed bars) in PBS. The relative turbidity at 450 nm against that in 0.3 M mannitol solution is shown. (B) Similar procedures as used in A were performed, except that the liposomes were incubated at $37°$ for 30 min in the presence of 50% fetal bovine serum (FBS). The net charge of liposomes is shown in parentheses. (C) Liposomes were incubated at $37°$ for 30 min in the presence of 50% FBS. Then the spin column procedure

pipetted off, and the wells are rinsed three times with ice-cold PBS for removal of unbound liposomes. Liposomes bound to J774 cells are measured fluorometrically after solubilization of the cells with reduced Triton X-100.

Figure 2 shows the time-dependent association (binding and internalization) of PGlcUA liposomes and dipalmitoylphosphatidylglycerol (DPPG) liposomes (control) to J774 cells. PGlcUA liposomes did not associate well with the macrophage-derived cells (Namba *et al.*, 1992).

In Vivo Behavior of Glucuronate-Modified Liposomes

Biodistribution

Long-circulating liposomes are expected to reside in the bloodstream for extended periods in normal animals and to accumulate in tumor tissues in tumor-bearing animals. Biodistribution of liposomes is determined by

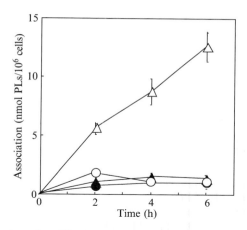

FIG. 2. Binding and uptake of liposomes containing DOCS by macrophage-derived J774 cells. Liposomes (500 nmol as phospholipids) composed of DPPC, cholesterol, and modifying lipids (PGlcUA or DPPG) (4:4:1 as molar ratio) were incubated at 4° (closed symbols) or 37° (open symbols) for the indicated times with J774 cells. Circles, PGlcUA liposomes; triangles, DPPG liposomes.

was performed for separating the liposomal fraction from unbound serum proteins. The amount of protein binding to liposomes was determined. PGlc, palmityl glucose; DCP, dicetyl phosphate; DPPG, dipalmitoylphosphatidylglycerol; PS, phosphatidylserine; SA, stearylamine; STAC, stearyltrimethylammonium chloride; DMRIE, 1,2-dimyristyloxypropyl-3-dimethylhydroxyethyl bromide; PGlcUA, palmityl glucuronide; PEG, poly(ethylene glycol)-conjugated distearoylphosphatidylethanolamine.

using radiolabeled liposomes, employing either radioactive lipophilic compounds as a membrane marker or hydrophilic compounds as internal aqueous space markers. Cholesteryl ether or ester is usually used as a liposomal
membrane marker, since they do not exchange rapidly with blood or tissue
lipids.

To determine the biodistribution of PGlcUA liposomes, [^{14}C]cholesteryl
oleate and [^{3}H]inulin are used as membrane and internal aqueous space
markers, respectively. To determine the biodistribution of PGlcUA and
control DPPG liposomes in normal mice, ddY male mice are injected in the
tail vein with liposomes (0.3 mmol lipids). At a selected time after administration, mice are sacrificed under anesthesia, and the radioactivity in each organ is
determined. To determine the biodistribution of the liposomes in tumor-
bearing mice, ddY male mice are inoculated subcutaneously into the hind
leg with 10^7 S180 tumor cells, and mice are injected with liposomes similarly to
normal mice when the local tumor weight reaches 1–2 g. Figure 3 shows
the distribution of [^{3}H]inulin encapsulated in liposomes 12 h after injection
(Oku *et al.*, 1992). As expected, liposomes stayed in the bloodstream longer in

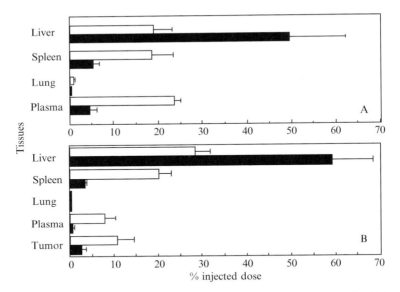

FIG. 3. Biodistribution of liposomal inulin in normal and tumor-bearing mice 12 h after
injection. [^{3}H]Inulin was encapsulated in liposomes composed of DPPC, cholesterol, and
modifying lipids (PGlcUA or DPPG) (4:4:1 as a molar ratio). Normal mice (A) or tumor-
bearing mice (B) were injected with PGlcUA liposomes (open bar) or DPPG liposomes
(closed bar), and biodistribution of [^{3}H]inulin was determined 12 h after injection. Data show
the percent injected dose per tissue and SD.

normal mice and accumulated in tumor tissue in tumor-bearing mice. Therefore, PGlcUA-modified long-circulating liposomes are expected to be useful
as a passive targeting tool for anticancer agents.

Liposomal Stability in the Bloodstream

Liposomal stability in the bloodstream is determined with double-
labeled liposomes using both a membrane marker and an internal aqueous
marker. If the liposomal integrity is maintained, the ratio of both markers
would not be changed.

PGlcUA and DPPG liposomes are double-labeled with [14C]cholesteryl
oleate and [3H]inulin and injected into ddY mice. Blood of the animals is
collected at selected times, and the ratio of both radioisotopes is determined. The ratio did not change at least up to 12 h, indicating that the
structural integrity of liposomes is maintained during this time.

In Vivo Trafficking

Liposomal behavior *in vivo* has usually been evaluated invasively in
sacrificed animals after injection of radiolabeled liposomes. A method for
studying the real-time trafficking of liposomes *in vivo* was developed by use
of positron emission tomography (PET) (Oku *et al.*, 1995). This technique
enabled real-time trafficking of liposomes to be evaluated, especially during the early time points just after administration. A scheme of the PET
study is shown in Fig. 4.

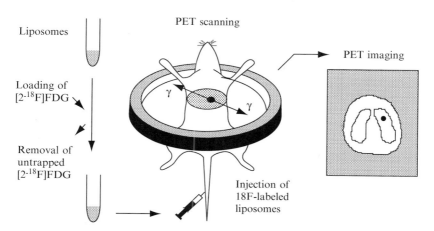

FIG. 4. PET analysis for liposomal trafficking. Liposomes are loaded with [2-18F]fluor-
odeoxyglucose ([2-18F]FDG) and injected into mice. Then the change in the distribution of
the positron is monitored by a PET camera.

Liposomes are positron labeled as follows: A thin lipid film is hydrated with 1.0 ml of 0.45 M sodium gluconate. This liposomal solution is then mixed with 2.0 ml of a [2-^{18}F]-2-fluoro-2-deoxy-D-glucose ([2-^{18}F]FDG) solution and freeze-thawed for three cycles using liquid nitrogen to encapsulate the positron-emitter–containing chemical into the liposomes. Next, the liposomes are extruded three times through a polycarbonate membrane filter (100-nm pore size), washed by centrifugation at 180,000g for 20 min after dilution with saline to remove the untrapped [2-^{18}F]FDG, and resuspended in saline. The encapsulation efficiency is about 10% in all preparations. Liposomal trafficking in normal or tumor-bearing mice is analyzed by PET. In the case of tumor-bearing mice, syngeneic mice are subcutaneously inoculated into their right posterior flank with 10^6 Meth A sarcoma cells 7 days before analysis. Liposomes are injected via the tail vein at a dose of about 1.85 MBq as [2-^{18}F]FDG radioactivity. The emission scan is started immediately after injection and performed for 120 min. The radioactivity in the form of coincidence γ photons is measured and converted to Bq/cm^3 of tissue volume by calibration after correction for decay and attenuation. After composition of the PET images, time–activity curves are obtained from the mean pixel radioactivity in the region of interest (ROI) of the images, where ROIs covered the image of specific organs.

Figure 5 shows the PET images composed during the first 30-min accumulation of PGlcUA liposomes; the corresponding X-ray images taken with age-matched mice are also shown. Intense accumulation shown in the eighth slice indicates the accumulation of liposomes in the heart, suggesting the high retention of the liposomes in the bloodstream. By use of this method, *in vivo* trafficking of PGlcUA and PEG liposomes as well as DPPG liposomes as a control was examined (Oku *et al.*, 1996b). The images of tumor accumulation of long-circulating liposomes and control DPPG liposomes are shown in Fig. 5. Figure 6 shows the time–activity curves of liposomal accumulation in liver, spleen, and tumor, as well as the heart, which may reflect the blood pool. The trafficking of long-circulating liposomes was different from that of conventional liposomes. For example, hepatic accumulation of long-circulating liposomes decreased time dependently, whereas that of the conventional DPPG liposomes increased for 40 min after injection. The initial increase in the hepatic accumulation of DPPG liposomes might have been due to RES trapping, and the following decrease might be due to the release of [2-^{18}F]FDG by metabolic breakdown of the liposomes. PGlcUA liposomes could avoid liver trapping most efficiently, although splenic accumulation of these liposomes was the highest among the three long-circulating liposomes. Tumor accumulation of liposomes was obvious immediately after injection for both long-circulating liposomes. The accumulation of liposomes in the heart may reflect the

PET

X-ray

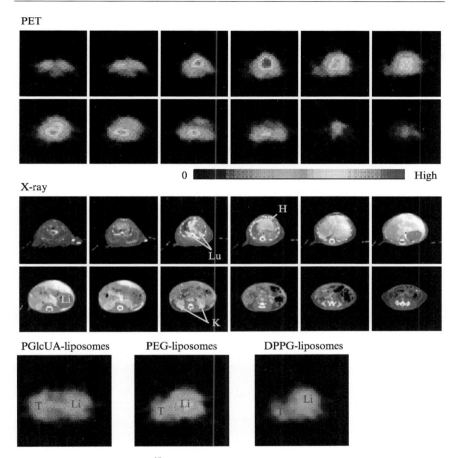

FIG. 5. Biodistibution of [2-^{18}F]FDG-labeled PGlcUA liposomes imaged by PET. PGlcUA liposomes composed of DPPC, cholesterol, and PGlcUA (4:4:1 as a molar ratio) were labeled with [2-^{18}F]FDG and extruded through a polycarbonate membrane with 100-nm pores. The [2-^{18}F]FDG-labeled liposomes were injected intravenously into 7-week-old BALB/c male mice, and the emission scan was performed immediately after injection. PET images (upper panel) and corresponding X-ray images (middle panel) are shown where the slice aperture is 3.25 mm from the head (upper left) to the tail (lower right). Lu, lung; H, heart; Li, liver; K, kidney. PGlcUA-and PEG-modified liposomes, as well as DPPG liposomes, were labeled with [2-^{18}F]FDG and injected into BALB/c male mice bearing Meth A sarcoma. The tumor accumulation of these liposomes is shown in the lower panel. (See color insert.)

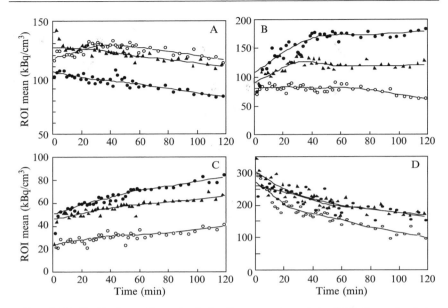

FIG. 6. Biodistribution of long-circulating liposomes imaged by PET. Long-circulating liposomes, namely, PGlcUA- and PEG-modified liposomes, as well as DPPG liposomes, were labeled with [2-^{18}F]FDG and injected into BALB/c male mice bearing Meth A sarcoma. Emission scan was performed for 120 min immediately after injection. Time–activity curves of ^{18}F in liver (A), spleen (B), tumor (C), and heart (D) were obtained after injection of PGlcUA liposomes (●), PEG liposomes (▲), and conventional DPPG liposomes (○). Fitting curves were calculated based on the method of weighed least squares.

liposomal distribution in the blood. In fact, high radioactivity was observed in blood but not in the residual heart after removal of the blood once the animal had been sacrificed and the radioactivity of each organ was determined by a γ-counter. Figure 6 indicates that long-circulating liposomes reside in the bloodstream longer than conventional liposomes. As a result, their accumulation in tumor would be distinct after a longer period of time, although their initial accumulation in the tumor was higher than that of conventional liposomes.

Glucuronate-Modified Liposome-Mediated Cancer Treatment

Liposomal encapsulation promises long circulation of drugs together with reduced side effects. Furthermore, long-circulating liposomes tend to accumulate in tumor tissues by passive targeting. The therapeutic effect of Adriamycin (ADM) encapsulated in PGlcUA liposomes has been examined by use of glucuronate-modified liposomes.

Adriamycin-Loading Method

The remote-loading method, which enables the encapsulation of drugs in preformed liposomes with almost 100% trapping efficiency, was originally developed by Mayer *et al.* (1990). A lipid film (DPPC/cholesterol/PGlcUA or DPPG = 4:4:1) is hydrated with 0.3 M sodium citrate (pH 4.0). After the liposomes are freeze-thawed for three cycles and sized with a 100-nm-pore filter, the pH outside of the liposomes is adjusted to 7.5 by the addition of sodium bicarbonate, and the suspension is then diluted with HEPES-buffered saline (pH 7.5). ADM loading is performed by incubation of the liposomal solution with ADM for 1 h at 60°. Encapsulation efficiency was more than 90% as determined by the absorbance of the supernatant at 480 nm after centrifugation of the liposomal suspension (100,000g for 5 min) in the presence or absence of reduced Triton X-100 (Aldrich, Milwaukee, WI). Untrapped ADM is removed by gel filtration through PD-10 (Pharmacia).

Stability of ADM-Encapsulated Liposomes

Release of ADM from liposomes in the presence or absence of serum is examined to evaluate liposomal stability. ADM-encapsulating liposomes are incubated in PBS (pH 7.2) or in 90% fetal bovine serum at 37°. At selected times, a 0.4-ml sample is removed and centrifuged at 100,000g for 5 min to pellet the liposomes. An aliquot of the supernatant is collected, and its ADM content is determined (absorbance at 480 nm). ADM release was not observed in PBS at least up to 48 h, and about 10% of ADM was released from PGlcUA and DPPG liposomes at the beginning of the incubation. However, further release of ADM was not observed for at least 48 h.

Biodistribution of Adriamycin Delivered by Glucuronate-Modified Liposomes

Meth-A sarcoma cells (1×10^6 cells/0.2 ml) are carefully injected subcutaneously into the posterior flanks of 5-week-old BALB/c male mice. Ten days after tumor implantation, PGlcUA liposomes encapsulating ADM or ADM alone (0.2 mg as ADM/mouse) are injected into the tail vein. At selected times after injection, the animals are sacrificed, and 100 mg of plasma and various organs are homogenized with 3 ml of 0.3 M HCl/50% methanol. After centrifugation (850g for 10 min and 8000g for 15 min), the fluorescence intensity of the supernatant (excitation 470 nm and emission 590 nm) is determined. As shown in Fig. 7, ADM concentration in the tumor was increased and that in heart, the main site for the side effect of ADM, was decreased when ADM was delivered by PGlcUA liposomes.

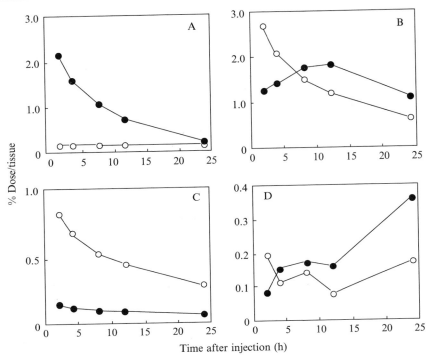

FIG. 7. Biodistribution of Adriamycin delivered by glucuronate-modified liposomes. PGlcUA liposomes encapsulating ADM (●) or ADM alone (○) were injected into the tail vein of Meth A sarcoma-bearing mice. At a selected time after injection, plasma and the removed organ were homogenized, and ADM was extracted. ADM concentration in plasma and various organs was determined fluorometrically.

Therapeutic Efficacy of Adriamycin Delivered by Glucuronate-Modified Liposomes

Since long-circulating liposomes tend to accumulate passively in tumor tissues, the therapeutic efficacy of anticancer agents is expected to increase with the long-circulating liposomal formulation. However, various reports and our experience indicate that the advantage of the liposomalization of anticancer drugs is more obvious in reducing side effects than in enhancing therapeutic efficacy.

Meth A sarcoma-bearing mice and ADM-loaded PGlcUA liposomes are prepared as described above. Liposomes and drugs are injected intravenously with various administration schedules into the tumor-bearing mice. The weight of each mouse and the size of the tumor are monitored every day after administration of the drugs. To determine the tumor volume, two bisecting diameters of each tumor are measured with slide calipers, and the formula $0.4 \, (a \times b^2)$ is employed where a is the largest diameter and b is the smallest. In several experiments, the animals are sacrificed, and the tumors are removed for weighing. Tumor volume calculated by means of the formula provides a good correlation with actual tumor weight ($r = 0.980$).

As shown in Fig. 8, ADM encapsulated in long-circulating liposomes effectively reduced tumor growth and prolonged the survival time of Meth A sarcoma-bearing mice after a two-dose treatment (Oku *et al.*, 1994). We used 15 mg/kg ADM in each injection, which is slightly lower than the LD_{50}. The reduction in toxicity of ADM due to liposomal formulation was obvious, since all mice in the group treated with free ADM died after the second administration, whereas none of the mice in the groups treated with liposomal formulation did. We next tried to obtain a complete cure of the animals with multidose administration of a PGlcUA liposomal formulation. The tumor volumes transiently increased until day 8 and then tended to decrease; some of them actually reached zero after day 20. Forty percent of the animals that were apparently tumor free at day 33 remained alive throughout the experimental period, which lasted 180 days. Mean survival of the other 60% of the ADM-PGlcUA liposome-treated animals was 67.2 ± 12.6 days, which was 2.3-fold longer than the survival of the saline treatment group (28.8 ± 2.7 days). These data show that encapsulation of ADM in PGlcUA liposomes that have long-circulating ability significantly increased

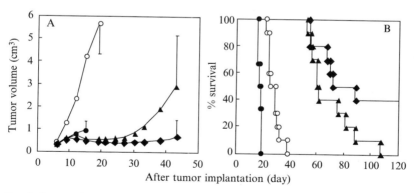

FIG. 8. Suppression of tumor growth by treatment with PGlcUA liposomal Adriamycin. Meth A sarcoma was implanted subcutaneously into the posterior flanks of BALB/c mice. These mice were then injected intravenously with saline (○), free ADM (●), or ADM encapsulated in PGlcUA liposomes (▲) at days 5 and 12. In each injection, 15 mg/kg ADM was given. Data for the four-dose treatment of ADM encapsulated in PGlcUA liposomes at days 5, 12, 19, and 29 are also shown (◆). (A) Tumor volume change; (B) survival.

the therapeutic efficacy of the drug against growing tumors compared with the free drug or drug encapsulated in conventional liposomes.

Reduced side effects and enhanced therapeutic efficacy of anticancer drugs by PGlcUA liposomalization were also observed when vincristine was used (Tokudome *et al.*, 1996). Taken together, PGlcUA liposomes have the potential of practical usage as drug carriers of anticancer agents in the field of cancer treatment.

Conclusions

Liposomes are thought to be useful carriers of drugs in drug delivery systems. However, the trapping of liposomes in the reticuloendothelial system mainly causes the clearance of liposomes from the circulation after opsonization with serum proteins. Therefore, liposomes with reduced opsonization are expected to have long-circulating characteristics and to accumulate in tumor tissues, due to the leaky endothelium of these tissues. The evaluation of the *in vivo* behavior of liposomes by PET analysis indicates the passive targeting ability of long-circulating liposomes. Glucuronate-modified liposomes having long-circulating characteristic are actually useful for the delivery of anticancer agents to the tumor site, and they can reduce the toxic side effects of the agents.

Acknowledgments

The authors thank Ms. Kanako Doi Kawaguchi and Dr. Yoshihiro Tokudome for technical assistance, and Dr. Hideo Tsukada and members of the PET Center at Hamamatsu Photonics, K.K. for supporting PET analysis.

References

Allen, T. M., and Chonn, A. (1987). Large unilamellar liposomes with low uptake into the reticuloendothelial system. *FEBS Lett.* **223**, 42–46.

Allen, T. M., Hansen, C., Martin, F., Redeman, C., and Yau-Yong, A. (1991). Liposomes containing synthetic lipid derivatives of poly(ethylene glycol) show prolonged circulation half-lives *in vivo. Biochim. Biophys. Acta* **1066**, 29–36.

Blume, G., and Cevc, G. (1990). Liposomes for the sustained drug release *in vivo. Biochim. Biophys. Acta* **1029**, 91–97.

Bradley, A. J., Brooks, D. E., Norris- Jones, R., and Devine, D. V. (1999). C1q binding to liposomes is surface charge dependent and is inhibited by peptides consisting of residues 14–26 of the human C1qA chain in a sequence independent manner. *Biochim. Biophys. Acta* **1418**, 19–30.

Cabanes, A., Briggs, K. E., Gokhale, P. C., Treat, J. A., and Rahman, A. (1998). Comparative *in vivo* studies with paclitaxel and liposome-encapsulated paclitaxel. *Int. J. Oncol.* **12**, 1035–1040.

Chonn, A., Semple, S. C., and Cullis, P. R. (1991). Separation of large unilamellar liposomes from blood components by a spin column procedure: Towards identifying plasma proteins which mediate liposome clearance *in vivo*. *Biochim. Biophys. Acta* **1070**, 215–222.

Colbern, G. T., Dykes, D. J., Engbers, C., Musterer, R., Hiller, A., Pegg, E., Saville, R., Weng, S., Luzzio, M., Uster, P., Amantea, M., and Working, P. K. (1998). Encapsulation of the topoisomerase I inhibitor GL147211C in pegylated (STEALTH) liposomes: Pharmacokinetics and antitumor activity in HT29 colon tumor xenografts. *Clin. Cancer Res.* **4**, 3077–3082.

Derksen, J. T. P., Morselt, H. W. M., and Scherphof, G. L. (1988). Uptake and processing of immunoglobulin-coated liposomes by subpopulations of rat liver macrophages. *Biochim. Biophys. Acta* **971**, 127–136.

Doi, K., Oku, N., Toyota, T., Shuto, S., Sakai, A., Ito, H., and Okada, S. (1994). Therapeutic effect of reticuloendothelial system (RES)-avoiding liposomes containing a phospholipid analogue of 5-fluorouracil, dipalmitoylphosphatidylfluorouridine, in Meth A sarcoma-bearing mice. *Biol. Pharm. Bull.* **17**, 1414–1416.

Embree, L., Gelmon, K., Tolcher, A., Hudon, N., Heggie, J., Dedhar, C., Logan, P., Bally, M. B., and Mayer, L. D. (1998). Pharmacokinetic behavior of vincristine sulfate following administration of vincristine sulfate liposome injection. *Cancer Chemother. Pharmacol.* **41**, 347–352.

Gabizon, A., Goren, D., Cohen, R., and Barenholz, Y. (1998). Development of liposomal anthracyclines: from basics to clinical applications. *J. Control Release* **53**, 275–279.

Harashima, H., Sakarta, K., Funato, K., and Kiwada, H. (1994). Enhanced hepatic uptake of liposomes through complement activation depend on the size of liposomes. *Pharm. Res.* **11**, 402–406.

Klibanov, A. L., Maruyama, K., Torchilin, V. P., and Huang, L. (1990). Amphipathic polyethyleneglycols effectively prolong the circulation time of liposomes. *FEBS Lett.* **268**, 235–237.

Krishna, R., and Mayer, L. D. (1997). Liposomal doxorubicin circumvents PSC 833-free drug interactions, resulting in effective therapy of multidrug-resistant solid tumors. *Cancer Res.* **57**, 5246–5253.

Mayer, L. D., Tai, L. C. L., Bally, M. B., Mitilenes, G. N., Ginsberg, R. S., and Cullis, P. R. (1990). Characterization of liposomal systems containing doxorubicin entrapped in response to pH gradients. *Biochim. Biophys. Acta* **1025**, 143–151.

Namba, Y., Oku, N., Ito, F., Sakakibara, T., and Okada, S. (1992). Liposomal modification with uronate, which endows liposomes with long circulation *in vivo*. Reduces the uptake of J774 cells *in vitro*. *Life Sci.* **50**, 1773–1779.

Oku, N. (1999). Anticancer therapy using glucuronate modified long-circulating liposomes. *Adv. Drug Delivery Rev.* **40**, 63–73.

Oku, N., and Namba, Y. (1994). Long-circulating liposomes. *Crit. Rev. Ther. Drug Carrier Syst.* **11**, 231–270.

Oku, N., Namba, Y., and Okada, S. (1992). Tumor accumulation of novel RES-avoiding liposomes. *Biochim. Biophys. Acta* **1126**, 255–260.

Oku, N., Doi, K., Namba, Y., and Okada, S. (1994). Therapeutic effect of adriamycin encapsulated in long-circulating liposomes on Meth-A-sarcoma-bearing mice. *Int. J. Cancer* **58**, 415–419.

Oku, N., Tokudome, Y., Tsukada, H., and Okada, S. (1995). Real-time analysis of liposomal trafficking in tumor-bearing mice by use of positron emission tomography *Biophys. Acta* **1238**, 86–90.

Oku, N., Tokudome, Y., Namba, Y., Saito, N., Endo, M., Hasagawa, Y., Kawai, M., Tsukada, H., and Okada, S. (1996a). Effect of serum protein binding on real-time trafficking of liposomes with different charges analyzed by positron emission tomography. *Biochim. Biophys. Acta* **1280**, 149–154.

Oku, N., Tokudome, Y., Tsukada, H., Kosugi, T., Namba, Y., and Okada, S. (1996b). *In vivo* trafficking of long-circulating liposomes in tumour-bearing mice determined by positron emission tomography. *Biopharm. Drug Dispos.* **17**, 435–441.

Oku, N., Tokudome, Y., Asai, T., and Tsukada, H. (2000). Evaluation of drug targeting strategies and liposomal trafficking. *Curr. Pharm. Des.* **6**, 1669–1691.

Sadzuka, Y., and Hirota, S. (1998). Does the amount of an antitumor agent entrapped in liposomes influence its tissue distribution and cell uptake? *Cancer Lett.* **131**, 163–170.

Sadzuka, Y., Hirotsu, S., and Hirotu, S. (1998). Effect of liposomalization on the antitumor activity, side-effects and tissue distribution of CPT-11. *Cancer Lett.* **127**, 99–106.

Scialli, A. R., Waterhouse, T. B., Desesso, J. M., Rahman, A., and Goeringer, G. C. (1997). Protective effect of liposome encapsulation on paclitaxel developmental toxicity in the rat. *Teratology* **56**, 305–310.

Symon, Z., Peyser, A., Tzemach, D., Lyass, O., Sucher, E., Shezen, E., and Gabizon, A. (1999). Selective delivery of doxorubicin to patients with breast carcinoma metastases by stealth liposomes. *Cancer* **86**, 72–78.

Tokudome, Y., Oku, N., Doi, K., Namba, Y., and Okada, S. (1996). Antitumor activity of vincristine encapsulated in glucuronide-modified long-circulating liposomes in mice bearing Meth A sarcoma. *Biochim. Biophys. Acta* **1279**, 70–74.

Vaage, J., Donovan, D., Wipff, E., Abra, R., Colbern, G., Uster, P., and Working, P. (1999). Therapy of a xenografted human colonic carcinoma using cisplatin or doxorubicin encapsulated in long-circulating pegylated stealth liposomes. *Int. J. Cancer* **80**, 134–137.

Vail, D. M., Chun, R., Thamm, D. H., Garrett, L. D., Cooley, A. J., and Obradovich, J. E. (1998). Efficacy of pyridoxine to ameliorate the cutaneous toxicity associated with doxorubicin containing pegylated (Stealth) liposomes: A randomized, double-blind clinical trial using a canine model. *Clin. Cancer Res.* **4**, 1567–1571.

Valero, V., Buzdar, A. U., Theriault, R. L., Azarnia, N., Fonseca, G. A., Willey, J., Ewer, M., Walters, R. S., Mackay, B., Podoloff, D., Booser, D., Lee, L. W., and Hortobagyi, G. N. (1999). Phase II trial of liposome-encapsulated doxorubicin, cyclophosphamide, and fluorouracil as first-line therapy in patients with metastatic breast cancer. *J. Clin. Oncol.* **17**, 1425–1434.

van Borssum Waalkes, M., Goris, H., Dontje, B. H., Schwendener, R. A., Scherphof, G., and Nijhof, W. (1998). Toxicity of liposomal 3′-5′-O-dipalmitoyl-5-fluoro-2′-deoxyuridine in mice. *Anticancer Drug Des.* **13**, 291–305.

Verdonck, L. F., Lokhorst, H. M., Roovers, D. J., and van Heugten, H. G. (1998). Multidrug-resistant acute leukemia cells are responsive to prolonged exposure of daunorubicin: Implications for liposome-encapsulated daunorubicin. *Leuk. Res.* **22**, 249–256.

Webb, M. S., Saxon, D., Wong, F. M., Lim, H. J., Wang, Z., Bally, M. B., Choi, L. S., Cullis, P. R., and Mayer, L. D. (1998). Comparison of different hydrophobic anchors conjugated to poly(ethylene glycol): Effects on the pharmacokinetics of liposomal vincristine. *Biochim. Biophys. Acta* **1372**, 272–282.

Working, P. K., Newman, M. S., Sullivan, T., Brunner, M., Podell, M., Sahenk, Z., and Turner, N. (1998). Comparative intravenous toxicity of cisplatin solution and cisplatin encapsulated in long-circulating, pegylated liposomes in cynomolgus monkeys. *Toxicol. Sci.* **46**, 155–165.

Zou, Y., Priebe, W., Stephens, L. C., and Perez-Soler, R. (1995). Preclinical toxicity of liposome-incorporated annamycin: Selective bone marrow toxicity with lack of cardio-toxicity. *Clin. Cancer Res.* **1**, 1369–1374.

Zucchetti, M., Boiardi, A., Silvani, A., Parisi, I., Piccolrovazzi, S., and D'Incalci, M. (1999). Distribution of daunorubicin and daunorubicinol in human glioma tumors after administration of liposomal daunorubicin. *Cancer Chemother. Pharmacol.* **44**, 173–176.

[9] Liposomalized Oligopeptides in Cancer Therapy

By Tomohiro Asai and Naoto Oku

Abstract

Organ-specific delivery of biofunctional agents is thought to enhance their activity and to reduce their side effects. Liposomes have been used as drug carriers in cancer chemotherapy, since they accumulate passively in tumor tissues due to an enhanced permeability and retention (EPR) effect. In addition, modification of liposomes with specific ligands enables active targeting. A small peptide having a high affinity for a certain antigen is suitable for modification of liposomes, since it is biocompatible, biodegradable, and less antigenic compared with antibody and other modifiers. Oligopeptide-modified liposomes are prepared by using lipophilic derivatives of the peptide, which are synthesized easily and incorporated readily into the liposomal bilayer. We describe two examples of the use of liposomal oligopeptides: one for antimetastatic therapy and the other for antineovascular therapy. Arg-Gly-Asp (RGD)-related peptides are known to contribute various cellular functions such as adhesion and invasion and to inhibit tumor metastasis. However, peptide drugs are generally rapidly hydrolyzed and eliminated from the bloodstream. Liposomal RGD enables the half-lives and affinity to be improved, resulting in enhancement of antimetastatic activity. We then describe the usefulness of liposomal Ala-Pro-Arg-Pro-Gly (APRPG) for tumor treatment, which is specific for tumor angiogenic vessels. APRPG is originally isolated by use of a phage-displayed peptide library. Adriamycin encapsulated in APRPG-modified liposomes accumulated specifically in and damage tumor neovessels, resulting in notable antitumor efficacy.

Introduction

Organ-selective targeting of therapeutic agents may enhance their activity and reduce their side effects. Liposomes are thought to be an ideal drug carrier in targeted cancer chemotherapy. Liposomes have a tendency to passively accumulate in tumor tissues due to the enhanced permeability and retention (EPR) effect (Asai *et al.*, 1998; Oku, 1999a,b; Oku *et al.*, 1992, 1994). In addition, modification of liposomes with specific ligands (such as antibody, glycoconjugate, and oligopeptide) enables active targeting to tumor tissues (Arap *et al.*, 1998; Huang *et al.*, 1997; Viti *et al.*, 1999).

In particular, small peptides having high affinity against certain antigens are suitable for modification of liposomes, since they are biocompatible, biodegradable, and less antigenic compared with antibody and other modifiers. Furthermore, liposomalization of oligopeptides requires a simple technique involving a lipid derivative of the oligopeptides that is synthesized easily and incorporated readily into the liposomal bilayer (Oku *et al.*, 1996; Takikawa *et al.*, 2000). Although most oligopeptides generally have short half-lives in the bloodstream, liposomalization of oligopeptides could enhance their stability and retention.

Here, we present the application of liposomal oligopeptides for antimetastatic therapy and for antineovascular therapy. First, we show the liposomalization of Arg-Gly-Asp (RGD)-containing peptides that contribute to various cellular functions such as adhesion and invasion (Ruoslahti and Pierschbacher, 1987). Then, we describe the inhibitory effect of these liposomes against experimental metastasis. We next present the isolation of oligopeptides specific for tumor angiogenic vessels by using a phage-displayed peptide library (Gho *et al.*, 1997; Ishikawa *et al.*, 1998; Koivunen *et al.*, 1993; Martens *et al.*, 1995; Scott and Smith, 1990), preparation of liposomal oligopeptides, and targeting efficacy of liposomal oligopeptides. Finally, we show data on antineovascular therapy using liposomal oligopeptides.

Inhibitory Effect of Liposomal RGD on Experimental Metastasis

The amino acid sequence RGD, which was originally found within the fibronectin molecule, has been found to be present in a number of extracellular molecules. The sequence inhibits adhesion, invasion, and tumor cell–induced platelet aggregation, which support metastasis and also inhibits microvessel development during angiogenesis (Eliceiri and Cheresh, 1999). Therefore, many attempts have been made to suppress tumor metastasis by use of RGD and related peptides, and a number of reports have shown the antimetastatic activity of these peptides (Kurohane *et al.*, 2000; Oku *et al.*, 1996). Since most peptides including RGD peptides generally have short half-lives in the circulation, which may result in a decrease in their biological and therapeutic properties, a high dose of RGD and related peptides is required to obtain acceptable antimetastatic effects. We have grafted hydrophobic molecules onto RGD-related peptides, incorporated the resulting RGD analogs into liposomal membranes, and investigated the antimetastatic property of such liposomes (Oku *et al.*, 1996). In this configuration, one liposome could carry several thousand core sequences. The results indicated that liposomal RGD-peptide analogues are effective in suppressing lung colonization of B16BL6 melanoma cells in experimental tumor metastasis.

Synthesis of RGD Analogues

The RGD analogues used in this experiment are shown in Fig. 1. These analogues are synthesized as follows. Protected (G)RGD(S) peptides are prepared by conventional methods of liquid phase synthesis. Phytanoic acid is prepared from phytol by hydrogenation and ruthenium oxidation. A fatty acid is grafted to benzylated (G)RGD(S) by the addition of dicyclocarbodiimide (DCC), and blocking agents are removed by hydrolysis. For the synthesis of the LAP-5 analogue, 1,2-O-diphytanyl-sn-glycerol (obtained by hydrogenolysis of 3-benzyl-sn-glycerol etherified with phytanyl iodide) and N-Boc ethanolamine are condensed with phenylphosphoryl di(3-nitro-1,2,4-triazole). After removal of the Boc group, RGD peptide is incorporated into the molecule by the DCC-HOBt method. The resulting RGD analogues were purified by silica gel chromatography, deprotected by TFA, hydrogenolyzed with Adams' catalyst, and analyzed by ^{1}H-NMR.

Preparation of Liposomal RGD

Distearoylphosphatidylcholine (DSPC) and cholesterol are used as the main liposomal components, since liposomes with rather solid membranes are known to have long circulation time in the bloodstream. For the assay of experimental metastasis, liposomes composed of DSPC, cholesterol,

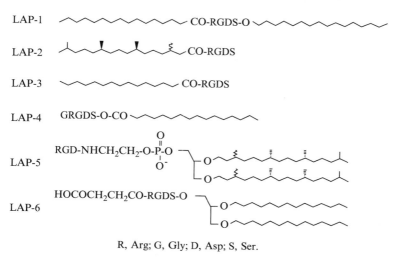

R, Arg; G, Gly; D, Asp; S, Ser.

FIG. 1. Structure of synthesized RGD analogues.

dipalmitoylphosphatidylglycerol (DPPG), and RGD-compound (20/20/5/6, as a molar ratio) are used. First, these components are dissolved in chloroform/methanol and dried under reduced pressure. The resulting thin lipid film is hydrated with 1 ml of phosphate-buffered saline (PBS) to produce RGD-bearing liposomes. After three cycles of freeze-thawing with liquid nitrogen, the liposomes are extruded through a 100-nm pore polycarbonate membrane filter to obtain rather homogeneously sized unilamellar RGD-liposomes. The resulting liposomes are centrifuged at 100,000g for 10 min to remove the unincorporated RGD analogs.

Inhibitory Effect of Liposomal RGD on Metastasis

One hundred microliters of B16BL6 (2×10^5 cells) suspension is coinjected with 100 μl of liposomal RGD solution (0.6 μmol RGD analogues equivalent to ~200 μg as RGD moiety) into the tail vein of 5-week-old C57BL/6 male mice. Fourteen days later, the mice are sacrificed under anesthesia, and the metastatic colonies in the lungs are counted. Figure 2

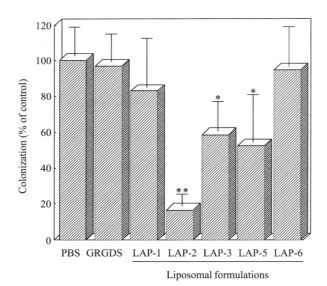

FIG. 2. Effect of liposomal RGD on experimental lung metastasis. Liposomal RGD (0.6 μmol RGD derivative equivalent to ~.0.2 mg as RGD moiety/mouse) and free GRGDS were coinjected with B16BL6 cells (1×10^5 cells/mouse) into the tail vein of male mice ($n = 5$). At day 14, the mice were sacrificed under anesthesia, and the metastatic colonies in the lungs were examined. Colony numbers are expressed as percent of control $\pm$ SD, where the colony number of control was 94.8 $\pm$ 14.9. *$p < 0.05$, **$p < 0.01$ vs. PBS control. A similar result was obtained in a separate experiment.

shows the inhibitory effect of liposomal RGD analogues on experimental lung metastasis. LAP-2, LAP-3, and LAP-5 liposomes significantly suppressed the number of tumor colonies in lungs to values that were 16.9, 58.6, and 52.3%, respectively, of the untreated controls. Furthermore, the mean size of colonies is also reduced. On the other hand, LAP-1 and LAP-6 liposomes fail to reduce the colonization. This result indicates that some, but not other, liposomal RGD analogues can inhibit experimental lung metastasis effectively. The availability of RGD on the liposomal surface may be an important factor, since anchoring both sides of RGD (in the case of LAP-1) to the liposomal membrane may reduce the availability of the peptide for suppression of metastasis. In the case of LAP-6, the RGD moiety is sterically hindered by the succinyl group, which may reduce the availability of the former for binding.

Isolation of Oligopeptides Homing to Angiogenic Vessels

Recently, neovascular targeting has become a focus of interest in cancer chemotherapy. This is because angiogenic vessels have properties different from those of the preexisting systemic vessels (Hanahan, 1997; St. Croix et al., 2000; Zetter, 1997), and certain drugs or drug carriers first meet angiogenic vessels before extravasation in the tumor. As one of their characteristics, the angiogenic endothelia express several marker molecules that are not expressed or only slightly expressed on that of preexisting vessels (St. Croix et al., 2000). Specific ligands against these molecules are considered to be useful for active targeting to tumor angiogenic vessels. It was suggested that cancer chemotherapy targeted to angiogenic endothelia induces endothelial cell apoptosis prior to tumor cell apoptosis in tumor tissues and eradicates even the drug-resistant tumor (Boehm et al., 1997; Browder et al., 2000). Furthermore, the metastatic pathway of tumor cells would be broken by the disruption of angiogenic vessels (Skobe et al., 1997). We have isolated peptides specific for tumor angiogenic vasculature using a phage-displayed peptide library. Our results indicate that the angiogenic vasculature-specific peptide, APRPG, might be useful as a tool for active targeting to tumor neovasculature and that liposomes modified with APRPG might be useful for antineovascular therapy.

Biopanning

Angiogenic vessels are formed on murine dorsal skin for *in vivo* biopanning (Pasqualini and Ruoslahti, 1996; Pasqualini et al., 1997). Highly metastatic murine B16BL6 melanoma cells (1×10^7 cells/ring) are loaded into a Millipore chamber ring. The chamber rings are dorsally implanted into 5-week-old C57BL/6 male mice (Kurohane et al., 2001). Five days after the implantation, these mice bearing angiogenic vessels on the dorsal skin are used for in vivo biopanning.

A phage-displayed random peptide library expressing pentadecamer amino acid residues at the N-terminus of pIII phage coat protein of M13 phage is used (kindly provided by Dr. Hideyuki Saya, Kumamoto University). In vivo biopanning is performed by a modified method as described by Pasqualini and Ruoslahti (1996) and Pasqualini et al. (1997). The phage-displayed peptide library (1×10^{13} cfu) is injected intravenously into angiogenic vessel-bearing mice. The mice are deeply anesthetized with pentobarbital sodium and snap frozen in liquid nitrogen 4 min after the injection. The skin attached to the Millipore chamber ring where the angiogenic vessels had been formed is dissected, minced, and homogenized with ice-cold Dulbecco's modified Eagle medium (DMEM) containing 1 mM phenylmethylsulfonylfluoride. This homogenate is washed three times (30,000g for 10 min) with ice-cold DMEM containing 1% bovine serum albumin (BSA), and the accumulated phages are recovered by infecting Escherichia coli K91KAN with them. A part of the phages in the homogenate is used for the titration of the accumulated phages, and the remaining phages are amplified in E. coli K91KAN and purified. Then, a second round of biopanning is performed similarly as for the first round. These biopanning steps are repeated for five cycles. At the fifth round of biopanning, the recovery rate of the phage (recovered phage titer to input phage titer) increased about 1000-fold over that of the first round, suggesting that selection of high-affinity phage clones capable of accumulating in the angiogenic site is successful.

Screening

Before affinity screening, the selected phages are cloned, and the sequence of presented peptides is determined. For in vivo screening, 1.0×10^6 B16BL6 cells are implanted subcutaneously into the posterior flank of 5-week-old C57BL/6 male mice. Each sample of phage clones (1.0×10^{11} cfu) is injected into tumor-bearing mice via the tail vein when the tumor size is about 10 mm in diameter. Four minutes after injection, the phages that had accumulated in tumor tissue are recovered and titrated. Similar experiments are performed in Meth A sarcoma-bearing mice. We have demonstrated that PRPGAPLAGSWPGTS-, DRWRPALPVVLFP-LH-, and ASSSYPLIHWRPWAR-presented phage clones accumulate extensively in two types of murine tumor.

Characterization of Peptides

Pentadecamer peptides are synthesized using of Rink amide resin (0.4–0.7 mmol/g) and a peptide synthesizer ACT357, resulting in an amide at the carboxyl-terminus. To confirm the capability of the synthetic peptides to accumulate in the tumor, 0.25 μmol of synthetic peptide (PRPGA-PLAGSWPGTS, DRWRPALPVVLFPLH, and ASSSYPLIHWRPWAR) and 5 × 10^8 cfu of the corresponding phage clone are coinjected into B16BL6 melanoma-bearing mice. Four minutes after injection, the titer of phages that accumulate in the tumor tissue is determined as described above. Tumor accumulation of each phage clone is suppressed in the presence of the corresponding peptide, and a random peptide, GLDLLGD VRIPVVRR, does not affect phage accumulation.

To determine the epitope sequences of the peptides, various short peptides based on original 15-mer sequences are synthesized, and the inhibitory effect of these peptides against tumor accumulation of the corresponding phage clones is examined. Our results indicated that PRP and WRP in original 15-mer sequences were important for their affinity.

Liposomalization of Peptides Homing to Angiogenic Vessels

Preparation of Angiogenic Vessel-Targeted Liposomes

Liposomes are modified with pentapeptides having PRP or WRP sequences, and the biodistribution of these liposomes in tumor-bearing mice is examined. For liposomalization of peptides, stearoyl 5-mer peptides (APRPG, RWRPA, or HWRPW) are synthesized using the DIPCI-HOBt coupling method. Liposomalization of stearoyl peptides is performed by the same procedure described in "Preparation of Liposomal RGD."

Biodistribution of Liposomes

C26 NL-17 carcinoma cells ($1.0 × 10^6$ cells/mouse) are injected subcutaneously into the posterior flank of 5-week-old BALB/c male mice, and a biodistribution study is performed when the tumor size becomes about 10 mm in diameter. Liposomes composed of DSPC, cholesterol, and stearoyl 5-mer peptides (10/5/2 as a molar ratio) are radiolabeled with [oleate-1-^{14}C]cholesteryl-oleate (74 kBq/mouse). Size-matched C26 NL-17 carcinoma-bearing mice are injected with the radiolabeled liposomes via the tail vein. Three hours after the injection, the mice are sacrificed under diethyl ether anesthesia for the collection of blood. After the mice are bled from the carotid artery, the heart, lung, liver, spleen, kidney, and tumor are removed, washed with saline, and weighed. The radioactivity in each organ

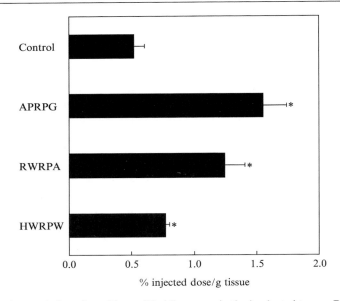

FIG. 3. Accumulation of peptide-modified liposomes in the implanted tumor. Biodistribution study of various peptide-modified liposomes 3 h after injection revealed that APRPG-modified liposomes accumulated in the tumor compared with RWRPA-modified liposomes, HWRPA-modified liposomes, or control liposomes in Meth A sarcoma-bearing mice. Data are presented as the percentage of the injected dose per gram tissue ± SD. Significant differences from the control are indicated (*$p < 0.05$).

is determined with a liquid scintillation counter. All peptide-modified liposomes showed high accumulation in tumor tissue compared with control liposomes (Fig. 3). In particular, APRPG-modified liposomes (PRP-Lip) showed the highest accumulation in tumors and had a tendency to avoid trapping by the reticuloendothelial system such as the spleen and liver.

Antitumor (Antineovascular) Activity of ADM-Encapsulated Neovasculature-Homing Liposomes

Adriamycin (ADM)-encapsulated liposomes are prepared by a modification of the remote-loading method (Oku *et al.*, 1994), with an encapsulation efficiency of more than 90%. Liposomes composed of DSPC/cholesterol (2:1) (control liposomes) or DSPC/cholesterol/stearoyl APRPG (20:10:1) (PRP-Lip) are prepared and sized to 100 nm. Tumor implantation is performed as described above. ADM encapsulated in the control liposomes (LipADM) or in liposomes modified with stearoyl APRPG (PRP-LipADM) (10 mg/kg as ADM), ADM alone (10 mg/kg), or 0.3 M glucose (control) are injected intravenously into tumor-bearing mice on days 6, 9, and 12 after implantation of Meth A sarcoma, or on days 9, 12, and 15 after implantation of C26 NL-17 carcinoma cells. The sizes of the tumor and body weight of each mouse are monitored every day. Tumor volume is calculated using the formula 0.4 ($a \times b^2$), where a is the largest and b is the smallest diameter of the tumor. Variance in a group is evaluated by the F test, and differences in mean tumor volume are evaluated by Student's t test. The results showed that administration of PRP-LipADM suppressed tumor growth most efficiently in Meth A sarcoma-bearing mice (Fig. 4A). The superior effect of PRP-LipADM compared with conventional LipADM may be explained by the increase of local concentration of ADM in tumor tissues. Alternatively, PRP-LipADM may have effectively damaged angiogenic endothelial cells resulting in the indirect suppression of tumor growth. The results also indicate that liposomalization itself is useful for decreasing side effects, since both PRP-LipADM and LipADM did not cause noticeable body weight change, but free ADM-treated mice decreased in body weight progressively. In fact, all free ADM-treated animals died of acute toxicity at the given dose. The mean survival times for the control, LipADM-treated, and PRP-LipADM-treated mice were 38.8, 47.7, and 50.2 days, respectively. PRP-LipADM was most efficient for prolonging survival time with an increased life span of about 30%. We also examined the suppression of tumor growth by PRP-LipADM in C26 NL-17–bearing mice. PRP-LipADM also showed marked suppression against C26 NL-17 (Fig. 4B). The free ADM treatment group lost weight, and three of six mice died of toxicity. The mean survival times for the control, LipADM-treated, and PRP-LipADM-treated mice were 55.8, 69.7, and 72.3 days, respectively. Again, PRP-LipADM was the most efficient (about 30% increase in life span), and one of six mice was completely cured.

For assay of antineovascular activity, a chamber ring loaded with C26 NL-17 cells (1×10^7 cells/ring) is dorsally inoculated into BALB/c male mice. At 2 days after inoculation, LipADM, PRP-LipADM (10 mg/kg as ADM), ADM alone (10 mg/kg), or 0.3 M glucose is injected intravenously into DAS model mice. Four days after inoculation, 1% Evans Blue solution is injected intravenously into them. After 1 min, they are scarified, and the pigment in the skin attached to the ring is extracted for the measurement of absorbance at 620 nm. PRP-LipADM markedly damaged newly forming blood vessels (Fig. 4C). These results suggest that modification of liposomes with APRPG enhance the antitumor activity of ADM and reduce the toxicity of ADM due to a targeting effect.

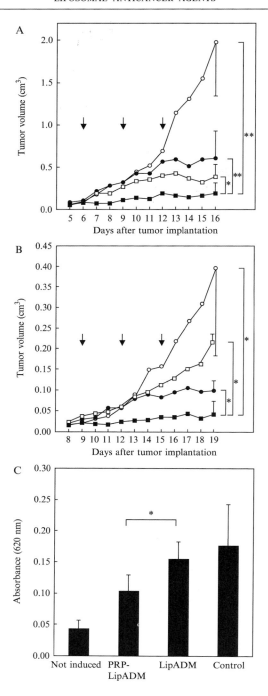

Confocal Observation

NBD-phosphatidylethanolamine (NBD-PE)–labeled liposomes composed of DSPC/cholesterol/NBD-PE (20:10:1) (control liposomes) or DSPC/cholesterol/stearoyl APRPG/NBD-PE (20:10:2:1) (PRP-Lip) are prepared. Human umbilical cord endothelial cells (HUVECs) pretreated with 100 ng/ml VEGF for 24 h are incubated with control liposomes or PRP-Lip for 30 min. To confirm the specific binding of PRP-Lip, HUVECs are incubated with PRP-Lip in the presence of 20-fold excess of free APRPG-peptides. Then, HUVECs are washed with phosphate-buffered saline (PBS) and fixed with 2% paraformaldehyde–0.2% glutaraldehyde in PBS. Localization of liposomes is monitored using an LSM510 confocal system. The binding capacity of PRP-Lip and control liposomes to HUVECs is determined by confocal microscopy. NBD-labeled liposomes bound to VEGF-activated HUVECs only when liposomes were modified with APRPG (Fig. 5A and B). This binding was cancelled in the presence of excess APRPG peptide.

Conclusion

RGD-related peptides must be administered in milligram doses to mice to suppress tumor metastasis *in vivo* because of rapid hydrolysis and elimination of the peptides. In contrast, liposomal RGD presented here may show both reduced hydrolysis due to steric factors and reduced elimination due to the size of the liposomes, since such liposomes have a long circulating time and are stable in the bloodstream. Furthermore, liposomes can present a number of RGD sequences on the same surface, which enables cooperative binding to metastatic tumor cells, i.e., clustering of RGD-reactive molecules on the cell surface. A 100-nm liposome containing 10 mol% RGD-analogue can be calculated to present more than 8000 RGD moieties on the liposomal surface, and a 150-nm liposome, more than

FIG. 4. Suppression of tumor growth and angiogenesis *in vivo* by PRP-LipADM. (A, B) Tumor-bearing BALB/c mice were injected iv with 0.3 M glucose (open circles), ADM solution (closed circles), LipADM (open squares), or PRP-LipADM (closed squares). PRP-LipADM markedly suppressed tumor growth. (A) Meth A sarcoma-bearing mice ($n = 6$; $*p < 0.05$; $**p < 0.01$). (B) C26 NL-17-bearing mice ($n = 6$; $*p < 0.001$). Data are presented as mean tumor volume ± SD. Arrows show the day of treatment. (C) Suppression of *in vivo* angiogenesis by PRP-LipADM was significantly different from LipADM treatment ($n = 4$–6; $*p < 0.05$), suggesting the antineovascular effect of PRP-LipADM. The relative amount of angiogenic vessels is indicated as Evans blue accumulation at the angiogenic site produced by the dorsal air sac method ± SD.

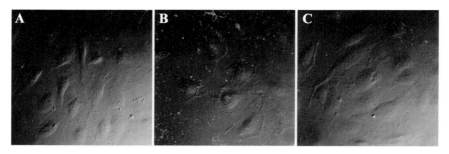

FIG. 5. Specific binding of APRPG-modified liposomes to VEGF-stimulated human umbilical endothelial cells. (B) Confocal microscopic observation indicates that the binding pattern of APRPG-modified liposomes to VEGF-stimulated HUVECs is specific, but not (A) that of control liposomes. (C) Binding of PRP-Lip was inhibited in the presence of excess APRPG, suggesting the presence of specific molecule(s) on the VEGF-stimulated HUVECs that have affinity to APRPG. (See color insert.)

18,000 RGD moieties. Therefore, liposomalization of oligopeptides such as RGD improves the short half-lives and affinity, which results in enhancement of the antimetastatic activity.

Liposomal APRPG might be useful as tools for active targeting to tumor neovasculature. In therapeutic experiments, modification of liposomes with APRPG enhanced the antitumor activity of ADM and reduced its toxicity. These effects of APRPG were independent of tumor type. The antitumor activity of PRP-LipADM may be explained partly by the increase of the local concentration of ADM in the tumor. In addition, we speculate that ADM damaged neovascular endothelial cells, since PRP-LipADM is expected to bind these growing cells efficiently. Taken together, it would be expected that PRP-Lip could deliver anticancer agents for antineovascular therapy.

Acknowledgments

The authors thank Mr. Yoshihiro Tokudome, Mr. Koh Watanabe, and Mr. Koichi Kuromi for technical assistance, and Dr. Koichi Ogino and Dr. Takao Taki for their collaboration in the phage library project.

References

Arap, W., Pasqualini, R., and Ruoslahti, E. (1998). Cancer treatment by targeted drug delivery to tumor vasculature in a mouse model. *Science* **279,** 377–380.
Asai, T., Kurohane, K., Shuto, S., Awano, H., Matsuda, A., Tsukada, H., Namba, Y., Okada, S., and Oku, N. (1998). Antitumor activity of 5'-*O*-dipalmitoylphosphatidyl 2'-*C*-cyano-2'-

deoxy-1-β-D-*arabino*-pentofuranosylcytosine is enhanced by long-circulating liposomalization. *Biol. Pharm. Bull.* **21,** 766–771.

Boehm, T., Folkman, J., Browder, T., and O'Reilly, M. S. (1997). Antiangiogenic therapy of experimental cancer does not induce acquired drug resistance. *Nature* **390,** 404–407.

Browder, T., Butterfield, C. E., Kraling, B. M., Shi, B., Marshall, B. M., O'Reilly, S., and Folkman, J. (2000). Antiangiogenic scheduling of chemotherapy improves efficacy against experimental drug-resistant cancer. *Cancer Res.* **60,** 1878–1886.

Eliceiri, B. P., and Cheresh, D. A. (1999). The role of αv integrins during angiogenesis: Insights into potential mechanisms of action and clinical development. *J. Clin. Invest.* **103,** 1227–1230.

Gho, Y. S., Lee, J. E., Oh, K. S., Bae, D. G., and Chae, C. B. (1997). Development of antiangiogenin peptide using a phage-displayed peptide library. *Cancer Res.* **57,** 3733–3740.

Hanahan, D. (1997). Signaling vascular morphogenesis and maintenance. *Science* **277,** 48–50.

Huang, X., Molema, G., King, S., Watkins, L., Edgington, T. S., and Thorpe, P. E. (1997). Tumor infarction in mice by antibody-directed targeting of tissue factor to tumor vasculature. *Science* **275,** 547–550.

Ishikawa, D., Kikkawa, H., Ogino, K., Hirabayashi, Y., Oku, N., and Taki, T. (1998). GD1α-replica peptides functionally mimic GD1α, an adhesion molecule of metastatic tumor cells, and suppress the tumor metastasis. *FEBS Lett.* **441,** 20–24.

Koivunen, E., Gay, D. A., and Ruoslahti, E. (1993). Selection of peptides binding to the $\alpha 5\beta 1$ integrin from phage display library. *J. Biol. Chem.* **268,** 20205–20210.

Kurohane, K., Namba, Y., and Oku, N. (2000). Liposomes modified with a synthetic Arg-Gly-Asp mimetic inhibit lung metastasis of B16BL6 melanoma cells. *Life Sci.* **68,** 273–281.

Kurohane, K., Tominaga, A., Sato, K., North, J. R., Namba, Y., and Oku, N. (2001). Photodynamic therapy targeted to tumor-induced angiogenic vessels. *Cancer Lett.* **167,** 49–56.

Martens, C. L., Cwirla, S. E., Lee, R. Y., Whitehorn, E., Chen, E. Y., Bakker, A., Martin, E. L., Wagstrom, C., Gopalan, P., Smith, C. W., Tate, E., Koller, K. J., Schatz, P. J., Dower, W. J., and Barrett, R. W. (1995). Peptides which bind to E-selectin and block neutrophil adhesion. *J. Biol. Chem.* **270,** 21129–21136.

Oku, N. (1999a). Delivery of contrast agents for positron emission tomography imaging by liposomes. *Adv. Drug Deliv. Rev.* **37,** 53–61.

Oku, N. (1999b). Anticancer therapy using glucuronate modified long-circulating liposomes. *Adv. Drug Deliv. Rev.* **40,** 63–73.

Oku, N., Namba, Y., and Okada, S. (1992). Tumor accumulation of novel RES-avoiding liposomes. *Biochim. Biophys. Acta* **1126,** 255–260.

Oku, N., Doi, K., Namba, Y., and Okada, S. (1994). Therapeutic effect of adriamycin encapsulated in long-circulating liposomes on Meth-A-sarcoma-bearing mice. *Int. J. Cancer* **58,** 415–419.

Oku, N., Tokudome, Y., Koike, C., Nishikawa, N., Mori, H., Saiki, I., and Okada, S. (1996). Liposomal Arg-Gly-Asp analogs effectively inhibit metastatic B16 melanoma colonization in murine lungs. *Life Sci.* **58,** 2263–2270.

Pasqualini, R., and Ruoslahti, E. (1996). Organ targeting *in vivo* using phage display peptide libraries. *Nature* **380,** 364–366.

Pasqualini, R., Koivunen, E., and Ruoslahti, E. (1997). αv integrins as receptors for tumor targeting by circulating ligands. *Nat. Biotechnol.* **15,** 542–546.

Ruoslahti, E., and Pierschbacher, M. D. (1987). New perspectives in cell adhesion: RGD and integrins. *Science* **238,** 491–497.

Scott, J. K., and Smith, G. P. (1990). Searching for peptide ligands with an epitope library. *Science* **249,** 386–390.

Skobe, M., Rockwell, P., Goldstein, N., Vosseler, S., and Fusenig, N. E. (1997). Halting angiogenesis suppresses carcinoma cell invasion. *Nat. Med.* **3,** 1222–1227.

St. Croix, B., Rago, C., Velculescu, V., Traverso, G., Romans, K. E., Montgomery, E., Lal, A., Riggins, G. J., Lengauer, C., Vogelstein, B., and Kinzler, K. W. (2000). Genes expressed in human tumor endothelium. *Science* **289,** 1197–1202.

Takikawa, M., Kikkawa, H., Asai, T., Yamaguchi, N., Ishikawa, D., Tanaka, M., Ogino, K., Taki, T., and Oku, N. (2000). Suppression of GD1α ganglioside tumor metastasis by liposomalized WHW-peptide. *FEBS Lett.* **446,** 381–384.

Viti, F., Tarli, L., Giovannoni, L., Zardi, L., and Neri, D. (1999). Increased binding affinity and valence of recombinant antibody fragments lead to improved targeting of tumoral angiogenesis. *Cancer Res.* **59,** 347–352.

Zetter, B. R. (1997). On target with tumor blood vessel markers. *Nat. Biotechnol.* **15,** 1243–1244.

[10] Separation of Liposome-Entrapped Mitoxantrone from Nonliposomal Mitoxantrone in Plasma: Pharmacokinetics in Mice

By ATEEQ AHMAD, YUE-FEN WANG, and IMRAN AHMAD

Abstract

A method is described for quantification of the liposomal and non-liposomal forms of mitoxantrone (MTO) in mouse plasma after intravenous administration of liposome-entrapped MTO Easy-to-Use (LEM-ETU) formulation. This is based on the property of liposome-entrapped MTO (LEM) to pass through reversed-phase C_{18} silica gel cartridges, while nonliposomal MTO or free MTO is retained with strong hydrophobicity and later is eluted with acidic methanol. Extraction of LEM and free MTO from plasma is performed in two steps. This technique is rapid and sensitive and can be used for a large series of sample preparation. The plasma samples are found stable after one freeze–thaw cycle. The recovery of MTO, as well as the precision, linearity, and accuracy of the method for both free and liposomal MTO, appears satisfactory for pharmacokinetic studies. The pharmacokinetic results in mice show a sustained release of MTO from LEM-ETU.

Introduction

The therapeutic benefits of encapsulating anticancer drugs such as doxorubicin and vincristine are well documented (Alving and Richards, 1983; Bonte and Juliano, 1986; Cullis *et al.*, 1989; Ostro and Cullis, 1989; Scherphof *et al.*, 1984; Senior, 1987). Presently, many anthracycline molecules are in clinical trials for treatment of several cancers. Efforts are being made to develop reliable methods for the delivery of anthracycline molecules, such as the use of liposomes as carrier systems (Bellott *et al.*, 2001; Chonn *et al.*, 1991; Druckman *et al.*, 1989; Mayer and St.-Onge, 1995; Srigritsanapol and Chan, 1994; Thies *et al.*, 1990).

Several factors contribute to the complexity of the pharmacokinetics of liposome-entrapped drugs. In the circulation, drugs may exist as liposome-associated, protein-bound nonliposomal drug or completely free drug. The plasma clearance of liposomal drugs occurs as a result of three processes with different elimination rates. These are (1) tissue uptake of drug-containing liposome, (2) release of drug from the liposome, or (3) clearance of nonliposomal drug. Obtaining a thorough understanding of the pharmacokinetic and pharmacodynamic relationship of liposome-entrapped anticancer drugs such as mitoxantrone (MTO, Fig. 1) will rely on the ability to accurately separate and quantify nonliposomal and liposome-associated drug fractions in plasma after administration of specific formulations. Methods reported in the literature include ultrafiltration (Krishna *et al.*, 2001) with limited recovery, cation exchange (Dowex, not reproducible with low recovery), ultracentrifugation (requiring prolonged sample process time for settling of drug protein complexes), and gel filtration (G-50) (which is unable to separate nonliposomal MTO from liposome-entrapped MTO). Liposome-entrapped MTO (LEM) Easy-to-Use (ETU) is one of NeoPharm's novel products currently in Phase I clinical trials. To understand the pharmacokinetics of MTO in LEM form, a simple and sensitive separation method is required to quantify MTO in both liposomal and nonliposomal (protein-bound and completely free) form in plasma.

The assay that we describe here can be used to monitor the concentration of MTO in different forms in mouse plasma after the separation of

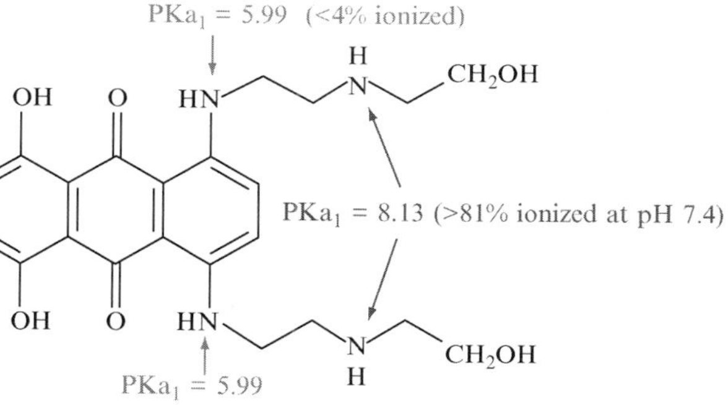

FIG. 1. Chemical structure of mitoxantrone.

liposome-associated MTO from nonliposomal MTO. This method was validated using a liquid chromatography (LC) method for determination of MTO concentration in mouse plasma. This assay is robust, reproducible from 10 to 5000 ng/ml, free from any interference of matrix or dilution effect, and meets the sensitivity and reproducibility criteria needed for pharmacokinetic studies of LEM in plasma. We have conducted a mouse pharmacokinetic study, after administration of LEM-ETU formulation, to compare pharmacokinetic profiles of liposome-entrapped with nonliposomal MTO and total (nonliposomal and liposomal) MTO.

Methods

Principle of the Assay

The quantification of LEM in mouse plasma is based on binding of nonliposomal MTO to silica-based C_{18} solid phase, with strong hydrophobicity and nonadsorption of charged liposomes to the solid phase. Samples containing LEM are spiked with an internal standard (IS), ametantrone, before processing. Drug to IS peak area ratios for the standards are plotted versus drug concentration to generate a linear calibration curve (Fig. 2).

Chemicals

The lipids 1,2-dioleoly-*sn*-glycero-3-phosphocholine, cholesterol, and 1,1′,2,2′-tetramyristoyl cardiolipin are obtained from Avanti Polar Lipids (Alabaster, AL). Mitoxantrone dihydrochloride (100%) is from Pharmacopeia (Rockville, MD). Ametantrone diacetate [1,4-bis([2-(2-hydroxyethyl)amino]ethylamino) 9,10-anthracenedione], the internal standard (IS), is a generous gift from the Drug Synthesis & Chemistry Branch, Development and Therapeutics Program, Division of Cancer Treatment and Diagnosis, National Cancer Institute (Bethesda, MD). Hexane is from Aldrich (Milwaukee, WI). Ammonium formate, tocopherol acid succinate, sodium hydroxide, phosphoric acid, ACS reagent, and ascorbic acid are from Sigma (St. Louis, MO); HPLC-grade acetonitrile and methyl alcohol and USP-grade sodium chloride are from EM Science (Gibbstown, NJ); USP-grade ethanol is from Aaper (Shelbyville, KY); formic acid is from Fisher Scientific (Hampton, NH); hexanesulfonic acid sodium salt is from Acros Organics (Pittsburgh, PA); and hydrochloric acid and potassium phosphate monobasic are from J.T. Baker (Phillipsburg, NJ). Milli-Q water is obtained from an in-house Millipore Milli-Q Synthesis System (Millipore, Bedford, MA).

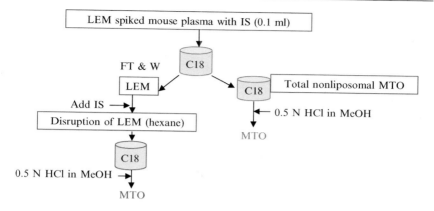

FIG. 2. Quantification of LEM in mouse plasma. FT & W, flow through and wash; IS, internal standard.

Preparation of Liposome-Entrapped Formulation of Mitoxantrone (LEM)

LEM is prepared as detailed previously (Gokhale *et al.*, 2001), but without the sonication step. Briefly, lipids and tocopherol acid succinate are solubilized in ethanol, diluted into an aqueous solution of mitoxantrone, sucrose, and saline, and then lyophilized. Reconstitution of the lyophilized cake is accomplished by adding 5 ml of water and shaking well to yield LEM with a mean particle size less than 200 nm and an approximate mitoxantrone concentration of 1 mg/ml.

Stock Solutions

Stock solutions of mitoxantrone and IS are prepared at 100 μg/ml in a saline/ascorbate solution ("diluent," 8.0 g/liter sodium chloride and 10.0 g/liter ascorbic acid) and stored at $-70 \pm 10°$. Due to the known adherence of mitoxantrone to glass (Priston and Sewell, 1994), all solutions are prepared and stored in polypropylene tubes. LEM is reconstituted in sterile in-house Milli-Q water, and a stock solution of LEM (100 μg/ml) is then prepared in diluent and stored at $-70 \pm 10°$. The reconstituted LEM is further diluted with HPLC mobile phase as described below to prepare test solutions of 50, 500, 1000, and 5000 ng/ml for determination of MTO in the LEM-ETU formulation. The exact concentration of MTO in the formulation is determined by comparing the peak areas of the test solutions to the standard curve made from USP MTO.

Preparation of Standards and Quality Control Solutions

CD1 mice are obtained from Bioreclamation, Inc. (Hicksville, NY). Mouse plasma is collected with sodium heparin and is stored at $-70 \pm 10°$ until use. Thawed plasma is spiked with MTO stock solution to prepare a 1000 ng/ml plasma standard. This solution is diluted further to prepare standards between 10 and 5000 ng/ml. A separate MTO stock solution is used to prepare quality control (QC) samples between 30 and 800 ng/ml plasma. A disruption procedure is needed for quantification of LEM. After hexane extraction, LEM becomes completely free MTO. Quantification of MTO in LEM truly is based on a standard curve generated from MTO/IS peak area ratios plotted against standard concentrations.

Chromatographic Equipment and Conditions

An Agilent 1100 Series HPLC system controlled by ChemStation A.8.04 software is used for chromatographic analysis. The system is composed of a quaternary pump (Model G1311A), vacuum degasser (Model G1322A), thermostatted autosampler (Model G1329A or G1367A), thermostatted column compartment (Model G1316A), and diode array detector (Model G1315B), all from Agilent Technologies (Palo Alto, CA). The autosampler is maintained at 4° and the column compartment at 30°.

The HPLC separations are achieved using a Nucleosil C_{18}, 250×4 mm i.d., 5-μm particle size column from Macherey-Nagel (Easton, PA). A precolumn filter with a 2-μm PEEK frit (Upchurch, Oak Harbor, WA) and a guard column (Macherey-Nagel CC Nucleosil C_{18}, 4×8 mm) are installed ahead of the analytical column. The isocratic mobile phase is 29:71 (v:v) acetonitrile/ammonium formate (160 mM) with hexanesulfonic acid (35 mM), adjusted to pH 2.7 with formic acid, running at a flow rate of 1.0 ml/min. The IS and MTO are detected by their absorbance at 655 nm.

Processing of Plasma Samples

1. The C_{18} cartridge (Waters Sep-pak Vac C_{18}, 1cc, Milford, MA) is conditioned with 1.0 ml methanol.
2. The cartridge is equilibrated with 1.0 ml of 0.067 M phosphate-buffered saline (PBS) buffer (KH_2PO_4, pH 7.4, in 0.9% normal saline).
3. A mouse plasma sample (0.1 ml), spiked with LEM or MTO and AMT (IS, 250 ng/ml), is loaded.
4. The flow through (0.1 ml) and the washing (2×0.5 ml) PBS buffer are collected if mouse plasma contains LEM. The fraction contains MTO in LEM form.

Processing of Nonliposomal MTO

1. The C_{18} column is eluted with 2×0.15 ml of 0.5 N HCl in methanol, and the eluent is collected.
2. Nonliposomal MTO is included in this fraction.

Processing of LEM Fraction

1. Concentrated phosphoric acid (6 μl) is added to the collected flow through and washes from steps 3 and 4 to obtain approximately pH 3.0, and the mixture is vortexed.
2. Hexane (0.4 ml) is added, the tubes are vortexed vigorously and centrifuged at 5000g for 10 min, and the aqueous phased is transferred to a microcentrifuge tube.
3. Forty microliters of 5.0 N sodium hydroxide solution are added to the aqueous solution to adjust pH to approximately 7.4 or more, then mixed well.
4. The aqueous solution is loaded onto a new C_{18} cartridge, which was already conditioned and equilibrated with methanol and PBS buffer.
5. The cartridge is washed with 2×0.5 ml PBS buffer.
6. The column is eluted with 2×0.15 ml of 0.5 N HCl in methanol, and the eluent is collected. This fraction contains MTO in LEM form.

Analysis of Mitoxantrone

Standard curves generate acceptable data over the concentration range of 10 to 5000 ng MTO/ml in mouse plasma spiked with MTO. A standard curve of peak area ratios of MTO to IS versus concentration gives acceptable results and is used for all concentration calculations. Typical chromatograms for MTO from LEM or MTO standard and IS spiked in plasma are shown in Fig. 3. A typical MTO standard curve is presented in Fig. 4. The developed and validated LC bioanalytical method for MTO in mouse plasma is robust and reproducible from 10 to 5000 ng/ml. The assay range may be extended up to 40,000 ng/ml by dilution. This method is free from any interference of matrix or dilution effects and meets the sensitivity and reproducibility criteria needed for pharmacokinetic studies of LEM-ETU in mouse plasma.

Pharmacokinetics of Nonliposomal MTO and LEM in Mice

The pharmacokinetics of MTO in both nonliposomal and liposomal forms following a single intravenous (iv) dose of LEM-ETU is determined in male CD2F1 mice. Following each iv administration, one blood sample per animal (three mice per time point) is collected, and plasma is prepared.

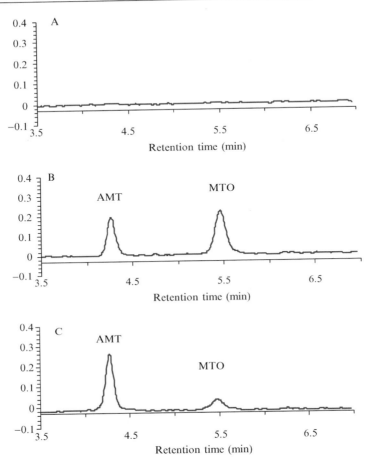

Fig. 3. Representative of HPLC chromatograms. (A) Blank mouse plasma; (B) MTO at 50 ng/ml with IS at 250 ng/ml in mouse plasma spiked with LEM; (C) MTO standard at LLOQ (10.0 ng/ml) level in mouse plasma.

The obtained plasma samples are immediately processed using the method described here. Concentrations of MTO in plasma are determined by HPLC. The plasma concentrations versus time profiles are shown in Fig. 5. The pharmacokinetic parameters of MTO in both forms and total MTO are calculated using the WinNonlin program (version 4.0 Pharsight Corporation, Mountain View, CA) and are summarized in Table I.

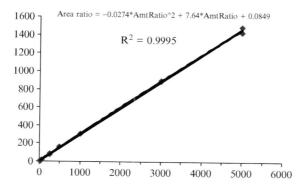

FIG. 4. Standard curve of mitoxantrone in mouse plasma.

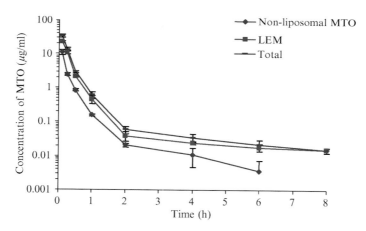

FIG. 5. Mean plasma concentration–time profiles in CD2F1 mice (5 mg/kg).

Conclusion

The method described here separately measures MTO concentrations in either liposome-entrapped form or nonliposomal form in mouse plasma. It can be used to determine individual pharmacokinetics profiles of LEM and nonliposomal MTO *in vivo*.

TABLE I

PHARMACOKINETIC PARAMETERS OF MTO IN LEM AND NONLIPOSOMAL FORMS IN CD2F1 MICE

Treatment	MTO form	Dose (mg/kg)	$t_{1/2}$ (h)	C_{max}[a] (μg/ml)	V_z (liters/kg)	V_{ss} (liters/kg)	MRT (h)	CL (liters/h/kg)	AUC_{0-last}[b] (μg·h/ml)	$AUC_{0-\infty}$ (μg·h/ml)
LEM-ETU	Total	4.71	3.39	52.9	2.10	0.149	0.348	0.429	10.9	11.0
	Nonliposomal	4.71	1.53	24.4	3.04	0.302	0.220	1.38	3.41	3.42
	LEM	4.71	6.23	31.7	5.45	0.347	0.573	0.606	7.63	7.77

[a] Extrapolated values at zero time point.

[b] The last time points of AUCs are 6 h for nonliposomal MTO and 8 h for LEM and total MTO.

Acknowledgment

The authors acknowledge Dr. Aqel Abu-Qare for his assistance in method development.

References

Alving, C. R. (1991). Liposomes as carriers of antigens and adjuvants. *J. Immunol. Methods* **140**, 1–13.

Bellott, R., Pouna, P., and Robert, J. (2001). Separation and determination of liposomal and non-liposomal daunorubicin from the plasma of patients treated with Daunoxome. *J. Chromatogr. B.* **757**, 257–267.

Bonte, F., and Juliano, R. L. (1986). Interactions of liposomes with serum proteins. *Chem. Phys. Lipids* **40**, 359–372.

Chonn, A., Semple, S. C., and Cullis, P. R. (1991). Separation of large unilamellar liposomes from blood components by a spin column procedure: Towards identifying plasma proteins which mediate liposome clearance *in vivo*. *Biochim. Biophys. Acta* **1070**, 215–222.

Cullis, P. R., Mayer, L. D., Bally, M. B., Madden, T. D., and Hope, M. J. (1989). Generating and loading of liposomal systems for drug-delivery applications. *Adv. Drug Delivery Rev.* **3**, 267–282.

Druckman, S., Gabizon, G., and Barenholz, Y. (1989). Separation of liposome-associated doxorubicin from non-liposome-associated doxorubicin in human plasma: Implications for pharmacokinetic studies. *Biochim. Biophys. Acta* **980**, 381–384.

Gokhale, P. C., Pei, J., Zhang, C., Ahmad, I., Rahman, A., and Kasid, U. (2001). Improved safety, pharmacokinetics, and therapeutic efficacy profiles of a novel liposomal formulation of mitoxantrone. *Anticancer Res.* **21**, 3313–3321.

Johnson, J. L., Ahmad, A., Khan, S., Wang, Y. F., Abu-Qare, A. W., Ayoub, J. E., Zhang, A., and Ahmad, I. (2004). Improved liquid chromatographic method for mitoxantrone quantification in mouse plasma and tissues to study pharmacokinetics of liposome entrapped mitoxantrone formulation. *J. Chromatography B.* **799**, 149–155.

Krishna, R., Webb, M., St.-Onge, G. S., and Mayer, L. D. (2001). Liposomal and nonliposomal drug pharmacokinetics after administration of liposome-encapsulated vincristine and their contribution to drug tissue distribution properties. *J. Pharmacol. Exp. Ther.* **298**, 1206–1212.

Mayer, L. D., and St.-Onge, G. (1995). Determination of Free and Liposome-associated doxorubicin and vincristine levels in plasma under equilibrium conditions employing ultrafiltration techniques. *Anal. Biochem.* **232**, 149–157.

Ostro, M. J., and Cullis, P. R. (1989). Use of liposomes as injectable-drug delivery systems. *Am. J. Hosp. Pharm.* **46**, 1576–1587.

Priston, M. J., and Sewell, G. J. (1994). Improved LC assay for the determination of mitozantrone in plasma: Analytical considerations. *J. Pharm. Biomed. Anal.* **12**, 1153–1162.

Senior, J. (1987). Fate and behavior of liposomes *in vivo*: A review of controlling factors. *Crit. Rev. Ther. Drug Carrier Syst.* **3**, 123–193.

Srigritsanapol, A. A., and Chan, K. K. (1994). A rapid method for the separation and analysis of leaked and liposomal entrapped phosphoramide mustard in plasma. *J. Pharm. Biomed. Anal.* **12**, 961–968.

Thies, R. L., Cowens, D. W., Cullis, P. R., Bally, M. B., and Mayer, L. D. (1990). Method for rapid separation of liposome-associated doxorubicin from free doxorubicin in plasma. *Anal. Biochem.* **188**, 65–71.

[11] Methodology and Experimental Design for the
Study of Liposome-Dependent Drugs

By TIMOTHY D. HEATH

Abstract

This chapter describes the concept of liposome-dependent drugs and the rationale for using them. Subsequently, procedures for studying and identifying liposome-dependent drugs are given. The first procedure described is a simple endpoint assay, and methods are given for both adherent and nonadherent cells. To establish in such a system that a drug is liposome dependent, it is necessary to demonstrate an IC_{50} for the encapsulated drug that is less than that of the free drug, preferably with continuous exposure of the cells to drug. Subsequently, a second procedure is described, which is a more rigorous approach able to identify liposome dependency for a drug that is less effective in a carrier system than it is in the free form. This procedure is a multicompartment growth inhibition assay, wherein two cell populations are separated by a semipermeable membrane, through which free drug but not the liposomal carrier system may diffuse. The first population is adherent and is directly exposed to the liposomal or free drug. The second cell population is nonadherent and is exposed only to the drug that diffuses through the membrane. In addition to the methodology, experimental design is discussed and also the calculations needed to determine percent leakage, percent processing, percent metabolism, and the delivery factor, a parameter equivalent to a therapeutic index in an *in vivo* study.

Introduction

The concept of liposome-dependent drugs developed from a simple question. How can it be demonstrated that a carrier system designed to interact selectively with a specific target cell population is delivering its contents to that target population without collateral effects on the adjacent cells? In the early 1980s when this question was being explored, many investigators were interested in the idea of using carrier systems to deliver drugs to specific cells. Efforts to do this with many widely accepted therapeutic agents had proved inconclusive, because drug that is extensively released from the carrier system can be cytotoxic independently of the carrier. It is typical with agents such as cytosine arabinoside, for example, to observe a similar potency for free and encapsulated drug when cells were continuously exposed to the drug (Allen *et al.*, 1981). Consequently, it cannot be determined whether the liposomes actually deliver the drug to the cells or simply release the drug into the medium, from where it will act independently of the carrier system. Some investigators have argued that such release may be beneficial in an *in vivo* situation, suggesting that local release of a drug might still produce a carrier-mediated targeted effect (Mayhew and Papahadjopoulos, 1983). Others refuted such claims on theoretical grounds. Blumenthal *et al.* (1982), for example, argued that at the cellular level, drug released from liposomes at the cell surface could not be expected to diffuse preferentially into the cell, a process referred to as appositional transfer. Both Levy (1987) and Stella and Himmelstein (1980) argued that the possibility of a beneficial effect from local drug release *in vivo* from a carrier system was not realistic based on pharmacokinetic considerations.

Whatever the most desired mechanism for *in vivo* drug delivery may be, *in vitro* drug delivery studies can be used only to demonstrate that the delivery system preferentially associates with target cells and that drug potency is increased by the delivery system. The latter phenomenon, of necessity, requires the use of a liposome-dependent drug. The carrier system is required to deliver the drug to cells, such that the free drug will be less potent than the encapsulated drug. Moreover, the drug must be more effective in a carrier directed to interact with the target cell population than it is in a carrier that has not been directed to interact with the target cell. Related to this, a carrier system must be more potent against a cell population with which it interacts than it is against a similar cell population with which it does not interact. The most effective liposome-dependent drugs make such proofs of efficacy easier to achieve by being less able to penetrate cell membranes than most commonly used cytotoxic agents (Heath *et al.*, 1985a). The result of this property, which might be called transport negative, is that the potency of the free drug is very low, resulting in a larger difference in potency between the free drug and the carrier system. It can also be observed that the difference in potency between the encapsulated and free drug depends on the efficiency of the carrier system. As a result, a transport-negative drug will be more likely to allow for the detection of the effect of the carrier system if that effect is small.

The list of compounds known to be liposome dependent is quite short. The first compound named this way was methotrexate-γ-aspartate (Heath *et al.*, 1985a), a transport-negative derivative of methotrexate, whose ability to inhibit dihydrofolate reductase is comparable to that of methotrexate, but whose potency is over 100 times less than that of methotrexate owing to its inability to penetrate cells. Subsequently, other derivatives of methotrexate were shown to be liposome dependent (Heath *et al.*, 1986). Methotrexate itself was also shown to be liposome dependent in some circumstances, although its higher potency makes an increase harder to achieve (Heath *et al.*, 1986; Leserman *et al.*, 1981). Other compounds have also proven liposome dependent, including *N*-(phosphonacetyl)-L-aspartate (PALA) (Heath and Brown, 1990), phosphonoformate (Heath and Brown, 1990; Szoka and Chu, 1988), phosphonoacetate (Heath and Brown, 1990), clodronate (Mönkkönen and Heath, 1983; van Rooijen *et al.*, 1988), gallium (Mönkkönen *et al.*, 1993), fluoroorotate (Heath *et al.*, 1985b, 1987), and hygromycin B (Vidal *et al.*, 1985). In addition to these drugs, there are also a number of biologically active molecules that function similarly to the liposome-dependent drugs, including toxin A chains (McIntosh and Heath, 1982) and viral DNA (Fraley *et al.*, 1980) and RNA (Wilson *et al.*, 1979). Indeed, the idea of a liposome-dependent drug was stimulated by work showing that polio RNA (Wilson *et al.*, 1979) or SV40 DNA (Fraley *et al.*, 1980) was infectious when the nucleic acid was encapsulated in liposomes.

Along with the liposome-dependent drugs, an experimental framework was developed, the description of which is the purpose of this chapter. This experimental framework enables the investigator to work methodically and unambiguously to document carrier-mediated drug delivery and build a clear picture of how the drug and the delivery system work. A positive result in such studies proves two things simultaneously, that the drug is liposome dependent and that the carrier system is effective. Therefore, once a particular drug has been identified as carrier dependent, it may be used together with new carrier systems to determine whether they are effective for drug delivery. Similarly, a proven carrier system can be used to discover other liposome-dependent drugs.

Simple Endpoint Assay

A liposome-dependent drug is likely to exhibit its selective delivery in a simple endpoint assay. The form of this can vary, but the most basic form will be described here, the growth inhibition assay. The simple rule that is followed is to establish that the endpoint of the assay (50% inhibitory concentration, IC_{50}) is lower for the drug in the carrier system than it is for the free drug. Drug concentration is expressed in terms of the average concentration in the medium. Hence, if the drug is released from the carrier sysystem into the medium, it will be present at this concentration. As a result, a lower IC_{50} for the encapsulated drug provides unambiguous evidence for its carrier-mediated delivery, because such an IC_{50} could not be achieved through the action of drug released from the carrier system.

The most rigorous proof of carrier-dependent delivery in such a simple system comes from the use of continuous exposure, where no attempt is made to remove drug until the cells are counted at the end of the growth period. "Wash" formats, in which cell exposure to the drug is for a period of time less than the entire growth period, can be useful to demonstrate that a targeted system is likely to be effective *in vivo*, where exposure times may be short. However, wash experiments should be viewed cautiously as proof for delivery on their own, particularly if the IC_{50} observed for the carrier system under these conditions is much greater than the IC_{50} of the free drug in continuous cell exposure. A delivery system designed to interact specifically with cells may cause the retention of a significant fraction of encapsulated drug by the cells. This drug could act independently of the delivery system by leaking into the medium after the wash has taken place. It is, therefore, safest in a simple endpoint assay to rule that a drug has been delivered by the carrier system only if it is more potent in continuous exposure.

Experimental Design

Assay Format. We have generally found 24-well plates to be a robust format for growth inhibition assays, and the methods described all use them. There are methods that use plates with smaller wells, going even as far as to use 96-well plates, particularly when using a colorimetric assay for cell concentration. Such a format has some clear advantages in terms of the number of samples per plate and the possibilities of automation of reading. Whatever well size is chosen, care is required when working with adherent cells. Disturbance caused by shaking or rocking of the plates prior to cell adhesion can result in an accumulation of cells at the center of the wells, and gowth will not be uniform, because the clustered cells will quickly reach confluency and will migrate to other areas of the well with difficulty.

Cell Lines. Almost any cell line that divides at least every 16 h can be used effectively for growth inhibition assays. Slower growing lines can also be used, although it is necessary to adjust the period of the assay if these are used. We have typically used either a 48-h growth assay for very rapidly dividing cells or a 72-h assay for slower growing cells such as CV1-P. Occasionally, a 96-h growth period has been necessary. Adherent cells are harvested from confluent cultures by whatever means is customary for the cell line (trypsin, EDTA, scraping, etc.) After harvesting and washing, a suspension is counted in a cell counter, the cells are diluted, and then plated in 24-well plates to give between 1.5×10^4 and 4×10^4 cells per well, 1 ml medium per well. The initial cell count may vary from one line to another, but allow for at least three doublings during the assay. To do this for an adherent cell line, it is necessary to know the total cell number in a confluent well for the cell line in question. This can readily be determined in trial studies where cells are plated and observed over a few days of growth, and ultimately counted to determine the confluent cell count. Nonadherent cells can be diluted from active cultures to 1×10^5 cells per ml in fresh medium, and plated by adding 1 ml per well to 24-well plates. For both adherent and nonadherent cell lines, the cells are allowed to grow overnight before treatment with drug.

Drug Treatment. It is best to use at least triplicate wells for each concentration of drug used. In addition, a half-logarithmic series of drug concentrations (3, 10, 300, 1000, etc.) will cover the range adequately. A half-logarithmic dilution series works well, because growth inhibition tends to be a threshold phenomenon, such that cell growth will almost always fall from 100% to 0% within a 10- to 30-fold drug concentration range. A half-logarithmic series always gives three or four experimental points on the inhibition curve, thereby allowing accurate measurement of the IC_{50}. Typically, a growth curve is generated from a single plate, which allows for seven different drug concentrations. If the drug potency is completely unknown, a range-finding experiment may be necessary, using either a logarithmic dilution series or, alternatively, using multiple plates. Once the range is known, a single plate is sufficient to generate a growth curve using a half-log drug series. Typically, drug may be added to the wells in 10 μl of buffered saline solution. However, it is possible to add as much as 100 μl of buffered saline without noticeably affecting cell growth.

Counting and Controls. To measure the inhibition of cell growth, it is necessary to determine two control values. The first is the cell count at time zero, and at least three wells are harvested from each batch of plates to determine this value prior to drug treatment. The second is the cell count of wells allowed to grow without drug treatment, and it is important to include this control on each plate used in the growth inhibition study. These two controls may be referred to as the "0%" and "100%" values and are used to determine the percent growth in drug-treated wells in the following formula:

$$\% \text{ Growth} = \frac{(\text{Cell Count} - 0\% \text{ Count})}{(100\% \text{ Count} - 0\% \text{ Count})}$$

Method

To help illustrate the general points made above, two specific methods that have been used will be provided, one for CV1-P cells, an adherent cell line that has been extensively used in liposome drug delivery studies, and EL4, a nonadherent murine T-lymphoma cell line.

CV1-P Cells. CV1-P cells are grown in Dulbecco's modified Eagle medium containing pyruvate and 1 g/liter glucose. The medium is supplemented with 10% fetal calf serum and 100 units/ml penicillin and streptomycin. Cells are grown at 37° in a humidified incubator, with a 7% CO_2 atmosphere. Cells are grown to confluency in 25-ml flasks, the medium is removed by washing with 5 ml of Dulbecco's phosphate-buffered saline with 2 mM calcium chloride, 2 mM magnesium chloride (PBS-CM), and harvested by treatment with 1 ml of 0.2% trypsin at 37° for 15 min. After release, the cells are suspended in 5 ml of growth medium, pelleted, and resuspended at 2×10^4 cells/ml in growth medium. The number of cells is measured using a Coulter model ZM or equivalent counter, both before and after dilution with medium. After dilution, the cells are transferred to 24-well plates, with 1 ml of suspension per well. Once the cells are transferred to the plates, care is taken to place the plates in the incubator without shaking so as to allow the cells to settle uniformly on the bottom surface of the well. The plates are allowed to incubate overnight before beginning the growth inhibition study. On the next day, three wells are freed of medium by aspiration, and 1 ml of 1:5 diluted trypsin solution in PBS (no calcium/magnesium) is added to each well to remove the cells. After 15 min in the incubator, the cells are resuspended by repeated pipetting of the trypsin solution with an Eppendorf style pipette. The suspension (0.2 ml) is diluted in 9.8 ml of isotonic counting fluid, and the cells are counted to obtain the 0% or original cell count. For a growth inhibition assay, triplicate wells are used for each drug concentration. One set of three wells on each plate is treated with PBS only (100% control), while the rest are treated with various concentrations of drug. The cells are allowed to grow for 72 h, the medium is aspirated, and the cells are harvested by treatment with 1 ml of 1:5 diluted trypsin solution in PBS for 15 min at 37°. After trypsin treatment, the cells are released from the plate by repeated pipetting, 0.2 ml is diluted in 9.8 ml of isotonic counting fluid, and the cells are counted. The percent growth is determined from the counts, as described above, and the mean for each drug concentration is calculated. The values are then plotted on semilog paper (percent growth vs. log drug concentration), and the IC_{50}, which is the concentration of drug that inhibits the cell count by 50%, is determined from the growth curve.

If, instead of continuous exposure, an exposure length shorter than the period of growth is required, it is possible, with CV1-P and other adherent cells, to remove the drug at the desired time and to return the cells to growth afterward. After the desired incubation period, usually between 1 and 24 h, plates are removed from culture, and the medium is aspirated using a Pasteur pipette connected to a water aspirator. Some care is required in this step to avoid removing cells from the plate. The plate is tilted at about 30° from horizontal, and the Pasteur is carefully applied to the lower side of the well. As the medium is withdrawn from the top surface, the pipette is gradually moved toward the bottom, touching it briefly against the intersection of the well side and bottom. Once a plate has been aspirated in this way, each well is immediately filled with 1 ml of fresh warm medium, and the plate is returned to culture conditions. Experience has shown that it is not necessary to wash the cell monolayer to further eliminate residual drug, as aspiration removes more than 99% of the medium. In fact, trying to wash the wells with PBS-CM results in considerable cell loss, something that also occurs if care is not taken in the way that the medium is aspirated. Therefore, it is important to monitor the 100% control to ensure that cell growth is equivalent to what would be observed in a continuous exposure study. The experiment is completed as described above for continuous exposure.

EL4 Cells. EL4 cells are grown in Dulbecco's modified Eagle medium containing pyruvate and 3 g/liter glucose. The medium is supplemented with 10% fetal calf serum and 100 units/ml penicillin and streptomycin. Cells are grown at 37° in a humidified incubator, with a 7% CO_2 atmosphere. Cells are grown to a density of approximately 1×10^6 cells/ml in 25-ml flasks and then counted and diluted to 3×10^4 cells/ml in fresh warm medium. The number of cells is determined both before and after dilution with medium. After dilution, the cells are transferred to 24-well plates, with 1 ml of suspension per well. The plates are allowed to incubate overnight before beginning the growth inhibition study. On the next day, three wells are resuspended by repeated pipetting of the medium with an Eppendorf style pipette. The cells are counted to obtain the 0% or original cell count. For a growth inhibition assay, triplicate wells are used for each drug concentration. One set of three wells on each plate is treated with PBS only (100% control), while the rest are treated with various concentrations of drug. The cells are allowed to grow for 48 h, resuspended by repeated pipetting, and are counted. The percent growth is determined from the counts, as described above, and the mean for each drug concentration is calculated. The values are then plotted on semilog paper (percent growth vs. log drug concentration), and the IC_{50}, which is the concentration of drug that inhibits the cell count by 50%, is determined from the growth curve.

Transient exposure of nonadherent cells is not possible in the 24-well plate format without some modification. Removing the cells efficiently from the drug requires centrifugation in either conical tubes or plates with conical wells. The simplest way is to substitute 24 conical centrifuge tubes for the 24-well plate and to spin the tubes at a low speed when removal of the drug is required. The medium is then carefully aspirated from the cell pellet, the pellet is loosened by gentle mixing, warm medium is added with further mixing, and the cells are returned to culture. The experiment is completed as described above for continuous exposure. Cells in such an experimental format can be susceptible to damage or loss. It is therefore important to monitor the 100% control to ensure that cell growth is equivalent to what would be observed in a continuous exposure study.

Multicompartment Growth Inhibition Assays

Karl Popper (1968) is usually credited with pointing out that negative assertions are impossible to prove in science. In this regard, the liposome dependency of a drug, or rather the lack thereof, is no exception. While the simple assay described above will prove that a drug is liposome dependent or that a carrier system effectively delivers such a drug, it will not prove with absolute certainty that the drug is not liposome dependent or that the carrier system is not at least partly effective. This arises because it is possible that a drug in liposomes may be delivered to cells selectively, yet its effect is not sufficient to exceed that of free drug. This is what occurs in many cases with methotrexate, for example (Leserman et al., 1981). To probe liposome dependency more thoroughly, we developed a two-compartment system described below. This system is based on a very simple idea. Two cell populations are separated by a semipermeable membrane that will allow passage of free drug without allowing passage of the carrier system. Population A is treated with the carrier system and reports the total effects of the system. Population B is exposed only to drug that passes through the semipermeable membrane and, therefore, reports only the effects of the drug that has leaked from the liposomes or carrier system. Population B can also report the loss of delivered drug from population A and, in some cases, will detect whether this drug has been converted to a more potent form. Evidence for this phenomenon is obtained by including parallel studies, in which population A is left out. The result gives a "leakage only" effect on population B, because the liposomes are still placed in compartment A, but there are no cells to process them. Comparison of this result with those obtained from the standard format will readily reveal whether cell-induced leakage is occurring. Cell-induced processing to a more potent metabolite is evident when the apparent leakage is much greater than 100%. This occurs when, with population A present, the IC_{50} of the encapsulated drug for population B is lower than the IC_{50} of the free drug. If the likely metabolite is known, it is possible to make estimates of the degree of metabolism and release by comparing the IC_{50} values with those of the metabolite. This is discussed below in more detail, and the equations for all relevant calculations are provided.

Method for Two-Compartment Assay

To ilustrate the use of the two-compartment system, the procedure that was developed for CV1-P cells as population A and EL4 cells as population B will be described. These assays are carried out using Costar transwells (Costar, Cambridge, MA) in 24-well plates, and it should be noted that in this experimental format population B must be a suspension cell culture, because adherent cells will grow on the transwell membrane and block diffusion of drug into the transwell compartment. The Costar transwells were fabricated with 0.1-μm pore size polycarbonate membranes. Validation studies showed that this pore size prevents the passage of liposomes prepared by reverse-phase evaporation from either egg phosphatidylglycerol and cholesterol (67:33) or distearoylphosphatidylglycerol and cholesterol (67:33) (Ng and Heath, 1989). In contrast, low-molecular-weight substances such as carboxyfluorescein diffuse to equilibrium in 4 h (Ng and Heath, 1989). CV1-P cells, if present, are grown in the wells of the plates directly, while the EL4 cells grow in suspension in the transwells. Drug is always introduced directly into the plate well and reaches the EL4 cells only by diffusion through the polycarbonate membrane of the transwell. Transwells are designed to allow direct transfer of material into the wells through small slots above the transwell compartment.

CV1-P cells and EL4 cells are both grown for the two-compartment assay under conditions described above, and they are both grown in the medium used for EL4 cells, namely Dulbecco's modified Eagle medium supplemented with 3 g/liter glucose and sodium pyruvate. CV1-P cells are diluted to 2×10^4 cells/ml, and 0.7 ml of this suspension is added to each well (1.4×10^4 cells/well). Plates are allowed to grow overnight, and a transwell is placed in each well of the plate. EL4 cells are diluted to 3×10^4 cells/ml, and 0.15 ml of cell suspension is added to each transwell. The plates are returned to culture conditions for 1 h before adding the drug to the compartment containing the CV1-P cells. As before, triplicate wells are treated with drug in a half-logarithmic concentration series. The cells are allowed to grow for 72 h before counting. For EL4 cell counting, the transwells are carefully removed from the plate before resuspending, diluting, and counting the cells. CV1-P cells are then released from the plate with trypsin, resuspended, and counted as described above.

Special mention should be made of the growing conditions used for the two-compartment system. The need for optimal diffusion of free drug through the polycarbonate membrane makes it necessary to put the plates on an orbital shaker during the 72-h growth period. The orbital shaker, which is placed in the incubator for this purpose, releases sufficient heat to cause an evaporation problem if no further steps are taken. Consequently, a 1-in.-thick slab of expanded polystyrene is placed on the shaker top to insulate the plates from the heat released. The plates are placed in a 4-in.-deep plastic container, which contains a 1-in.-thick layer of sponge saturated with water. This prevents loss of volume from the plates, especially the wells at the edge of the plate, and renders cell growth uniform in all wells.

Data Treatment for Two-Compartment Assay

The two-compartment assay comes with an array of useful calculations, described briefly above. These calculations can be confusing, because of the use of IC_{50} values derived for free drug, encapsulated drug, and, in certain cases, potential metabolite. Moreover, there are IC_{50} values for population A (CV1-P), for population B (EL4), and also for population B in the absence of population A. In short, six different IC_{50} values derived from at least four related studies are being used. To clarify this, the following describes the necessary studies and the abbreviations for their respective IC_{50} values that are subsequently used in the equations. These experiments are generally each carried out at least five times to demonstrate the reproducibility of the results. Typically, the mean of all determinations is used in the calculations.

Experiment 1. A two-compartment study with both cell lines present to determine the IC_{50} of free drug on both population A and population B.

$$FA = IC_{50} \text{ of the free drug for cell population A}$$

$$FB = IC_{50} \text{ of the free drug for cell population B}$$

Experiment 2. A two-compartment study with both cell lines present to determine the IC_{50} of encapsulated drug on both population A and population B.

$$EA = IC_{50} \text{ of the encapsulated drug for cell population A}$$

$$EB = IC_{50} \text{ of the encapsulated drug for cell population B}$$

Experiment 3. A two-compartment study with only population B present to determine the IC_{50} of encapsulated drug on population B in the absence of population A.

$$EB_0 = IC_{50} \text{ of the free drug for cell population B}$$
$$\text{in the absence of population A}$$

Experiment 4. A two-compartment study with only population B present to determine the IC_{50} of the putative metabolite on population B.

$MB_0 = IC_{50}$ of the putative metabolic for cell population B

From the six values defined above, the following calculations can be carried out as necessary.

Percent Leakage. The initial intention in designing the two-compartment assay was to permit the measurement of the extent of liposome leakage under the conditions of simple drug delivery experiments, calculated as follows:

$$\% \text{ Leakage} = \frac{FB}{EB} \times 100$$

Percent Spontaneous Leakage. It is useful to determine to what extent leakage occurs spontaneously through the interaction of the liposomes with serum or medium components. This calculation gives that information and is the primary reason for carrying out experiment 2. The value of spontaneous leakage is calculated as follows:

$$\% \text{ Spontaneous Leakage} = \frac{FB}{EB_0} \times 100$$

Percent Processing. Using percent leakage and percent spontaneous leakage, it is possible to calculate the percentage of drug that is released as a result of the interaction of the liposomes with cell population A. This value is calculated as follows:

$$\% \text{ Processing} = \% \text{ Leakage} - \% \text{ Spontaneous Leakage}$$

Percent Metabolism. The results obtained in the above three calculations may suggest that the target cell population is processing the drug and releasing it in a more potent form. This is particularly so if the percent leakage value is greater than 100%, which is clearly impossible. This conclusion might also be drawn if the value for percent processing seems very large. Usually, if a drug is being metabolized, a fairly specific idea as to the nature of the metabolite produced is possible. For example, in the case of methotrexate-γ-aspartate (Ng and Heath, 1989), the work of Baugh *et al.* (1970) showed that methotrexate was a very likely product of its metabolism. Of course, by analysis, the structure of the metabolite might also be demonstrated. Therefore, it is possible to calculate percent metabolism by reference to the IC_{50} of the putative metabolite. This calculation assumes that if the metabolite is produced, it is produced rapidly and is the sole source of the growth inhibitory effect. These two assumptions are good ones, because the logarithmic nature of growth inhibition is such that the most potent compound in the system will typically dominate in terms of its effects. Percent metabolism is calculated as follows:

$$\% \text{ Metabolism} = \frac{MB_u}{EB} \times 100$$

Delivery Factor

All the above calculations use only the data derived form the bystander cell population, population B. The delivery factor is the only value to use the IC_{50} values obtained for the target cell population, population A. The delivery factor is the *in vitro* equivalent of a therapeutic index, and it shows by how much the effect of the liposome-encapsulated drug is greater on the target cell population than it is on the nontarget population. This calculation is the prime accomplishment of the two-compartment system, and it provides the most rigorous proof for the liposome dependency of a drug–drug delivery system combination. Methotrexate-γ-aspartate (Ng and Heath, 1989) was 92 times more effective for the target than for the nontarget cells based on this value. Fluorodeoxyuridine or fluorodeoxyuridine monophosphate (Heath and Brown, 1989) in a variety of negatively charged and neutral liposome formulations exhibited delivery factors close to one, showing convincingly that these two drugs were not liposome dependent in these delivery systems. The delivery factor is calculated as follows:

$$\text{Delivery Factor} = \frac{FA \times EB}{EA \times FB}$$

Precision

As mentioned above, each of the component studies is generally carried out at least five times, producing, for each IC_{50} value, a mean and standard deviation. The fractional standard deviation for percent leakage, percent spontaneous leakage, percent metabolism, or delivery factor is the square root of the sum of variance/mean^2 for each IC_{50} value included in the caculation. For all calculations except the delivery factor, which requires no further calculation, it is convenient to multiply this answer by 100 so that it is given as a percent value rather than a fractional value.

Other Endpoint Assays

The principles applied here may be used with any endpoint assay. For example, we have measured the ability of drugs to inhibit cell contractility (Heath *et al.*, 1990) and have shown that certain agents, principally fluoroorotate, are more effective in this regard in liposomes than they are in the free form (Heath *et al.*, 1987). We have also used 96-well colorimetric assays for cell growth to show that gallium (Mönkkönen *et al.*, 1993) and clodronate (Mönkkönen and Heath, 1993) are liposome dependent. The reader should refer to our publications for more details (Mönkkönen and Heath, 1993; Mönkkönen et al., 1993).

Conclusions

While space does not permit a description of all the experimental methods that have been used for the study of liposome-dependent drug delivery, it is hoped that what has been described above will provide sufficient conceptual insight and ideas to anyone wishing to carry out their own studies in this area and that they will be able to design experiments to suit their own particular experimental situation. The idea of using a carrier system to deliver a drug to cells is a very simple one. However, experimental proof that a carrier is delivering its contents in the desired way comes only if appropriate drugs are chosen for study. The experimental formats described here provide the means to identify effective carrier systems and the compounds that may effectively be delivered by them. It is hoped that this chapter will encourage anyone who chooses to work in this area of study to do so with rigor and thoroughness.

Acknowledgments

While the above chapter has been written from my perspective, it would be remiss of me not to recognize the contributions of the many colleagues with whom I have worked in this field of study. My thanks go to the late Demetri Papahadjopoulos, Jukka Mönkkönen, Keith Bragman, Kate Matthay, Bob Debs, Carolyn Brown, Ninfa Lopez Straubinger, Anna Abai, Steve Comiskey, Lawrence Ng, Maria Bruno, Jin-Seok Kim, Tormaine Thompson, and Ketan Amin. Finally, my special thanks go to Bob Straubinger for many valuable hours of discussion of liposome-dependent drug delivery.

References

Allen, T. M., McAllister, L., Mausolf, S., and Hyoffry, E. (1981). Liposome-cell interactions. A study of the interactions of liposomes containing entrapped anti-cancer drugs with the EMT6, S49 and AE1 (transport-deficient) cell lines. *Biochim. Biophys. Acta* **643**, 346–362.

Baugh, C. M., Stevens, J. C., and Krumdieck, C. L. (1970). Studies on gamma-glutamyl carboxypeptidase. I. The solid phase synthesis of analogs of polyglutamates of folic acid and their effects on human liver gamma-glutamyl carboxypeptidase. *Biochim. Biophys. Acta* **212**, 116–125.

Blumenthal, R., Ralston, E., Dragsten, P., Leserman, L. D., and Weinstein, J. N. (1982). Lipid vesicle-cell interactions: Analysis of a model for transfer of contents from adsorbed vesicles to cells. *Membr. Biochem.* **4**, 283–303.

Fraley, R., Subramani, S., Berg, P., and Papahadjopoulos, D. (1980). Introduction of liposome-encapsulated SV40 DNA into cells. *J. Biol. Chem.* **255**, 10431–10435.

Heath, T. D., and Brown, C. S. (1989). Targeted drug delivery: A two-compartment growth inhibition assay demonstrates that fluorodeoxyuridine and fluorodeoxyuridine monophosphate are liposome-independent drugs. *Sel. Cancer Ther.* **5,** 179–184.

Heath, T. D., and Brown, C. S. (1990). Liposome-dependant delivery of N-(phosphonactyl)-L-aspartic acid to cells *in vitro. J. Liposome Res.* **1,** 303.

Heath, T. D., Lopez, N. G., and Papahadjopoulos, D. (1985a). The effects of liposome size and surface charge on liposome-mediated delivery of methotrexate-gamma-aspartate to cells *in vitro. Biochim. Biophys. Acta* **820,** 74–84.

Heath, T. D., Lopez, N. G., Stern, W. H., and Papahadjopoulos, D. (1985b). 5-Fluoroorotate: A new liposome-dependent cytotoxic agent. *FEBS Lett.* **187,** 73–75.

Heath, T. D., Lopez, N. G., Piper, J. R., Montgomery, J. A., Stern, W. H., and Papahadjopoulos, D. (1986). Liposome-mediated delivery of pteridine antifolates to cells *in vitro*: Potency of methotrexate, and its alpha and gamma substituents. *Biochim. Biophys. Acta* **862,** 72–80.

Heath, T. D., Lopez, N. G., Lewis, G. P., and Stern, W. H. (1987). Antiproliferative and anticontractile effects of liposome encapsulated fluoroorotate. *Invest. Ophthalmol. Vis. Sci.* **28,** 1365–1372.

Leserman, L. D., Machy, P., and Barbet, J. (1981). Cell-specific drug transfer from liposomes bearing monoclonal antibodies. *Nature* **293,** 226–228.

Levy, G. (1987). Targeted drug delivery—some pharmacokinetic considerations. *Pharm. Res.* **4,** 3–4.

Mayhew, E., and Papahadjopoulos, D. (1983). Therapeutic applications of liposomes. *In* "Liposomes" (M. Ostro, ed.), p. 289. Marcel Dekker, New York.

McIntosh, D. P., and Heath, T. D. (1982). Liposome-mediated delivery of ribosome inactivating proteins to cells *in vitro. Biochim. Biophys. Acta* **690,** 224–230.

Mönkkönen, J., and Heath, T. D. (1993). The effects of liposome-encapsulated and free clodronate on the growth of macrophage-like cells *in vitro*: The role of calcium and iron. *Calcif. Tissue Int.* **53,** 139–146.

Mönkkönen, J., Brown, C. S., Thompson, T. T., and Heath, T. D. (1993). Liposome-mediated delivery of gallium to macrophage-like cells *in vitro*: Demonstration of a transferrin-independent route for intracellular delivery of metal ions. *Pharm. Res.* **10,** 1130–1135.

Ng, K.-Y., and Heath, T. D. (1989). Liposome-dependent delivery of pteridine antifolates: A two-compartment growth inhibition assay for evaluating drug leakage and metabolism. *Biochim. Biophys. Acta* **981,** 261–268.

Popper, K. R. (1968). "The Logic of Scientific Discovery." Hutchinson, London.

Stella, V. J., and Himmelstein, K. J. (1980). Prodrugs and site-specific drug delivery. *J. Med. Chem.* **23,** 1275–1282.

Szoka, F. C., Jr., and Chu, C. J. (1988). Increased efficacy of phosphonoformate and phosphonoacetate inhibition of herpes simplex virus type 2 replication by encapsulation in liposomes. *Antimicrob. Agents Chemother.* **32,** 858–864.

van Rooijen, N., Kors, N., ter Hart, H., and Claassen, E. (1988). *In vitro* and *in vivo* elimination of macrophage tumor cells using liposome-encapsulated dichloromethylene diphosphonate. *Virchows Archiv. B Cell Pathol. Incl. Mol. Pathol.* **54,** 241–245.

Vidal, M., Sainte-Marie, J., Philippot, J. R., and Bienvenue, A. (1985). LDL-mediated targeting of liposomes to leukemic lymphocytes *in vitro. EMBO J.* **4,** 2461–2467.

Wilson, T., Papahadjopoulos, D., and Taber, R. (1979). The introduction of poliovirus RNA into cells via lipid vesicles (liposomes). *Cell* **17,** 77–84.

[12] Liposome-Mediated Suicide Gene Therapy in Humans

By REGINA C. RESZKA, ANDREAS JACOBS, and JÜRGEN VOGES

Abstract

The LIPO-HSV-1-*tk* gene transfer system was developed for a 3-day pump application in a first prospective Phase I/II clinical study. Eight patients suffering from recurrent glioblastoma multiforme were treated intratumorally on the basis of convection-enhanced delivery using the nonviral vector system. It was possible to identify the target tissue together with assessment of vector distribution and gene product expression, as well as the metabolic effect of ganciclovir treatment, noninvasively, by the combination of magnetic resonance imaging and positron emission tomography as a multimodal molecular imaging system. The therapy was well tolerated without major side effects. In two of eight patients, we observed a greater than 50% reduction of tumor volume and in six of eight patients focal treatment effects. The noninvasive visualization of therapeutic effects on tumor metabolism and documentation of gene expression will be important for the further successful development and implementation of patient individual gene therapy.

Introduction

The most widely used suicide gene transfer system in preclinical investigations, as well as human clinical trials, is the one based on herpes simplex virus type 1 thymidine kinase/ganciclovir (HSV-1-*tk*/GCV) (Fillat *et al.*, 2003; O'Malley, 2000). HSV-1-*tk* encodes a viral enzyme that is foreign for human cells and is able to convert the inactive and relatively less toxic prodrug ganciclovir (GCV) into its monophospate form. As shown in Fig. 1, intracellular host kinases metabolize this GCV monophosphate into di- and triphosphates (Matthews and Boehme, 1988). The GCV triphosphate can be incorporated into dividing cells and inhibits DNA polymerase, which results in chain termination, termination of DNA synthesis, and cell death and can induce cytotoxicity in neighboring HSV-1-*tk*-negative (bystander) cells (Asklund *et al.*, 2003; Fillat and Sangro, 2003).

The primary tumor entities in HSV-1-*tk*/GCV gene therapy are brain, ovarian, and liver tumors (O'Malley, 2000). HSV-1-*tk*/retrovirus–producing cells (Boviatsis and Chiocca, 1994; Chen and Wao, 1994; Culver and Blaese, 1992; Ezzedine and Breakefield, 1991; Ram and Oldfield, 1993) were used

METHODS IN ENZYMOLOGY, VOL. 391

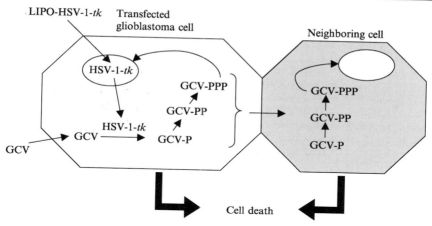

Fig. 1. Mode of action of the HSV-1-*tk*/GCV system and bystander effect. Expression of the transfected HSV-1-*tk*-gene in the presence of ganciclovir (GCV) results in the formation of GCV phosphate derivatives by the activities of HSV-1-thymidine kinase and cellular kinases. The purinic nucleotide 2′-deoxyguanosine triphosphate analogue GCV triphosphate can either incorporate into DNA or inhibit DNA polymerase. When transferred in untransduced cells that have direct contact with the transfected cell, the GCV phosphate derivatives will additionally cause cell death in neighboring cells (bystander effect).

for the treatment of rat glioblastoma models of HSV-1-*tk* carrying adeno- and herpes viruses. In particular, the use of adenoviral vectors encoding HSV-1-*tk* may give rise to liver toxicity (van der Eb and Hoeben, 1998).

Based on issues of immunogenicity, safety, and inflammation (Marshall, 1995), the development of secondary tumors (Check, 2002; Twombly, 2003), and the death (Lehrmann, 1999) of one patient associated with the use of viral vectors, nonviral *in vivo* gene therapy strategies (Ogris and Wagner, 2002; Schätzlein, 2001) based on cationic lipid delivery systems for cancer treatment gained considerable attention over the past years (Anwer and Sullivan, 2000; Felgner and Felgner, 1994; Felgner and Cheng, 1995; Gao and Huang, 1995; Nabel and Cho, 1994). The main advantages of cationic liposomes are no safety requirements, uncomplicated production under GMP conditions and upscaling of the pharmaceutical formulation, the possibility of repeated injection or infusion with pump systems due to stability at room temperature, low immunogenicity, and long-term stability after lyophilization of the DNA–lipid complexes. Safety, efficacy, and long-term stability at body temperature had made the LIPO-HSV-1-*tk* formulation attractive for a protracted intracerebral infusion via pump systems over several days (Reszka and Walther, 1995; Von Eckardstein and Reszka, 2001; Zhu and Reszka, 1996). According to the experimental data, we have chosen the LIPO-HSV-1-*tk* gene transfer system for a first nonviral clinical Phase I/II study in patients with recurrent glioblastomas.

HSV-1-*tk* Gene Therapy for the Treatment of Brain Tumors

Glioblastoma multiforme (GBM) is the second most frequent cause of neurological-related death in a population ranging in age from 45 to 70 years (Preston-Martin, 1999). Without treatment, the median survival is 4–6 months after diagnosis. Following surgery, radiotherapy, and adjuvant chemotherapy, the median survival times range from 10 to 12 months (Tatter and Harsh, 1998).

During the past years, several novel therapeutic concepts have been developed (Takeshima and Van Meier, 2000) to improve the prognosis for these patients. For the treatment of GBM, different clinical trials using the retroviral-mediated HSV-1-*tk*/GCV strategy have been applied via vector-producing cells (VPCs) (Klatzmann and Philippan, 1998; Rainov, 2000; Ram and Oldfield, 1997; Shand and Baubier, 1999).

In these brain tumor studies, an important key limiting step of gene therapy was found to be the limited efficacy of local delivery and vector particle distribution within the heterogeneous target tissue. Therefore, individual monitoring of both delivery of vector-gene constructs and gene expression is of great importance for the estimation of therapeutic effects. Based on molecular imaging technology employing positron emission tomography (PET) and a specific radioactively labeled nucleoside analogue, 2′-fluoro-5-iodo-1-β-D-arabinofuranosyl-uracil ([^{124}I]FIAU) (Jacobs and Heiss, 2001a; Tjuvajev and Blasberg, 1995, 1996, 1998), it was possible for the first time to identify the viable and appropriate target tissue and to noninvasively visualize the transduced "tissue dose" of vector-mediated HSV-1-*tk* gene expression in patients (Jacobs and Heiss, 2001b). In a prospective Phase I/II clinical study, eight patients suffering from recurrent GBM were treated with stereotactically guided intratumoral convection-enhanced delivery (CED) of the LIPO-HSV-1-*tk*-gene followed by a systemic application of ganciclovir (Voges and Kapp, 2002; Voges and Jacobs, 2003). Figure 2 shows the gene construct and the lipids used for the generation of the cationic lipid–DNA complex in LIPO-HSV-1-*tk* administered to patients. For regional assessment, MR and multitracer PET images were coregistered (Jacobs and Heiss, 2001a; Voges and Kapp, 2002; Voges and Jacobs, 2003).

The aims of the study were to determine the safety of the intratumoral pump application over a 48-h period of LIPO-HSV-1-*tk* by CED and the feasibility of multimodal molecular imaging technology for the noninvasive monitoring of CED-mediated HSV-1-*tk* gene expression. The study design including clinical considerations, preparation of LIPO-HSV-1-*tk*

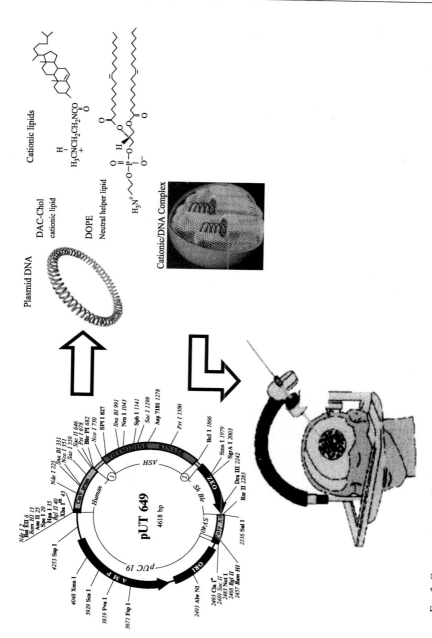

Fig. 2. Gene construct and cationic lipids used for generation of LIPO-HSV-1-*tk*. The plasmid pUT649 was complexed with the cationic lipid DAC-Chol (3β-(*N,N,N'*-dimethylaminoethane-carbamoxyl cholesterol) in the presence of the helper lipid DOPE (1,2-dioleoylphosphatidylethanolamine) and administered to patients with glioblastoma multiforme.

gene complexes, and the protocol scheme have been published by Voges and Kapp (2002) and are summarized in Fig. 3.

Monitoring

The general and neurological status of the patients, as well as standard laboratory parameters (hematology, blood biochemistry, urinalysis), were assessed daily. For the examination of cytokines and hematopoietic growth factors (TNF-α, IFN-γ, IL-2R, IL-4, IL-6, sCD-14, GM-CSF, and G-CSF), blood samples were collected during screening, after CED of the LIPO-HSV-1-*tk* gene system, and after completion of GCV application. To detect plasmid DNA in peripheral blood, Taqman-PCR (Becker and Whitley, 1999; Kawai and Saisho, 1999) was used as described previously (Voges and Kapp, 2002). Upon completion of the 29 days of treatment, patients were evaluated at months 2, 3, 4, 7, and 12 after LIPO-HSV-1-*tk* gene infusion.

In summary, we were able to show the following:

- The cationic liposome LIPO-HSV-1-*tk* gene vector system appears to be safe. Intratumoral infusion with a total volume of 30 or 60 ml caused no major adverse effects.
- Neither the cytokine level of TNF-α, IFN-γ, IL-2R, IL-4, IL-6, and sCD-14 nor the level of the hematopoietic growth factors GM-CSF and G-CSF changed after the treatment of patients using the LIPO-HSV-1-*tk* gene/GCV system.
- Via Taqman-PCR, persistent HSV-1-*tk*-plasmid DNA (at an average up to 1 pg/ml serum) was detected in individual patients at several days up to day 70 after LIPO-HSV-1-*tk* treatment.
- Local tumor responses (MET-PET) have been documented for six of eight patients.
- Noninvasive imaging of vector-mediated HSV-1-*tk* gene expression by FIAU-PET is feasible in the clinical setting.
- The extent of the FIAU-PET signal may predict therapeutic outcome.

Conclusions

Gene therapy for cancer is one of the most intriguing and complex applications of gene transfer technology. Together with developments of molecular targeted therapies including immunotherapy, further improvements in gene therapy are of utmost importance in improving the prognosis for patients with cancer. Based on the complexity and heredity of the cancer, different molecular concepts and gene delivery systems are currently being developed. A growing number of therapeutic concepts such as

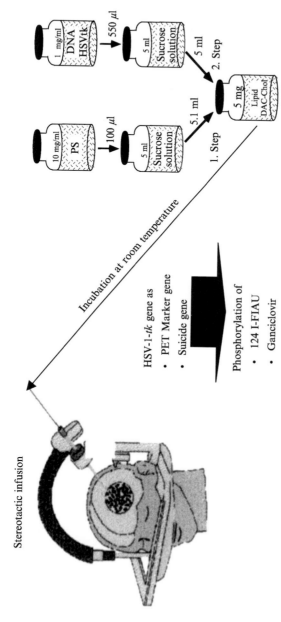

FIG. 3. Formation and administration of the LIPO-HSV-1-*tk* complex. Thirty minutes prior to intratumoral infusion, the DNA–liposome complexes were prepared according to the scheme presented at the right. First, a protamine/sucrose solution is mixed with DAC-Chol/DOPE. Following complete resuspension of the lipid film after 30 s of sonication, the DNA–sucrose solution is gently added. After incubation at room temperature for 30 min, the LIPO-HSV-1-*tk* complex is formed, and the sterile suspension fills the pump application system. Sucrose solutions, 10%; PS, protamine sulfate.

immunomodulation, different suicide and gene replacement strategies, and antiangiogenetic or cell death gene therapies, viral oncolysis or drug resistance gene therapy is now an essential part of current preclinical and clinical anticancer trials. With respect to clinical gene therapy trials, more than 70% of these are directed toward solid tumors.

Most important for the further successful development and implementation of gene therapy will be the improved determination and understanding of molecular and genetic characteristics, tumor architecture, metabolism, and various barrier functions. These will enable the following:

1. The efficient and targeted delivery of the transgene (improvement of vector distribution) and controlled expression of the gene product according to the delivery schedule (pump application versus a designed single treatment schedule).
2. The further implementation of methodologies suitable for noninvasive monitoring of vector application and gene transfer by molecular imaging.
3. The development of individual treatment schedules with the prodrug, depending of the expression of the gene product after noninvasive monitoring.
4. Controlled delivery of the prodrug or its formulation to the target cells.

Acknowledgments

The project was funded in part by the German Ministry of Education and Research (BMBF GE 9620-22). We thank Jana Richter for liposome preparation, Dipl. Ing. Ingrid Berger for HPLC analysis of the lipids and liposomes, Barbel Pohl for the *in vitro* quality control experiments, and Dr. Brigitte Hoch for help in preparing the manuscript.

References

Anwer, K., and Sullivan, S. M. (2000). Cationic lipid-based delivery system for systemic cancer gene therapy. *Cancer Gene Ther.* **7,** 1156–1164.

Asklund, T., and Almqvist, P. M. (2003). Gap junction-mediated bystander effect in primary cultures of human malignant gliomas with recombinant expression of the HSVtk gene. *Exp. Cell Res.* **284,** 185–195.

Becker, K., and Whitley, C. B. (1999). Real-time quantitative polymerase chain reaction to assess gene transfer. *Hum. Gene Therapy* **10,** 2559–2566.

Boviatsis, E. J., and Chiocca, E. A. (1994). Long-term survival of rats harboring brain neoplasms treated with ganciclovir and a herpes simplex virus vector that retains an intact thymidine kinase gene. *Cancer Res.* **54,** 5745–5751.

Check, E. (2002). Safety panel backs principle of gene-therapy trials. *Nature* **420,** 595.

Chen, S. H., and Woo, S. L. (1994). Gene therapy for brain tumors: Regression of experimental gliomas by adenovirus-mediated gene transfer *in vivo*. *Proc. Natl. Acad. Sci. USA* **91**, 3054–3057.

Culver, K. W., and Blaese, R. M. (1992). *In vivo* gene transfer with retroviral vector-producer cells for treatment of experimental brain tumors. *Science* **256**, 1550–1552.

Ezzedine, Z. D., and Breakefield, X. O. (1991). Selective killing of glioma cells in culture and *in vivo* by retrovirus transfer of the herpes simplex virus thymidine kinase gene. *New Biol.* **3**, 608–614.

Felgner, J. H., and Felgner, P. L. (1994). Enhanced gene delivery and mechanism studies with a novel series of cationic lipid formulations. *J. Biol. Chem.* **269**, 2550–2561.

Felgner, P. L., and Cheng, S. H. (1995). Improved cationic lipid formulations for *in vivo* gene therapy. *Ann. N.Y. Acad. Sci.* **772**, 126–139.

Fillat, C., and Sangro, B. (2003). Suicide gene therapy mediated by the herpes simplex virus thymidine kinase gene/ganciclovir system: Fifteen years of application. *Curr. Gene Ther.* **3**, 13–26.

Gao, X., and Huang, L. (1995). Cationic liposome-mediated gene transfer. *Gene Ther.* **2**, 710–722.

Jacobs, A., and Heiss, W. D. (2001a). Quantitative kinetics of [124I]FIAU in cat and man. *J. Nucl. Med.* **42**, 467–475.

Jacobs, A., and Heiss, W. D. (2001b). Positron-emission tomography of vector-mediated gene expression in gene therapy for gliomas. *Lancet* **358**, 727–729.

Kawai, S., and Saisho, H. (1999). Quantification of hepatitis C virus by TaqMan PCR: Comparison with HCV Amplicor Monitor assay. *J. Med. Virol.* **58**, 121–126.

Klatzmann, D., and Philippon, J. (1998). A phase I/II study of herpes simplex virus type 1 thymidine kinase "suicide" gene therapy for recurrent glioblastoma. Study Group on Gene Therapy for Glioblastoma. *Hum. Gene Ther.* **9**, 2595–2604.

Lehrmann, S. (1999). *Nature* **401**, 517.

Marshall, E. (1995). Gene therapy's growing pains. *Science* **269**, 1052–1055.

Matthews, T., and Boehme, R. (1988). Antiviral activity and mechanism of action of ganciclovir. *Rev. Infect. Dis.* **10**, S490–494.

Nabel, G. J., and Cho, K. (1994). Immunotherapy for cancer by direct gene transfer into tumors. *Hum. Gene Ther.* **5**, 57.

Ogris, M., and Wagner, E. (2002). Tumor-targeted gene transfer with DNA polyplexes. *Somat. Cell Mol. Genet.* **27**, 85–95.

O'Malley, B. W. (2000). Suicide gene therapy. *In* "Gene Therapy for Cancer: Therapeutic Mechanisms and Strategies" (N. Templeton, ed.), pp. 353–370. Marcel Dekker, New York.

Preston-Martin, S. (1999). Role of medical history in brain tumour development. Results from the international adult brain tumour study. *In* "The Gliomas" (M. S. Berger and C. B. Wilson, eds.), pp. 2–6. W. B. Saunders Company, Philadelphia, PA.

Rainov, N. G. (2000). A phase III clinical evaluation of herpes simplex virus type 1 thymidine kinase and ganciclovir gene therapy as an adjuvant to surgical resection and radiation in adults with previously untreated glioblastoma multiforme. *Hum. Gene Ther.* **11**, 2389–2401.

Ram, Z., and Oldfield, E. H. (1993). *In situ* retroviral-mediated gene transfer for the treatment of brain tumors in rats. *Cancer Res.* **53**, 83–88.

Ram, Z., and Oldfield, E. H. (1997). Therapy of malignant brain tumors by intratumoral implantation of retroviral vector-producing cells. *Nat. Med.* **3**, 1354–1361.

Reszka, R., and Walther, W. (1995). Liposome mediated transfer of marker and cytokine genes into rat and human glioblastoma cells *in vitro* and *in vivo*. *J. Liposome Res.* **5**, 149–167.

Schätzlein, A. G. (2001). Non-viral vectors in cancer gene therapy: Principles and progress. *Anti-Cancer Drugs* **12,** 275–304.

Shand, N., and Barbier, N. (1999). A phase 1–2 clinical trial of gene therapy for recurrent glioblastoma multiforme by tumor transduction with the herpes simplex thymidine kinase gene followed by ganciclovir. GLI328 European-Canadian Study Group. *Hum. Gene Ther.* **10,** 2325–2335.

Takeshima, H., and Van Meir, E. G. (2000). Application of advances in molecular biology to the treatment of brain tumors. *Curr. Oncol. Rep.* **2,** 425–433.

Tatter, S. B., and Harsh, G. H. (1998). Neurosurgical management of brain tumors. *In* "Gene Therapy for Neurological Disorders" (E. A. Chiocca and X. O. Breakfield, eds.), pp. 161–170. Humana Press, Totowa, NJ.

Tjuvajev, J. G., and Blasberg, R. G. (1995). Imaging the expression of transfected genes *in vivo. Cancer Res.* **55,** 6126–6132.

Tjuvajev, J. G., and Blasberg, R. G. (1996). Noninvasive imaging of herpes virus thymidine kinase gene transfer and expression: A potential method for monitoring clinical gene therapy. *Cancer Res.* **56,** 4087–4095.

Tjuvajev, J. G., and Blasberg, R. (1998). Imaging herpes virus thymidine kinase gene transfer and expression by positron emission tomography. *Cancer Res.* **58,** 4333–4341.

Twombly, R. (2003). For gene therapy, now-quantified risks are deemed troubling. *J. Natl. Cancer Inst.* **95,** 1032–1033.

van der Eb, M. M., and Hoeben, R. C. (1998). Severe hepatic dysfunction after adenovirus-mediated transfer of the herpes simplex virus thymidine kinase gene and ganciclovir administration. *Gene Ther.* **5,** 451–458.

Voges, J., and Kapp, J. F. (2002). Clinical protocol. Liposomal gene therapy with the herpes simplex thymidine kinase gene/ganciclovir system for the treatment of glioblastoma multiforme. *Hum. Gene Ther.* **13,** 675–685.

Voges, J., and Jacobs, A. (2003). Imaging-guided convection-enhanced delivery and gene therapy of glioblastoma. *Ann. Neurol* **55,** 479–487.

Von Eckardstein, and Reszka, R. (2001). Short-term neuropathological aspects of *in vivo* suicide gene transfer to the F98 rat glioblastoma using liposomal and viral vectors. *Histol. Histopathol.* **16,** 735–744.

Zhu, J., and Reszka, R. (1996). A continuous intracerebral gene delivery system for *in vivo* liposome-mediated gene therapy. *Gene Ther.* **3,** 472–476.

Section II

Liposomal Antibacterial, Antifungal, and Antiviral Agents

[13]   Use of Liposomes to Deliver Bactericides to Bacterial Biofilms

By Malcolm N. Jones

Abstract

Methods are described for the preparation of anionic and cationic liposomes and proteoliposomes with covalently linked lectins or antibodies by the extrusion technique (vesicles by extrusion, VETs). The liposomes are prepared from the phospholipid dipalmitoylphosphatidylcholine (DPPC), together with the anionic lipid phosphatidylinositol (PI) or the cationic amphiphile dioctadecyldimethylammonium bromide (DDAB) together with the reactive lipid DPPE-MBS, the *m*-maleimidobenzoyl-*N*-hydroxysuccinimide (MBS) derivative of dipalmitoylphosphatidylethanolamine (DPPE). Proteins (lectin or antibody), after derivatization with *N*-succinimidyl-*S*-acetylthioacetate (SATA), can be covalently linked to the surface of the liposomes by reaction with the reactive lipid, DPPE-MBS. The physical and chemical characterization of the liposomes and proteoliposomes by photon correlation spectroscopy (PCS) and protein analysis, to determine the number of chemically linked protein molecules (lectin or antibody) per liposome, are described. The liposomes can be used for carrying oil-soluble bactericides (e.g., Triclosan) or water-soluble antibiotics (e.g., vancomycin or benzylpenicillin) and targeted to immobilized bacterial biofilms of oral or skin-associated bacteria adsorbed on microtiter plates. Techniques for the preparation of immobilized bacterial biofilms, applicable to a wide range of bacterial suspensions, and for the analysis of the adsorption (targeting) of the liposomes to the bacterial biofilms are given. The mode of delivery and assessment of antibacterial activity of liposomes encapsulating bactericides and antibiotics, when targeted to the bacterial biofilms, by use of an automated microtiter plate reader, are illustrated, with specific reference to the delivery of the antibiotic benzylpenicillin encapsulated in anionic liposomes to biofilms of *Staphylococcus aureus*. The methods have potential application for the delivery of oil-soluble or water-soluble bactericidal compounds to a wide range of adsorbed bacteria responsible for infections in implanted devices such as catheters, heart valves, and artificial joints.

Introduction

Liposomes have been used to deliver bactericides to a relatively wide range of bacterial species. One example is the skin-associated bacterium *Staphylococcus epidermidis* (Sanderson and Jones, 1996), which is normally a nonpathogenic organism but can cause infections when implanted devices such as catheters (Raad *et al.*, 1993), heart valves (Etienne *et al.*, 1988), and artificial joints (Verheyen *et al.*, 1993) are used. Methicillin-resistant *Staphylococcus aureus* (MRSA) is a major nosocomial pathogen in the hospital environment (Kerr *et al.*, 1990). Liposomes have been used to deliver vancomycin, one of the few antibiotics effective against *S. aureus* (Kim *et al.*, 1999; Onyeji *et al.*, 1994). Bactericide delivery to oral bacteria such as *Streptococcus mutans* and *oralis* (formerly *sanguis*) is an approach to improved dental hygiene by use of proteoliposomes with surface-bound lectins (Jones *et al.*, 1993) or immunoliposomes with surface-bound antibodies (Robinson *et al.*, 2000). Phagocytic cells of the reticuloendothelial system (RES) are the natural targets of intravenously injected liposomes used to carry antibiotics to infections due to intracellular microbes such as *Salmonella, Mycobacterium*, and *Leishmania* (Gupta and Haq, 2004; Salem *et al.*, 2004; Szoka, 1990). The antibiotics ciprofloxacin and gentamicin when encapsulated in poly(ethylene glycol)-coated liposomes have enhanced therapeutic potential against rat lung infected with *Klebsiella pneumoniae* (Bakker-Woudenberg *et al.*, 2001; Schiffelers *et al.*, 2001). Ciproflaxacin-loaded liposomes are also effective against *Pseudomonas aeruginosa* (Nicholov *et al.*, 1993). Cationic liposomes prepared from the pure amphiphile dioctadecyldimethylammonium bromide (DDAB) show bactericidal properties toward *Escherichia coli, Salmonella thyphimurium, P. aeruginosa*, and *S. aureus* (Campanha *et al.*, 1999; Martins *et al.*, 1997).

Preparation of Liposomes by Extrusion

Liposomes are prepared by the vesicle extrusion technique (VET) (Hope *et al.*, 1985; Mayer *et al.*, 1986; Olson *et al.*, 1979). This method gives largely unilamellar liposomes, which, when two stacked polycarbonate filters of pore size 100 nm are used (Poretics, Livermore, CA), have diameters of 100 ± 15 nm. For larger pore sizes (e.g., 200 nm) the resulting VET diameters are generally smaller than the pore size.

VETs are prepared from the required lipid composition using dipalmitoylphosphatidylcholine (DPPC) as the major lipid component. Anionic liposomes are prepared by incorporation of phosphatidylinositol (PI) in a mol% range of 0–20%. Cationic liposomes are prepared by incorporation of a cationic amphiphile [e.g., stearylamine or preferably DDAB] in a mol% range of 0–20% and cholestrol (~18 mol%). If poly(ethylene glycol)-coated liposomes are required, dipalmitoylphosphatidylethanola-mine with covalently linked poly (ethylene glycol) of the required molecu-lar mass (DPPE-PEG, Avanti, Polar Lipids Inc. Alabaster, AL) in a mol% range of 0–10% can be used (Nicholas *et al.*, 2000). If liposomes are required incorporating a reactive lipid (DPPE-MBS, see below), this can be added to the lipid mixture at the required mol% in a range up to 15 mol%. For analytical purposes, the VETs are radiolabeled with either [^{3}H]DPPC on [^{14}C]DPPC, depending on whether they are used to entrap a radiolabeled drug that is ^{14}C or ^{3}H labeled, respectively.

The required lipid mixture (total mass ~30 mg) together with 5 μCi [^{3}H]- or [^{14}C]DPPC and any required oil-soluble drug are dissolved in either chloroform–methanol (4:1 by volume) or 20 ml of *tert*-butyl alcohol in the case of cholesterol-containing (cationic) liposomes in a 100-ml round-bottom flask. The solvents are freshly distilled and dried using molecular sieves. The solvent is removed by rotary evaporation at 60° to leave a thin lipid film. The film is hydrated by addition of 3 ml of phosphate-buffered saline (PBS, pH 7.4) at 60° with vigorous mixing using a vortex mixer. Water-soluble drugs for encapsulation are added at this stage. The temper-ature of 60° is chosen to be above the chain-melting temperature of all the lipids present (e.g., DPPC has a chain-melting temperature of approximate-ly 41°). The resulting suspension after vortex mixing is turbid and contains multilamellar vesicles (MLVs). To enhance drug encapsulation, the MLVs are transferred to a thick-walled glass tube, then frozen in liquid nitrogen (−196°), and thawed (60°). The freeze–thaw cycle is repeated five times. The resulting suspension of frozen and thawed MLVs (FATMLV) is ex-truded at 60° through two stacked polycarbonate filters (100-nm pore size) under nitrogen pressure (200–500 psi) with a Lipex Biomembranes, Inc extruder (Lipex Biomembranes, Inc. Vancouver, B.C., Canada). The extru-sion is repeated 5–10 times as required to produce an opalescent suspension of approximately 100-nm-diameter VETs.

Preparation of Proteoliposomes

Lectin Attachment

The covalent linking of lectins [e.g., wheat germ agglutinin (WGA) or succinyl concanavalin A (sCon A)] to the surface of liposomes involves three steps (Jones and Chapman, 1994; Kaszuba and Jones, 1997).

1. The preparation of liposomes incorporating a reactive lipid [e.g., the *m*-maleimidobenzoyl-*N*-hydroxysuccinimide (MBS) derivative of dipalmitoylphosphatidylethanolamine (DPPE-MBS)].

2. The derivatization of the lectin with N-succinimidyl-S-acetylthio-acetate (SATA), which on deacetylation generates free − SH groups.
3. The conjugation of derivatized lectin to the liposome. Conjugation occurs by reaction of the free − SH groups on the lectin and the maleimido group of the DPPE-MBS.

Preparation of Reactive Lipid, DPPE-MBS

DPPE (40 mg) is dissolved in a mixture of dry chloroform (16 ml), dry methanol (2 ml), and dry triethylamine (20 mg). MBS (20 mg, Pierce and Warriner, Chester, UK) is added, and the reaction mixture is stirred under nitrogen at room temperature for 24 h, after which the organic phase is washed three times with PBS (pH 7.4) to remove unreacted MBS. The DPPE-MBS derivative is recovered from the organic phase by rotary evaporation and analyzed by thin-layer chromatography (TLC) using a silica plate and a solvent mixture containing chloroform, methanol, and glacial acetic acid (65:25:13 v/v). The TLC should confirm that there is negligible contamination of the DPPE-MBS (R_f = 0.78) with DPPE (R_f = 0.56). The DPPE-MBS is stored in a chloroform–methanol (9:1 v/v) mixture at 4°.

Preparation of the SATA Derivative of Lectin

Lectin (e.g., sCon A or WGA) is derivatized using SATA at a molar ratio of SATA to lectin in a range up to 35 (Duncan et al., 1983). A volume of stock solution (in a range of 0–8.75 μl) containing 9 mg SATA in 50 μl dimethylformamide is added to sCon A [10 mg in 2.5 ml phosphate (50 mM)–ethylenediaminetetraacetate (EDTA) (1 mM) buffer, pH 7.5] at room temperature. After reaction for 15 min, the derivatized sCon A is separated from unreacted SATA by gel filtration on a Sephadex G-50 column (15 cm × 2 cm diameter). Fractions (2 ml) of derivatized sCon A are separated from unreacted SATA (eluted at a larger elution volume). The absorbance of the fractions is monitored spectroscopically at 280 nm, where both lectin and SATA absorb. The derivatized sCon A is assayed using a Lowry microassay (Lowry et al., 1951) with sCon A as the standard. The deacetylation of the SATA derivative of sCon A is done using 0.1 M hydroxylamine. Derivatized sCon A fractions (volume 2 ml) containing 0.9–1.1 mg ml^{-1} protein are reacted with 200 μl of hydroxylamine stock solution for 1 h (concentration range 0–1.25 M made up in 2.5 mM EDTA with sufficient solid Na_2HPO_4 added to bring the pH to 7.5). After deace-tylation, the sulfhydryl content of the derivatized sCon A is determined by the Ellman method (Ellman, 1959; Kaszuba and Jones, 1997). The sCon A is dimeric at pH 7 (molecular mass 52,000) (Gunther *et al.*, 1973). The number of − SH groups per dimer increases with the SATA to sCon A molar ratio (Fig. 1) and with the concentration of hydroxylamine used for deacetylation (Fig. 2) (Francis *et al.*, 1992).

Conjugation of Liposomes Containing Reactive Lipid (DPPE-MBS) with Derivatized Lectin (sCon A)

Conjugation is carried out at room temperature for 2 h or overnight at 4°. The liposomes incorporating DPPE-MBS (2 ml, total lipid concentration 1 mg ml^{-1}) are incubated with 2 ml of the deactylated SATA derivative of sCon A at a protein concentration of 1 mg ml^{-1}. After conjugation the reaction mixture is applied to a Sepharose 4B column to separate the proteoliposomes from the unreacted SATA derivative of sCon A. Protein analysis of the fractions (Lowry assay, Lowry *et al.*, 1951) and scintillation counting of the lipid show the coelution of protein and lipid that confirms conjugation. The extent of conjugation, provided the molar ratio of SATA to lectin and the concentration of hydroxylamine are not limiting, increases linearly with the mol% DPPE-MBS in the liposome. The number of lectin molecules per liposome depends on the liposome size and size of the lectin molecule. For example, for sCon A conjugated to liposomes incorporating

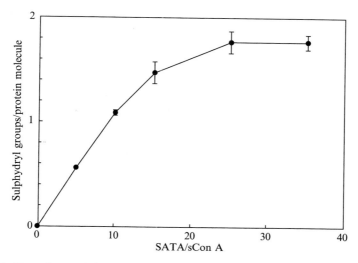

Fig. 1. Dependence of the sulfydryl content of sCon A (dimer) (i.e., number of − SH groups per dimer) on the SATA to sConA molar ratio. The concentration of hydroxylamine used for deacetylation was 0.1 *M*. Reproduced from Francis *et al.* (1992) with permission from Elsevier Science.

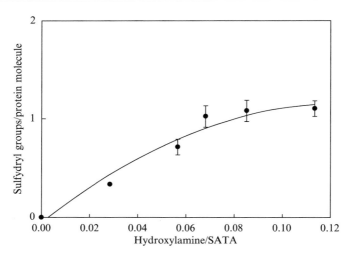

FIG. 2. Dependence of deacetylation of the SATA derivative of sConA on hydroxylamine concentration. The concentration of sConA was approximately constant at 1 mg ml^{-1}. Reproduced from Francis *et al.* (1992) with permission from Elsevier Science.

15 mol% DPPE-MBS, a small liposome (diameter ∼80 nm) conjugates ∼900 sCon A molecules and ∼175 WGA molecules (Francis *et al.*, 1992).

Immunoliposomes

The preparation of liposomes with covalently linked antibodies (immunoliposomes) is carried out as for lectins as previously described (Robinson *et al.*, 2000). Immunoglobulins (IgGs) generally have a larger molecular mass than lectins (∼160,000) so the numbers of conjugated antibodies per liposome are smaller. The methods described above may be adapted to the linking of any protein to the liposome surface. SATA reacts with the terminal amine groups on lysine residues so that the number of potential sites for derivatization will vary from one protein to another. In principle, only one lysine residue is required for derivatization and subsequent conjugation. For the antibody raised to the antigenic determinants on the surface of *S. oralis*, the numbers of antibody molecules per 100-nm-diameter liposomes incorporating 16 mol% DPPE-MBS is approximately 25.

Characterization of Liposomes and Proteoliposomes

Liposomes are characterized in terms of size by photon correlation spectroscopy (PCS). The Malvern Zetasizer 3000 or a similar instrument (e.g., Coulter N4 Plus Submicron Particle Sizer) is very convenient for obtaining the size and size distributions of dilute suspensions of liposomes. The sample cell is a square-sided cuvette. A 50-μl sample of liposomes (approximate lipid concentration 10 mg ml^{-1}) is added to 2 ml of PBS and scattering data are processed with the Malvern software to yield the z-average diameter of the liposomes ($\bar{d}_z$) and the standard deviation from fitting the scattering data to a normal distribution by cummulants analysis. The z-average diameter is defined by the following relations:

$$\bar{d}_z = \frac{\sum_i n_i d_i^3}{\sum_i n_i d_i^2} = \frac{\sum_i w_i d_i^2}{\sum_i w_i d_i} \qquad (1)$$

where n_i and w_i are the number and mass of species of diameter d_i. PCS measures the z-average diffusion coefficient ($\bar{D}_z$) of the particles, and the diameters are derived from the assumption that the particles are spherical and obey the Stokes–Einstein equation

$$\bar{D}_z = \frac{kT}{3\pi \eta \bar{d}_z} \qquad (2)$$

where k, T, and η are the Boltzmann constant, absolute temperature, and solvent viscosity, respectively. For VETs this is a valid assumption.

The Malvern Zetasizer 3000 is also used to obtain the ζ-potential of the liposomes from measurements of their electrophoretic mobilites (u), which are related to ζ-potentials (ζ) by the Smoluchowski equation,

$$u = \frac{\varepsilon_0 \varepsilon_r \zeta}{\eta} \qquad (3)$$

where ε_0 and ε_r are the permittivity of vacuum and relative permittivity, respectively. The Smoluchowski equation is applicable at high ionic strength when the electrical double-layer thickness is small relative to the size of the liposomes. For liposomes of diameter approximately 100 nm, the ionic strength of the solvent used to measure ζ-potentials should be in excess of 0.02 M. This corresponds to approximately a 1/10 dilution of PBS (Reboiras et al., 2001). Any concentration of PBS greater than 1/10 dilution will be satisfactory for the determination of liposome ζ-potentials based on the Smoluchowski equation. An electrolyte concentration greater than 10^{-2} M will also eliminate problems arising from the surface conductivity of the particles (Dulklin and Derjaguin, 1974).

The characterization of proteoliposomes in terms of size is done using PCS as for "naked" liposomes, but also a measurement of the number of protein molecules linked to the liposome surface is required. The calculation of this parameter is made from the molar ratio of protein to lipid (P/L) derived from the assay of protein by the Lowry method (Lowry et al., 1951) and the assay of lipid by scintillation counting. The number of protein molecules (p_n) in the liposomes sample is calculated. The liposome diameter (d) is used to calculate the number of lipid molecules per liposome (N^L). The number of lipid molecules per liposome is given by

$$N^L = \frac{4\pi}{a^L}\left(\frac{d}{2}\right)^2 + \frac{4\pi}{a^L}\left[\left(\frac{d}{2}\right) - h\right]^2 \tag{4}$$

where a^L is the area per lipid molecule in the liposomal bilayer (taken as 0.5 nm^2) and h is the bilayer thickness (taken as 7.5 nm) (Hutchinson et $al.$, 1989). The first term gives the number of lipid molecules in the outer monolayer of the bilayer and the second term the number in the inner monolayer. The number of protein molecules per liposome (P) is given by

$$P = \left(\frac{p_n}{l_n}\right)N^L \tag{5}$$

where l_n is the number of lipid molecules in the sample.

Unless the liposome sample is monodisperse, a distribution of liposome sizes will lead to a distribution of P values. Thus we may define number ($\bar{P}_n$), weight ($\bar{P}_w$), and z-average ($\bar{P}_z$) values of P calculated from the corresponding number $N(d_i)$, weight $W(d_i)$, and z-average $Z(d_i)$ distributions as follows:

$$\bar{P}_n = \frac{\sum_i P_i n_i}{\sum_i n_i} = \frac{\sum_i P_i N(d_i)}{\sum_i N(d_i)} \tag{6}$$

$$\bar{P}_w = \frac{\sum_i P_i w_i}{\sum_i w_i} = \frac{\sum_i P_i W(d_i)}{\sum_i W(d_i)} \tag{7}$$

$$\bar{P}_z = \frac{\sum_i P_i n_i d_i^2}{\sum_i n_i d_i^2} = \frac{\sum_i P_i Z(d_i)}{\sum_i Z(d_i)} \tag{8}$$

For a monodisperse liposome sample $\bar{P}_n = \bar{P}_w = \bar{P}_z$, but in practice a liposomal sample will not be completely homogeneous in terms of liposome size, so that $\bar{d}_n < \bar{d}_w < \bar{d}_z$ and $\bar{P}_n < \bar{P}_w < \bar{P}_z$. The computation of $\bar{P}_n$ and $\bar{P}_w$ from a normal weight distribution $W(d_i)$ of liposome diameters is carried out as previously described (Hutchinson et $al.$, 1989). It should be noted that a normal weight distribution is a symmetrical bell-shaped curve identical to the Gaussian law of errors, but the corresponding number distribution becomes progressively more asymmetrical with the increase in the distribution width (Hutchinson et $al.$, 1989). The number of protein molecules per liposome increases linearly with the mol% of DPPE-MBS with a slope that increases with the average size of the liposomes (Fig. 3).

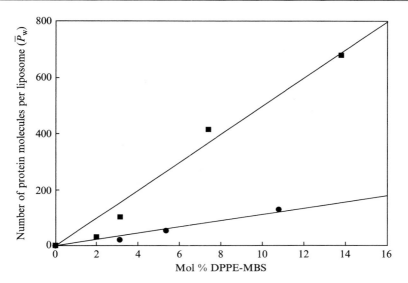

Fig. 3. Weight average number of proteins (WGA) per liposome (composition DPPC/ phosphatidylinositol 9:1 by wt.) as a function of DPPE-MBS incorporation. ■, Liposome weight average diameter 163 ± 6 nm; ●, Liposome weight average diameter 83 ± 4 nm. Reproduced from Hutchinson *et al.* (1989) with permission from Elsevier Science.

Targeting (Adsorption) of Liposomes to Bacterial Biofilms

Preparation of Bacterial Biofilms

Bacterial biofilms are prepared on the surface of microtiter plate wells made of irradiated polystyrene. Immulon 2HB flat-bottom 96-well microtiter plates (Dynex Technologies Inc, Chantilly, VA) are used. The required bacteria are grown in suspension culture to obtain a suspension of bacteria in PBS with an absorbance of 0.5. The precise conditions of growth may vary with the particular bacterium to be used, but the following is representative. The bacteria are taken from storage in glycerol at −70° or as supplied to prepare agar or blood agar plates [e.g., agar plus 5% (v/v) defibrinated horse blood]. The agar plates are prepared from 3.7 g of brain heart infusion (BHI) (code CM 255 Oxoid Ltd., Basingstoke, Hants., UK) in 100 ml of double-distilled water plus 1.5 g of bacteriological agar No. 1 (code L11 Oxoid Ltd.). The agar plates are streaked with bacteria and inserted upside down at 37° for 18 h in a candle jar [if anaerobic growth conditions are required, e.g., for *S. oralis* (*sanguis*)] or in a sterile environment (if aerobic conditions are required, e.g., for *S. epidermidis*). Streaked agar plates can be stored for a maximum of 1 month at 4°. The colonies of bacteria are used to inoculate nutrient broth prepared from 3.7 g BHI, 0.3 g yeast extract powder (code L21, Oxoid Ltd.), plus 1% (w/v) sucrose. The nutrient broth is divided into 10-ml aliquots in glass screw-cap bottles and sterilized (15 lb pressure for 15 min) prior to inoculation. After growth at 37° for 18 h, the cells are harvested by centrifugation (2000 rpm for 15 min) and washed four times in sterile PBS prior to suspension in PBS to give an absorbance of 0.5 at 550 nm. The final suspension is used to load microtiter plate wells (200 μl per well), which are then left overnight at room temperature for the bacterial cells to adhere. The choice of an absorbance of 0.5 is based on the observation that adsorption of the cells to the plate wells increases steeply from suspension of bacteria with absorbance in the range of 0–0.3 and more slow thereafter (Kaszuba et al., 1995). After cell adhesion, the supernatants are removed, and the biofilms are washed three times with 300-μl aliquots of PBS prior to use. The method yields biofilms of close-packed bacteria as shown by electron microscopy (Kaszuba et al., 1995). The bacterial cell density on the surface of the microtiter plate wells is measured radiochemically by labeling the cells with [^{3}H]thymidine. Table I shows data for the cell densities on the surface of microtiter plate wells of strains of S. epidermidis selected on the basis of their hydrophobicity using their affinity for partitioning into hexadecane.

Targeting Assay

Targeting of liposomes or proteoliposomes to immobilized bacterial biofilms is carried out following biofilm preparation. The biofilm and the microtiter plates can be stored at 4° for short periods (days) prior to use. Microtiter plate wells of bacterial biofilms are "blocked" with 300 μl per well of bovine serum albumin solution [1% (w/v) in PBS] or 0.02% (w/v) β-casein in PBS for 15–60 min at room temperature. They are then washed

TABLE I
ADSORPTION OF STRAINS OF STAPHYLOCOCCUS EPIDERMIDIS OF DIFFERING HYDROPHOBICITIES ON WELLS OF MICROTITER PLATES (IMMULON 2)[a]

S. epidermidis subpopulation	Mean % hydrophobicity	Cell density (cells m^{-2})
Wild type	88.7 ± 4.2	2.34 ± 0.11 × 10^{10}
Hydrophobic subpopulation	92.8 ± 1.3	2.34 ± 0.15 × 10^{10}
Hydrophilic subpopulation	61.2 ± 4.8	1.76 ± 0.08 × 10^{10}
M3 (adhesion defective) mutant	2.1 ± 0.6	2.15 ± 0.07 × 10^9

[a] Data from Sanderson et al. (1996).

three times with 300 μl of PBS and blotted dry by inversion on paper tissues followed by sharp "slapping" to remove any adhering droplets prior to inoculation with radiolabeled liposome suspensions (200 μl) at a liposomal lipid concentration in a range of 0–5 mM for 2 h at 37°. After incubation, the wells are washed three times with 300 μl PBS, blotted dry, and 200 μl of 5% (w/v) sodium n-dodecyl sulfate is added to each well, incubated for 30 min, and briefly sonicated (2 min) to fully disperse the biofilm with adsorbed liposomes. Aliquots (180 μl) of the dispersed biofilm are taken for scintillation counting. Control wells containing only bacteria, only PBS, and only liposomes are used to correct for background activity.

The extent of adsorption of liposomes to the bacterial biofilms is assessed from the measurement of the moles of adsorbed lipid (N_{obs}) relative to the number of moles that would be adsorbed if the biofilm was covered with a close-packed monolayer of liposomes (L_a). The value of L_a depends on the size of the liposomes and is calculated from the equation

$$L_a = \frac{A_{bf}\overline{N}_w}{\pi\left(\frac{\overline{d}_w}{2}\right)^2} \tag{9}$$

where A_{bf} is the geometric area of the biofilm exposed to liposomes of weight average diameter ($\overline{d}_w$), and $\overline{N}_w$ is the weight-average number of moles of lipid per liposomes. $\overline{N}_w$ is calculated from Eq. (4); when $d = \overline{d}_w$, $N^L = \overline{N}_w$. The area of the biofilm (A_{bf}) is measured from the geometric dimensions of the microtiter plate wells, which can be measured by loading the wells with 200 μl of a suitably colored liquid (e.g., copper sulfate solution) and using a cathatometer to determine the width and height of the liquid in the wells. For 96-well microtiter plates, A_{bf} is of the order of 2×10^{-4} m^2 per well (Chapman $et\ al.$, 1990). The true surface area of the biofilm will exceed the geometric area because of the approximately hemispherical shape and surface roughness of the cells. Also there is "dead" space between cells in a closely packed monolayer that is neglected in taking the geometric area. However, this is corrected for, as it is easily shown that the true area covered by closely packed spherical cells is 0.785 A_{bf}. The targeting or adsorption is expressed as the apparent monolayer coverage (percent amc) and is calculated from the equation

$$\%\ amc = \frac{N_{obs}}{L_a} \times 100 \tag{10}$$

The percent amc increases with liposomal lipid concentration and often can be fitted to the Langmuir adsorption isotherm written in the form (Sanderson $et\ al.$, 1996)

$$\% \text{ amc} = \frac{(\% \text{ amc})_{max} [L]}{K_d + [L]} \tag{11}$$

where $(\% \text{ amc})_{max}$ is the maximum apparent monolayer coverage, K_d is the dissociation constant, and $[L]$ the free liposomal lipid concentration. The association constant K_a $(=1/K_d)$ is used to calculate the Gibbs energy of association of the liposomes to the biofilm. The $(\% \text{ amc})_{max}$ can exceed 100% for some systems. This implies that either a multilayer of liposomes is formed at the bacterial surface, or more often, values of >100% arise because of the assumption that the biofilm is geometrically flat, and surface roughness has been neglected.

Table II shows parameters derived by the application of the Langmuir isotherm to measurements of the adsorption of cationic liposomes on biofilms of *S. aureus*. The maximum adsorption increases with the mol% of the cationic lipid component dimethylaminoethanecarbamoyl cholesterol (DC-Chol). The average Gibbs energy of adsorption of -20.5 ± 0.6 kJ mol^{-1} is typical for the adsorption of liposomes of various compositions to bacterial biofilms.

Delivery of Bactericides to Biofilms and the Assessment of Antibacterial Activity

Liposomes are prepared as described above with encapsulated bactericidal agent. The agent is added, in the required concentration range, to the initial lipid mixture if it is oil soluble, e.g., Triclosan (Irgasan). If it is water

TABLE II
PARAMETERS FOR THE ADSORPTION OF CATIONIC LIPOSOMES (DPPC/Chol/DC-Chol)[a]
OF VARYING MOLAR RATIO (MOL% CATIONIC LIPID) TO BIOFILMS OF *STAPHYLCOCCUS AUREUS*
AT 25°, pH 7.4[b]

Molar ratio (DPPC/Chol/ DC-Chol)	DC-Chol (%)	$(\% \text{amc})_{max}$	K_a (M)	ΔG_a^{0c} (kJ mol^{-1})
(1:0.49:0.11)	6.4	35.1 ± 1.6	5525 ± 1136	−21.36 ± 0.52
(1:0.49:0.22)	11.9	43.1 ± 2.1	4193 ± 884	−20.68 ± 0.53
(1:0.49:0.33)	16.9	46.0 ± 2.6	2876 ± 679	−19.74 ± 0.60
(1:0.49:0.43)	21.4	75.2 ± 3.6	3228 ± 646	−20.03 ± 0.50
(1:0.49:0.65)	28.9	75.5 ± 5.2	4153 ± 1288	−20.65 ± 0.80

[a] Dipalmitoylphosphatidylcholine/cholesterol/dimethylaminoethanecarbamoyl cholesterol.
[b] Data from Kim *et al.* (1999).
[c] Calculated from $\Delta G_a^0 = -RT \ln K_a$ (liposome diameter 121 ± 11 nm).

soluble, it is added to the aqueous phase on hydration of the lipid film prior to preparation of the MLV, e.g., vancomycin, chlorohexidine, gentamicin, or benzylpenicillin. Ideally, radiolabeled drug should be used if available. The liposomal lipid has a tracer determined by the drug label, e.g., [^{14}C]DPPC is used with [^{3}H]Triclosan or [^{3}H]DPPC with [^{14}C]benzylpeni-cillin. Immediately before use, liposomes are separated from free drug by gel filtration using a Sepharose 4B column (24 cm × 1.8 cm diameter). The peak fraction of liposomes is identified by removal of 10 μl for scintillation counting of lipid and drug. The biofilms prepared as above are incubated with aliquots (250 μl) of liposomes. The liposome lipid concentrations are usually in a range up to 5 mM and the incubation times in a range up to 3 h, although for some applications, e.g., delivery to oral bacteria, a short time range up to 10 min may be appropriate. The drug concentration in the liposome samples is measured [e.g., radiochemically or using an appropri-ate wet assay such as a Lowry assay (Lowry et al., 1951) for vancomycin (Kim et al., 1999) and gentamicin (Sanderson and Jones, 1996)], and con-trols are done with the equivalent concentrations of free drug. Other controls with wells containing only bacteria and only PBS are run to define the maximum and minimum growth of bacteria, respectively.

After incubation (targeting) of the liposomes with the biofilms, the liposome suspension and controls are removed from the wells, and the wells are washed three times with 300 μl PBS and blotted. Sterile nutrient broth (200 μl) (as used for preparation of the cell cultures above) is added to the wells, and they are then incubated for 18 h at 37° for cell growth to occur. After incubation, the biofilms are dispersed by sonication, and the absorbance of each well is measured at 630 nm using a suitable microtiter plate reader (e.g., Dynatech MR60 coupled to a suitable computer). The increase in absorbance over the 18-h period is taken as a measure of bacterial growth. The absorbance of wells containing only bacteria (A^{b}) and only PBS (A^{0}) are taken as 100% and 0% growth, respectively. The percent growth of the bacteria is given by,

$$\% \text{ Growth } = \frac{(A - A^{0})}{A^{b} - A^{0}} \times 100 \qquad (12)$$

A preferable method is to monitor the growth of bacteria in each well as a function of time up to 24 h using an automated programmable plate reader such as the DIAS Dynatech Laboratories plate reader, which has software capable of monitoring wells up to 24 h and recording either the time for each well to reach a specified absorbance or the time to reach maximum growth rate. Absorbance time profiles for each well are also produced. The use of this method enables a direct comparison to be made between the growth profiles of biofilms exposed to liposomes carrying bactericide over a range of concentration, free drug at equivalent concentrations, and bacterial growth in the absence of bactericide.

Figure 4 shows the absorbance (A)–time (t) profiles of biofilms of $S.$ *aureus* and the effects of liposomes [composition DPPC/Chol/DC-Chol (molar ratio 1:0.40:0.49)] carrying benzylpenicillin (PenG) and an equivalent concentration of free drug on the growth profiles. The times for maximum growth rate $(dA/dt)_{max}$ are computed from the growth curves by the Dynatech software and are shown plotted in Fig. 5. Figures 4 and 5 demonstrate that the targeting of liposomes and the delivery of PenG inhibit the growth of bacteria from the biofilms more effectively than "free" drug, i.e., it takes longer to reach the maximum growth rate after liposome delivery than it does for biofilms exposed to an equivalent concentration of the free drug. A suitable measure of the effectiveness of liposome delivery as measured from the inhibition of bacterial growth is the expression

$$\text{Growth inhibition factor} \quad = \quad \frac{(dA/dt)_{max}^{liposomes}}{(dA/dt)_{max}^{free\ drug}} \tag{13}$$

In general for the delivery of PenG using cationic liposomes with DC-Chol as the targeting lipid, the growth inhibition factor is in a range up to 3,

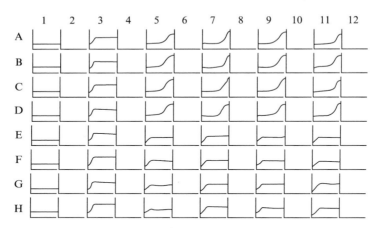

FIG. 4. Growth profiles (absorbance) of biofilms of *Staphylococcus aureus* as a function of time. The biofilms were incubated with liposomes of composition DPPC/DC-chol/cholesterol molar ratio 1:0.43:0.49 carrying benzylpenicillin at an overall concentration of 0.19 mM for different lengths of time as follows: column 5 A–D (1 h 45 min), column 7 A–D (1 h 30 min), column 9A–D (1 h 15 min), column 11A–D (1 h). Column 1 is a control containing no bacteria. Column 3 is a control containing only biofilms. Rows E–H are biofilms treated with free benzylpenicillin at a concentration of 0.19 mM for the same times as used for the liposome treatments. The profiles cover a range of absorbance from −0.5 to 1.5 and a time scale from 0 to 24 h. Reproduced from Kim and Jones (2004) with permission from Marcel Dekker, Inc.

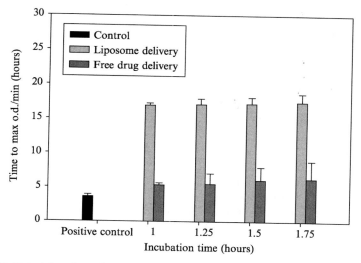

Fig. 5. Data taken from the profiles in Fig. 4 for the time to reach maximum growth rate [o.d. (abs) per minute] for biofilms treated with liposomes carrying benzylpenicillin and free benzylpenicillin. The conditions of the experiments are described in the legend to Fig. 4.

depending on the conditions of targeting (time of incubation of biofilm with liposomes) and the concentration of PenG in the liposomes (Kim and Jones, 2004). The growth inhibition factor decreases with PenG concentration. Clearly, the advantage of liposome delivery of drugs is most notably seen at low overall drug concentrations (i.e., low concentrations of encapsulated drug), where the targeting and delivery of the drug depend on the effectiveness of adsorption of the liposomes to the biofilm. Clearly, at sufficiently high drug concentrations, the free drug is an effective inhibitor of bacterial growth.

Advantages and Limitations of the Methods

The advantages of liposomes for drug delivery to bacterial biofilms may be summarized as follows:

- Liposomes are biocompatible, and their lipid composition can be varied widely.
- Liposome surfaces can be chemically modified by the introduction of site-directing molecules (lectins, antibodies, etc.) for targeting. The introduction of poly(ethylene glycol) to prevent uptake by the

reticuloendothelial system also inhibits uptake by bacterial biofilms (Ahmed *et al.*, 2001).

- Oil- or water-soluble drugs can be carried to bacterial biofilms by liposomes and are similarly effective in inhibiting bacterial biofilm growth (Robinson *et al.*, 2000).
- The effectiveness of liposome delivery is greatest with low drug concentrations in the liposomes, where equivalent concentrations of free drug do not inhibit cell growth as effectively. Liposome delivery is particularly advantageous when reduction of the toxic side effects of drugs is required.
- The methods are applicable to the targeting of drugs to, and the inhibition of, the cell growth of any type of bacterial cell or other cell types that form monolayers on a solid substrate.
- The methods are limited to the study of biofilms of bacteria immobilized on solid supports. The results obtained under these conditions may not necessarily reflect the behavior of the bacteria grown in suspension or *in vivo*, although they will give a useful indication of the possible value of liposome delivery under these conditions.

References

Ahmed, K., Muiruri, P. W., Jones, G. H., Scott, M. J., and Jones, M. N. (2001). The effect of grafted poly(ethylene glycol) on the electrophoretic properties of phospholipid liposomes and their adsorption to bacterial biofilms. *Colloids Surf. A* **194**, 287–296.

Bakker-Woudenberg, I. A. J. M., ten Kate, M. T., Guo, L., Working, P., and Mouton, J. W. (2001). Improved efficiency of ciprofloxacin in polyethylene glycol-coated liposomes for treatment of *Klebsiella pneumoniae* pneumonia in rats. *Antimicrob. Agents Chemother.* **45**, 1487–1492.

Campanha, M. T. N., Mamizukar, E. M., and Carmona-Ribeiro, A. M. (1999). Interactions between cationic liposomes and bacteria: The physical chemistry of bactericidal action. *J. Lipid Res.* **40**, 1495–1500.

Chapman, V., Fletcher, S. M., and Jones, M. N. (1990). A simple theoretical treatment of a competitive immunosorbent assay (ELISA) and its application to detection of human blood group antigens. *J. Immunol. Methods* **131**, 91–98.

Dulklin, S. S., and Derjaguin, B. V. (1974). *In* "Surface and Colloid Science" (E. Matijevic, ed.), p. 210. Wiley-Interscience, New York.

Duncan, R. J. S., Weston, P. D., and Wrigglesworth, R. (1983). A new reagent which may be used to introduce sulfhydryl groups into proteins and its use in the preparation of conjugates for immunoassay. *Anal. Biochem.* **132**, 68–73.

Ellman, G. L. (1959). Tissue sulphydryl groups. *Arch. Biochem. Biophys.* **82**, 70–77.

Etienne, J., Brun, Y., Elsolh, N., Delorme, V., Mouren, C., Bes, M., and Fleurette, J. (1988). Characterization of clinically significant isolates of *Staphylococcus epidermidis* from patients with endocarditis. *J. Clin. Microbiol.* **26**, 613–617.

Francis, S. E., Hutchinson, F. J., Lyle, I. G., and Jones, M. N. (1992). The control of protein surface density on proteoliposomes. *Colloids Surf. A* **62**, 177–184.

Gunther, G. R., Wang, J. L., Yahara, I., Cunningham, B. A., and Edelman, G. H. (1973). Concanavalin A derivatives with altered biological-activity. *Proc. Natl. Acad. Sci. USA* **70**, 1012–1016.

Gupta, C. M., and Haq, W. (2004). Tuftsin-bearing liposomes as a antibiotic carriers in treatment of macrophage infections. *Methods Enzymol.* **391**, 291–304.

Hope, M. J., Bally, M. B., Webb, G., and Cullis, P. R. (1985). Production of large unilamellar vesicles by a rapid extrusion procedure—Characterisation of size distribution, trapped volume and ability to maintain a membrane potential. *Biochim. Biophys. Acta* **812**, 55–65.

Hutchinson, F. J., Francis, S. E., Lyle, I. G., and Jones, M. N. (1989). The characterization of liposomes with covalently attached proteins. *Biochim. Biophys. Acta* **978**, 17–24.

Jones, M. N., and Chapman, D. (1994). "Micelles, Monolayers and Biomembranes," pp. 117–142. Wiley-Liss, New York.

Jones, M. N., Francis, S. E., Hutchinson, F. J., Handley, P. S., and Lyle, I. G. (1993). Targeting and delivery of bactericide to adsorbed oral bacteria by use of proteoliposomes. *Biochim. Biophys. Acta* **1147**, 251–261.

Kaszuba, M., and Jones, M. N. (1997). The use of lectins for targeting in drug delivery. *In* "Methods in Molecular Medicine: Lectin Methods and Protocols" (J. M. Rhodes and J. D. Milton, eds.). Vol. 9, Chapter 49. Humana Press Inc., Totowa, NJ.

Kaszuba, M., Lyle, I. G., and Jones, M. N. (1995). The targeting of lectin-bearing lipsomes to skin-associated bacteria. *Colloids Surf. B. Biointerfaces* **4**, 151–158.

Kerr, S., Kerr, G. E., Mackintosh, C. A., and Marples, R. R. (1990). A survey of methicillin-resistant *Staphylococcus aureus* affecting patients in England and Wales. *J. Hosp. Infect.* **16**, 35–48.

Kim, H.-J., and Jones, M. N. (2004). The delivery of benzyl pencillin to *Staphylococcus aureus* biofilms by use of liposomes. *J. Liposome Res.* **14**, 123–139.

Kim, H.-J., Gias, E. L. M., and Jones, M. N. (1999). The adsorption of cationic liposomes to *Staphylococcus aureus* biofilms. *Colloids Surf. A* **149**, 561–570.

Lowry, O. H., Roseburgh, N. J., Farr, A. L., and Randall, R. J. (1951). Protein measurement with the folin phenol reagent. *J. Biol. Chem.* **193**, 265–275.

Martins, L. M. S., Mamizukar, E. M., and Carmona-Ribeiro, A. M. (1997). Cationic vesicles as bactericides. *Langmuir* **13**, 5583–5587.

Mayer, L. D., Hope, M. J., and Cullis, P. R. (1986). Vesicles of variable sizes produced by a rapid extrusion procedure. *Biochim. Biophys. Acta* **858**, 161–168.

Nicholas, A. R., Scott, M. J., Kennedy, N. I., and Jones, M. N. (2000). Effect of grafted polyethylene glycol (PEG) on the size, encapsulation efficiency and permeability of vesicles. *Biochim. Biophys. Acta* **1463**, 167–178.

Nicholov, R., Khoury, A. E., Bruce, A. W., and DiCosmo, F. (1993). Interaction of ciprofloxacin loaded cells with *Pseudimonas aeruginosa* cells. *Cells Mater.* **3**, 321–325.

Olson, F., Hunt, C. A., Szoka, F. C., Vail, W. J., and Papahadjopoulos, D. (1979). Preparation if liposomes of defined size distribution by extrusion through polycarbonate membranes. *Biochim. Biophys. Acta* **557**, 9–23.

Onyeji, C. O., Nightingale, C. H., and Marangos, M. N. (1994). Enhanced killing of methicillin-restant *Staphylococcus aureus* in human macrophages by liposome-entrapped vancomycin and teicoplanin. *Infection* **22**, 1–5.

Raad, I., Costerton, W., Sabharwal, U., Sacilowski, M., Anaissie, E., and Bodey, G. P. (1993). Ultrastructure analysis of indwelling vascular catheters: a quantitative relationship between luminal colonization and duration of placement. *J. Infect. Dis.* **168**, 400–407.

Reboiras, M. D., Kaszuba, M., Connah, M. T., and Jones, M. N. (2001). Measurement of wall potentials and their time-dependent changes due to adsorption processes: Liposome adsorption on glass. *Langmuir* **17**, 5314–5318.

Robinson, A. M., Creeth, J. E., and Jones, M. N. (2000). The use of immunoliposomes for specific delivery of antimicrobial agents to oral bacteria immobilized on polystyrene. *Biomater. Sci.* **11,** 1381–1393.

Salem, I. I., Flasher, D. L., and Düzgüneş, N. (2004). Liposome-encapsulated antibiotics. *Methods Enzymol.* **391,** 261–291.

Sanderson, N. M., and Jones, M. N. (1996). Encapsulation of vancomycin and gentamycin within cationic liposomes for inhibition of growth of *Staphylococcus epidermidis. J. Drug Target.* **4,** 181–189.

Sanderson, N. M., Guo, B., Jacob, A. E., Handley, P. S., Cunniffe, J. E., and Jones, M. N. (1996). The interaction of cationic liposomes with the skin-associated bacterium *Staphylococcus epidermidis*: Effects of ionic strength and temperature. *Biochim. Biophys. Acta* **1283,** 207–214.

Schiffelers, R. M., Storm, G., and Bakker-Woudenberg, I. A. J. M. (2001). Therapeutic efficacy of liposomal gentamycin in clinically relevant rat models. *Int. J. Pharm.* **214,** 103–105.

Szoka, F. C., Jr. (1990). The future of liposomal drug delivery. *Biotechnol. Appl. Biochem.* **12,** 496–500.

Verheyen, C. C. P. M., Dhert, W. J. A., Petit, P. L. C., Rozing, P. M., and De Groot, K. (1993). *In vitro* study on the integrity of a hydroxyapatite coating when challenged with staphylococci. *J. Biomed. Mater. Res.* **27,** 775–781.

[14]   Long-Circulating Sterically Stabilized Liposomes in the Treatment of Infections

By IRMA A. J. M. BAKKER-WOUDENBERG, RAYMOND M. SCHIFFELERS, GERT STORM, MARTIN J. BECKER, and LUKE GUO

Abstract

The administration of antimicrobial agents encapsulated in long-circulating sterically stabilized liposomes results in a considerable enhancement of therapeutic efficacy compared with the agents in the free form. After liposomal encapsulation, the pharmacokinetics of the antimicrobial agents is significantly changed. An increase in circulation time and reduction in toxic side effects of the agents are observed. In contrast to other types of long-circulating liposomes, an important characteristic of these sterically stabilized liposomes is that their prolonged blood circulation time is, to a high degree, independent of liposome characteristics such as liposome particle size, charge and lipid composition (rigidity) of the bilayer, and lipid dose. This provides the opportunity to manipulate antibiotic release from these liposomes at the site of infection, which is important in view of the differences in pharmacodynamics of different antibiotics and can be done without compromising blood circulation time and degree of target localization of these lipoosomes. Depending on the liposome characteristics and the agent encapsulated, antibiotic delivery to the infected site is achieved, or the liposomes act as a microreservoir function for the antibiotic. In experimental models of localized or disseminated bacterial and fungal infections, the sterically stabilized liposomes have successfully been used to improve antibiotic treatment using representative agents of various classes of antibacterial agents such as the β-lactams, the aminoglycosides, and the quinolones or the antifungal agent amphotericin B. Extensive biodistribution studies have been performed. Critical factors that contribute to liposome target localization in infected tissue have been elucidated. Liposome-related factors that were investigated were poly(ethylene glycol) density, particle size, bilayer fluidity, negative surface charge, and circulation kinetics. Host-related factors focused on the components of the inflammatory response.

Introduction

The rationale for using liposomes as carriers of antimicrobial agents is their ability to modulate the pharmacokinetics of the encapsulated agents (Allen, 1997, 1998; Bakker-Woudenberg, 1995; Pinto-Alphandry *et al.*, 2000; Wasan and Lopez-Berestein, 1995). As a result, the biodistribution and clearance of the compounds are substantially changed, leading to significant effects on the therapeutic efficacy, as well as the toxicity, of the antibiotics. One evident limitation of "classical"/"conventional" liposomes is their fast elimination by cells of the mononuclear phagocyte system (MPS) (Abra and Hunt, 1981; Gregoriadis, 1988). Various strategies have been followed to reduce the recognition of the liposomes by the MPS and consequently to increase the blood residence time of the liposomes without substantial drug leakage, allowing for interaction with tissues beyond the MPS (Huang *et al.*, 1995; Storm and Woodle, 1998). Small, neutral liposomes with bilayers composed of "rigid" phospholipids of high-phase-transition temperature, administered at relatively high lipid dose, are required to obtain increased vascular circulation (Allen, 1981; Hwang *et al.*, 1980; Senior, 1987). MiKasome and AmBisome (NeXstar Pharmaceuticals Inc., San Dimas, CA) are liposome types with such characteristics, containing amikacin and amphotericin B, respectively. These formulations have been studied extensively in animals and humans with respect to their pharmacokinetics and therapeutic efficacy (Fielding *et al.*, 1998, 1999; Petersen *et al.*, 1996; Walsh *et al.*, 1998; Whitehead *et al.*, 1998). In particular, AmBisome has been widely used with success in patients with severe fungal infections and a variety of underlying diseases refractory to conventional amphotericin B (Ellis *et al.*, 1998; Tollemar *et al.*, 2001; Walsh *et al.*, 1998). The long blood residence time of these liposomes is strongly dependent on the lipid dose administered.

Long circulation of liposomes can also be achieved by incorporation of specific glycolipids such as monosialoganglioside (G_{M1}) and hydrogenated phosphatidylinositol (HPI) without the constraint of high lipid dose but still limited to the use of rigid lipids (Allen, 1994; Allen et al., 1989; Gabizon and Papahadjopoulos, 1992). The development of long-circulating so-called sterically stabilized liposomes (SSL) creates new possibilities in this respect. These SSL are prepared by using a lipid derivative of the hydrophilic polymer poly(ethylene glycol) (PEG) such as PEG-destearoyl-phosphatidylethanolamine (PEG-DSPE) as the stabilizing component. The presence of PEG provides a steric barrier against a variety of interactions at the bilayer surface with components in the biological environment that could destabilize the liposomes or could serve as opsonins resulting in decreased uptake by the MPS (Allen et al., 1991; Klibanov et al., 1990; Papahadjopoulos et al., 1991; Storm et al., 1995; Woodle, 1998; Woodle and Lasic, 1992). An important characteristic of SSL is that they show prolonged blood residence time and localization in infectious targets without requiring a small particle size, rigid nature of the bilayer, or administration of a high lipid dose (Allen and Hansen, 1991; Woodle et al., 1995). The bilayer fluidity of the liposomes, which is the determinant of the release of the encapsulated agent, can be manipulated; this is of great value in view of the differences in pharmacodynamics of the different classes of antibiotics (Bakker-Woudenberg et al., 1989; Schiffelers et al., 2001a; Silvander et al., 1998; Woodle et al., 1992). SSL show dose-independent, first-order pharmacokinetics (Allen et al., 1995).

In the treatment of infectious diseases, SSL are applied as carriers of antimicrobial agents to achieve drug localization at the infected site, to reduce side effects of potentially toxic drugs, or to serve as a microreservoir of drug in the circulation. The preparation and use of SSL-encapsulating antimicrobial agents of different classes such as aminoglycosides, β-lactams, quinolones, and polyenes are described in this chapter.

Preparation and Characterization of SSL

Radiolabeled SSL

The pharmacokinetics and biodistribution of intact liposomes are determined by the use of radioactively labeled liposomes. For the localization studies in bacterial pneumonia described in this chapter, a high-affinity ^{67}Ga-deferoxamine-mesylate (^{67}Ga-DF) complex as an aqueous liposomal marker has been used. As shown by Gabizon et al. (1988), this complex is appropriate for in vivo tracing of intact liposomes because of the advantages of minimal translocation of radioactive labels to plasma proteins and the rapid renal clearance rate when the label is released from the liposomes. Liposomes were prepared as described by Bakker-Woudenberg *et al.* (1993). In brief, appropriate amounts of partially hydrogenated egg phosphatidylcholine (PHEPC, iodination value 40), cholesterol, 1,2-distearoyl-*sn*-glycero-3-phosphoethanolamine-*N*-[poly(ethylene glycol)-2000] (PEG-DSPE), egg L-α-phosphatidylcholine (EPC), egg L-α-phosphatidylglycerol (EPG), L-(-phosphatidyl-L-serine) (PS), or distearoylphosphatidylcholine (DSPC) are dissolved in chloroform/methanol (1:1; v/v) in a round-bottom flask. The solvent is evaporated under reduced pressure in a rotary evaporator, and the lipid mixture is dried under nitrogen for 15 min, redissolved in 2-methyl-2-propanol, and freeze-dried overnight. The resulting lipid film is hydrated for 2 h in HEPES/NaCl buffer, pH 7.4 [10 mM N-(2-hydroxy ethyl)piperazine-N'-ethane sulfonic acid (HEPES) and 135 mM NaCl containing 5 mM of the chelator deferoxamine mesylate (Desferal)]. The chelator was added to enable labeling with [67]Ga.

SSL are sized in various ways. Liposomes of approximately 100 nm (range, 80–120 nm) are obtained by sonication of the hydrated lipid dispersion for 8 min with an amplitude of 8 μm using a 9.5-mm probe in an MSE Soniprep 150 (Sanyo Gallenkamp PLC, Leicester, UK). Other particle sizes of liposomes are obtained by multiple extrusion of the hydrated lipid dispersion through two stacked polycarbonate membranes, with pore sizes of 100 and 50 nm for the 100-nm liposomes, 400 and 200 nm for the 280-nm liposomes, and 600 and 400 nm for the 360-nm liposomes.

Particle size distribution is measured using dynamic light scattering detected at an angle of 90° to the laser beam on a Malvern 4700 System (Malvern Instruments Ltd., Malvern, UK). The polydispersity of the liposome population is reported by the system with an index ranging from 0.0 for an entirely monodisperse dispersion up to 1.0 for a complete polydisperse dispersion. For all liposome preparations used in the experiments described here, the polydispersity index value is below 0.3.

Nonencapsulated DF is removed by gel filtration of the liposomes on a Sephadex G-50 column (Pharmacia, Uppsala, Sweden) using HEPES/NaCl buffer as an eluent. The SSL are subsequently concentrated via ultracentrifugation at 365,000g for 2 h at 4°. Liposomes are labeled with ^{67}Ga according to Gabizon *et al.* (1988). ^{67}Ga-citrate (1 mCi/ml) is diluted 1:10 in aqueous 5 mg/ml 8-hydroxyquinone and incubated for 1 h at 52° to obtain ^{67}Ga-oxine. One milliliter of this mixture is added per 1000 μmol total lipid (TL) of liposomes. Since ^{67}Ga-oxine can pass the liposomal membrane and has a high affinity for the encapsulated chelator DF, the radioactive label becomes entrapped, resulting in the formation of a ^{67}Ga-DF complex in the aqueous interior of the liposomes. Unencapsulated ^{67}Ga is removed by gel filtration. Radiolabeled liposomes are concentrated by ultracentrifugation at 365,000g for 2 h at 4°. Resulting specific activities are between 4×10^4 and 2×10^5 cpm/μmol TL. Phosphate concentration is determined colorimetrically according to Bartlett (1959).

For the SSL localization studies in fungal pneumonia described in this chapter, ^{99m}Tc labeling of liposomes has been applied. ^{99m}Tc-SSL are prepared as described by Laverman *et al.* (1999a). The liposomes are composed of EPC, PEG-DSPE, cholesterol, and the hydrazinonicotinamide derivative of distearoylphosphatidylethanolamine (HYNIC-DSPE) in a molar ratio of 1.85:0.15:1:0.07. The mean diameter of the liposomes is 85 nm with a polydispersity index of 0.1. Preformed HYNIC-PEG liposomes are labeled with ^{99m}Tc as described by Laverman *et al.* (1999a). The radiochemical purity of the labeled SSL is determined using instant thin-layer chromatography on ITLC-SG strips with 0.15 M sodium citrate (pH 5.0) as the mobile phase and verified by elution on a PD-10 column.

SSL-Gentamicin

Gentamicin belongs to the aminoglycosides, a class of antimicrobial agents clinically used for treating severe (nosocomial) gram-negative and gram-positive infections, especially in immunocompromised patients, and for treating mycobacterial infections (Kumana and Yuen, 1994). Its chemical structure is shown in Fig. 1. Due to their polycationic nature, they

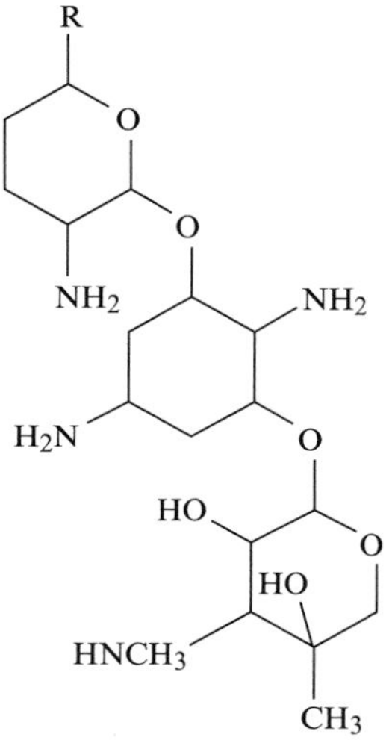

FIG. 1. Chemical structure of gentamicin. The amine moieties are protonated at physiological pH.

require parenteral administration. Plasma concentrations should be maintained within a narrow therapeutic window due to the dose-related adverse effects on kidneys and audiovestibular apparatus (Begg and Barclay, 1995). Liposomal encapsulation may help to increase the therapeutic index of aminoglycosides by site-specific delivery of the drug and/or reducing nephrotoxicity and ototoxcity (Schiffelers et al., 2001b).

So far, approaches to actively load gentamicin in liposomes have met with limited succes. Ion pairing and pH gradient methods did not improve gentamicin loading (Ruijgrok et al., 1999). For passive encapsulation we use 500 mg gentamicin in 2.5 ml of distilled water to hydrate a lipid film for 2 h. The lipid film is made by rotary evaporation of 1000 μmol TL composed of PHEPC/Chol/PEG-DSPE (molar ratio 1.85:1:0.15, respectively) dissolved in a mixture of chloroform/methanol (1:1). Higher concentrations of gentamicin result in the irreversible formation of aggregates. The total volume is brought to 10 ml by the addition of HEPES/NaCl buffer, pH 7.4. Subsequently, liposome size is reduced to approximately 100 nm by sonication for 8 min at an amplitude of 8 μm (MSE Sonyprep 150, Sanyo Gallenkamp PLC, Leicester, UK). Interestingly, separation of the unencapsulated gentamicin from the liposomal gentamicin by size-exclusion gel chromatography over a Sephadex G-50 column (Pharmacia, Uppsala, Sweden) with HEPES/NaCl buffer as the eluent results in a 2-fold lower encapsulation efficiency than separation by three consecutive ultracentrifugation steps at 60,000 rpm for 2 h per run (Beckman ultracentrifuge L-70, Beckman, Palo Alto, CA) in changes of fresh buffer. The mechanism behind this difference is as yet unclear. The specific activity of the liposomal gentamicin formulation purified by the ultracentrifugation method is approximately 80–90 μg gentamicin/μmol lipid. The amount of gentamicin added is approximately linearly related to the encapsulated amount, as is shown in Fig. 2.

Drug concentrations are determined using a diagnostic sensitivity agar diffusion test using Staphylococcus aureus Oxford strain (ATCC 9144) as the indicator organism. Total (liposome-encapsulated and free) and free (unencapsulated) drug concentrations are measured after disintegration of the liposomes using 0.1% (v/v) Triton X-100. By measuring the inhibitory zones of bacterial growth in the agar, the gentamicin concentration can be determined. After the ultracentrifugation step, less than 10% of the drug is shown to be unencapsulated. Gentamicin remains stably encapsulated in the liposomes for at least 1 month when stored at 4°.

The patent literature describes a way to increase loading of gentamicin into liposomes by using negatively charged phospholipids (such as phosphatidylinositol) (Meirinha da Cruz et al., 1993). It is based on the natural affinity of aminoglycosides for negatively charged phospholipids, which has

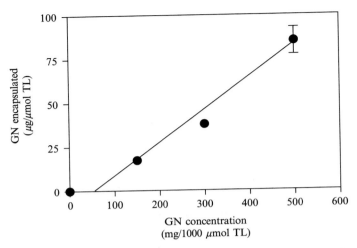

Fig. 2. Specific activity (μg gentamicin/μmol TL) as a function of the amount of gentamicin added per 1000 μmol TL.

also been implicated in their toxicity (Mingeot-Leclercq and Tulkens, 1999). This approach is valid only for a limited number of applications, as the high mol% of charged phospholipids needed will undoubtedly compromise circulation times.

SSL-Ceftazidime

Ceftazidime belongs to the cephalosporins. Its chemical structure is shown in Fig. 3. The cephalosporins are a class of antimicrobial agents with optimal pharmacodynamics when the bacteria are exposed for a prolonged period of time. Liposomal encapsulation may be beneficial to form a local depot at the target site or a circulating depot from which the encapsulated drug can be gradually released.

For passive encapsulation we use 750 mg ceftazidime in 2.5 ml of physiological aqueous NaCl solution to hydrate a lipid film for 2 h. The lipid film is made by rotary evaporation of 1000 μmol TL composed of PHEPC/Chol/PEG-DSPE (molar ratio 1.85:1:0.15, respectively) dissolved in a mixture of chloroform/methanol (1:1). Total volume is brought to 10 ml by the addition of HEPES/NaCl buffer, pH 7.4. Higher concentrations of ceftazidime appear too viscous for liposome formation. Subsequently, liposome size is reduced to approximately 100 nm by sonication for 8 min at an amplitude of 8 μm (MSE Sonyprep 150, Sanyo Gallenkamp PLC, Leicester, UK). The specific activity of the liposomal ceftazidime

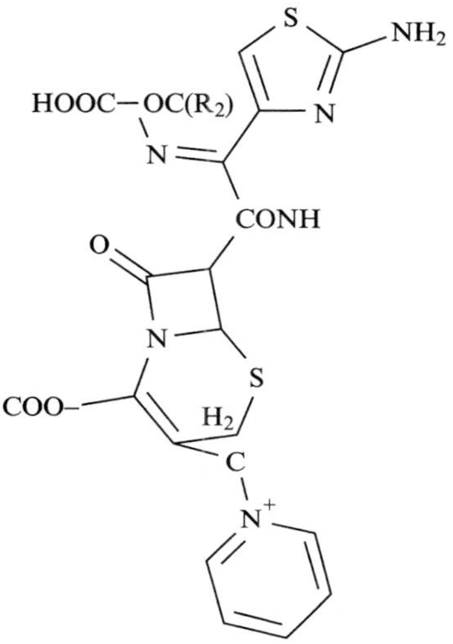

Fig. 3. Chemical structure of ceftazidime. Ceftazidime contains the four-membered β-lactam ring.

formulation purified is approximately 80–90 μg ceftazidime/μmol lipid irrespective of purification method. The amount of ceftazidime encapsulated is related linearly to the initial drug concentration (Fig. 4). Drug concentrations are determined using a diagnostic sensitivity agar diffusion test using the *Escherichia coli* (clinical isolate) strain as the indicator organism. Total (liposome-encapsulated and free) drug concentrations are measured after disintegration of the liposomes using 0.1% (v/v) Triton X-100. After ultracentrifugation or gel chromatography, less than 10% of the drug is shown to be unencapsulated. Ceftazidime remained stably encapsulated in the liposomes for at least 1 month when stored at 4°.

SSL-Ciprofloxacin

Ciprofloxacin is a synthetic fluoroquinolone antibiotic with a broad spectrum of activity. Like other fluoroquinolones, it contains a fluorine atom at position 6 of the 4-quinolone nucleus, a structural feature that results in enhanced activity and potency compared with nonfluorinated quinolone (Fig. 5). Although it has proven to be a highly effective antimicrobial agent, the potential exists to improve its therapeutic efficacy and expand its clinical use. One possible method of achieving this is through the use of liposomes as a drug carrier system. Liposomes that have been sterically stabilized

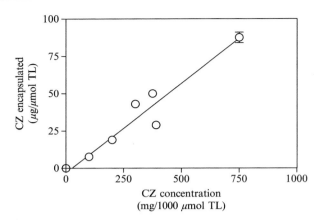

FIG. 4. Specific activity (μg ceftazidime/μmol TL) as a function of the amount of ceftazidime (CZ) added per 1000 μmol TL.

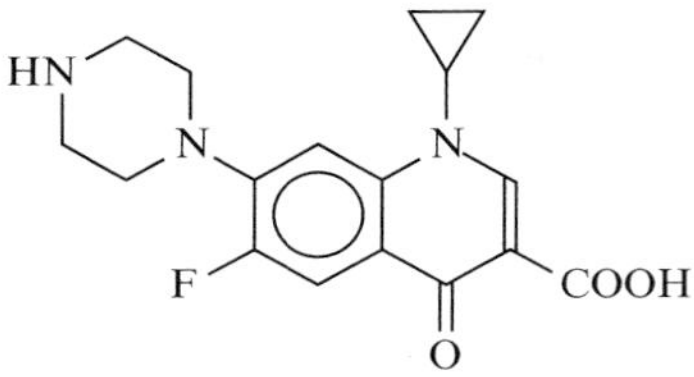

FIG. 5. Chemical structure of ciprofloxacin.

by a synthetic PEG-derivatized phospholipid (MPEG-DSPE) are especially promising (Papahadjopoulos *et al.*, 1991).

The HCl salt of ciprofloxacin is freely soluble in water at acidic pH (<5.0) but is poorly soluble at neutral pH or in the presence of sulfate ions. As shown in Fig. 6, the solubility of ciprofloxacin HCl in a 10% sucrose solution at ambient temperature is approximately 40 mg/ml, whereas the solubility decreases significantly in 250 mM ammonium sulfate solutions at either pH 5.7 or 4.5. Because of this solubility profile and the existence of titratable amines in the molecule, the drug is amenable to loading into preformed liposomes using a transmembrane ammonium sulfate gradient system (Haran *et al.*, 1993).

Liposomes are prepared using a thin lipid film hydration method. Lipid components (HSPC/cholesterol/MPEG-DSPE/49.97:45.18:4.83 mol%) are dissolved with chloroform in a round-bottom flask, and a thin lipid film is formed by rotary evaporation under vacuum. To hydrate, an appropriate

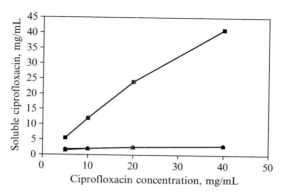

FIG. 6. Solubility of ciprofloxacin HCl in 10% sucrose (■), 250 mM ammonium sulfate, pH 5.7 (▲) or 4.5 (●).

ammonium sulfate solution is added, and the flask is rotated in a 60–65° waterbath, alternating with vigorous shaking by hand. Liposomes are extruded using a stainless-steel thermbarrel extruder (Lipex Biomembranes, Inc. Vancouver, B.C., Canada) at 60–65° through sequential polycarbonate membranes of decreasing pore size (0.4, 0.2, 0.1, and 0.05 μm) until the particle size is reduced to 100 ± 10 nm. They are then dialyzed against a 200-fold or greater volume of 10% sucrose 5 mM NaCl to remove the external ammonium sulfate (four exchanges over 36 h). To load, ciprofloxacin HCl is first dissolved in 10% sucrose at concentrations indicated below and then is incubated with the liposomes at 60–65° for 30 min with occasional mixing. Loaded liposomes are immediately placed in an ice-water bath to cool to room temperature. Unencapsulated drug is removed by dialysis with 10% sucrose 5 mM NaCl (200-fold or greater volume, one exchange over 12 h), followed by 10% sucrose 10 mM histidine, pH 6.5 (200-fold or greater volume, two exchanges over 24 h). The final product is sterilized by filtering through a 0.2-μm membrane (Gelman Sciences, Ann Arbor, MI).

To study the effect of ammonium sulfate concentration on ciprofloxacin loading, liposomes are prepared with an internal ammonium sulfate concentration of 250, 300, 350, or 400 mM and then loaded at drug concentrations ranging from 6.25 to 40 mg/ml. Varying drug concentrations allow an additional comparison of the encapsulated drug/lipid ratio to determine optimal loading conditions. As shown in Fig. 7, loading efficiency at each drug concentration increases with greater ammonium sulfate concentrations, confirming that loading capacity does, in fact, increase with higher ammonium sulfate concentrations. However, the difference between loading efficiencies at 350

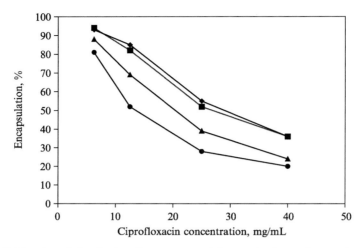

FIG. 7. Ciprofloxacin loading into liposomes containing varying internal ammonium sulfate concentrations: 250 mM (●), 300 mM (▲), 350 mM (■), and 400 mM (◆).

and 400 mM ammonium sulfate is minimal. This suggests a limit to the loading capacity, possibly due to the size of the liposomes. Since an increase to 400 mM ammonium sulfate did not yield a significant advantage, 350 mM ammonium sulfate was selected as the target concentration for SSL-Cipro preparation.

Figure 8 illustrates the relationships between drug concentration at loading, percent encapsulation, and encapsulated drug/lipid ratio for liposomes containing an ammonium sulfate concentration of 350 mM. While the encapsulated drug/lipid ratio increases with greater drug concentration at loading, the loading efficiency decreases significantly. For example, loading at 40 mg/ml ciprofloxacin achieves the highest encapsulated drug/lipid ratio (0.41 mg/mg), but loading efficiency is only 36%, with the majority of drug left unencapsulated. Based on the criteria of minimizing the loss of unencapsulated drug and maximizing the encapsulated drug/lipid ratio, the intermediate condition (12.5 mg/ml drug at loading) was chosen for loading SSL-Cipro at a total lipid concentration of 35 mg/ml (50 mM). At this loading condition, the weight ratio of drug/lipid in the loading mixture is 0.36 with a loading efficiency of 82% and an encapsulated drug/lipid ratio of 0.30.

The same data in Fig. 8 are used to calculate the encapsulated drug concentrations in liposomes, based on a theoretical entrapping volume of 2 µl per mmol of liposomes (Lasic, 1993). As seen in Table I, the encapsulated drug concentration increases with greater drug concentration

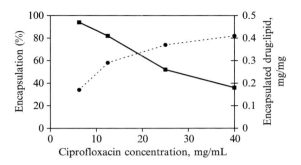

FIG. 8. Ciprofloxacin loading into liposomes containing 350 mM ammonium sulfate. The effect of ciprofloxacin concentration at loading on percent encapsulation (●) and encapsulated drug/lipid ratio (■).

TABLE I

CALCULATION OF ENCAPSULATED CIPROFLOXACIN CONCENTRATION IN LIPOSOMES

Cipro conc. at loading (mg/ml)	Encapsulation (%)	Lipid concentration (mM)	Calculated liposome internal volume (ml)[a]	Encapsulated Cipro concentration in liposomes (mg/ml)
6.25	94	50	0.1	59
12.5	82	50	0.1	103
25	52	50	0.1	130
40	36	50	0.1	144

[a] A theoretical entrapping volume of 2 μl per mmol lipids was used for unilamellar liposomes with mean particle diameter of 100 nm.

at loading. Loading at 6.25, 12.5, 25, and 40 mg/ml results in encapsulated ciprofloxacin concentrations of 59, 103, 130, and 144 mg/ml, respectively. Because ciprofloxacin is poorly soluble in sulfate ions (Fig. 6), the encapsulated ciprofloxacin in these liposomes is likely to form insoluble drug complexes. In fact, cryoelectron micrographic examination of SSL-Cipro revealed small globular crystals in the interior of these liposomes (Fig. 9) (Lasic et al., 1995).

A pilot study was conducted to evaluate the stability of SSL-Cipro stored at 2–8° and 30° (Table II). Over a 3-month period, SSL-Cipro appears to be stable at 2–8° but is less so at 30°. While the percent free drug at 2–8° shows no significant change, it increases to the 5–7% range at 30°, indicating that some leakage has occurred. However, the amount of free drug at 30° does

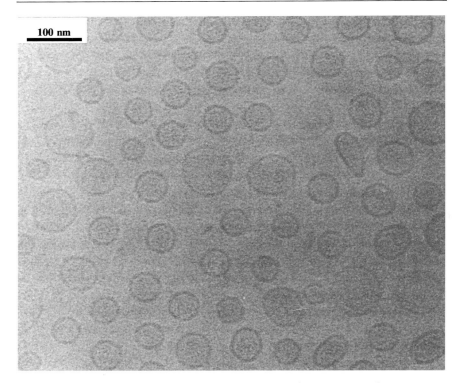

FIG. 9. Cryoelectron microscopy of SSL-Cipro (Lasic *et al.*, 1995).

not appear to increase further after 1 month. The data from the lysopho-sphatidylcholine (LPC) content indicate that the greatest concern for insta-bility is, in fact, degradation of HSPC. SSL-Cipro shows minor amounts of LPC (5% hydrolysis of HSPC) under 2–8° conditions at the 3-month time point, but significant degradation (15% hydrolysis of HSPC) at 30°. The results support a storage temperature of 2–8° for SSL-Cipro.

SSL-Amphotericin B

Amphotericin B is a macrolide polyene antobiotic. Its chemical struc-ture is shown in Fig. 10. Liposomal encapsulation of amphotericin B aims to reduce its toxicity, allowing higher doses to be given and therefore improving clinical efficacy.

Amphotericin B–containing SSL are prepared as described by van Etten *et al.* (1995a). The liposomes consist of PEG-DSPE, hydrogenated soybean phosphatidylcholine (HSPC), and cholesterol in a molar ratio of

TABLE II
PILOT STABILITY OF SSL-CIPRO AT 2–8° AND 30°

Time point (months)	Cipro concentration[a] (mg/ml)	Free Cipro[b] (%)	LPC[c] (mg/ml)	Mean particle diameter[d] (nm)
2–8°				
0	5.43	1.2	0.09	111
1	5.01	2.0	0.20	108
2	5.76	1.0	0.24	105
3	4.97	1.6	0.28	105
30°				
0	5.43	1.2	0.09	111
1	5.05	7.0	0.29	108
2	5.66	6.0	0.64	104
3	4.97	5.3	0.90	105

[a] The ciprofloxacin concentration was determined by reverse-phase HPLC.
[b] The percentage of free drug was measured by separating free drug from liposomes on a size-exclusion column and quantifying the amount of free and encapsulated drugs by UV spectrophotometry.
[c] The lysophosphatidylcholine (LPC) was determined by reverse-phase HPLC.
[d] The mean particle size of liposomes was determined by dynamic light-scattering analysis.

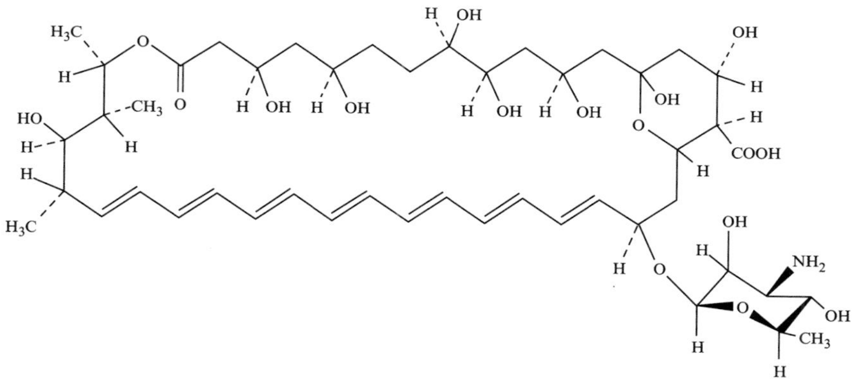

FIG. 10. Chemical structure of amphotericin B.

0.21:1.79:1. Amphotericin B is very poorly soluble in chloroform/methanol (1:1; v/v). Therefore, as a first step in liposome preparation, amphotericin B is complexed to PEG-DSPE by adding small volumes (20–50 μl) of 1 N HCl to a suspension of amphotericin B and PEG-DSPE in 2 ml of chloroform/methanol (1:1; v/v). This is followed by heating at 65° and vortex mixing until the solution clears. Small volumes (10–50 μl) of 1 N NaOH are added. This is followed by the addition of HSPC and cholesterol. When precipitation of amphotericin B is observed, small volumes of 1 N HCl are again added until the solution clears. This lipid mixture is evaporated to dryness in a round-bottom flask at 65°. The lipid film is hydrated by vortex mixing in a buffer solution containing 10 mM sodium succinate and 10% sucrose (pH 5.5) at 65°. Liposomes are sonicated resulting in an average particle size of 100 nm (range, 95–105 nm). Liposomes are separated from nonentrapped amphotericin B by gel filtration on a Sephadex G-50 column and are concentrated by using 300-kDa Microsep filters (Filtron, Breda, The Netherlands). The phospholipid concentration is determined by a phosphate assay (Ames and Dubin, 1960). The concentration of amphotericin B is determined spectrophotometrically at 405 nm after destruction of the liposomes in DMSO/methanol (1:1 v/v).

Infection Models in Experimental Animals

Bacterial Pneumonia

The experimental model of bacterial pneumonia applied in this chapter is a unilateral (left-sided) pneumonia caused by *Klebsiella pneumoniae* in rats, as described by Bakker-Woudenberg *et al.* (1982). RP/Aeur/RijHsd strain albino rats with a specified pathogen-free status (18–25 weeks of age, weighing 185–225 g) (Harlan, Horst, The Netherlands) are used. Rats are housed individually with free access to sterilized water and SRMA chow. In brief, rats are anesthetized by intramuscular injection of fluanisone and fentanyl citrate (Hypnorm), followed by intraperitoneal injection of pentobarbital (Nembutal). The left primary bronchus is intubated, a cannula is passed through the tube, and the left lung lobe is inoculated with 0.02 ml of a saline suspension containing 10^6 *K. pneumoniae* bacteria in the logarithmic phase of growth. Following bacterial inoculation, the narcotic antagonist nalorphine bromide is administered intramuscularly. After bacterial inoculation, the development of the infection in the left lung is progressive, whereas in the right lung of the same animal, signs of an infection are not observed. Untreated rats develop septicemia and pleuritis and die between day 3 and 6 after bacterial inoculation. At 64 h after inoculation of the left lung, a lobar pneumonia has developed. Three zones characteristic of this type of pneumonia can be clearly distinguished. The consolidated zone is the lower part of the lung where the bacterial inoculum is deposited and the infectious process starts. As the active infection gradually moves upward, the consolidated area is characterized by gray hepatization.

Microscopic evaluation indicates disintegration of alveolar walls, cellular debris, limited blood flow, and the deposition of connective tissue. The hemorrhagic zone is the active area of the infection featured by a dark red appearance. Microscopic evaluations indicate the presence of edema fluid, a large number of bacteria, infiltrating leukocytes, and hemorrhagic areas. The early-infected zone appears macroscopically as normal lung tissue; however, microscopic evaluation reveals the presence of bacteria and leukocytes involved in the infectious process. As the three zones of the lobar pneumonia can be clearly distinguished at 64 h after inoculation, and these zones show limited variation in size and macroscopic appearance between the animals, this time point is chosen to examine SSL localization, bacterial counts and capillary permeability

Antimicrobial treatment is started at 24 h after bacterial inoculation, when the bacterial count in the left lung has increased approximately 10^3-fold, to 3×10^9 viable *K. pneumoniae* (range, 5×10^8 to 8×10^9; $n = 10$), and 7 of 10 rats have developed positive blood cultures.

Studies are performed in immunocompetent or leukopenic rats. Leukopenia is induced by intraperitoneal injections of 60 mg/kg cyclophosphamide every 4 days starting at 5 days before bacterial inoculation. Leukopenia is ascertained by measuring leukocyte counts in fresh blood samples, obtained by retroorbital bleeding. Leukocytes are counted on a Cobas Minos Stex (Roche Haematology, Montpllier, France) using Minotrol 16 standards. As a result of cyclophosphamide treatment, the number of leukocytes in the blood is reduced 6-fold from $5.8 \times 10^9 \pm 1 \times 10^9$/liter (controls) to $1 \times 10^9 \pm 8 \times 10^8$/liter (mean $\pm$ SD, $n = 3$). Leukocyte counts remain reduced throughout the study period.

Fungal Disseminated Infection

The experimental model of disseminated fungal infection applied in this chapter is a systemic infection caused by *Candida albicans* in persistently leukopenic mice, as described by van Etten *et al.* (1995a). In brief, leukopenia is induced by intraperitoneal administration of cyclophosphamide at 100 mg/kg 4 days before *C. albicans* inoculation, followed by additional doses of 75 mg/kg on the day of fungal inoculation and the same dose at 3-day intervals thereafter. This treatment results in persistent leukopenia ($<1 \times 10^8$/liter) from the time of fungal inoculation to the termination of the study. Leukopenic mice are infected by inoculation of 3×10^4 *C. albicans* into the tail vein. Antifungal treatment is given intravenously and started at various times after inoculation.

In vitro antifungal activities of amphotericin B formulations in terms of effective killing ($>99.9\%$) of an inoculum of 1.3×10^7 viable *C. albicans*/

liter during 6 h of incubation are determined. A logarithmic-growth-phase culture of *C. albicans* is prepared and exposed for 6 h to increasing concentrations of each agent. At various intervals, the numbers of viable *C. albicans* are determined by plate counts of 10-fold serial dilutions of the washed specimens on Sabouraud dextrose agar.

The toxicities of amphotericin B formulations are measured in mice treated intravenously for 5 consecutive days once daily, with doses ranging from 0.1 to 1.0 mg/kg/day in steps of 0.1 mg/kg/day and above 1 mg/kg/day in steps of 2 mg/kg/day. Acute mortality is assessed directly following injection of the preparation. Blood urea nitrogen and serum creatinine levels as parameters for renal toxicity and aspartate aminotransferase and alanine aminotransferase levels as parameters for liver toxicity are determined by established methods in serum samples from mice 24 h after the termination of treatment.

Fungal Pneumonia

The model of fungal pneumonia in persistently leukopenic rats is based on the experimental bacterial pneumonia described in this chapter and adapted for *Aspergillus fumigatus* (Leenders *et al.*, 1996, 1999). In brief, leukopenia is induced by intraperitoneal administration of cyclophosphamide at 75 mg/kg 5 days before inoculation, followed by repeated doses of cyclophosphamide 60 mg/kg at 1 day before and 3 and 7 days after inoculation. This treatment results in granulocyte counts of less than 10^8 liter on the day of fungal inoculation and afterward. To prevent bacterial superinfections, rats receive ciprofloxacin (660 mg/liter) and polymyxin E (100 mg/liter) in their drinking water during the whole experiment. Starting 1 day before inoculation, daily amoxicillin (40 mg/kg/day), delivered intramuscularly, is added to this regimen for the remainder of the experiment. An *A. fumigatus* strain, originally isolated from an immunocompromised patient with invasive pulmonary aspergillosis, and once every month passed into rats to maintain its virulence, is used. To infect the left lung of the rats, 2×10^4 *A. fumigatus* conidia are used. In untreated rats, the mortality rate is $\pm 50\%$ on day 7 and 90–100% on day 12 after fungal inoculation. In approximately half of the rats fungal dissemination to extrapulmonary organs occurs, particularly to the liver. Blood cultures for aspergilli always remain negative.

Antifungal treatment is started at 24 h after inoculation. At that time, histopathological examination reveals short hyphae in the infected left lung. Hyphae increase in length over time and show radial growth and branching at later time points. Infiltrating leukocytes, predominantly granulocytes, are seen in low numbers, as a thin granulocytic rim around the fungal foci. There is no clear increase in the amount of leukocytes over time. Tissue invasion is seen at 48 h and later, showing especially broncho-invasion at earlier stages, progressing to increasing angioinvasion at later stages. Small hemorrhagic infarcts are seen in some histological sections at 48 h, increasing over time to extensive tissue infarction at 168 h. Tissue necrosis is seen relatively late in the course of the disease, first observed at 72 h and increasing at later time points to significant tissue necrosis with widespread fungal growth.

Localization of Radiolabeled SSL at Sites of Infection

Blood Circulation Kinetics and Biodistribution

Radiolabeled liposomes are administered intravenously at the selected dose via the tail vein at selected times after inoculation of the left lung. Blood samples of approximately 0.3 ml are taken from rats by retro-orbital bleeding using heparinized capillaries at selected times after liposome injection. Blood sample volume is measured, and radioactivity was counted in a Minaxi autogamma 5000 gamma counter (Packard Instrument Company, Meriden, CT) allowing calculation of the radioactivity present in the blood.

Prolonged blood residence time of the SSL in rats was observed. Circulation half-life is 20 h in uninfected rats and was accompanied by a relatively low hepatosplenic uptake of approximately 18% at 1 h after liposome administration (Bakker-Woudenberg et al., 1993; Schiffelers et al., 1999). In rats with K. pneumoniae pneumonia, the circulation time of the SSL is not different from uninfected control rats (Bakker-Woudenberg et al., 1993).

Biodistribution of liposomes is determined by sacrificing rats by CO_2 inhalation, and at selected times after liposome injection, various organs such as infected left lung, right lung, spleen, liver, kidneys, and heart were dissected. The organs are weighed, and radioactivity is counted to assess the biodistribution of the liposomes. The contribution of radioactivity in the blood to the radioactivity measured in the organs was subtracted, based on the total blood volume of the injected rats being 5.3% of the body weight. This percentage is determined by labeling syngeneic erythrocytes with ^{111}In-oxine according to Kurantsin-Mills et al. (1989). In an independent experiment, blood samples are taken 10 min after injection of the labeled erythrocytes, assuming that all erythrocytes were still present in the circulation. The dilution factor of the radioactive ^{111}In label allows calculation of the total blood volume, as well as the blood content of the various organs.

In the experimental pneumonia models, it has been observed that in infected rats the degree of localization of SSL in the infected left lung tissue is higher than localization of SSL in the left lung of uninfected rats (Bakker-Woudenberg *et al.*, 1993; Schiffelers *et al.*, 1999). The localization of SSL at the infected site is related to the local infected process, since in the right lung of the same infected rat (in which lung infection is not developed), the degree of localization of SSL was not increased compared with the localization observed in uninfected rats.

Localization in Bacterial Pneumonia: Influence of Liposome Characteristics

The degree of SSL localization in the infected left lung and uninfected right lung is determined in rats with varying intensity of infection (Bakker-Woudenberg *et al.*, 1993). The target localization of SSL appears to be strongly positively correlated with the severity of the infection. Up to 9% of the liposomes was recovered from the infected left lung tissue in severely infected rats. In contrast, classical short-circulating liposomes localize to a significantly lower degree compared with the SSL. The liposome-related factors that influenced the target localization of SSL have been investigated (Schiffelers *et al.*, 1999, 2000, 2001c). The liposome characteristics studied are particle size (100, 280, 360 nm), negative surface charge (0, 1, 10% PS), PEG density (1, 5, 10%), and bilayer fluidity (EPC, DSPC, PHEPC). In addition, the SSL are administered at various doses. The data are shown in Fig. 11 and Table III. The prolonged blood residence time and target localization of SSL appears to be relatively independent of the physicochemical characteristics of the SSL. All SSL preparations studied localize to a high degree at the site of infection compared with conventional liposomes. However, by increasing the particle size from 100 to 280 or 360 nm, or by reducing the PEG density from 5 to 1 mol%, the uptake of SSL by the MPS is increased, and as a result the SSL blood circulation time is decreased, leading to a reduced degree of target localization. Variation of liposome bilayer fluidity has no effect on blood residence time or on SSL localization at the infected site. Manipulation of the circulation kinetics by using different amounts of PS, or by varying the lipid dose, also has an effect on circulation kinetics of SSL. It is concluded that the degree of SSL localization at the infected site is positively linearly correlated with the area under the blood–concentration time curve (AUC) of the liposome formulations ($r = 0.92$, $p < 0.001$) (Fig. 11). In addition, the degree of SSL localization is remarkably independent of the physicochemical characteristics of the liposomes.

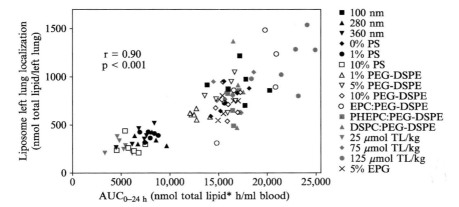

FIG. 11. Relationship between AUC over 24 h postinjection of a liposome formulation and corresponding degree of infected left lung localization of individual animals for indicated liposome formulations. ^{67}Ga-labeled liposomes were injected iv at 40 h after bacterial inoculation of the left lung. The lipid composition, particle size, or lipid dose are shown in Table III when they differ from the standard liposome composition PHEPC/Chol/PEG-DSPE 1.85:1.0:0.15 mol/mol, with a liposome size of 100 nm administered at a dose of 75 μmol TL/kg.

The general mechanism behind SSL localization at the site of infection is based on an equation by Kedem and Katchalsky (1958) that describes the plasma protein influx at sites of increased capillary permeability resulting from infection or inflammation.

Localization in Bacterial Pneumonia: Influence of Host Factors

The critical factors on the side of the host that may influence localization of SSL at the site of infection have also been investigated (Schiffelers *et al.*, 2001c). The highest number of *K. pneumoniae* organisms are present within the active hemorrhagic zone of the infection compared with the consolidated zone. High bacterial numbers correlate well with a pronounced inflammatory response, featured by a significantly increased capillary permeability as reflected by Evans blue dye levels and wet-to-dry weight ratio. In this zone, SSL localization also appears highest compared with the consolidated zone of infection. A study in rats with increasing severity of infection (established by using increasing inocula of *K. pneumoniae*) revealed that both SSL localization and capillary permeability correlate positively with the severity of the infection (Fig. 12). It can be concluded that increased capillary permeability allows extravasation of SSL.

TABLE III
LIPOSOMES: LIPID COMPOSITION, PARTICLE SIZE, AND LIPID DOSE[a]

Liposome property studied	Lipid composition (mol:mol)	Size (nm)	Dose (μmol TL/kg)
100 nm		100	
280 nm		280	
360 nm		360	
	PHEPC:Chol:PEG-DSPE:PS		
0% PS	1.85:1.0:0.15:0		
1% PS	1.82:1.0:0.15:0.03		
10% PS	1.55:1.0:0.15:0.30		
	PHEPC:Chol:PEG-DSPE		
1% PEG-DSPE	1.95:1.0:0.05		
5% PEG-DSPE	1.85:1.0:0.15		
10% PEG-DSPE	1.70:1.0:0.30		
	Phospholipid:PEG-DSPE		
EPC:PEG-DSPE	2.85:0.15		
PHEPC:PEG-DSPE	2.85:0.15		
DSPC:PEG-DSPE	2.85:0.15		
25 μmol TL/kg			25
75 μmol TL/kg			75
125 μmol TL/kg			125
	PHEPC:Chol:PEG-DSPE:EPG		
5% EPG	1.70:1.0:0.15:0.15		

[a] See the legend to Fig. 11 for details.

The contribution of circulating leukocytes to SSL localization seems of minor importance, as concluded from the data obtained in leukopenic rats. Apparently, drug delivery by SSL could also be feasible in the leukopenic host, in which severe infections frequently occur.

SSL target localization is also observed when inflammatory stimuli other than viable *K. pneumoniae* organisms are used, including lipopolysaccharide, 0.1 *M* HCl, or other infectious agents effecting an increase in capillary permeability (Boerman *et al.*, 2000; Laverman *et al.*, 1999b; Schiffelers *et al.*, 2001c).

Localization in Fungal Pneumonia

The SSL localization in the infected left lung of leukopenic rats with invasive pulmonary aspergillosis (IPA) at different stages of the disease has been investigated (Becker *et al.*, 2002). Liposomal uptake in the infected left lung increases over time (Table IV) and is significantly correlated with other parameters of progression of the fungal infection such as the size of

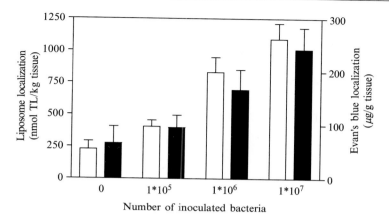

FIG. 12. Localization of SSL (open bars) and Evans blue (solid bars) in rats with increasing severity of *K. pneumoniae* left lung infection induced by increasing numbers of inoculated bacteria ($n = 6$, mean $\pm$ SD). Liposomes composed of PEG-DSPE/PHEPC/Chol (0.15:1.85:1) at a dose of 75 μmol TL/kg or Evans blue were administered intravenously at 24 h after bacterial inoculation of the left lung. Liposome localization and the amount of Evans blue were determined 24 h after injection of liposomes and Evans blue. From Schiffelers *et al.* (2001c), with permission.

TABLE IV

LOCALIZATION OF SSL IN THE INFECTED LEFT LUNG, RIGHT LUNG, AND BLOOD OF RATS WITH INVASIVE PULMONARY ASPERGILLOSIS AT DIFFERENT STAGES OF THE DISEASE[a]

	Localization of SSL (% ID/g tissue) at times after inoculation		
	24 h	72 h	168 h
Left lung	2.47 ± 0.47	3.84 ± 1.03	4.94 ± 0.63
Right lung	1.75 ± 0.41	1.37 ± 0.12	0.96 ± 0.09
Blood	5.51 ± 0.16	3.76 ± 0.34	2.71 ± 0.18
Left/right lung ratio	1.61 ± 0.28	2.83 ± 0.75	5.91 ± 0.97
Left lung/blood ratio	0.46 ± 0.08	0.93 ± 0.20	1.94 ± 0.26

[a] At 0, 48, and 144 h after fungal inoculation of the left lung, SSL were injected intravenously into groups of rats. Tissues were dissected at 24 h after injection of liposomes. Data are shown as mean $\pm$ SEM, $n = 8$ (24 h), $n = 7$ (72 h), and $n = 15$ (168 h).

the pulmonary hemorrhagic lesion and the levels of serum galactomannan. As shown in Table IV, a significant increase over time was found in the left/right lung ratios ($p < 0.001$). The mean left/right lung ratios are approximately 1 in uninfected rats and significantly elevated in infected animals at

72 h and 168 h postinoculation ($p = 0.04$ and $p = 0.0001$, respectively). At 168 h after inoculation, the mean left/right lung ratio, the mean percent ID/g left lung, and the mean left lung/blood ratio are highest. It is concluded that uptake of SSL in the infected lung is strongly associated with the severity of the disease.

Therapeutic Efficacy of SSL Formulations

SSL-Gentamicin in Bacterial Pneumonia

In rats, SSL composed of PEG-DSPE/PHEPC/Chol (molar ratio, 0.15:1.85:1) and containing the aminoglycoside gentamicin appears to be stable in the circulation after intravenous administration. The half-life of SSL-gentamicin is approximately 20 h, whereas the half-life of gentamicin in the free form is 20 min.

In the model of left-sided *K. pneumoniae* pneumonia in rats with intact host defense, the therapeutic efficacy of gentamicin administered at 24 h after bacterial inoculation is substantially increased when it was administered in the SSL-form (Bakker-Woudenberg *et al.*, 1995; Schiffelers *et al.*, 2001a). Increased survival of the rats and decreased bacterial counts are observed in the infected left lung at day 7 after bacterial inoculation compared with the effects of free gentamicin. As intact SSL-gentamicin does not kill the *K. pneumoniae* organisms *in vitro*, it can be concluded that the superior therapeutic effect of SSL-gentamicin is the result of localization and subsequent degradation of liposomes and release of encapsulated gentamicin at the infected site.

In leukopenic rats in which bacteremia as a complication of the pneumonia rapidly occurs, the antibiotic treatment should be directed toward the bacteria at the infectious focus (left lung), as well as the bacteria in blood. Table V shows that in these animals free gentamicin at the low dose of 5 mg/kg/day, which effected therapeutic efficacy in rats with intact host defense, is not effective anymore and had to be administered at the maximum tolerated dose (MTD) (40 mg/kg/day) for 5 days to obtain complete therapeutic efficacy. However, with the addition of a single dose of SSL-gentamicin (5 mg/kg) to low-dose gentamicin treatment (5 mg/kg/day), a therapeutic effect is achieved and results in a 7-fold lower cumulative amount of gentamicin compared with free gentamicin treatment alone. A single dose of SSL-gentamicin alone has no effect. Determination of the bacterial counts in the infected tissue reveals that local bacterial killing is effected by the administration of SSL-gentamicin, whereas the free gentamicin doses prevent the occurrence of bacteremia and hence mortality. Thus, free gentamicin and SSL-gentamicin complement each other.

TABLE V
SURVIVAL OF LEUKOPENIC RATS INFECTED IN THE LEFT LUNG WITH A HIGH OR A LOW
GENTAMICIN-SUSCEPTIBLE *K. PNEUMONIAE*[a]

Infected with high gentamicin-susceptible *K. pneumoniae*

GEN (5 days q 12 h)		SSL-GEN (5 days q 24 h)		GEN + SSL-GEN (5 days q 12 h + one single dose)	
Dose (mg/kg/day)	Survival (%)	Dose (mg/kg/day)	Survival (%)	Dose (mg/kg/day)	Survival (%)
5	10	5^b	0	$5 + 5^b$ (day 1)	100
10	30	10^b	0		
20	70	20^b	0		
40	100				

Infected with low gentamicin-susceptible *K. pneumoniae*

GEN (5 days q 12 h)		SSL-GEN (5 days q 24 h)		GEN + SSL-GEN (5 days q 12 h + 5 days q 24 h)	
Dose (mg/kg/day)	Survival (%)	Dose (mg/kg/day)	Survival (%)	Dose (mg/kg/day)	Survival (%)
40	0	20^b	0	$40 + 20^b$	50
>40	Toxicity			$40 + 20^c$	100

[a] Rats were treated with free gentamicin (GEN), SSL-gentamicin (SSL-GEN), or the combination; 10–15 rats were used per experimental group.
[b] A liposome formulation was used consisting of PEG-DSPE/PHEPC/Chol.
[c] A liposome formulation was used consisting of PEG-DSPE/EPC.

When the infection in the leukopenic rats is caused by a low gentamicin-susceptible *K. pneumoniae*, free gentamicin, even when administered at the MTD, is no longer effective (Table V). Combination with SSL-gentamicin is needed for therapeutic success (50% survival of rats). In this case, increasing the fluidity of the liposome bilayer by omitting Chol from the liposome formulation, which consisted of PEG-DSPE/EPC, effects an increased gentamicin release from the liposomes and hence a maximum therapeutic effect (100% survival of rats).

Other investigators have encapsulated the aminoglycoside, streptomycin, in SSL composed of PEG-DSPE/DSPC/Chol (molar ratio, 1:9:6.7). These liposomes were applied in a model of *Mycobacterium avium* complex infection in beige mice in which after intravenous inoculation the infection is disseminated to various organs such as liver, spleen, lungs, and lymph nodes (Gangadharam et al., 1995). The streptomycin-containing SSL, administered twice weekly for a period of 2 weeks, effect bactericidal activity in the spleen and a bacteriostatic effect in the liver and lung. During this treatment course, which is relatively short in view of the chronic M. avium infection, a superior efficacy of the SSL containing streptomycin over streptomycin encapsulated in classic liposomes composed of EPG/EPC/Chol (molar ratio, 1:9:6.7) was not observed. However, the authors speculate that a superior efficacy of SSL as a carrier of antibiotic may be expected when treatment is continued for a relatively long period of time, as in that case SSL have been shown to localize in deep-seated tissue macrophages.

SSL-Ceftazidime in Bacterial Pneumonia

Ceftazidime encapsulated in SSL composed of PEG-DSPE/PHEPC/Chol (molar ratio, 0.15:1.85:1) and administered intravenously into rats also appear to be relatively stable in the circulation (Bakker-Woudenberg et al., 1995). For β-lactam antibiotics (such as ceftazidime), it is known that continuously maintained plasma concentrations are important for successful treatment of severe infections. Therefore, we compared the efficacy of SSL-ceftazidime as a single dose with a continuous 2-day infusion of free ceftazidime at the same total dose. In the model of K. pneumoniae left-sided lung infection, the therapeutic efficacy of SSL-ceftazidime at a single dose is as effective as free ceftazidime at a continuous 2-day infusion. This is observed over a dose range varying from 1 mg/kg (not effective) up to 30 mg/kg (fully effective) (Bakker-Woudenberg et al., 1995).

Another study in the same infection model in rats with intact host defense revealed that a single dose of SSL-ceftazidime (12 mg/kg) given 24 h after bacterial inoculation is as effective as 10 doses of free ceftazidime administered over a period of 5 days at a total dose of 500 mg/kg (Schiffelers et al., 2001d). The superior efficacy of SSL-ceftazidime is even more pronounced when the rats with intact host defense are infected with a low ceftazidime-susceptible K. pneumoniae strain. SSL-ceftazidime administered over a 2-day period is effective at a total dose fo 25 mg/kg, which is 80-fold lower than the total dose of 2000 mg/kg needed to obtain the same therapeutic efficacy with free ceftazidime. It is concluded that maintaining ceftazidime at active concentrations, which is of critical importance for therapeutic efficacy, can also be achieved by single-dose administration of the antibiotic entrapped in SSL.

SSL-Ciprofloxacin in Bacterial Pneumonia

As stated in the Introduction to this chapter, SSL may also be used as a microreservoir of antimicrobial agents in the circulation. Experimental evidence to support the application of SSL in this way is provided by the encapsulation of ciprofloxacin in SSL to modulate the pharmacokinetics of ciprofloxacin after intravenous administration. As a result, extended ciprofloxacin activity in the blood and tissues is achieved. As shown in Table VI, 30 min after intravenous injection of ciprofloxacin, 77% of the injected dose is still circulating in the blood when it is administered in the SSL-encapsulated form compared with only 1.3% after administration in the free form. A substantial increase in $AUC_{0-6\ h}$, 800 μg/h/ml was obtained for SSL-ciprofloxacin, compared with 15.8 μg/h/ml for free ciprofloxacin. In addition, a decrease in toxic side effects is observed, facilitating the use of relatively high doses and decreasing dose frequency (Bakker-Woudenberg *et al.*, 2001). In these studies, the lipid composition of the SSL is PEG-DSPE/HSPC/Chol (molar ratio, 5:50:45). In the model of *K. pneumoniae* left-sided pneumonia, SSL-ciprofloxacin is superior to free ciprofloxacin (Table VII), achieving a 90% rat survival rate, which cannot be achieved with once-daily free ciprofloxacin. At a once-daily schedule, SSL-ciprofloxacin is slightly, but not significantly, more active than free ciprofloxacin at a twice-daily schedule.

SSL-Amphotericin B in Fungal Disseminated Infection

Intravenous administration of amphotericin B encapsulated in SSL composed of PEG-DSPE/HSPC/Chol (molar ratio, 0.21:1.79:1) in mice results in a substantially reduced toxicity of amphotericin B. The MTDs

TABLE VI
CONCENTRATIONS OF TOTAL CIPROFLOXACIN IN BLOOD AFTER INTRAVENOUS ADMINISTRATION OF SSL-CIP OR CIP IN RATS WITH *K. PNEUMONIAE* LUNG INFECTION[a]

	Concentration[b] (μg/ml) (% of injected dose) at 30 min after treatment	$AUC_{0-6\ h}$ (μg/h/ml)
SSL-CIP	291 ± 32 (77)	800
CIP	5.0 ± 1.2 (1.3)	15.8

[a] SSL-CIP (20 mg/kg; total lipid, 86 μmol/kg) or CIP (20 mg/kg) was injected as a single dose into rats at 24 h after bacterial inoculation of the left lung.
[b] Data are expressed as mean ± SD of six rats.

TABLE VII

THERAPEUTIC EFFECT OF TREATMENT OF RATS WITH *K. PNEUMONIAE* LEFT LUNG INFECTION[a]

	ED_{50}[b] (mg/kg/day)	ED_{90}[b] (mg/kg/day)
SSL-CIP, once daily	3.3 (1.7–5.7)	15.0 (8.0–78.9)
CIP, once daily	18.9 (10.5–37.8)	> MTD
CIP, twice daily	5.1 (2.6–8.6)	36.0 (18.3–164)

[a] Rats were inoculated with 10^6 viable *K. pneumoniae* in the left lung. Untreated rats died within 5 days. Treatment was started at 24 h after inoculation and continued for 3 days. The doses were escalated by 2-fold increases (*n*, 8–10 per dosage), ranging from 0.3 to 80 mg/kg/day. Survival of rats was assessed for a period of 21 days.

[b] ED_{50} and ED_{90}, dosages that effect 50% and 90% survival of rats, respectively. Values in parentheses are 95% confidence intervals.

TABLE VIII

PARAMETERS FOR THERAPEUTIC INDEX OF VARIOUS AMPHOTERICIN B FORMULATIONS

	Antifungal killing (>99.9%) capacity[a]	MTD (mg/kg/day)[b]	
		Acute toxicity	Chronic toxicity
AMB-DOC	0.2	0.8	>0.8
SSL-AMB	0.4	15	>15
AmBisome	12.8	>31	>31

[a] The minimal amphotericin B concentrations required to kill (>99.9%) the initial inoculum of 1.3×10^7 viable *C. albicans* within 6 h of incubation.

[b] Maximum tolerated dosage (MTD) determined in mice that were treated iv for 5 consecutive days once daily with amphotericin B-DOC, SSL-amphotericin B, or AmBisome. Amphotericin B dosages ranged from 0.1 to 1.0 mg/kg/day in steps of 0.1 mg/kg/day, and above 1 mg/kg/day in steps of 2 mg/kg/day. Six mice per group. Acute toxicity in terms of death during treatment and chronic toxicity in terms of more than a 3-fold increase in the indices for renal function (BUN and CREAT) and liver function (ALAT and ASAT) compared with placebo-treated mice, determined at 24 h after termination of treatment.

of SSL-amphotericin B versus conventional amphotericin B determined in mice to which the drugs are administered intravenously once daily for 5 consecutive days are 15 mg/kg and 0.8 mg/kg, respectively (Table VIII). With respect to *in vitro* antifungal activity, the killing capacity of SSL-amphotericin B measured in terms of >99.9% killing of *C. albicans* during a 6-h incubation is similar to that of conventional amphotericin B (Table VIII). So, in contrast to liposomal amphotericin B (AmBisome), which has reduced toxicity as well as reduced antifungal activity, with

SSL-amphotericin B it is possible to reduce toxicity without reducing the drug's antifungal activity.

Long circulation of SSL-amphotericin B after intravenous administration is observed, with a half-life of around 20 h, which is dose independent over a lipid dose range of 5–85 μmol TL/kg (van Etten *et al.*, 1995a). This is in contrast to AmBisome, for which long blood residence time is dependent on the use of high dosages, the half-life being about 8 h at a lipid dose of 70 μmol TL/kg (van Etten *et al.*, 1995b).

Evidence for an increased therapeutic index of amphotericin B following administration in the SSL form is obtained in the model of severe disseminated *C. albicans* infection in persistently leukopenic mice (van Etten *et al.*, 1995a,c, 1998) (Table IX). When treatment is started early (at 6 h after fungal inoculation), SSL-amphotericin B and AmBisome were similarly fully effective (100% survival of mice) when administered as a single dose (5 mg/kg). Delay of start of treatment of 16 h results in diminished therapeutic efficacy of AmBisome administered at a single dose in contrast to a single dose of SSL-amphotericin B, which is still fully effective. SSL-amphotericin B retains its therapeutic efficacy even when conventional amphotericin B treatment is postponed to 20 h after fungal inoculation, whereas AmBisome effects only 30% survival. These data demonstrate clearly that high antifungal activity as present in SSL-amphotericin B is of great importance in the treatment of severe fungal infection.

TABLE IX

THERAPEUTIC EFFECT OF EARLY OR DELAYED TREATMENT WITH A SINGLE DOSE OF SSL-AMPHOTERICIN B VERSUS SINGLE OR MULTIDOSE AMBISOME ON SURVIVAL OF LEUKOPENIC MICE WITH DISSEMINATED CANDIDIASIS[a]

Start of treatment (h postinoculation)	Treatment	Daily dose (mg/kg)	Number of doses	Survival (%)	
				Day 3	Day 7
	None			50	30
6	SSL-AMB	5	1	100	100
	AmBisome	5	1	100	100
	AmBisome	5	5	100	100
16	SSL-AMB	5	1	100	100
	AmBisome	5	5	90	90
20	SSL-AMB	5	1	100	100
	AmBisome	5	5	60	30

[a] Leukopenic mice were inoculated iv at zero time with 3×10^4 *C. albicans*. Untreated mice died between 24 h and 8 days after inoculation. Mice were treated iv as indicated.

SSL-Amphotericin B in Fungal Pneumonia

In the model of invasive *Aspergillus fumigatus* left-sided pneumonia in persistently leukopenic rats, the therapeutic efficacy of SSL-amphotericin B administered at 30 h after fungal inoculation, the time that mycelial growth was firmly established, was investigated. Intravenous single- or multiple-dose treatment with SSL-amphotericin B was compared with that of conventional amphotericin B (van Etten *et al.*, 2000). A single dose of SSL-amphotericin B at 10 mg/kg appeared as effective as a 10-day treatment with 1 mg/kg/day in terms of a significantly increased survival rate of rats, as well as reduced fungal dissemination to the liver. Prolongation of SSL-amphotericin B to 5-day treatment did not further prolong survival of rats.

Concluding Remarks

The techniques of encapsulation of antimicrobial agents from four different classes in SSL and the characterization of the formulations are described in this chapter. Biodistribution studies in models of bacterial or fungal infections show that radioactively labeled SSL substantially localize at the infectious focus. This target localization of SSL is strongly associated with the severity of the infectious disease. The effect of the encapsulation in SSL of the antimicrobial agents on their therapeutic efficacy was determined in clinically relevant models of serious, difficult-to-treat infection addressing the issues of impaired host defense, low antimicrobial susceptibility of the infectious agent, or subeffective antibiotic concentrations at the site of infection. With the use of SSL-encapsulated antomicrobial agents, a considerable enhancement of therapeutic efficacy is achieved. This is the consequence of site-specific and/or site-avoidance antibiotic delivery by SSL, or a microreservoir function of SSL for antibiotics.

References

Abra, R. M., and Hunt, C. A. (1981). Liposome disposition *in vivo*. III. Dose and vesicle-size effects. *Biochim. Biophys. Acta* **666**, 493–603.

Allen, T. M. (1981). A study of phospholipid interactions between high-density lipoproteins and small unilamellar vesicles. *Biochim. Biophys. Acta* **640**, 385–397.

Allen, T. M. (1994). Long circulating (sterically stabilized) liposomes for targeted drug delivery. *Adv. Drug Deliv. Rev.* **13**, 285–309.

Allen, T. M. (1997). Liposomes. Opportunities in drug delivery. *Drugs* **54**(Suppl. 4), 8–14.

Allen, T. M. (1998). Liposomal drug formulations. Rationale for development and what we can expect for the future. *Drugs* **56**, 747–756.

Allen, T. M., and Hansen, C. (1991). Pharmacokinetics of stealth versus conventional liposomes: Effect of dose. *Biochim. Biophys. Acta* **1068**, 133–141.

Allen, T. M., Hansen, C., and Rutledge, J. (1989). Liposomes with prolonged circulation times: Factors affecting uptake by reticuloendothelial and other tissues. *Biochim. Biophys. Acta* **981**, 27–35.

Allen, T. M., Hansen, C., Martin, F., Redemann, C., and Yau-Young, A. (1991). Liposomes containing synthetic lipid derivatives of poly(ethylene glycol) show prolonged circulation half-lives *in vivo*. *Biochim. Biophys. Acta* **1066**, 29–36.

Allen, T. M., Hansen, C. B., and Lopes de Menezes, D. E. (1995). Pharmacokinetics of long-circulating liposomes. *Adv. Drug Deliv. Rev.* **16**, 267–284.

Ames, B. N., and Dubin, D. T. (1960). The role of polyamines in the neutralization of bacteriophage deoxyribonucleic acid. *J. Biol. Chem.* **235**, 769–775.

Bakker-Woudenberg, I. A. J. M. (1995). Liposomes in the treatment of parasitic, viral, fungal and bacterial infections. *J. Liposome Res.* **5**, 169.

Bakker-Woudenberg, I. A. J. M., van den Berg, J. C., and Michel, M. F. (1982). Therapeutic activities of cefazolin, cefotaxime, and ceftazidime against experimentally induced *Klebsiella pneumoniae* pneumonia in rats. *Antimicrob. Agents Chemother.* **22**, 1042–1050.

Bakker-Woudenberg, I. A. J. M., Lokerse, A. F., and Roerdink, F. H. (1989). Antibacterial activity of liposome-entrapped ampicillin *in vitro* and *in vivo* in relation to the lipid composition. *J. Pharmacol. Exp. Ther.* **251**, 321–327.

Bakker-Woudenberg, I. A. J. M., Lokerse, A. F., ten Kate, M. T., Mouton, J. W., Woodle, M. C., and Storm, G. (1993). Liposomes with prolonged blood circulation and selective localization in *Klebsiella pneumoniae*-infected lung tissue. *J. Infect. Dis.* **168**, 164–171.

Bakker-Woudenberg, I. A. J. M., ten Kate, M. T., Stearne-Cullen, L. E. T., and Woodle, M. C. (1995). Efficacy of gentamicin or ceftazidime entrapped in liposomes with prolonged blood circulation and enhanced localization in *Klebsiella pneumoniae*-infected lung tissue. *J. Infect. Dis.* **171**, 938–947.

Bakker-Woudenberg, I. A. J. M., ten Kate, M. T., Guo, L., Working, P., and Mouton, J. W. (2001). Improved efficacy of ciprofloxacin administered in polyethylene glycol-coated liposomes for treatment of *Klebsiella pneumoniae* pneumonia in rats. *Antimicrob. Agents Chemother.* **45**, 1487–1492.

Bartlett, G. R. J. (1959). Phosphorus assay in column chromatography. *J. Biol. Chem.* **234**, 466–468.

Becker, M. J., Dams, E. Th. M., de Marie, S., Oyen, W. J. G., Boerman, O. C., Fens, M. H. A. M., Verbrugh, H. A., and Bakker-Woudenberg, I. A. J. M. (2002). Scintigraphic imaging using 99mTc-labeled PEG liposomes allows early detection of experimental invasive pulmonary aspergillosis in neutropenic rats. *Nucl. Med. Biol.* **29**, 177–184.

Begg, E. J., and Barclay, M. L. (1995). Aminoglycosides–50 years on. *Br. J. Clin. Pharmacol.* **39**, 597–603.

Boerman, O. C., Laverman, P., Oyen, W. J. G., Corstens, F. H. M., and Storm, G. (2000). Radiolabeled liposomes for scintigraphic imaging. *Prog. Lipid Res.* **39**, 461–465.

Ellis, M., Spence, D., de Pauw, B., Meunier, F., Marinus, A., Collette, L., Sylvester, R., Meis, J., Boogaerts, M., Selleslag, D., Krcmery, V., von Sinner, W., MacDonald, P., Doyen, C., and VanderCam, B. (1998). An EORTC international multicenter randomized trial (EORTC number 19923) comparing two dosages of liposomal amphotericin B for treatment of invasive aspergillosis. *Clin. Infect. Dis.* **27**, 1406–1416.

Fielding, R. M., Lewis, R. O., and Moon-McDermott, L. (1998). Altered tissue distribution and elimination of amikacin encapsulated in unilamellar, low-clearance liposomes (MiKasome). *Pharm. Res.* **15**, 1775–1781.

Fielding, R. M., Moon-McDermott, L., Lewis, R. O., and Horner, M. J. (1999). Pharmacokinetics and urinary excretion of amikacin in low-clearance unilamellar liposomes after a single or repeated intravenous administration in the rhesus monkey. *Antimicrob. Agents Chemother.* **43,** 503–509.

Gabizon, A., and Papahadjopoulos, D. (1992). The role of surface charge and hydrophilic groups on liposome clearance *in vivo. Biochim. Biophys. Acta* **1103,** 94–100.

Gabizon, A., Huberty, J., Straubinger, R. M., Price, D. C., and Papahadjopoulos, D. (1988). An improved method for *in vivo* tracing and imaging of liposomes using a gallium 67-deferoxamine complex. *J. Liposome Res.* **1,** 123–135.

Gangadharam, P. R. J., Ashtekar, D. R., Flasher, D. L., and Düzgüneş, N. (1995). Therapy of *Mycobacterium avium* complex infections in beige mice with streptomycin encapsulated in sterically stabilized liposomes. *Antimicrob. Agents Chemother.* **39,** 725–730.

Gregoriadis, G., (ed.) (1988). "Liposomes as Drug Carriers. Recent Trends and Progress." John Wiley & Sons, New York.

Haran, G., Cohen, R., Bar, L. K., and Barenholz, Y. (1993). Transmembrane ammonium sulfate gradients in liposomes produce efficient and stable entrapment of amphipathic weak bases. *Biochim. Biophys. Acta* **1151,** 201–215.

Huang, S. K., Martin, F. J., Friend, D. S., and Papahadjopoulos, D. (1995). Mechanism of Stealth Liposomes Accumulation in Some Pathological Tissues. *In* "Stealth Liposomes" (D. Lasic and F. Martin, eds.). pp. 119–125. CRC Press, Boca Raton, FL.

Hwang, K. J., Luk, K. F., and Beaumier, P. L. (1980). Hepatic uptake and degradation of unilamellar sphingomyelin/cholesterol liposomes: A kinetic study. *Proc. Natl. Acad. Sci. USA* **77,** 4030–4034.

Kedem, O., and Katchalsky, A. (1958). Thermodynamic analysis of the permeability of biological membranes to non-electrolytes. *Biochem. Biophys. Acta* **27,** 229–246.

Klibanov, A. L., Maruyama, K., Torchilin, V. P., and Huang, L. (1990). Amphipathic polyethyleneglycols effectively prolong the circulation time of liposomes. *FEBS Lett.* **268,** 235–237.

Kumana, C. R., and Yuen, K. Y. (1994). Parenteral aminoglycoside therapy. Selection, administration and monitoring. *Drugs* **47,** 902–913.

Kurantsin-Mills, J., Jacobs, H. M., Siegel, R., Cassidy, M. M., and Lessin, L. S. (1989). Indium-111 oxine labeled erythrocytes: Cellular distribution and efflux kinetics of the label. *Int. J. Rad. Appl. Instrum.* **B16,** 821–827.

Lasic, D. D. (1993). "Liposomes: From Physics to Applications." Elsevier Science Publishers, Amsterdam.

Lasic, D. D., Ceh, B., Stuart, M. C. A., Guo, L., Frederik, P. M., and Barenholz, Y. (1995). Transmembrane gradient driven phase transitions within vesicles: Lessons for drug delivery. *Biochim. Biophys. Acta* **1239,** 145–156.

Laverman, P., Dams, E. T., Oyen, W. J., Storm, G., Koenders, E. B., Prevost, R., van der Meer, J. W., Corstens, F. H., and Boerman, O. C. (1999a). A novel method to label liposomes with 99mTc by the hydrazino nicotinyl derivative. *J. Nucl. Med.* **40,** 192–197.

Laverman, P., Boerman, O. C., Oyen, W. J. G., Dams, E. Th. M., Storm, G., and Corstens, F. H. M. (1999b). Liposomes for scintigraphic detection of infection and inflammation. *Adv. Drug Deliv. Rev.* **37,** 225–235.

Leenders, A. C., de Marie, S., ten Kate, M. T., Bakker-Woudenberg, I. A. J. M., and Verbrugh, H. A. (1996). Liposomal amphotericin B (AmBisome) reduces dissemination of infection as compared with amphotericin B deoxycholate (Fungizone) in a rate model of pulmonary aspergillosis. *J. Antimicrob. Chemother.* **38,** 215–225.

Leenders, A. C. A. P., van Etten, E. W. M., and Bakker-Woudenberg, I. A. J. M. (1999). *In* "Handbook of Animal Models of Infection" (O. Zak and A. Sande, eds.), pp. 693–696. Academic Press, San Diego, CA.

Meirinha da Cruz, M. E., Ataide de Carvalhosa, N. D., and Gameiro Francisco, A. P. (1993). Patent no. WO9323015.

Mingeot-Leclercq, M. P., and Tulkens, P. M. (1999). Aminoglycosides: Nephrotoxicity. *Antimicrob. Agents Chemother.* **43,** 1003–1012.

Papahadjopoulos, D., Allen, T. M., Gabizon, A., Mayhew, E., Matthay, K., Huang, S. K., Lee, K. D., Woodle, M. C., Lasic, D. D., and Redemann, C. (1991). Sterically stabilized liposomes: Improvements in pharmacokinetics and antitumor therapeutic efficacy. *Proc. Natl. Acad. Sci. USA* **88,** 11460–11464.

Petersen, E. A., Grayson, J. B., Hersh, E. M., Dorr, R. T., Chiang, S. M., Oka, M., and Proffitt, R. T. (1996). Liposomal amikacin: Improved treatment of *Mycobacterium avium* complex infection in the beige mouse model. *J. Antimicrob. Chemother.* **38,** 819–828.

Pinto-Alphandry, H., Andremont, A., and Couvreur, P. (2000). Targeted delivery of antibiotics using liposomes and nanoparticles: Research and applications. *Int. J. Antimicrob. Agents* **13,** 155–168.

Ruijgrok, E. J., Vulto, A. G., and van Etten, E. M. (1999). *J. Liposome Res.* **9,** 291–300.

Schiffelers, R. M., Bakker-Woudenberg, I. A. J. M., Snijders, S. V., and Storm, G. (1999). Localization of sterically stabilized liposomes in *Klebsiella pneumoniae*-infected rat lung tissue: Influence of liposome characteristics. *Biochim. Biophys. Acta* **1421,** 329–339.

Schiffelers, R. M., Bakker-Woudenberg, I. A. J. M., and Storm, G. (2000). Localization of sterically stabilized liposomes in experimental rat *Klebsiella pneumoniae* pneumonia: Dependence on circulation kinetics and presence of poly(ethylene)glycol coating. *Biochim. Biophys. Acta* **1468,** 253–261.

Schiffelers, R. M., Storm, G., ten Kate, M., and Bakker-Woudenberg, I. A. J. M. (2001a). Therapeutic efficacy of liposome-encapsulated gentamicin in rat *Klebsiella pneumoniae* pneumonia in relation to impaired host defense and low bacterial susceptibility to gentamicin. *Antimicrob. Agents Chemother.* **45,** 464–470.

Schiffelers, R., Storm, G., and Bakker-Woudenberg, I. (2001b). Liposome-encapsulated aminoglycosides in pre-clinical and clinical studies. *J. Antimicrob. Chemother.* **48,** 333–344.

Schiffelers, R. M., Storm, G., and Bakker-Woudenberg, I. A. J. M. (2001c). Host factors influencing the preferential localization of sterically stabilized liposomes in *Klebsiella pneumoniae*-infected rat lung tissue. *Pharm. Res.* **18,** 780–787.

Schiffelers, R. M., Storm, G., ten Kate, M. T., Stearne-Cullen, L. E., den Hollander, J. G., Verbrugh, H. A., and Bakker-Woudenberg, I. A. J. M. (2001d). *In vivo* synergistic interaction of liposome-coencapsulated gentamicin and ceftazidime. *J. Pharmacol. Exp. Ther.* **298,** 369–375.

Senior, J. H. (1987). Fate and behavior of liposomes *in vivo*: A review of controlling factors. *Crit. Rev. Ther. Drug Carrier Syst.* **3,** 123–193.

Silvander, M., Johnsson, M., and Edwards, K. (1998). Effects of PEG-lipids on permeability of phosphatidylcholine/cholesterol liposomes in buffer and in human serum. *Chem. Phys. Lipids* **97,** 15–26.

Storm, G., and Woodle, M. C. (1998). *In* "Long Circulating Liposomes: Old Drugs, New Therapeutics" (M. C. Woodle and G. Storm, eds.), pp. 3–16. Springer-Verlag, Berlin.

Storm, G., Belliot, S. O., Daemen, T., and Lasic, D. D. (1995). Surface modification of nanoparticles to oppose uptake by the mononuclear phagocyte system. *Adv. Drug Deliv. Rev.* **17,** 31–48.

Tollemar, J., Klingspor, L., and Ringdén, O. (2001). Liposomal amphotericin B (AmBisome) for fungal infections in immunocompromised adults and children. *Clin. Microbiol. Infect.* **7**(Suppl. 2), 68–79.

Torchillin, V. P. (1996). How do polymers prolong circulation time of liposomes? *J. Liposome Res.* **6,** 99.

van Etten, E. W. M., ten Kate, M. T., Stearne, L. E. T., and Bakker-Woudenberg, I. A. J. M. (1995a). Amphotericin B liposomes with prolonged circulation in blood: *In vitro* antifungal activity, toxicity, and efficacy in systemic candidiasis in leukopenic mice. *Antimicrob. Agents Chemother.* **39,** 1954–1958.

van Etten, E. W. M., Otte-Lambillion, M., van Vianen, W., ten Kate, M. T., and Bakker-Woudenberg, I. A. J. M. (1995b). Biodistribution of liposomal amphotericin B (AmBisome) and amphotericin B-desoxycholate (Fungizone) in uninfected immunocompetent mice and leucopenic mice infected with *Candida albicans. J. Antimicrob. Chemother.* **35,** 509–519.

van Etten, E. W. M., van Vianen, W., Tijhuis, R. H. G., Storm, G., and Bakker-Woudenberg, I. A. J. M. (1995c). Sterically stabilized amphotericin B-liposomes: Toxicity and biodistribution in mice. *J. Control. Rel.* **37,** 123–129.

van Etten, E. W. M., Snijders, S. V., van Vianen, W., and Bakker-Woudenberg, I. A. J. M. (1998). Superior efficacy of liposomal amphotericin B with prolonged circulation in blood in the treatment of severe candidiasis in leukopenic mice. *Antimicrob. Agents Chemother.* **42,** 2431–2433.

van Etten, E. W. M., Stearne-Cullen, L. E. T., ten Kate, M. T., and Bakker-Woudenberg, I. A. J. M. (2000). Efficacy of liposomal amphotericin B with prolonged circulation in blood in treatment of severe pulmonary aspergillosis in leukopenic rats. *Antimicrob. Agents Chemother.* **44,** 540–545.

Walsh, T. J., Yeldandi, V., McEvoy, M., Gonzalez, C., Chanock, S., Freifeld, A., Seibel, N. I., Whitcomb, P. O., Jarosinski, P., Boswell, G., Bekersky, I., Alak, A., Buell, D., Barret, J., and Wilson, W. (1998). Safety, tolerance, and pharmacokinetics of a small unilamellar liposomal formulation of amphotericin B (AmBisome) in neutropenic patients. *Antimicrob. Agents Chemother.* **42,** 2391–2398.

Wasan, K. M., and Lopez-Berestein, G. (1995). The past, present, and future uses of liposomes in treating infectious diseases. *Immunopharmacol. Immunotoxicol.* **17,** 1–15.

Whitehead, T. C., Lovering, A. M., Cropley, I. M., Wade, P., and Davidson, R. N. (1998). Kinetics and toxicity of liposomal and conventional amikacin in a patient with multidrug-resistant tuberculosis. *Eur. J. Clin. Microbiol. Infect. Dis.* **17,** 794–797.

Woodle, M. C. (1998). Controlling liposome blood clearance by surface-grafted polymers. *Adv. Drug Deliv. Rev.* **32,** 139–152.

Woodle, M. C., and Lasic, D. D. (1992). Sterically stabilized liposomes. *Biochim. Biophys. Acta* **1113,** 171.

Woodle, M. C., Matthay, K. K., Newman, M. S., Hidayat, J. E., Collins, L. R., Redemann, C., Martin, F. J., and Papahadjopoulos, D. (1992). Versatility in lipid compositions showing prolonged circulation with sterically stabilized liposomes. *Biochim. Biophys. Acta* **1105,** 193–200.

Woodle, M. C., Newman, M. S., and Working, P. K. (1995). Biological Properties of Sterically Stabilized Liposomes. *In* "Stealth Liposomes" (D. D. Lasic and F. J. Martin, eds.), pp. 103–117. CRC Press, Boca Raton, FL.

[15] Liposome-Encapsulated Antibiotics[1]

By Isam I. Salem, Diana L. Flasher, and Nejat Düzgüneş

Abstract

Encapsulation of certain antibiotics in liposomes can enhance their effect against microorganisms invading cultured cells and in animal models. We describe the incorporation of amikacin, streptomycin, ciprofloxacin, sparfloxacin, and clarithromycin in a variety of liposomes. We delineate the methods used for the evaluation of their efficacy against *Mycobacterium avium-intracellulare* complex (MAC) infections in macrophages and in the beige mouse model of MAC disease. We also describe the efficacy of pH-sensitive liposomes incorporating sparfloxacin or azithromycin. We summarize studies with other antibiotics, including rifampicin, rifabutin, ethambutol, isoniazid, clofazimine, and enrofloxacin, and their use against MAC, as well as other infection models, including *Mycobacterium tuberculosis*.

Introduction

Drug association to an effective transportation system, or "vehicle," offers several advantages over conventional dosage forms, overcoming some bioavailability limitations and guaranteeing drug effectiveness and safety. Liposomes, in this context, are by far the most studied and suitable vehicles to carry and target active substances to the site of action, due to their ability to localize in certain cell types, thus producing optimum therapeutic results with a minimum of side effects caused by the drug (Gregoriadis and Florence, 1993). These vehicles are suitable to carry hydrophobic or hydrophilic drugs, either in their lipid bilayer or in the liposome interior, at concentrations several-to-many-fold above their aqueous solubility (Lasic, 1993). Liposomes exhibit different biodistribution and pharmacokinetics than free drug molecules, thereby improving the therapeutic efficacy of the encapsulated drug molecules (Dupond, 2002; Gabizon *et al.*, 1989a; Gill *et al.*, 1995; Sparano and Winer, 2001).

In the treatment of infectious diseases, liposomes are applied as carriers of antimicrobial agents to achieve localization at the infected site, to reduce side effects, or to serve as a microreservoir of the drug in circulation. Selective delivery of drugs to cells of the reticuloendothelial system by liposomes

[1]This chapter is dedicated to the memory of Professor Pattisapu R. J. Gangadharam.

METHODS IN ENZYMOLOGY, VOL. 391

has resulted in greater efficacy against various intracellular pathogens. Encapsulation in liposomes enhances the antimicrobial activities of a number of antibiotics against infections both in cultured cells and *in vivo* (Düzgüneş, 1998; Emmen and Storm, 1987; Popescu *et al.*, 1987). Delivery of ampicillin (Bakker-Woudenberg *et al.*, 1985), streptomycin (Tadakuma *et al.*, 1985), and amphotericin B (Lopez-Berestein *et al.*, 1987) in liposomes has markedly enhanced the activities of these drugs against experimental infections induced by *Listeria monocytogenes, Salmonella enteritidis*, and *Candida albicans*, respectively. Several laboratories have reported increased activities of liposome-encapsulated antimycobacterial agents against *Mycobacterium avium-intracellulare* complex (MAC) infections *in vivo* (Bermudez *et al.*, 1990; Cynamon *et al.*, 1989; Düzgüneş *et al.*, 1988, 1991a, 1995; Gangadharam *et al.*, 1989, 1991). One of the advantages of liposome-encapsulated drugs is the significant reduction of drug toxicity. Administration of amphotericin or adriamycin in liposomes substantially reduces drug toxicity in patients (Gabizon *et al.*, 1989b; Lopez-Berestein *et al.*, 1987).

Our laboratories have been using liposome-encapsulated antibiotics for the therapy of MAC infections *in vitro* and *in vivo* and have developed new liposome-encapsulated drugs and formulations for this purpose (Ashtekar *et al.*, 1991; Düzgüneş, 1998; Düzgüneş *et al.*, 1988, 1991a, 1995, 1996; Gangadharam *et al.*, 1991, 1995; Kesavalu *et al.*, 1990; Majumdar *et al.*, 1992; Pedroso de Lima *et al.*, 1990; Salem and Düzgüneş, 2003). MAC causes serious pulmonary and disseminated infection in AIDS and non-AIDS patients (Bartley *et al.*, 1999; Chin *et al.*, 1994; Horsburgh, 1991). Most patients with MAC and AIDS survive no more than 1 year after diagnosis. MAC is resistant to most antituberculosis drugs (Hawkins *et al.*, 1986). Many drugs that have *in vitro* effects against MAC are not effective *in vivo*. Furthermore, AIDS patients infected with MAC do not respond to therapy as well as patients without AIDS (Horsburgh *et al.*, 1985).

It is well known that MAC invades resident macrophages in the lungs, liver, spleen, and lymph nodes and is also present in the bone marrow and blood (Armstrong *et al.*, 1985; Eng *et al.*, 1989). Our studies indicate that some of the components of multidrug regimens currently administered to patients, such as amikacin and ciprofloxacin, are relatively ineffective against MAC inside macrophages (Majumdar *et al.*, 1992), although they may be bactericidal against the organism in culture. It is, therefore, imperative to target the drugs to the intercellular sites of MAC infection to achieve *in vivo* efficacy (Düzgüneş *et al.*, 1988, 1991a; Gangadharam *et al.*, 1991; Perronne *et al.*, 1991).

Liposomes are avidly phagocytosed by macrophages both *in vitro* (Dijkstra *et al.*, 1984; Daleke *et al.*, 1970) and *in vivo* (Poste *et al.*, 1982; Roerdink *et al.*, 1986; Scherphof, 1991). Thus, liposomes are naturally targeted to the major organs infected with MAC and may be ideal vehicles for directing antibiotics to sites of infection (Düzgüneş *et al.*, 1988, 1991a). The therapeutic efficacy of liposome-encapsulated antibiotics can be estimated from studies with cultured macrophages infected with MAC. The superior efficacy of liposome-encapsulated amikacin, streptomycin, and ciprofloxacin versus the free antibiotics against MAC has been demonstrated in infected macrophages *in vitro* (Ashtekar *et al.*, 1991; Bermudez *et al.*, 1987; Majumdar *et al.*, 1992).

In this chapter, we describe methods used to investigate the efficacy of liposomal antimycobacterials against MAC in infected macrophages derived from human peripheral blood monocytes and in infected beige mice. We also summarize liposomal formulations used in other infection models, including *Mycobacterium tuberculosis* and *Salmonella dublin*.

Human Monocyte–Derived Macrophages

Human macrophages to be used in the evaluation of the antimycobacterial action of liposomal antibiotics are prepared from peripheral blood obtained from healthy donors (ascertained to be HIV-, hepatitis B-, and hepatitis C-negative at the blood bank). Peripheral blood mononuclear cells are isolated by Histopaque-1077 (Sigma, St. Louis, MO) density gradient centrifugation. The cells are then cultured at a density of 2×10^6 per well in 48-well tissue culture plates (Falcon) at 37° and 5% CO_2 in a cell culture incubator. The medium and nonadherent cells are aspirated after 24 h, and the adherent monocyte monolayer is washed twice with 1 ml of Dulbecco's modified Eagle medium with high glucose (DME-HG). The cell layer is incubated in 1 ml of cell differentiation medium [DME-HG-L-glutamine + 20% fetal bovine serum (FBS, Sigma) + 10% human serum (Advanced Biotechnologies, Columbia, MD)], with a change to a maintenance medium (DME-HG-L-glutamine + 20% FBS) on day 3. The Trypan Blue exclusion technique is used to determine the viability of the cells after washing. Adherent cells are expected to develop morphological characteristics of macrophages on day 4.

Microorganisms

Different MAC strains are used in the experiments described below. Strain 101 came from an AIDS patient in California and was supplied by Dr. Clark Inderlied. Strain LR541 was isolated from an AIDS patient at the Memorial Sloan Kettering Institute in New York and was provided by Dr. Jack Crawford and Dr. Joseph Bates from Little Rock, Arkansas. Strains SK32, SK47, and SK52 are more recent isolates from the Sloan

Kettering Memorial Institute from different AIDS patients and were provided by Dr. Sheldon Brown. Strain 571-8 was from a non-AIDS patient undergoing treatment at the National Jewish Center for Immunology and Respiratory Medicine, Denver, CO. For clarithromycin experiments, the MAC strain MAC 11 obtained from Drs. K. Hadley and D. Yajko (San Francisco General Hospital) is used. A single-cell suspension of a predominantly (>95%) transparent colony type is obtained from these isolates and stored in aliquots at $-7°$. The use of the virulent strain 101 allows for comparison with earlier results obtained with this organism.

MAC is cultured in Middlebrook 7H9 broth (Difco). After 7 days of incubation, the bacterial suspension is subcultured 24 h before the infection of the macrophage monolayer, and the bacterial suspension is adjusted to 10^7 ml by using a McFarland standard.

Infection of Macrophages by MAC

Macrophages are inoculated with the MAC suspension at a ratio of 5 to 1. The cells are incubated for 24 h at $37°$ in 5% CO_2 and moist air. After 24 h, the monolayer is washed three times with Hanks' balanced salt solution (HBSS; UCSF Cell Culture Facility, San Francisco, CA, or Life Technologies, GIBCO/BRL, Rockville, MD) to remove the extracellular MAC. After the first day of infection, the number and viability of the cells are ascertained by Naphthol Blue Black staining of nuclei (Nakagawara and Nathan, 1983) and by Trypan Blue exclusion, respectively. The cultures are also monitored for cell loss by visual inspection under a phase-contrast microscope.

In general, the infected macrophages are incubated for 7 days in RPMI 1640 with 10% FBS without antibiotics. The colony-forming units (CFU) of MAC are determined after lysis of the macrophages with 0.25% sodium dodecyl sulfate (SDS). The lysate is diluted serially (10^4- to 10^7-fold) in sterile water and plated in duplicate on 7H10 agar plates. The colonies are enumerated after 2–3 weeks of incubation at $37°$. The initial infection of the macrophages is assessed after 24 h of incubation with MAC, following removal of extracellular bacteria. Triplicate wells are used for each experimental condition tested.

Infection of Beige Mice

Four- to six-week-old male or female beige mice (C57BL/6/bg^j/ bg^j) are used for our experiments. This species is a suitable animal model for disease due to MAC infections (Gangadharam *et al.*, 1983). Each experimental animal receives approximately 10^6–10^7 viable units of MAC by intravenous (iv) injection. The animals receive injections of free or liposome-encapsulated antibiotics iv, via tail vein injection. Treatment is started either 1 day or 3 weeks after infection. The injections are normally given weekly or twice weekly for 4 weeks. As controls, mice are injected with (1) empty (buffer loaded) liposomes administered at the same lipid concentration as the antibiotic-liposomes; (2) buffer only; (3) empty liposomes plus free drug; (4) iv free drug at the same dose as the liposome-encapsulated drug; or (5) intramuscular free drug, six times a week, at a conventional dose, usually 10-fold higher than the weekly liposomal dose.

After the animals are sacrificed, the CFU counts are determined in the blood, liver, spleen, lungs, kidney, bone marrow, and pooled lymph nodes (superficial inguinal, mesenteric, superficial deep cervical, and renal). Following homogenization, cells are lysed with 0.25% SDS, and the lysis medium is neutralized with 10% bovine serum albumin (BSA). The bacteria are diluted serially (10-fold dilutions) in water, and an appropriate aliquot (50–100 μl) is plated on 7H11 agar medium, ensuring that the medium is spread evenly. The inoculated plates are incubated (with the lids at the bottom, to prevent condensation) in a humidified 37° incubator. In addition, representative samples from these tissues are taken to determine tissue levels of drugs, the localization of liposomes, and histopathological observations. Standard randomization procedures are employed in selecting animals to be sacrificed at scheduled intervals.

MAC Growth Determination by BACTEC

Growth of MAC is monitored radiometrically by means of a BACTEC 460-TB or 760 instrument (Becton Dickinson Diagnostic Systems, Sparks, MD). Growth is measured as a function of the release of ^{14}C-labeled CO_2 resulting from the metabolism of ^{14}C-labeled palmitate in Middlebrook 7H12 broth (Siddiqi et al., 1981). Growth is then expressed as a numerical value referred to as the "growth index" (GI) that ranges from 1 to 900. The data are presented in each case as the mean of three determinations.

The minimal inhibitotory concentrations (MIC) of the drugs are determined after 7 days of incubation of MAC in 7H9 broth, following the adjustment of the bacterial suspension by using a McFarland standard. From this suspension, 10-fold dilutions are made in BACTEC medium from 10^{-1} to 10^{-5}. Drug-containing vials and drug-free control vials are injected with 200 μl of the suspension from the 10^{-3} dilution. A 1/100 control vial is inoculated with 200 μl of the 10^{-5} dilution. The vials are read at the same time every day until the 1/100 control shows a GI > 30, with an increase in GI > 10 for 3 consecutive days.

Following the treatment, the growth index of MAC in macrophages is determined after lysis of the macrophages with 0.25% SDS, which is subsequently neutralized with 10% BSA. The lysate is diluted serially to 10^3, and aliquots of the diluted lysates (200 μl) are injected into the BACTEC vials. The BACTEC vials are incubated at 37° and assayed every 24 h. All measurements are carried out in duplicate.

Liposome-Encapsulated Amikacin

Preparation of Liposomes

Amikacin is encapsulated in liposomes composed of phosphatidylglycerol (PG)/phosphatidylcholine (PC)/cholesterol (chol) (1:1:1 molar ratio) by reverse-phase evaporation, followed by extrusion through polycarbonate membranes (Düzgüneş, 2003; Düzgüneş et al., 1983, 1988). All phospholipids are obtained from Avanti Polar Lipids (Alabaster, AL), and cholesterol is purchased from Calbiochem (San Diego, CA) or Avanti Polar Lipids. A sterile solution (0.667 ml) of 50 mg/ml amikacin (in 10 mM KCl, 5 mM glycine, pH 9.6, adjusted to an osmolality of 300 mOsm with NaCl) is added to a solution of lipids (5.0 μmol) in 2 ml diethyl ether (previously washed with distilled water and allowed to phase separate) in a sterile Kimax test tube with a Teflon-lined screw-cap. The mixture is sonicated for 5 min under an argon atmosphere in a bath-type sonicator (Laboratory Supply Co., Hicksville, NY) to form an emulsion. The tube is covered with Teflon tape and placed in a larger tube that fits onto a rotary evaporator (Büchi, Flawil, Switzerland). The emulsion is evaporated under controlled vacuum (about 350 mm Hg, to avoid excessive effervescence) until a gel is formed, the gel is broken up by vortexing briefly, and the mixture is evaporated further until an opalescent aqueous solution is formed. The vortexing process may be repeated if the gel is not broken completely. To the aqueous solution, 1.32 ml of the amikacin stock solution is added, and the sample is placed back into the rotary evaporator for an additional 20 min at maximal vacuum to remove any residual ether. The liposomes are extruded through polycarbonate membranes of 0.2 μm pore diameter (Nuclepore, Pleasanton, CA) under argon pressure in a high-pressure extrusion apparatus (Lipex Biomembranes, Vancouver, British Columbia, Canada, or equivalent). Unencapsulated amikacin is eliminated by chromatographing the liposome suspension through a Bio-Rad (Hercules, CA) column (20 cm × 1 cm) packed with sterilized Sephadex

G-75 (Pharmacia, Piscataway, NJ), using 140 mM NaCl, 10 mM KCl, 10 mM glycine, pH 9.6, as the elution buffer. The liposome suspension is filtered through a 0.22-μm filter (Schleicher and Schuell, Keene, NH) to ensure sterility. The size distribution of the liposomes is ascertained in a dynamic light-scattering instrument (NP-4, Coulter Electronics, Inc., Hialeah, FL).

The amount of encapsulated amikacin is determined by an enzyme-linked immunosorbent assay (ELISA) after lysing the liposomes with 0.5% Triton X-100 (Sigma) using a DuPont clinical analyzer. The assay is based on the activity of glucose-6-phosphate dehydrogenase conjugated to amikacin. The enzyme is inhibited when the conjugate is bound to anti-amikacin antibody. The amount of bound conjugate is determined by the concentration of free amikacin that competes for the antibody. The amikacin concentration can be determined also by using the procedure outlined below for streptomycin, as long as the liposomes do not contain amino lipids. Phospholipid concentrations are determined by phosphate analysis (Bartlett, 1959; Düzgüneş, 2003). Control liposomes containing buffer only without amikacin are prepared with 140 mM NaCl, 10 mM KCl, 5 mM glycine, pH 9.6, as the aqueous medium.

Efficacy against MAC In Vivo

Free or liposome-encapsulated amikacin is administered to mice iv at a dose of 100 μg per mouse per week (5 mg/kg once a week). Treatment is started 1 day after infection and is continued for 3 weeks (a total of four injections). In addition, one group of animals receives a daily dose of 1 mg per mouse (50 mg/kg daily) im 6 days a week for the entire 8-week period of the experiment. Two control groups are included, one receiving empty (buffer-loaded) liposomes administered at the same lipid concentration as the amikacin-liposome group and the other receiving buffer. In a variation of this experiment, different doses of liposome-encapsulated amikacin may be administered in five weekly injections. Free amikacin is given iv at a dose of 10 mg/kg. Three randomly selected mice are sacrificed on day 1 following infection (baseline) and at 2-week intervals thereafter for up to 8 weeks.

Administration of four weekly low doses of amikacin (5 mg/kg) encapsulated in liposomes prevent *in vivo* multiplication of MAC. The same dose of the free drug given under the same conditions does not arrest the growth of the organisms. At the 8-week point, CFU counts in the livers of the liposome-encapsulated amikacin group are more than three orders of magnitude below those of the buffer or empty-liposome controls and the free-amikacin group (Düzgüneş *et al.*, 1988). The effect of liposome-encapsulated amikacin persists for at least 5 weeks after the termination of treatment.

Essentially similar results are obtained in the spleen, liposome-encapsulated amikacin being more effective than the free drug given iv. By 2 weeks, the CFU counts in animals treated with liposomal amikacin are about 1.5 orders of magnitude lower than the CFU in untreated animals and those that receive the free drug. At 8 weeks, i.e., 5 weeks after the end of treatment, the CFU counts are approximately three orders of magnitude lower in animals treated with liposome-encapsulated amikacin than in controls or those receiving free amikacin iv. In the lymph nodes, the results are essentially the same as with the lungs, in that neither free nor encapsulated amikacin is effective for 8 weeks. The im treatment group, at a dose 60-fold higher than that of the free or liposome-encapsulated amikacin treatment group, shows a reduction in CFU counts below the baseline (day 1) level after 4 weeks.

The results demonstrate that iv delivery of amikacin encapsulated in large unilamellar liposomes arrests the growth of MAC in the livers and spleens of beige mice. Liposome-encapsulated amikacin is substantially more effective than comparable amounts of free amikacin administered iv. These observations suggest that the negatively charged liposomes are taken up by the infected macrophages of the liver and spleen. It is also likely that the liposomes, and hence the encapsulated amikacin, are localized at intracellular sites where the mycobacteria reside, since it is known that liposomes are avidly phagocytosed by macrophages *in vitro* and *in vivo* (Daemen *et al.*, 1986; Poste *et al.*, 1982; Roerdink *et al.*, 1986). Liposomal amikacin was found to be effective against MAC even when administered once a month (Leitzke *et al.*, 1998).

A liposome preparation composed of hydrogenated soy PC/chol/distearoyl PG (2:1:0.1), and with average diameter 45–65 nm, appears to be more effective in reducing MAC infection in the spleen and liver (Petersen *et al.*, 1996) compared with studies employing liposomes of larger mean diameter (Düzgüneş *et al.*, 1988). It should be noted, however, that the study with the smaller liposomes (Petersen *et al.*, 1996) utilized much higher doses (8- to 24-fold) of liposomal amikacin, which was administered three times as frequently than that used in Düzgüneş *et al.* (1988). The same liposome formulation has also been tested against murine tuberculosis and shows 2.4- to 5-fold higher efficacy when given three times a week compared with free amikacin given im five times a week (Dhillon *et al.*, 2001). A proliposome method to encapsulate amikacin with high efficiency has been described (Zhang and Zhu, 1999).

Liposome-Encapsulated Streptomycin

Preparation of Liposomes

Streptomycin is encapsulated in unilamellar liposomes composed of PG/PC/chol (molar ratio, 1:9:5) prepared by reverse-phase evaporation, followed by extrusion through polycarbonate membranes (0.2 μm pore diameter) (Düzgüneş *et al.*, 1988, 1991a; Szoka *et al.*, 1980). Streptomycin is dissolved at a concentration of 140 mg/ml in pyrogen-free water containing 10 mM HEPES, pH 7.4. The osmolality of the solution is adjusted to 290 mOsm with NaCl, using a vapor pressure osmometer from Wescor (Logan, Utah). The solution is then sterilized by passage through a 0.22-μm Millex filter. Lipids (10–20 μmol) are dried from a chloroform (CHCl$_3$) solution and redissolved in 1 ml of diethyl ether (which is washed with water or HEPES-buffered saline just before use to eliminate any peroxidation products). An aliquot (0.34 ml) of the streptomycin solution is added to the lipid and sonicated for 2 min under an argon atmosphere in a sealed glass tube. The emulsion is placed in a rotary evaporator under controlled vacuum and temperature to achieve a liposome suspension (Düzgüneş, 2003). An additional 0.66 ml of streptomycin solution is added to the liposomes, and the mixture is vortexed and further evaporated for 20 min to eliminate any residual ether. The liposomes are extruded through polycarbonate membranes of 0.2 μm pore diameter (Nuclepore) under argon pressure in a high-pressure extrusion apparatus (Lipex Biomembranes). The unencapsulated drug is separated by passing the liposome preparation through a sterile Sephadex G-75 (Pharmacia) column eluted with 140 mM NaCl, 10 mM KCl, 10 mM HEPES buffer, pH 7.4. The preparation is then filter-sterilized by passage through a 0.22-μm Millex filter. The amount of encapsulated streptomycin is determined by a fluorescence assay (Düzgüneş *et al.*, 1991a; Gangadharam *et al.*, 1991; Hiraga and Kinoshita, 1981) after lysis of an aliquot of the liposome suspension with 10-fold excess methanol and sonication for 5 min to release the encapsulated antibiotic. Ninhydrin is added to a final conncentration of 5 mM, and the mixture is incubated for 10–20 min at 50°. Then, sodium hydroxide is added at a final concentration of 0.075 N, and the mixture is incubated for 10 min. A standard curve is established with 1–10 mg/ml streptomycin using the same procedure. The fluorescence is measured at 50° in a fluorometer (SLM 4000, Spex Fluorolog 2, Perkin-Elmer LS 50B, or equivalent), with the excitation wavelength at 395 nm, and either using a high-pass filter (KV470) in the emission channel or setting the emission monochromator to 505 nm. The phospholipid concentration in the preparation is determined by phosphate analysis (Bartlett, 1959; Düzgüneş, 2003).

For *in vivo* experiments, different formulations of liposomes are prepared: (1) Poly(ethylene glycol)-distearoylphosphatidylethanolamine (PEG-DSPE)/distearoylphosphatidylcholine (DSPC)/chol (molar ratio 1:9:6.7), (2) soybean phosphatidylinositol (PI)-DSPC-chol (molar ratio 1:9:6.7), and (3) PG/PC/chol (molar ratio, 1:9:6.7). Lipids (200 μmol of phospholipid or 334 μmol of total lipid) are dried from a chloroform solution onto the sides of a round-bottom glass flask. The dried film is then hydrated by the addition of 3.5 ml of a sterile streptomycin solution (168 mg/ml in 10 mM HEPES, pH 7.4), adjusted to an osmolality of 300 mOsm with NaCl. For negative-control liposomes, 100 μmol of phospholipid (or 167 mmol of total lipid) are dried and hydrated with 3.5 ml of HEPES buffer (10 mM HEPES, 140 mM NaCl, 10 mM KCl, pH 7.4). Full hydration is achieved by vortexing. The sterically stabilized liposomes are submitted to four cycles of freezing (in a dry ice–ethanol bath) and thawing (in a 21° waterbath), and are extruded four times through dual polycarbonate membranes (a 0.4-μm pore-diameter membrane on top of a 0.1-μm pore-diameter membrane) under argon pressure in a high-pressure extrusion apparatus. The extruder is preheated to 56° by circulating water through the water jacket around the chamber to ensure that the DSPC component is above its phase transition temperature. Unencapsulated streptomycin is removed by size-exclusion chromatography on sterile Sephadex G-75 equilibrated with argon-saturated HEPES buffer and eluting with the same buffer. The liposome suspensions are sterilized by passage through 0.22-μm pore-diameter filters.

The amount of encapsulated streptomycin varies in the range of 188–298 μg of drug per μmol of phospholipid for the different preparations. The liposomes containing PI or PEG-DSPE retain 100% of their contents upon storage for 2 weeks at 4° and are used for all of the injections over the 2-week period of therapy. Liposomes containing PG leak 38% of the encapsulated streptomycin upon storage at 4° for 2 weeks. Therefore, in the case of the PG-containing liposomes, unencapsulated streptomycin is removed from half of the initial preparation and is used for the injections during the first week. Unencapsulated antibiotic is removed from the remaining half of the preparation 1 week later to be used for the injections during the second week.

Efficacy against MAC in Macrophages

The effect of streptomycin encapsulated in liposomes has been compared with that of unencapsulated streptomycin against MAC inside human monocyte–derived macrophages in a concentration range achievable in the serum of treated patients (Majumdar *et al.*, 1992). The number of

MAC CFU decreases with increasing streptomycin concentration, in comparison with untreated controls, at the end of the 7-day incubation period following the treatment. Liposome-encapsulated streptomycin at 20 μg/ml reduces the CFU by about two orders of magnitude from the value obtained for untreated controls, and at 50 μg/ml the decrease is about three orders of magnitude. The CFU counts in cells treated with liposome-encapsulated streptomycin are consistently lower than those in cells treated with free streptomycin. Liposomal streptomycin at 10 μg/ml is as effective in reducing the MAC CFU as free streptomycin at 30 μg/ml. Compared with the level of the initial infection achieved at the end of the first 24 h of incubation with MAC, liposome-encapsulated streptomycin at 50 μg/ml reduces the CFU by 89%. Similar results have been obtained with MAC-infected murine peritoneal macrophages (Ashtekar *et al.*, 1991).

Enhancement of the antimycobacterial effect of liposome-encapsulated streptomycin, compared with the same concentration of free streptomycin, is similar over the concentration range of 10–40 μg/ml but is more pronounced at 50 μg/ml. It is possible that the higher intracellular concentration of liposomal streptomycin achieved with this dose reaches a critical value for higher antibacterial activity. Another explanation for this finding is that liposomes that do not encapsulate any streptomycin (buffer-loaded liposomes) are also found to inhibit MAC growth inside macrophages when added at a lipid concentration corresponding to that of 50 μg of liposome-encapsulated streptomycin per ml. Some lipids have a potent macrophage-activating function (Dijkstra *et al.*, 1984); thus, it is possible that the liposomes enhance host resistance to bacteria through an immunomodulatory effect.

Efficacy against MAC in Beige Mice

The transparent colony type of MAC (strain 101; serotype 1) is injected into 5- to 7-week-old male beige (C57BL/6/bg^j/bg^j) mice via the tail vein (approximately 7×10^6 viable units per animal). The infected animals are treated for 2 weeks by twice weekly injections of the liposome-encapsulated antibiotics (at doses of 15 or 30 mg/kg of body weight for the PG/PC/choesterol or PI/DSPC/chol liposomes and 15 mg/kg for the PEG-DSPE/DSPC/chol liposomes) via the tail vein. The treated animals are sacrificed at 2 and 4 weeks following infection to determine the CFU, as described above. The results are expressed as the mean and standard deviation of the CFU from three mice per experimental point.

Compared with the initial (1 day) infection, 15 or 30 mg/kg of streptomycin in either liposome composition reduces the CFU by about 2 log units at the end of the 4-week experiment, indicating a bactericidal effect

(Gangadharam *et al.*, 1995). Compared with the untreated control at the end of the 4-week period, this corresponds to about a 3.4-log-unit reduction in CFU. Streptomycin encapsulated in PEG-DSPE/DSPC/chol liposomes (at a dose of 15 mg/kg) causes a 1.3-log-unit reduction in the CFU compared with the CFU of the initial infection and a 2.7-log-unit reduction in the CFU compared with CFU in the untreated controls at week 4. The difference between the CFU obtained with 15 mg of streptomycin per kg in PEG-DSPE/DSPC/chol and either PG/PC/chol or PI/DSPC/chol liposomes is significant. Buffer-loaded control liposomes have no effect on the CFU in the case of PG/PC/chol liposomes, while liposomes of the two other compositions containing DSPC cause some reduction in the CFU compared with the CFU in untreated controls. Intramuscular injections of free streptomycin (at a dose of 150 mg/kg at 5 days/week for 4 weeks) as a control result in a reduction in CFU of 3.6 log units at the 4-week time point compared with the CFU in untreated controls. It should be noted that the total dose of antibiotic administered in the latter case is 25- to 50-fold higher than that given in liposomes. In the spleen, all three liposome compositions cause a 1.8- to 2.4-log-unit reduction in CFU at the 4-week point compared with the CFU at the initial infection.

Twice weekly administration of streptomycin encapsulated in PG/PC/chol or PI/DSPC/chol liposomes is bactericidal to MAC in the liver, while the antibiotic encapsulated in liposomes of all three compositions is bactericidal in the spleen. PEG-DSPE-containing liposomes may be less effective than the other liposomes in the liver because of increased levels of circulation in blood and uptake in other tissues (Allen *et al.*, 1991; Woodle and Lasic, 1992). While PI-containing liposomes are effective in the liver and spleen, they are the least effective liposome composition in terms of CFU reduction in the lungs. Studies with *Klebsiella pneumoniae*–infected rats indicate that the percentage of hydrogenated PI-containing liposomes that localize in the infected lungs is much lower than that of PEG-DSPE–containing liposomes (Bakker-Woudenberg *et al.*, 1992, 1993). However, the degree of localization of the former liposomes in infected lungs is significantly higher than that of PG-containing liposomes (Bakker-Woudenberg, 1992). It is also interesting to note that the enhanced localization of streptomycin in multilamellar liposomes in the lung compared with that of unilamellar liposomes does not result in enhanced efficacy against MAC (Düzgüneş *et al.*, 1991a).

It is possible that improved liposome stability in plasma may prevent the release of encapsulated antibiotics once the liposomes are endocytosed by macrophages. Our results demonstrate, however, that encapsulated streptomycin is available in MAC-infected tissues and cells and can inhibit the growth of MAC. Thus, sterically stabilized liposomes with prolonged circulation times, including those composed of PEG-DSPE/DSPC/chol, can be used effectively to deliver antibiotics to tissues and infected macrophages. Twice weekly administration of streptomycin encapsulated in PG/PC/chol or PEG-DSPE/DSPE/chol liposomes can reduce the level of MAC infection in the lungs by several orders of magnitude. Thus, liposome-encapsulated aminoglycosides can inhibit MAC growth in the lungs at doses much lower than the free im doses required to achieve the same effect (Düzgüneş *et al.*, 1991a; Gangadharam *et al.*, 1991). The ability to reduce levels of mycobacterial infection in the lungs with liposomes may also provide the opportunity to treat *M. tuberculosis* infections by the same technique. Vladimirsky and Ladigina (1982) have shown that administration of streptomycin in liposomes to mice infected with *M. tuberculosis* causes a significant reduction of colony counts in the spleen and increases survival compared with the free drug.

Liposome-Encapsulated Ciprofloxacin

Preparation of Liposomes

Ciprofloxacin is encapsulated in multilamellar vesicles composed of PG/PC/chol (molar ratio, 1:9:5). The lipids are first dried from chloroform onto the sides of a glass tube as a thin film by using a rotary evaporator (Büchi) and then hydrated with a sterile ciprofloxacin solution. Since ciprofloxacin crystallizes at a neutral pH but is soluble at an acidic pH, solutions of the antibiotic at 25 mg/ml are prepared in either (1) unbuffered water, (2) 40 mM glycine buffer at pH 3.5, or (3) 10 mM acetate buffer at pH 5.6. These solutions are adjusted to an osmolality of 290 mOsm by addition of sucrose crystals and then sterilized by filtration through 0.22-μm Millex syringe filters.

The hydrated lipid is mixed thoroughly by vortexing, and the mixture is subjected to four cycles of freezing (using an ethanol–dry ice bath) and thawing (using a water bath at room temperature). To enable pelleting and washing of the liposomes made in a solution containing sucrose, the prepared liposomes are diluted 30-fold in one of the following "dilution media": (1) 140 mM NaC, 10 mM KCl, (2) 100 mM NaCl, 10 mM KCl, 40 mM glycine, pH 3.5, or (3) 140 mM NaCl, 10 mM KCl, 10 mM acetate, pH 5.6. The unencapsulated drug is eliminated by centrifugation at 15,000 rpm for 15 min at 4° followed by washing with the respective dilution medium; this process is repeated three times. The liposomes are then suspended in 140 mM NaCl, 10 mM KCl, 10 mM HEPES, pH 7.4. The amount of encapsulated ciprofloxacin is determined by measuring the absorbance of methanol-lysed vesicles at 276 nm (methanol/buffer ratio of 9:1).

Antimycobacterial Effects of Liposome-Encapsulated Ciprofloxacin

A liposomal ciprofloxacin concentration as low as 0.1 μg/ml reduces the CFU in MAC-infected human macrophages by 2 log units compared with untreated controls, while the free drug at this concentration is ineffective (Majumdar *et al.*, 1992). Liposome-encapsulated ciprofloxacin at 2.5 μg/ml reduces the CFU by four orders of magnitude compared with the CFU obtained with the free drug. The CFU are reduced below 4 log units at a concentration of 5 μg/ml of the encapsulated antibiotic. Buffer-loaded (empty) liposomes result in no significant reduction in CFU compared with untreated controls, while incubation of infected macrophages with empty liposomes plus 5 μg of free ciprofloxacin per ml reduces the CFU to 0.27 log unit below the value obtained with the same concentration of the drug alone. Oh *et al.* (1995) have demonstrated that ciprofloxacin encapsulated in distearoylphosphatidylglcerol (DSPG/DSPC/chol (1:1:1) or DSPG/chol (2:1) liposomes, using a remote-loading procedure, is taken up to a much greater extent by murine J774 macrophages compared with the free drug. The antimycobacterial effect of liposomal ciprofloxacin is also drastically enhanced compared with the free drug.

Antibacterial Effects of Liposomal Ciprofloxacin In Vivo

The *in vivo* efficacy of liposomal ciprofloxacin has been evaluated in a murine model of *Salmonella dublin* infection (Magallanes *et al.*, 1993). A single injection of liposomal ciproflaxin is 10 times more effective in preventing mortality than a single iv injection of free drug. The enhanced effect of liposomal ciprofloxacin may be attributed to the high concentration and persistence of the antibiotic in infected organs (Magallanes *et al.*, 1993). Liposome-encapsulated ciprofloxacin is also highly effective against respiratory *Francisella tularensis* infection when delivered to the lungs in aerosolized form, with mean mass aerodynamic diameters ranging from 3.1 to 3.8 μm (Conley *et al.*, 1997). While the mortality rate is 100% in untreated or free ciprofloxacin-treated animals, all the animals given aerosolized liposomal ciprofloxacin survive the infection. These experiments point to the potential for the treatment of lower respiratory tract infections with liposome-encapsulated ciprofloxacin (Conley *et al.*, 1997). Ciprofloxacin encapsulated in sterically stabilized liposomes composed of hydrogenated soy PC/chol/PEG-DSPE (50:45:5) is superior to the free drug in the treatment of *K. pneumoniae* in rats (Bakker-Woudenberg *et al.*, 2001).

Liposome-Encapsulated Sparfloxacin

Preparation of Liposomes

Due to the limited aqueous solubility of sparfloxacin, it is initially dissolved together with the phospholipids and cholesterol in chloroform (Düzgüneş *et al.*, 1996). Following the evaporation of the solvent and hydration of the dried lipid in HEPES-buffered saline, sparfloxacin is incorporated in the membrane phase of the liposomes. Multilamellar liposomes are prepared as follows. Chloroform solutions of PG, PC, chol, and sparfloxacin are mixed at a 1:1:1:0.4 molar ratio in a glass tube (total phospholipid, 40–80 μmol), dried to a thin film in a Büchi rotary evaporator, and then placed in a vacuum oven at room temperature to remove any residual $CHCl_3$. The dried film is hydrated with 1–2 ml of HEPES buffer (140 mM NaCl, 10 mM KCl, 10 mM HEPES, pH 7.4) by vortexing under an argon atmosphere. Unencapsulated sparfloxacin is removed by centrifugation of the liposomes at 10,000 rpm in an Eppendorf centrifuge equilibrated at 4°, followed by resuspension in ice-cold HEPES buffer, and this process is repeated four times. The encapsulation efficiency of sparfloxacin in these multilamellar liposomes is between 45 and 62%, as determined by its A_{304} after an aliquot of the liposomes is dissolved in a 10-fold excess of methanol. The lipid concentration is determined by phosphate assay (Bartlett, 1959; Düzgüneş, 2003).

Antimycobacterial Effect of Liposomal Sparfloxacin in Macrophages

The murine macrophage-like cell line J774 is maintained in Dulbecco's modified Eagle medium supplemented with 10% FBS and 50 μg of gentamicin per ml in a 5% CO_2 incubator at 37°. MAC is added to the cells at a ratio between 10 and 20 to 1 and incubated for 2 h at 37°. The cells are washed three times with HBSS to remove free bacteria and incubated further in Dulbecco's modified Eagle medium containing only 1% FBS to minimize the growth of the cells. Free or liposome-encapsulated sparfloxacin is added to the medium for either 24 h or 4 days and is then removed by washing. Seven days after the initial infection, the macrophages are lysed with 0.5 ml of 0.25% SDS for 10 min, and the lysate is neutralized by the addition of 0.5 ml of 10% BSA. The viable bacteria are quantitated by the radiometric BACTEC method, as described above.

Since the MICs of sparfloxacin for the MAC strains SK12 and 101 are 0.5 μg/ml, higher concentrations of sparfloxacin are tested in the MAC-infected macrophage model. J774 cells infected with MAC SK12 are treated for 24 h with free or liposome-encapsulated sparfloxacin at concentrations of 1.5 or 3 μg/ml. Free sparfloxacin at 1.5 μg/ml has no effect on the growth of MAC, while at the same concentration, the liposome-encapsulated drug is slightly effective, reducing the growth index to 80% of that of the untreated control (Düzgüneş et al., 1996). At 3 μg/ml, free and liposome-encapsulated sparfloxacin reduce the growth index to about 80 and 66% of that of the untreated control, respectively. The reduction in growth index by free drug is not statistically significant, while that caused by liposomal sparfloxacin is. At 6 μg/ml, the growth index is reduced to about 25 and 30% of the untreated controls for free and liposome-encapsulated sparfloxacin, respectively. When the treatment time is extended to 4 days, free sparfloxacin reduces the growth index to 6% of that of the untreated control, while liposome-encapsulated sparfloxacin reduces it to 8% of that of the control. The treatment of the infected macrophages for 4 days, and the resulting decrease in the growth index of the MAC, also reduces the toxicity of the MAC to the macrophages they infect.

Free and liposome-encapsulated sparfloxacin have similar effects on the growth of intracellular MAC, particularly at the higher drug concentrations used. The lack of enhancement of efficacy by liposome-encapsulation is most likely due to the efficient uptake of the free drug by the infected macrophages, thereby obviating the dependence on liposome-mediated uptake. This may be attributed partially to the lipophilic nature of sparfloxacin and its ability to partition readily into cellular membranes. These results may also be explained by the low uptake of large multilamellar liposomes by macrophages, coupled with the slow release of the antibiotic from the inner lamellae of the liposomes. The slightly greater anti-MAC effect of liposomal sparfloxacin at the lowest concentration used, compared with that of the free antibiotic, could be the result of a greater accumulation and availability of the drug in the compartments containing MAC. Although liposome-encapsulated aminoglycosides show a modest increase in efficacy against MAC in macrophages in culture (Ashtekar et al., 1991; Kesavalu et al., 1990; Majumdar et al., 1992), they are effective at much lower doses in vivo than that required for the free drug (Düzgüneş et al., 1988, 1991a; Gangadharam et al., 1991). Analogous to the aminoglycosides, it is likely that liposome-encapsulated sparfloxacin will have an enhanced effect against MAC infection in vivo because of its ability to localize in the mononuclear phagocyte (reticuloendothelial) system.

Liposome-Encapsulated Clarithromycin

Liposome Preparation

Clarithromycin encapsulation in liposomes is achieved by either incorporating the drug into the lipid solution before formation of a thin dried film, by including the drug in the hydration solution, or by utilizing both methods at the same time. In the first case, the drug is dissolved initially in $CHCl_3$ at three different ratios together with the phospholipids and cholesterol because of the low aqueous solubility limit of clarithromycin. Chloroform solutions of PG, PC, and cholesterol at a molar ratio of 1:9:5 are mixed with 10, 20, or 30% of clarithromycin (Table I) in glass tubes, dried to a thin film on a Büchi rotary evaporator, and then placed in a vacuum oven at room temperature to remove any residual $CHCl_3$. The dried films are hydrated with 1–2 ml of 0.05 M citrate buffer at pH 5.2 by vortexing under an argon atmosphere. Unincorporated clarithromycin is removed by centrifugation of the multilamellar liposomes at 3000 rpm for 10 min in an Eppendorf centrifuge equilibrated at 4° and resuspended in ice-cold citrate buffer (3 times). The liposomes are finally resuspended in HEPES-buffered saline (HBS, 140 mM NaCl, 10 mM KCl, 10 mM HEPES, pH 7.4, and 303 mOsm).

In the second method of preparation, clarithromycin corresponding to 10, 20, or 30 mol% with respect to the phospholipids is incorporated into the hydration solution, rather than the liposome membrane. These liposomes are submitted to five freeze–thaw cycles at −80° and 37°, respectively. In the third method, 30 mol% clarithromycin is incorporated in the lipid film and in the hydration solution at the same time. Unencapsulated drug is removed in both cases by centrifugation as described above. In all these methods, the amount of encapsulated clarithromycin is quantitated by

TABLE I

ENCAPSULATION OF CLARITHROMYCIN BY LIPID FILM HYDRATION

Composition PG/PC/CH	Drug added (% phospholipid)	Drug in liposomes (μg/μmol phospholipid)	Drug in solution (μg/μmol phospholipid)
1:9:5	10	4.99	0
1.9:5	20	5.33	1.49
1:9:5	30	17.74	6.72
5:5:5	30	143.72	—
10:0:5	30	185.22	—

high-pressure liquid chromatography (HPLC) after dissolving an aliquot of the liposomes in an excess of methanol and normalizing to the lipid concentration determined by phosphate assay. Control liposomes are prepared similarly, but without the clarithromycin.

Of all the methods employed to encapsulate clarithromycin in liposomes, the lowest efficiency is obtained when the drug is incorporated into the hydration solution, between 0 and 2%. The hydration of drug-containing thin lipid films (composed of PG/PC/chol at a molar ratio of 1:9:5) yields encapsulation efficiencies between 3.6 and 6.7%. A slightly higher percentage (7.2%) is obtained when 30% of the drug is included in both the lipid film and the hydration solution. In contrast, when 30% of the drug is incorporated in a lipid film composed of PG/PC/chol at a ratio of 5:5:5, the encapsulation efficiency increases up to 47.9%. Furthermore, when the composition is changed to 10:5 (PG/chol), the efficiency increases up to 61.7%, suggesting that electrostatic interactions between PG and clarithromycin stabilize the drug in the liposome membrane.

Antimycobacterial Activity of Liposomal Clarithromycin

Since the aqueous solubility limit of clarithromycin is very low (Piscitelli *et al.*, 1992), the drug is initially dissolved in dimethyl sulfoxide (DMSO) (Mor *et al.*, 1994) for MIC determinations and then diluted in distilled water to at least 40 times the required final concentration (as BACTEC vials contain 4 ml of medium). The accurate determination of clarithromycin concentrations for MIC assessment and encapsulation rates is realized using a Rabbit-HP (Rainin, Emeryville, CA) liquid chromatograph, precolumn, and a prepacked 30 cm × 3.9-mm i.d. C_{18} Mirosorb column (Rainin Instruments, Woburn, MA). Sample analyses are performed at room temperature using a mobile phase consisting of 65% methanol and 35% (v/v) 0.05 M monobasic sodium phosphate buffer. The pH is adjusted to 4.0 using orthophosphoric acid. The separation is monitored at 210 nm.

After 7 days of incubation of MAC in 7H9 broth, the bacterial suspension is adjusted by using a McFarland standard. From this, 10-fold dilutions are made in BACTEC medium from 10^{-1} to 10^{-5}. Clarithromycin-containing vials (1.0, 2.0 and 4.0 μg/ml) and drug-free control vials are injected with 200 μl of the suspension from the 10^{-3} dilution. A 1/100 control vial is inoculated with 200 μl of the 10^{-5} dilution. The vials are read at the same time every day until the 1/100 control shows a GI > 30, with an increase in GI > 10 for 3 consecutive days.

Human peripheral blood monocyte–derived macrophages are prepared as described above. The cells are inoculated with the MAC suspension at a ratio of 5 to 1. The cells are incubated for 24 h at 37° in 5% CO_2 and moist air. After 24 h, the monolayer is washed three times with HBSS to remove the extracellular MAC. After the first day of infection, the number and viability (Nakagawara and Nathan, 1983) of the cells are ascertained by Naphthol Blue Black staining of nuclei and by Trypan Blue exclusion, respectively.

The effects of free and clarithromycin-containing liposomes are compared at three different concentrations clinically achievable in the serum of treated patients. Generally, peak plasma concentrations of clarithromycin in HIV-infected adults receiving 0.5 or 1 g doses of the drug orally every 12 h range from 2 to 4 or 5 to 10 μg/ml, respectively (Abbott Laboratories, 1996). Thus, infected macrophages are treated with 2, 4, and 8 μg/ml of either free or liposomal clarithromycin for 24 h. The medium is then removed, and the cells are washed three times with HBS and incubated for 7 days in DME-high glucose-L-glutamine containing 20% FBS. Seven days after treatment, the growth index of MAC is determined as described above.

The effects of different concentrations of clarithromycin-loaded liposomes on intracellular MAC growth are presented in Fig. 1 (Salem and Düzgüneş, 2003). The liposome composition used for these studies, PG/PC/CH (1:9:5), is similar to that employed in our *in vivo* studies with other

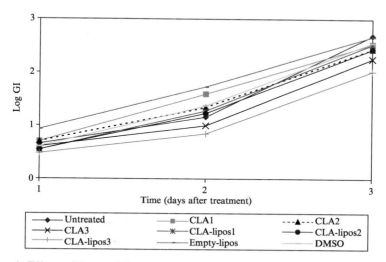

FIG. 1. Effects of free and liposome-encapsulated clarithromycin against MAC growth inside human macrophages. Infected cells were incubated for 24 h with free clarithromycin (CLA1, 2, and 3, corresponding to 2, 4, and 8 μg/ml of drug, respectively) or liposome-encapsulated clarithromycin (CLA-Lipos1, 2, and 3, corresponding to 2, 4, and 8 μg/ml of drug, respectively). After 7 days, cells were lysed to determine MAC growth by BACTEC.

antibiotics (Düzgüneş *et al.*, 1991a; Gangadharam *et al.*, 1991, 1995). Free and liposome-encapsulated clarithromycin have similar effects on the growth of intracellular MAC at the lower concentrations. Liposome-encapsulated clarithromycin at 8 μg/ml reduces the GI by 0.3 log units compared with the GI obtained with the free drug. Buffer-loaded (empty) liposomes have no effect on the intracellular growth of MAC or macrophage viability.

The effect of the various treatments on macrophage viability has also been investigated (Fig. 2). On day 7 postinfection, macrophage viability is comparable in infected but untreated controls, and MAC-infected cells treated with free drug, clarithromycin-loaded liposomes, or empty liposomes. Thus, the decrease in the MAC growth index observed in macrophages treated with clarithromycin formulations cannot be attributed to the decrease in the number of macrophages in the cultures due to cell death or removal of the cells from the monolayer.

Despite clarithromycin's capacity to concentrate into phagocytes, liposome-encapsulated clarithromycin at 8 μg/ml is slightly more effective against MAC than the free drug. This observation is partially in agreement with previous results obtained with ciprofloxacin, streptomycin, and amikacin, showing that the liposome-encapsulated drugs are more

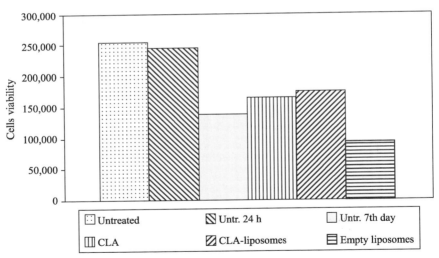

FIG. 2. Relative viability of MAC-infected macrophages using the Trypan Blue exclusion technique. Viability was determined by counting viable cells in infected but untreated control wells, as well as in wells containing MAC-infected cells treated with free or liposome-loaded clarithromycin, or empty liposomes. Counts were performed just before cell lysis for growth determination by BACTEC.

effective than the free drug against MAC inside murine peritoneal or human monocyte–derived macrophages (Ashtekar et al., 1991; Kesavalu et al., 1990; Majumdar et al., 1992). The slightly higher anti-MAC effect of clarithromycin-loaded liposomes, compared with that of the free drug, could be the result of a greater accumulation and availability of the drug in the endosomes or phagosomes containing MAC. The enhancement in efficacy is not as expected, however, probably due to the higher efficiency of the uptake of free clarithromycin by the infected macrophages compared with aminoglycoside antibiotics and ciprofloxacin. The high lipid solubility of clarithromycin may be responsible for this effect, as in the case of sparfloxacin described above. An additional factor that plays a role in the activities of the free and encapsulated clarithromycin is the relative inactivity of clarithromycin in the acidic milieu within phagolysosomes (Mor et al., 1994).

Clarithromycin encapsulated in the aqueous phase of liposomes composed of PC/dicetylphosphate/chol (7:2:1) is slightly more efficacious than free drug in reducing MAC CFU in macrophages (Onyeji et al., 1994); in this study, however, no growth of intracellular MAC was observed in untreated controls. Although liposomal clarithromycin shows a modest enhancement in efficacy against MAC in macrophages in culture, additional factors, including the stability of liposomes during circulation in the bloodstream and their passive targeting to infected resident macrophages in the liver and spleen, may influence the bioavailability of this antibiotic.

Antibiotics in pH-Sensitive Liposomes

Although antibiotics such as azithromycin are thought to have a particular advantage because they accumulate in tissues, they primarily localize in lysosomes (Schlossberg, 1995) and are thus not entirely "bioavailable" in the phagosomes harboring MAC and may be degraded. Following endocytosis, liposomes are also transported to lysosomes. For liposomal antibiotics to be effective, it is essential for endosomes containing liposomes to colocalize with MAC-containing phagosomes or for encapsulated antibiotics to exit the endosomes and localize in the infected phagosomes. If endosome–phagosome fusion is impaired in macrophages infected with mycobacteria (Xu et al., 1994), the likelihood of liposomes colocalizing in phagosomes may be relatively low. Therefore, enhancing the ability of liposomal antibiotics to leak out of liposomes, as well as out of endosomes via the destabilization of the endosome membrane, before the liposome is delivered to lysosomes may present a therapeutic advantage. pH-sensitive liposomes destabilize at the mildly acidic pH inside endosomes in which they are internalized. They also destabilize the endosome membrane and facilitate the cytoplasmic delivery of encapsulated molecules (Connor and Huang, 1985; Düzgüneş et al., 1991b; Straubinger, 1993; Straubinger et al., 1985). The destabilization and the delivery of the encapsulated contents into the cytoplasm may thus provide an escape mechanism for liposome-encapsulated antibiotics that would otherwise localize in lysosomes and get degraded.

Treatment of MAC-infected macrophages with pH-sensitive liposomes composed of dioleoylphosphatidylethanolamine (DOPE)/cholesterylhemisuccinate (CHEMS)/sparfloxacin (6:4:2) reduces the growth index of the mycobacteria to a level considerably lower than that obtained with free sparfloxacin at the same concentration (Table II). Sparfloxacin in conventional liposomes (PG/PC/sparfloxacin, 1:1:1:0.4) reduces the growth index to levels comparable to that obtained with free sparfloxacin, as described above. Delivery of sparfloxacin in pH-sensitive liposomes to M. tuberculosis-infected macrophages also results in a decrease in the growth index below that observed with free sparfloxacin (Table III).

Azithromycin encapsulated in pH-sensitive liposomes is also more effective than the antibiotic encapsulated in conventional liposomes composed of PG/PC/chol (1:1:1) or in highly negatively charged PG/chol (2:1) liposomes (Table IV). These results indicate the potential for the use of pH-sensitive liposomes for the intracellular delivery of antimycobacterial drugs. Gentamicin encapsulated in pH-sensitive liposomes of a composition different from that used in our studies was shown to be significantly more effective than the free drug in killing vacuole-resident Salmonella typhimurium in macrophages (Lutwyche et al., 1998).

TABLE II

ACTIVITY OF SPARFLOXACIN ENCAPSULATED IN pH-SENSITIVE LIPOSOMES AGAINST
MYCOBACTERIUM AVIUM COMPLEX (STRAIN 101) IN J774 MACROPHAGES

Treatment	Growth index[a]
Day 0 control	55.5 ± 1.5
Day 4 control	92.3 ± 11.8
Free sparfloxacin[b]	52.5 ± 2.5
pH-sensitive sparfloxacin liposomes[c]	36.5 ± 3.5
Conventional sparfloxacin liposomes[c]	55.0 ± 1.0
Empty pH-sensitive liposomes	143 ± 0.5

[a] Mean and standard deviation of growth indexes from triplicate determinations from duplicate wells. Data are representative of three independent experiments.
[b] The sparfloxacin concentration was 3 μg/ml.
[c] The liposomes were dialyzed to remove any sparfloxacin released from the liposomes. The sparfloxacin concentration was identical to that of free sparfloxacin (3 μg/ml) (D. Flasher, M. V. Reddy, P. R. J. Gangadharam, and N. Düzgüneş, unpublished data).

TABLE III
EFFICACY OF pH-SENSITIVE SPARFLOXACIN LIPOSOMES AGAINST
MYCOBACTERIUM TUBERCULOSIS (STRAIN H37Ry) IN J774 MACROPHAGES

Treatment	Growth index[a]
Day 0 control	4.0 ± 0.0
Day 4 control	104.0 ± 7.1
Free sparfloxacin[b]	18.5 ± 3.5
pH-sensitive sparfloxacin liposomes (1 μm)[b,c]	13.5 ± 3.5
pH-sensitive sparfloxacin liposomes (0.2 μm)[b,c]	7.5 ± 0.7
Empty pH-sensitive liposomes[d]	193.0 ± 14.1
DMSO control	100.5 ± 13.4

[a] Mean and standard deviation of growth indexes from triplicate samples from duplicate wells.
[b] The sparfloxacin concentration was 3 μg/ml.
[c] The liposomes were extruded through polycarbonate membranes with the indicated pore diameter.
[d] Buffer-loaded liposomes without the antibiotic, extruded through 1-μm pore-diameter polycarbonate membranes at the same lipid concentration as the liposomes containing sparfloxacin (M. V. Reddy, D. Flasher, N. Düzgüneş, and P. R. J. Gangadharam, unpublished data).

TABLE IV
EFFICACY OF LIPOSOME-ENCAPSULATED AZITHROMYCIN AGAINST
MAC 101 IN J774 A.1 MACROPHAGES

Treatment[a]	Growth index
Day 0 control	37.0 ± 2.8
Day 4 control	141 ± 7.0
Free azithromycin[b]	66.0 ± 4.2
Azithromycin liposomes (PG:PC:chol)[b]	42.0 ± 2.8
pH-sensitive azithromycin liposomes[b]	31.0 ± 2.8
Empty liposomes	131 ± 8.4

[a] The liposomes or the free drug were incubated with MAC-infected macrophages for 24 h and then removed, and the macrophages were incubated for another 3 days. The cells were lysed, and the growth index of MAC was measured in a BACTEC instrument (± SD).
[b] The concentration of azithromycin was 4 μg/ml (D. Flasher, M. V. Reddy, P. R. J. Gangadharam, and N. Düzgüneş, unpublished data).

One disadvantage of regular pH-sensitive liposomes is that they are cleared rapidly from circulation. This property would limit their use in delivering their contents to tissues other than the spleen and liver, or in delivering their contents over a prolonged period of time as an intravenous sustained-release system. Nevertheless, it is possible to generate pH-sensitive liposomes with prolonged circulation time *in vivo* that can deliver their contents into macrophages in a pH-dependent manner (Slepushkin *et al.*, 1997, 2004). Such sterically stabilized pH-sensitive liposomes are also highly effective in delivering functional antisense oligonucleotides and ribozymes to HIV-infected macrophages (Düzgüneş *et al.*, 2001). These liposomes are thus likely to be effective for the *in vivo* delivery of antimycobacterial drugs.

Other Liposomal Antibiotics

A number of other antibiotics have been encapsulated in liposomes and tested for their activity against MAC or *M. tuberculosis* infections. Gaspar *et al.* (2000) found that the rifampin derivative rifabutin is encapsulated with higher efficiency in fluid liposomes compared with rigid liposomes with higher transition temperature lipids. Rifabutin encapsulated in PC/phosphatidylserine (PS) (7:3) liposomes and administered iv is slightly more effective against MAC in the liver and spleen than an equivalent dose of free drug, reducing the CFU by about 0.4–0.8 log units, depending on the dose. It is likely that hydrophobic rifabutin forms micelles that are taken up by the reticuloendothelial system, thereby conferring an advantage to the free drug given by a rather nonconventional route. It will thus be of interest to compare iv liposomal rifabutine with a more conventional route of administration of rifabutine, such as subcutanous or oral.

Liposomal clofazimine is also effective against MAC *in vivo* under certain conditions, such as when treatment is started 3 weeks after infection (Kansal *et al.*, 1997). It is possible to inject considerably higher doses of liposomal clofazimine compared with free clofazimine before reaching the maximum tolerated dose. Significant reductions in CFU can be obtained in the lungs, spleen, and liver with clofazimine encapsulated in liposomes.

Isoniazid and rifampicin have been encapsulated in sterically stabilized liposomes containing *O*-stearoyl amylopectin, which appears to confer higher accumulation in the lungs when the liposomes are injected iv (Deol *et al.*, 1997). These liposomes can reduce the CFU of *M. tuberculosis* in the lungs, liver, and spleen of mice compared with untreated controls and free drug, but the level of reduction is generally less than half a log compared to the free drug. Nevertheless, once weekly administration of liposomal isoniazid or rifampicin in guinea pigs produces a reduction in CFU equivalent to that obtained with daily oral administration of the free drugs after 46 days of treatment (Pandey *et al.*, 2004). Rifampicin encapsulated in PC liposomes injected iv into *M. tuberculosis*–infected mice twice a week for 2 weeks reduces the CFU in the lungs by about 2.5 log units compared with untreated controls and by about 1.5 log units compared with free rifampicin (Agarwal et al., 1994). When the liposomes are targeted to macrophages via the tetrapeptide "Tuftsin," this effect is enhanced, facilitating a 5 log unit reduction in CFU in the lungs compared with untreated controls. A 3 log unit reduction compared with controls is obtained in the liver and spleen with this formulation.

Rifampicin liposomes have also been used for aerosol delivery to the lungs (Vyas et al., 2004). Higher drug concentrations are achieved in the lungs with negatively charged liposomes (PC/chol/dicetylphosphate) or liposomes containing macrophage-targeting ligands (maleylated bovine serum albumin or O-stearoyl amylopectin) than with free drug or PC:chol liposomes. If alveolar macrophages are isolated following aerosolization of the various formulations and infected with Mycobacterium smegmatis, rifampicin in the targeted liposomes causes an 89–93% reduction in the viability of the microorganism, while negatively charged liposomes, neutral liposomes, and free drug reduce the viability by 78, 68, and 54%, respectively (Vyas et al., 2004).

Liposomal ethambutol can be obtained using a "remote-loading" technique by first encapsulating 0.5 M H_2SO_4, and adjusting the external pH to 7.5 by $NaSO_4$–NaH_2PO_4, thereby generating a pH gradient across the membrane (Wiens et al., 2004). A large percentage (76–92%) of the added ethambutol can be encapsulated by this technique. Gentamicin and ceftazidime have been coencapsulated in liposomes composed of partially hydrogenated egg PC/chol/PEG-DSPE (1.85:1:0.15) (Schiffelers et al., 2001). When delivered in liposomes, the antibiotics show a synergistic interaction in the treatment of K. pneumoniae in a rat model, while the free antibiotics have only an additive effect. Enrofloxacin encapsulated in multilamellar PC/chol liposomes administered im in rabbits can increase the elimination half-life of the drug (Cabanes et al., 1995) and enhance the antibacterial effect of the drug against Staphylococcus aureus in canine neutrophils (Bas et al., 2000).

Azithromycin incorporated in the membrane phase of DSPG/chol (2:1) liposomes has an IC_{50} of 0.44 μM compared with an IC_{50} of 18 μM for the free drug in the treatment of MAC inside J774 cells (Oh et al., 1995). The efficacy of liposomal azithromycin increases with the negative charge of the liposomes.

Conclusions

The methods and studies described in this chapter indicate that numerous antibiotics can be encapsulated stably in liposomes and that liposomal formulations are more effective than the free drug in several infection models. The success of several liposome-encapsulated antimycobacterial agents in animal models suggests that they may be used effectively against MAC or *M. tuberculosis* infections. Other liposome-encapsulated antibiotics may also have significant advantages over the free drug in other infectious disease models. Further studies are necessary to find the most effective combination of liposome-encapsulated antibiotics and to identify other antibiotics that show increased efficacy when delivered in liposomes.

References

Abbott Laboratories. (1996). Biaxin (clarithromycin) Filmtab tablets and granules for oral suspension prescribing information. Abbott Park, IL.

Agarwal, A., Kandpal, H., Gupta, H. P., Singh, N. B., and Gupta, C. M. (1994). Tuftsin-bearing liposomes as rifampin vehicles in treatment of tuberculosis in mice. *Antimicrob. Agents Chemother.* **38,** 588–593.

Allen, T. M., Hansen, C., Martin, F., Redemann, C., and Yau-Young, A. (1991). Liposomes containing synthetic lipid derivatives of poly(ethylene glycol) show prolonged circulation half-lives *in vivo. Biochim. Biophys. Acta* **1066,** 29–36.

Armstrong, D., Gold, J. W. M., Drysanski, J., Whimbey, E., Polsky, B., Hawkins, C., Brown, A. E., Bernard, E., and Kiehn, T. K. (1985). Treatment of infections in patients with the acquired immunodeficiency syndrome. *Ann. Intern. Med.* **103,** 738–743.

Ashtekar, D., Düzgüneş, N., and Gangadharam, P. R. J. (1991). Activity of free and liposome-encapsulated streptomycin against *Mycobacterium avium* complex (MAC) inside peritoneal macrophages. *J. Antimicrob. Chemother.* **28,** 615–617.

Bakker-Woudenberg, I. A. J. M., Lokerse, A. F., Roerdink, F. H., Regts, D., and Michel, M. F. (1985). Free versus liposome-entrapped ampicillin in treatment of infection due to *Listeria monocytogenes* in normal and athymic (nude) mice. *J. Infect. Dis.* **151,** 917–924.

Bakker-Woudenberg, I. A. J. M., Lokerse, A. F., Ten Kate, M. T., and Storm, G. (1992). Enhanced localization of liposomes with prolonged blood circulation time in infected lung tissue. *Biochim. Biophys. Acta* **1138,** 318–326.

Bakker-Woudenberg, I. A. J. M., Lokerse, A. F., Ten Kate, M. T., Mouton, J. W., Woodle, M. C., and Storm, G. (1993). Liposomes with prolonged blood circulation and selective localization in *Klebsiella pneumoniae*-infected lung tissue. *J. Infect. Dis.* **168,** 164–171.

Bakker-Woudenberg, I. A. J. M., Ten Kate, M. T., Guo, L., Working, P., and Mouton, J. W. (2001). Improved efficacy of ciprofloxacin administered in polyethylene glycol-coated liposomes for treatment of *Klebsiella pneumoniae* pneumonia in rats. *Antimicrob. Agents Chemother.* **45,** 1487–1492.

Bartlett, G. R. (1959). Phosphorus assay in column chromatography. *J. Biol. Chem.* **234,** 466–468.

Bartley, P. B., Allworth, A. M., and Eisen, D. P. (1999). *Mycobacterium avium* complex causing endobronchial disease in AIDS patients after partial immune restoration. *Int. J. Tuberculosis Lung Dis.* **12,** 1132–1136.

Baş, A. L., Elmas, M., Simsek, A., Tras, B., Yazar, E., and Hadimli, D. D. (2000). Efficacies of free and liposome-encapsulated enrofloxacin (Baytril®) against *Staphylococcus aureus* infection in Turkish Shepard Dog neutrophils *in vitro. Rev. Méd. Vét.* **151,** 415–420.

Bermudez, L. E. M., Wu, M., and Young, L. S. (1987). Intracellular killing of *Mycobacterium avium* complex by rifapentine and liposome-encapsulated amikacin. *J. Infect. Dis.* **156,** 510–513.

Bermudez, L. E., Yau-Young, A. O., Lin, J. P., Cogger, J., and Young, L. S. (1990). Treatment of disseminated *Mycobacterium avium* complex infection in beige mice with liposome-encapsulated aminoglycosides. *J. Infect. Dis.* **161**, 1262–1268.

Cabanes, A., Reig, F., Anton, J. M., and Arboix, M. (1995). Sustained release of liposome-encapsulated enrofloxacin after intramuscular administration in rabbits. *Am. J. Vet. Res.* **56**, 1498–1501.

Chin, D. P., Reingold, A. L., Stone, E. N., Vittinghoff, E., Horsburgh, C. R., Simon, E. M., Yajko, D. M., Hadley, W. K., Ostroff, S. M., and Hopewell, P. C. (1994). The impact of *Mycobacterium avium* complex bacteremia and its treatment on survival of AIDS patients: A prospective study. *J. Infect. Dis.* **170**, 578–584.

Conley, J., Yang, H., Wilson, T., Blasetti, K., Di Ninno, V., Schnell, G., and Wong, J. P. (1997). Aerosol delivery of liposome-encapsulated ciprofloxacin: Aerosol characterization and efficacy against *Francisella tularensis* infection in mice. *Antimicrob. Agents Chemother.* **41**, 1288–1292.

Connor, J., and Huang, L. (1985). Efficient cytoplasmic delivery of a fluorescent dye by pH-sensitive immunoliposomes. *J. Cell Biol.* **101**, 582–589.

Cynamon, M. H., Swenson, C. S., Palmer, G. S., and Ginsberg, R. S. (1989). Liposome encapsulated amikacin therapy of *Mycobacterium avium* complex infection in beige mice. *Antimicrob. Agents Chemother.* **33**, 1179–1183.

Daemen, T., Veninga, A., Roerdink, F. H., and Scherphof, G. L. (1986). *In Vitro* activation of rat liver macrophages to tumoricidal activity by free or liposome encapsulated muramyl dipeptide. *Cancer Res.* **46**, 4330–4335.

Daleke, D. L., Hong, K., and Papahadjopoulos, D. (1990). Endocytosis of liposomes by macrophages: Binding, acidification and leakage of liposomes monitored by a new fluorescence assay. *Biochim. Biophys. Acta* **1024**, 352–366.

Deol, P., Khuller, G. K., and Joshi, K. (1997). Therapeutic efficacies of isoniazid and rifampin encapsulated in lung-specific stealth liposomes against *Mycobacterium tuberculosis* infection induced in mice. *Antimicrob. Agents Chemother.* **41**, 1211–1214.

Dhillon, J., Fielding, R., Adler-Moore, J., Goodall, R. L., and Mitchison, D. (2001). The activity of low-clearance liposomal amikacin in experimental murine tuberculosis. *J. Antimicrob. Chemother.* **48**, 869–876.

Dijkstra, J., van Galen, W. J. M., Hulstaert, C. E., Kalicharan, D., Roerdink, F. H., and Scherphof, G. L. (1984). Interaction of liposomes with Kupffer cells *in vitro*. *Exp. Cell Res.* **150**, 161–176.

Dupond, B. (2002). Overview of the lipid formulations of amphotericin B. *J. Antimicrob. Chemother.* **49**(Suppl. S1), 31–36.

Düzgüneş, N. (1998). Treatment of human immunodeficiency virus, *Mycobacterium avium* and *Mycobacterium tuberculosis* infections by liposome-encapsulated drugs. *In* "Medical Applications of Liposomes" (D. D. Lasic and D. Papahadjopoulos, eds.), pp. 189–219. Elsevier, Amsterdam.

Düzgüneş, N. (2003). Preparation and quantitation of small unilamellar liposomes and large unilamellar reverse-phase evaporation liposomes. *Methods Enzymol.* **367**, 23–27.

Düzgüneş, N., Wilschut, J., Hong, K., Fraley, R., Perry, C., Friend, D. S., James, T. L., and Papahadjopoulos, D. (1983). Physico-chemical characterization of large unilamellar phospholipid vesicles prepared by reverse-phase evaporation. *Biochim. Biophys. Acta* **732**, 289–299.

Düzgüneş, N., Perumal, V. K., Kesavalu, L., Goldstein, J. A., Debs, R. J., and Gangadharam, P. R. J. (1988). Enhanced effect of liposome-encapsulated amikacin on *Mycobacterium avium-M. intracellulare* complex in beige mice. *Antimicrob. Agents Chemother.* **32**, 1404–1411.

Düzgüneş, N., Ashtekar, D. R., Flasher, D. L., Ghori, N., Debs, R. J., Friend, D. S., and Gangadharam, P. R. J. (1991a). Treatment of *Mycobacterium avium-intracellulare* complex infection in beige mice with free and liposome encapsulated streptomycin: Role of liposome type and duration of treatment. *J. Infect. Dis.* **164,** 143–151.

Düzgüneş, N., Straubinger, R. M., Baldwin, P. A., and Papahadjopoulos, D. (1991b). pH-sensitive liposomes: Introduction of foreign substances into cells. *In* "Membrane Fusion" (J. Wilschut and D. Hoekstra, eds.), pp. 713–730. Marcel Dekker, New York.

Düzgüneş, N., Flasher, D., Pretzer, E., Konopka, K., Slepushkin, V. A., Steffan, G., Salem, I. I., Reddy, M. V., and Gangadharam, P. R. J. (1995). Liposome-mediated therapy of human immunodeficiency virus type-1 and *Mycobacterium* infections. *J. Liposome Res.* **5,** 669–691.

Düzgüneş, N., Flasher, D., Reddy, M. V., Luna-Herrera, J., and Gangadharam, P. R. J. (1996). Treatment of *Mycobacterium avium-intracellulare* complex infection by free and liposome-encapsulated sparfloxacin. *Antimicrob. Agents Chemother.* **40,** 2618–2621.

Düzgüneş, N., Simões, S., Slepushkin, V., Pretzer, E., Rossi, J. J., De Clercq, E., Antao, V. P., Collins, M. L., and Pedroso de Lima, M. C. (2001). Enhanced inhibition of HIV-1 replication in macrophages by antisense oligonucleotides, ribozymes and acyclic oligonucleotide phosphonate analogs delivered in pH-sensitive liposomes. *Nucleosides Nucleotides Nucleic Acids* **20,** 515–523.

Emmen, F., and Storm, G. (1987). Liposomes in treatment of infectious diseases. *Pharm. Weekbl.* **9,** 162–171.

Eng, R. H., Bishburg, E., Smith, S. M., and Mangia, A. (1989). Diagnosis of Mycobacterium bacteremia in patients with acquired immunodeficiency syndrome by direct examination of blood films. *J. Clin. Microbiol.* **27,** 768–769.

Gabizon, A., Sulkes, A., Peretz, T., Druckmann, S., Goren, D., Amselem, S., and Barenholz, Y. (1989a). Liposome-associated doxorubicin: Preclinical pharmacology and exploratory clinical phase. *In* "Liposomes in Therapy of Infectious Diseases and Cancer" (G. Lopez-Berenstein and I. J. Fidler, eds.), pp. 391–402. Alan R. Liss, New York.

Gabizon, A., Peretz, T., Sulkes, A., Amselem, S., Ben-Yosef, R., Ben-Baruch, N., Catane, R., Biran, S., and Barenholz, Y. (1989b). Systemic administration of doxorubicin-containing liposomes in cancer patients: A phase I study. *Eur. J. Cancer. Clin. Oncol.* **25,** 1795–1803.

Gangadharam, P. R. J., Edwards, C. K., III, Murthy, P. S., and Pratt, P. F. (1983). An acute infection model for *Mycobacterium intracellulare* disease using beige mice: Preliminary results. *Am. Rev. Respir. Dis.* **127,** 648–649.

Gangadharam, P. R. J., Perumal, V. K., Kesavalu, L., Debs, R. J., Goldstein, J., and Düzgüneş, N. (1989). Comparative activities of free and liposome encapsulated amikacin against *Mycobacterium avium* complex (MAC). *In* "Liposomes in the Therapy of Infectious Diseases and Cancer" (G. Lopez-Berestein and I. J. Fidler, eds.), pp. 177–190. Alan R. Liss, New York.

Gangadharam, P. R. J., Ashtekar, D. A., Ghori, N., Goldstein, J. A., Debs, R. J., and Düzgüneş, N. (1991). Chemotherapeutic potential of free and liposome-encapsulated streptomycin against experimental *Mycobacterium avium* complex infections in beige mice. *J. Antimicrob. Chemother.* **28,** 425–435.

Gangadharam, P. R. J., Ashtekar, D. R., Flasher, D., and Düzgüneş, N. (1995). Therapy of *Mycobacterium avium* complex infections in beige mice with streptomycin encapsulated in sterically stabilized liposomes. *Antimicrob. Agents Chemother.* **39,** 725–730.

Gaspar, M. M., Neves, S., Portaels, F., Pedrosa, J., Silva, M. T., and Cruz, M. E. M. (2000). Therapeutic efficacy of liposomal rifabutin in a *Mycobacterium avium* model infection. *Antimicrob. Agents Chemother.* **44,** 2424–2430.

Gill, P. S., Espina, B. M., Muggia, F., Cabriales, S., Tulpule, A., Esplin, J. A., Liebman, H. A., Forssen, E., Ross, M. E., and Levine, A. M. (1995). Phase I/II clinical and pharmacokinetic evaluation of liposomal daunorubicin. *J. Clin. Oncol.* **13**, 996–1003.

Gregoriadis, G., and Florence, A. T. (1993). Liposomes in drug delivery: Clinical, diagnostic and ophthalmic potential. *Drugs* **1**, 15–28.

Hawkins, C. C., Gold, J. W. M., Whimbey, E., Kiehn, T. E., Brannon, P., Cammarata, R., Brown, A. E., and Armstrong, D. (1986). *Mycobacterium avium* complex infections in patients with the acquired immunodeficiency syndrome. *Ann. Intern. Med.* **105**, 184–188.

Hiraga, Y., and Kinoshita, T. (1981). Post column derivatization of guanidino compounds in high performance liquid chromatography using ninhydrin. *J. Chromatogr.* **226**, 43–51.

Horsburgh, C. R. (1991). *Mycobacterium avium* complex infection in the acquired immunodeficiency syndrome. *N. Engl. J. Med.* **324**, 1332–1338.

Horsburgh, C. R., Mason, U. G., III, Farhi, D. C., and Iseman, M. D. (1985). Disseminated infection with *Mycobacterium avium-intracellulare*. A report of 13 cases and a review of the literature. *Medicine (Baltimore)* **64**, 36–48.

Kansal, R. G., Gomez-Flores, R., Sinha, I., and Mehta, R. T. (1997). Therapeutic efficacy of liposomal clofazimine against *Mycobacterium avium* complex in mice depends on size of initial inoculum and duration of infection. *Antimicrob. Agents Chemother.* **41**, 17–23.

Kesavalu, L., Goldstein, J. A., Debs, R. J., Düzgüneş, N., and Gangadharam, P. R. J. (1990). Differential effects of free and liposome encapsulated amikacin on the growth of *Mycobacterium avium* complex in mouse peritoneal macrophages. *Tubercle* **71**, 215–218.

Lasic, D. D. (1993). "Liposomes: From Physics to Applications." Elsevier, Amsterdam.

Leitzke, S., Bucke, W., Borner, K., Müller, R., Hahn, H., and Ehlers, S. (1998). Rationale for and efficacy of prolonged-intereval treatment using liposome-encapsulated amikacin in experimental *Mycobacterium avium* infection. *Antimicrob. Agent. Chemother.* **42**, 459–461.

Lopez-Berestein, G., Bodey, G. P., Frankel, L. S., and Mehta, K. (1987). Treatment of hepatosplenic candidiasis with liposomal-amphotericin *Br. J. Clin. Oncol.* **5**, 310–317.

Lutwyche, P., Cordeiro, C., Wiseman, D. J., St. Louis, M., Uh, M., Hope, M. J., Webb, M. S., and Finaly, B. B. (1998). Intracellular delivery and antibacterial activity of gentamicin encapsulated in pH-sensitive liposomes. *Antimicrob. Agents Chemother.* **42**, 2511–2520.

Magallanes, M., Dijkstra, J., and Fierer, J. (1993). Liposome-incorporated ciprofloxacin in treatment of murine salmonellosis. *Antimicrob. Agents Chemother.* **37**, 2293–2297.

Majumdar, S., Flasher, D., Friend, D. S., Nassos, P., Yajko, D., Hadley, W. K., and Düzgüneş, N. (1992). Efficacies of liposome-encapsulated streptomycin and ciprofloxacin against *Mycobacterium avium-M. intracellulare* complex infections in human peripheral blood monocyte/macrophages. *Antimicrob. Agents Chemother.* **36**, 2808–2815.

Mor, N., Vanderkolk, J., Mezo, N., and Heifets, L. (1994). Effects of clarithromycin and rifabutin alone and in combination on intracellular and extracellular replication of *Mycobacterium avium*. *Antimicrob. Agents Chemother.* **38**, 2738–2742.

Nakagawara, A., and Nathan, C. F. (1983). A simple method for counting adherent cells: Application to cultured human monocytes, macrophages and multinucleated giant cells. *J. Immmunol. Methods* **56**, 261–268.

Oh, Y. K., Nix, D. E., and Straubinger, R. M. (1995). Formulation and efficacy of liposome-encapsulated antibiotics for therapy of intracellular *Mycobacterium avium* infection. *Antimicrob. Agents Chemother.* **39**, 2104–2111.

Onyeji, C. O., Nightingale, C. H., Nicolau, D. P., and Quintiliani, R. (1994). Efficacies of liuposoome-encapsulated clarithromycin and ofloxacin against *Mycobacterium avium* complex in human macrophages. *Antimicrob. Agents Chemother.* **38**, 523–527.

Pandey, R., Sharma, S., and Khuler, G. K. (2004). Liposome-based antitubercular drug therapy in a guinea pig model of tuberculosis. *Int. J. Antimicrob. Agents* **23**, 414–415.

Pedroso de Lima, M. C., Chiche, B. H., Debs, R. J., and Düzgüneş, N. (1990). Interaction of antimycobacterial and anti-pneumocystis drugs with phospholipid membranes. *Chem. Phys. Lipids* **53**, 361–371.

Perronne, C., Gikas, A., Truffot-Pernot, C., Grosset, J., Vilde, J.-L., and Pocidalo, J. J. (1991). Activities of sparfloxacin, azithromycin, temafloxacin, and rifapentine compared with that of clarithromycin against multiplication of *Mycobacterium avium* complex within human macrophages. *Antimicrob. Agents Chemother.* **35**, 1356–1359.

Petersen, E. A., Grayson, J. B., Hersh, E. M., Dorr, R. T., Chiang, S.-M., Oka, M., and Proffitt, R. T. (1996). Liposomal amikacin: Improved treatment of *Mycobacterium avium* complex infection in the beige mouse model. *J. Antimicrob. Chemother.* **38**, 819–828.

Piscitelli, S. C., Danziger, L. H., and Rodvold, K. A. (1992). Clarithromycin and azithromycin: New macrolide antibiotics. *Clin. Pharm.* **11**, 137–152.

Popescu, M. C., Swenson, C. E., and Ginsberg, R. S. (1987). Liposome-mediated treatment of viral, bacterial and protozoal infections. *In* "Liposomes: From Biophysics to Therapeutics" (M. J. Ostro, ed.), pp. 219–251. Marcel Dekker, New York.

Poste, G., Bucana, C., Raz, A., Bugelski, P., Kirsh, R., and Fidler, I. J. (1982). Analysis of systemically administered liposomes and implications for their use in drug delivery. *Cancer Res.* **42**, 1412–1422.

Roerdink, F., Regts, J., Daemen, T., Bakker-Woudenberg, I., and Scherphof, G. (1986). Liposomes as drug carriers to liver macrophages: Fundamental and therapeutic aspects. *In* "Targeting of Drugs with Synthetic Systems" (G. Gregoriadis, J. Senior, and G. Poste, eds.), pp. 193–206. Plenum Press, New York.

Salem, I. I., and Düzgüneş, N. (2003). Efficacies of cyclodextrin-complexed and liposome encapsulated clarithromycin against *Mycobacterium avium* complex infection in human macrophages. *Int. J. Pharm.* **250**, 403–414.

Scherphof, G. L. (1991). *In vivo* behavior of liposomes: Interactions with the mononuclear phagocyte system and implications for drug targeting. *In* "Handbook of Experimental Pharmacology" (R. L. Juliano, ed.), Vol. 100, pp. 285–327. Springer-Verlag, Berlin.

Schiffelers, R. M., Storm, G., Ten Kate, M. T., Stearne-Cullen, L. E. T., Den Hollander, J. G., Verbrugh, H. A., and Bakker-Woudenberg, I. A. J. M. (2001). *In vivo* synergistic interaction of liposome-coencapsulated gentamicin and ceftazidime. *J. Pharm. Exp. Ther.* **298**, 369–375.

Schlossberg, D. (1995). Azithromycin and clarithromycin. *Med. Clin. North Am.* **79**, 803–815.

Siddiqi, S., Libonati, J. P., and Middlebrook, G. (1981). Evaluation of a rapid radiometric method for drug susceptibility testing of *Mycobacterium tuberculosis*. *J. Microbiol.* **13**, 908–913.

Slepushkin, V. A., Simões, S., Dazin, P., Newman, M. S., Guo, L. S., Pedroso de Lima, M. C., and Düzgüneş, N. (1997). Sterically stabilized pH-sensitive liposomes: Intracellular delivery of aqueous contents and prolonged circulation *in vivo*. *J. Biol. Chem.* **272**, 2382–2388.

Slepushkin, V., Simões, S., Pedroso de Lima, M. C., and Düzgüneş, N. (2004). Sterically stabilized pH-sensitive liposomes. *Methods Enzymol.* **387**, 134–147.

Sparano, J. A., and Winer, E. P. (2001). Liposomal anthracyclines for breast cancer. *Semin. Oncol.* **28**(suppl. 12), 32–40.

Straubinger, R. M. (1993). pH-sensitive liposomes for delivery of macromolecules into cytoplasm of cultured cells. *Methods Enzymol.* **221**, 361–376.

Straubinger, R. M., Düzgüneş, N., and Papahadjopoulos, D. (1985). pH-sensitive liposomes mediate cytoplasmic delivery of encapsulated macromolecules. *FEBS Lett.* **179**, 148–154.

Szoka, F., Olson, F., Heath, T., Vail, W., Mayhew, E., and Papahadjopoulos, D. (1980). Preparation of unilamellar liposomes of intermediate size (0.1–0.2 μm) by a combination of reverse phase evaporation and extrusion through polycarbonate membranes. *Biochim. Biophys. Acta* **601,** 559–571.

Tadakuma, T., Ikewaki, N., Yasuda, T., Tsutsumi, M., Saito, S., and Saito, K. (1985). Treatment of experimental salmonellosis in mice with streptomycin entrapped in liposomes. *Antimicrob. Agents Chemother.* **28,** 28–32.

Vladimirsky, M. A., and Ladigina, G. A. (1982). Antibacterial activity of liposome entrapped streptomycin in mice infected with *Mycobacterium tuberculosis. Biomedicine* **36,** 375–377.

Vyas, S. P., Kannan, M. E., Jain, S., Mishra, V., and Singh, P. (2004). Design of liposomal aerosols for improved delivery of rifampicin to alveolar macrophages. *Int. J. Pharmaceut.* **269,** 37–49.

Wiens, T., Redelmeier, T., and Av-Gay, Y. (2004). Development of a liposome formulation of ethambutol. *Antimicrob. Agents Chemother.* **48,** 1887–1888.

Woodle, M. C., and Lasic, D. D. (1992). Sterically stabilized liposomes. *Biochim. Biophys. Acta* **1113,** 171–199.

Xu, S., Cooper, C., Sturgill-Koszycki, S., van Heyningen, T., Chatterjee, D., Orme, I., Allen, P., and Russell, D. G. (1994). Intracellular trafficking in *Mycobacterium tuberculosis* and *Mycobacterium avium*-infected macrophages. *J. Immunol.* **153,** 2568–2578.

Zhang, J. H., and Zhu, J. B. (1999). A novel method to prepare liposomes containing amikacin. *J. Microencapsul.* **16,** 511–516.

[16] Tuftsin-Bearing Liposomes as Antibiotic Carriers in Treatment of Macrophage Infections

By C. M. Gupta and W. Haq

Abstract

Tuftsin is a tetrapeptide (Thr-Lys-Pro-Arg) that specifically binds monocytes, macrophages, and polymorphonuclear leukocytes and potentiates their natural killer activity against tumors and pathogens. The antimicrobial activity of this peptide is significantly increased by attaching at the C-terminus a fatty acyl residue through the ethylenediamine spacer arm. This activity is further augmented by incorporating the modified tuftsin in the liposomes. The tuftsin-bearing liposomes not only enhance the host's resistance against a variety of infections but also serve as useful vehicles for the site-specific delivery of drugs in a variety of macrophage-based infections, such as tuberculosis and leishmaniasis.

Introduction

Tuftsin is a macrophage natural killer activator tetrapeptide (Thr-Lys-Pro-Arg) that is an integral part of the Fc portion of the heavy chain of the leukophilic immunoglobulin G (residues 289–292) and is released physiologically as the free peptide after enzymatic cleavage (Najjar, 1987). It is known to possess specific receptors on monocytes, macrophages, and polymorphonuclear (PMN) leukocytes (Najjar, 1987) and potentiates the natural killer activity of these cells against tumors and, presumably, also against a variety of pathogens (Agrawal and Gupta, 2000). These properties of tuftsin make it an attractive candidate for use as a cell-specific ligand for targeting of drugs to the pathogens that reside and proliferate within macrophages and monocytes. Since these cells can be infected by a number of pathogens, including *Mycobacterium tuberculosis* and *Leishmania donovani*, and provide nonspecific resistance against a variety of infections, it is expected that grafting of tuftsin on the surface of the liposomes that are loaded with antibacterial/antiparasitic drugs should significantly increase their therapeutic efficacy against these infections.

Tuftsin, due to its hydrophilic character, cannot be grafted on the surface of liposomes without being attached to a sufficiently long hydrophobic anchor. Structure–function studies of this tetrapeptide indicate that its binding and subsequent activation of the mononuclear phagocyte system (MPS) is dependent upon rather strict conservation of its molecular structure. Thus, modifications of the peptide at its N-terminus or within the chain lead to a significant reduction or even loss of its biological activity (Fridkin and Gottlieb, 1981). However, the activity is largely retained if modifications are restricted only to the C-terminus (Gottlieb et al., 1982). All the modifications are, thus, limited to the carboxyl group of the Arg residue. Direct attachment of a fatty acyl group to the Arg residue, without any spacer arm, leads to modified tuftsin, which does not allow formation of liposomes, presumably due to perturbation of the phospholipid polar head group packing by the bulky Arg residue (Singhal et al., 1984). This problem is, however, circumvented by introducing an ethylenediamine spacer arm between the Arg residue and the hydrophobic anchor (Fig. 1).

Liposomes containing palmitoyl tuftsin (**I**) specifically recognize macrophages and PMN leukocytes (Singhal et al., 1984). Treatment of macrophages with these liposomes considerably increases their respiratory burst activity (Singh et al., 1992). Pretreatment of animals with tuftsin-bearing liposomes enables the animals to resist malaria (Gupta et al., 1986), leishmania (Guru et al., 1989), and fungal (Owais et al., 1993) infections. In addition, delivery of antileishmanial (Agrawal et al., 2002; Guru et al.,

$$\text{Thr-Lys-Pro-Arg-NH-(CH}_2)_2\text{-NH-C-C}_{15}\text{H}_{31}$$
$$\overset{\text{||}}{\text{O}}$$

Fig. 1. Structure of palmitoyl tuftsin (**I**).

1989), antitubercular (Agarwal *et al.*, 1994), antifungal (Owais *et al.*, 1993), and antifilarial (Owais *et al.*, 2003) drugs in liposomes containing **I** is shown to increase the therapeutic efficacy of drugs against these infections.

Here, we describe procedures for preparation of **I** as well as the liposomes that contain **I** in their bilayers, entrapment of various drugs in these liposomes, and their delivery to experimental animals with infected *L. donovani*, *M. tuberculosis*, or *Aspergillus*.

Synthesis of Palmitoyl Tuftsin (I)

Synthesis of **I** is carried out by the standard solution phase methods of peptide synthesis. The first intermediate, $H_2N-(CH_2)_2-NH-CO-C_{15}H_{31}$, is prepared by reacting palmitoyl chloride with an excess of ethylenediamine in a dilute ether solution. The amine thus obtained is coupled without purification with *t*-butyloxycarbonyl (Boc)-Arg(NC_2)-OH, using the N, N-dicyclohexylcarbodiimide/N-hydroxybenzotriazole (DCC/HOBt) procedure. The poduct is purified by silica gel chromatography to obtain Boc-Arg(NC_2)-NH-CH_2-CH_2-NH-CO-$C_{15}C_{31}$. The Boc group is removed by treatment with 4 N HCI in dioxane. The resulting amine hydrochloride is then coupled with Boc-Pro-OH using DCC/HOBt to obtain the dipeptide Boc-Pro-Arg(NO_2)-NH-CH_2-CH_2-NH-CO-$C_{15}H_{31}$ as a white powder. The Boc group is removed by treatment with 4 N HCI/dioxane, and the resulting amine is coupled with Boc-Lys[benzyloxycarbonyl (Z)]-OH using benzotriazole-1-yl-oxy-tris-(dimethylamino)-phosphonium hexafluorophosphate (BOP) reagent to obtain the tripeptide Boc-Lys(Z)-Pro-Arg(NO_2)-NH-CH_2-CH_2-NH-CO-$C_{15}H_{31}$. Removal of the Boc group from the tripeptide and subsequent coupling of the resulting amine with Boc-Thr[benzyl (Bz)]-OH using BOP reagent affords the fully blocked tetrapeptide Boc-Thr (Bz)-Lys(Z)-Pro-Arg(NO_2)-NH-CH_2-CH_2-NH-CO-$C_{15}H_{31}$. The crude tetrapeptide is purified by silica gel chromatography and crystallized from ethylacetate/hexane. The tetrapeptide is treated with 4 N HCl/dioxane to remove the Boc group, and the resulting product is subjected to catalytic hydrogenation over Pd black to obtain the desired compound as a white powder. The detailed synthetic procedure is given below.

Preparation of $H_2N-CH_2-CH_2-NH-CO-C_{15}H_{31}$ (II)

Palmitoyl chloride (2 ml) is added dropwise to a solution of ethylenediamine (5 ml; Sigma-Aldrich) in anhydrous ether (50 ml) at 0° under vigorous stirring. A white precipitate separates out in the reaction mixture. It is further stirred for 1.5 h, and then the precipitate is filtered and washed with cold ether. The solid is taken in chloroform and washed with saline.

The organic layer is dried over anhydrous sodium sulfate and concentrated under reduced pressure to obtain $H_2N-CH_2-CH_2-NH-CO-C_{15}H_{31}$ as a gummy material (2.1 g). This product is dried over P_2O_5 and KOH pellets under vacuum and used as such without further purification. Fast atom bombardment mass spectrometry (FAB-MS): nuclear magnetic resonanance MH^+, 299; (NMR) ($CDCl_3$) δ ppm: 0.88 [triplet (t), 3H, CH_3], 1.24 [broad singlet (bs), 26H, CH_2], 2.17 (t, 2H, $CO-CH_2$), 3.36 (bs, 4H, $NH-CH_2-CH_2-NH_2$), 5.7–5.84 [broad hump (bh), NH].

Preparation of Boc-Arg (NO₂)-HN-CH₂-CH₂-NH-CO-C₁₅H₃₁ (III)

To a solution of Boc-Arg(NO_2)-OH (2.1 g, 6.5 mM) in anhydrous dimethylformamide (DMF) (15 ml), HOBt (0.9 g, 6.5 mM; Janssen Chimica) is added, and the reaction mixture is cooled to 0°. To this solution DCC (1.4 g, 7 mM; Janssen Chimica) is added under vigorous stirring. After 10 min, a solution of **II** (2 g, 7 mM) in dichloromethane (10 ml) is added, and the reaction mixture is stirred for 1.5 h at 0° and then kept in a refrigerator overnight. Dicyclohexylurea (DCU) is filtered, and the filtrate is evaporated under reduced pressure. The residue is partitioned between ethylacetate (100 ml) and saline (25 ml). The organic layer is washed with 5% citric acid (3 × 20 ml), saline (3 × 20 ml), 5% $NaHCO_3$ (3 × 20 ml), and finally with saline. The organic layer is dried over anhydrous sodium sulfate and concentrated under reduced pressure to obtain the crude compound (3.2 g). The crude product is subjected to silica gel (Merck) chromatography using 2% methanol (MeOH) in chloroform as the eluent to obtain 2.5 g chromatographically homogeneous compound. It is characterized by mass and NMR spectroscopy. Melting point (M.P.) 120–121°; $[\alpha]_D^{25} - 3.5$ ($c = 1$, DMF); FAB-MS: MH^+, 599; NMR ($CDCl_3$) δ ppm: 0.88 (t, 3H, CH_3), 1.24 (bs, 26H, CH_2), 1.4 [singlet (s), 9H], 1.68–1.79 (bs, 4H, β and γ CH_2 Arg), 2.17 (t, 2H, $CO-CH_2$), 3.36 (bs, 6H, $NH-CH_2$-CH_2-NH and δ CH_2 Arg), 4.23 [multiplet (m), 1H, α CH], 5.7–8.24 (bh, NH).

Preparation of Boc-Pro-Arg(NO₂)-HN-CH₂-CH₂-NH-CO-C₁₅H₃₁ (IV)

Compound **III** (3.0 g, 5 mM) is dissolved in 4 N HCl/dioxane (25 ml), and the solution is kept at room temperature for 1.5 h. Solvent is removed under reduced pressure, and the residue is precipitated by addition of an excess of dry ether. The hydrochloride salt is collected by filtration and washed with anhydrous ether three times. It is dried over P_2C_5 and KOH pellets in a vacuum desiccator.

To a solution of Boc-Pro-OH (1.1 g, 5.2 mM) and HOBt (0.75 g, 5.5 mM) in dry DMF (25 ml) is added the hydrochloride salt of **III,** obtained as above, in dry DMF (15 ml) and N-methyl morpholine (1.2 ml) at 4° under vigorous stirring. To this vigorously stirred mixture, DCC (1.2 g, 6 mM) in dry dichloromethane (10 ml) is added dropwise. The reaction mixture is stirred further for 1.5 h at 0° and 8 h at room temperature. DCU is removed by filtration, and the filtrate is evaporated under reduced pressure. The residue is extracted in ethylacetate (150 ml) and the organic layer washed with 5% citric acid (3 × 20 ml), saline (3 × 20 ml), 5% NaHCO$_3$ (3 × 20 ml), and finally with saline. The organic layer is dried over anhydrous sodium sulfate and concentrated under reduced pressure to obtain an oily residue. The residue is crystallized with ethylacetate/hexane to obtain chromatographically homogeneous **IV** as a white solid. Yield: 2.8 g (83%). It is characterized by mass and NMR spectroscopy. M. P. 140–142°; $[\alpha]_D^{25}$ − 14.2 ($c = 1.1$, DMF); FAB-MS: MH$^+$, 697; NMR (CDCl$_3$) δ ppm: 0.88 (t, 3H, CH_3), 1.24 (bs, 26H, CH_2), 1.4 (s, 9H), 1.68–1.79 (bs, 4H, β and γ CH_2 Arg), 1.95–2.1 (bs, 4H, β and γ CH_2 Pro), 2.17 (t, 2H, CO-CH_2), 3.4 (bs, 2H, δ CH_2 Pro), 3.36 (bs, 6H, NH-CH_2CH_2NH and δ CH_2 Arg), 4.31 (m, 2H, α CHs), 5.7–8.24 (bh, NH).

Preparation of Boc-Lys(Z)-Pro-Arg(NO$_2$)-HN-CH$_2$-CH$_2$-NH-CO-C$_{15}$H$_{31}$ (V)

Boc-Lys(Z)-OH (1.6 g, 4 mM) and HOBt (0.68 g, 4.5 mM) are dissolved in dry DMF (25 ml), and the solution is cooled to 0°. A solution of dipeptide amine H-Pro-Arg(NO$_2$)-HN-CH$_2$-CH$_2$-NH-CO-C$_{15}$H$_{31}$, obtained by treating **IV** (4 mM) with 4 N HCl/dioxane and subsequent neutralization with N-methyl morpholine, is added to the above mixture. After 5 min, diisopropylethyl amine (0.5 ml) and BOP reagent (2.2 g, 4.5 mM; Novabiochem) are added to the reaction mixture, and the stirring is continued for another 4 h. Solvent is removed under reduced pressure, and the residue is extracted in ethylacetate (50 ml × 3). The organic layer is washed with 5% citric acid (3 × 20 ml), saline (3 × 20 ml), 5% NaHCO$_3$ (3 × 20 ml), and finally with saline. It is dried over anhydrous sodium sulfate and concentrated under reduced pressure to obtain an oily residue. The residue is crystallized with ethylacetate/hexane to obtain **V** as a white solid. Yield: 2.7 g (72%). It is characterized by mass and NMR spectroscopy. M.P. 103–105°; $[\alpha]_D^{25}$ −12.4 ($c = 1$, DMF); FAB-MS: MH$^+$, 960; NMR (CDCl$_3$) δ ppm: 0.88 (t, 3H, CH_3), 1.24 (bs, 26H, CH_2), 1.4 (s, 9H), 1.45–1.82 (bs, 6H, β, γ, and δ CH_2 Lys), 1.68–1.79 (bs, 4H, β and γ CH_2 Arg), 1.95–2.1 (bs, 4H, β and γ CH_2 Pro), 2.17 (t, 2H, CO-CH_2), 3.21 (bs, 2H, ε CH_2 Lys), 3.4 (bs, 2H, δ CH_2 Pro), 3.36 (bs, 6H, NH-CH_2-CH_2-NH and δ CH_2 Arg), 4.3–4.81 (m, 3H, α CH), 5 (s, 2H, CH_2), 7.3 (s, 5H, CH aromatic), 5.7–8.24 (bh, NH).

Preparation of Boc-Thr(Bz)-Lys(Z)-Pro-Arg(NO₂)-HN-CH₂-CH₂-NH-CO-C₁₅H₃₁ (VI)

Boc group from **V** (2.7 g, 2.8 mM) is removed using 4 N HCl/dioxane in the usual manner, and the resulting amine hydrochloride is neutralized with N-methyl morpholine (Sigma-Aldrich) in dry DMF (25 ml). To this solution, Boc-Thr(Bz)-OH (0.86 g, 2.8 mM) and HOBt (0.45 g, 3 mM) are added, and the reaction mixture is stirred at 0°. After 5 min, diisopropylethyl amine (0.5 ml; Sigma-Aldrich) and BOP reagent (1.7 g, 3.5 mM) are added, and stirring is continued for 4–5 h. Solvent is removed under reduced pressure, and the residue is extracted in ethylacetate (3 × 50 ml). The organic layer is washed with 5% citric acid (3 × 20 ml), saline (3 × 20 ml), 5% NaHCO₃ (3 × 20 ml), and finally with saline. It is dried over anhydrous sodium sulfate and concentrated under reduced pressure to obtain an oily residue. The crude product is subjected to silica gel column chromatography using 3% MeOH in chloroform as the eluent to obtain 2.1 g (62%) chromatographically homogeneous **VI**. It is characterized by mass and NMR spectroscopy. M.P. 82–83°; $[\alpha]_D^{25}$ −19.5 (c = 1.2, DMF); FAB-MS: MH⁺ 1150; NMR (CDCl₃) δ ppm: 0.88 (t, 3H, CH_3), 1.24 (bs, 26H, CH_2), 1–4 (s 9H), 1.44 (s, 3H, CH_3 Thr), 1.45–1.82 (bs, 6H, β, γ, and δ CH_2 Lys), 1.68–1.79 (bs, 4H, β and γ CH_2 Arg), 1.95–2.1 (bs, 4H, β and γ CH_2 Pro), 2.17 (t, 2H, CO-CH_2), 3.21 (bs, 2H, ε CH_2 Lys), 3.4 (bs, 2H, δ CH_2 Pro), 3.36 (bs, 6H, NH-CH_2-CH_2-NH and δ CH_2 Arg), 4.3–4.81 (m, 4H, α CH), 5–5.1 (2s, 4H, CH_2), 7.3–7.4 (2s, 10H, aromatic CH), 5.7–8.24 (bh, NH).

Preparation of Thr-Lys-Pro-Arg-NH-CH₂-CH₂-NH-CO-C₁₅H₃₁ (I)

Compound **VI** (0.92 g, 0.8 mM) is subjected to acidolysis using 4-N HCl/dioxane to remove the Boc group of the Thr residue. The hydrochloride salt is then subjected to catalytic hydrogenation over palladium black for 20 h in glacial acetic acid. Solvent is removed under reduced pressure, and the residue is crystallized using anhydrous methanol/ether two times to obtain a chromatographically homogeneous white powder. It is characterized by NMR and FAB-MS. Yield: 0.51 g, (80%). M.P. 149–151° (d); $[\alpha]_D^{25}$ −38.6 (c = 0.7, DMF); FAB-MS: MH⁺, 781; NMR (d_6-DMSO) δ ppm: 0.85 (t, 3H, CH_3), 1.24 (bs, 26H, CH_2), 1.44 (s, 3H, CH_3 Thr), 1.45–1.82 (bs, 6H, β, γ, and δ CH_2 Lys), 1.68–1.79 (bs, 6H, β and γ CH_2 Arg and β CH_2 alkyl), 1.95–2.1 (bs, 4H, β and γ CH_2 Pro), 2.04 (t, 2H, CO-CH_2), 3.08 (bs, 2H, ε CH_2 Lys), 3.4 (bs, 2H, δ CH_2 Pro), 3.36 (bs, 6H, NH-CH_2-CH_2-NH and δ CH_2 Arg), 4.18–4.51 (m, 4H, α CH), 7.24–8.80 (bh, NH).

Preparation of Liposomes

Liposomes are prepared from egg phosphatidylcholine (PC) and cholesterol (CH) by sonication except in the case of rifampin (RFP)-incorporated liposomes (Agarwal *et al.*, 1994). Tuftsin is grafted in the liposome bilayer by mixing 7–8% of **I** by weight of PC in the above mixture. Quantities greater than 7–8% are not well tolerated by the liposome bilayer, whereas lower quantities are inadequate to give optimal results. In case of RFP-incorporated liposomes, only PC is used as the lipid component, as the presence of CH in the lipid mixture reduces the amount of RFP intercalation within the bilayer (Agarwal *et al.*, 1994). Detailed procedures for preparation of tuftsin-bearing liposomes for studies of their interactions with PMN leukocytes, for modulating the nonspecific host resistance against infections, for delivering antileishmanial, antitubercular, and antifungal drugs to macrophages and monocytes are given below.

Preparation of Liposomes for Studies of Interactions with PMN Leukocytes

To monitor the interactions between the tuftsin-grafted liposomes and macrophages/monocytes/PMN leukocytes, the liposomal membrane is labeled with $[^{14}C]PC$ to determine the extent of exchange/transfer of lipids between the liposomes and the cells, and the fluorescent dye 6-carboxyfluorescein (6-CF) (Ralston *et al.*, 1981) is entrapped as the model solute to monitor the extent of solute leakage caused by liposome–cell interactions. Thus, small unilamellar liposomes are prepared from PC (15 μmol), CH (7.5 μmol), traces of $[^{14}C]PC$ (about 10 μCi), and 6-CF (0.2 M) with or without **I** (7–8% by PC weight) in 0.8 ml Tris-buffered saline (10 mM Tris containing 150 mM NaCl, pH 7.4) or sucrose-supplemented Tris-buffered saline [10 mM Tris containing 150 mM NaCl, 44 mM sucrose, and 5 mM ethylenediaminetetracetic acid (EDTA), pH 7.4] by probe sonication (Singhal *et al.*, 1984). Typically, a solution of lipids in chloroform is dried in a glass tube under a slow jet of oxygen-free nitrogen gas, resulting in the formation of a thin lipid film on the wall of the tube. Final traces of solvents are removed after leaving the tube *in vacuo* for 1–2 h. The lipid film is dispersed in 0.8 ml Tris-buffered saline. The tube is flushed with nitrogen and stoppered. It is vortexed at 35–40° for 20 min. The lipid dispersion so obtained is transferred to a cuvette and sonicated (1–1.5 h) under nitrogen using a probe-type sonicator to obtain an optically clear suspension. The sonicated preparations are centrifuged at 10,500g for 1 h at 5°. Only the liposomes found in the top two thirds of the supernatant are used. Free and liposomal 6-CF are separated by gel filtration over a Sephadex G-50 column (1.4 × 20 cm), using Tris-buffered saline as the eluting buffer. Liposome-rich fractions are pooled together and used in further experiments. The outer diameter of liposomes is determined by negative-staining electron microscopy.

The extent of incorporation of **I** in the liposome bilayer is estimated as follows. To an aliquot (20 μl) of PC/CH liposomes containing **I** are added 580 μl phosphate buffer (10 mM, pH 8.5), 20 μl trinitrobenzenesulfonic acid (1.5% in 0.8 M NaHCO$_3$, pH 8.5), and 200 μl Triton X-100 (1% in 0.8 M NaHCO$_3$, pH 8.5) or buffer. It is mixed and then incubated in the dark at 18–20° for 1 h. After this period, the reaction is stopped by adding 400 μl HCl (1.5 M). The absorbance for yellow color thus obtained is read at 410 nm. Quantities of **I** incorporated in the bilayers are calculated from the standard curve, which is drawn by reacting varying amounts of **I** with trinitrobenzenesulfonic acid under identical conditions. The absorbance at 410 nm is linear with concentration to at least 0.8 absorbance units. The amounts of **I** incorporated into the liposome bilayer are over 95%, and about 60% of this amount is localized in the outer monolayer.

Preparation of Liposomes to Increase Nonspecific Host Resistance against Parasitic and Fungal Infections

Liposomes are prepared from equimolar amounts of PC and CH with or without **I** (8% by weight of PC) in Tris-buffered saline (10 mM Tris containing 150 mM NaCl, pH 7.4) by probe sonication, as described above. The sonicated preparation is centrifuged at 12,000g for 30 min (10°) to sediment undispersed lipids and titanium particles. The amount of **I** incorporated in the liposome bilayer is estimated as described above.

Preparation of Liposomes as Vehicles to Deliver RFP in Mycobacterium tuberculosis-Infected Mice

Mycobacterium tuberculosis resides and proliferates primarily within mononuclear phagocytes. Therefore, delivery of antitubercular drugs, like RFP, through tuftsin-bearing liposomes should help to enrich this drug within the diseased cells, which in turn may result in reduction of the drug dose, leading to reduced RFP toxicity (Raleigh, 1972). Tuftsin-bearing liposomes are thus prepared by incorporating RFP in both the bilayers and internal aqueous compartment of the liposomes.

Typically, PC (62.5 μmol), RFP (0.61 μmol), and **I** (7–8% by PC weight) are dissolved in chloroform methanol (1:1, v/v). The solution is dried in a glass tube under a slow jet of oxygen-free nitrogen, resulting in the formation of a thin lipid film on the wall of the tube. Final traces of the solvent are removed after leaving the film *in vacuo* overnight at 4°.

The lipid film is dispersed in 1.5 ml of phosphate-buffered saline (PBS; 10 mM phosphate containing 150 mM NaCl, pH 7.4) and vortexed at 35–40° for 15–20 min. The lipid dispersion thus obtained is transferred to a cuvette and sonicated for 30 min under oxygen-free nitrogen, using a probe-type sonicator. The sonicated preparation is centrifuged at 12,000g for 30 min to remove titanium particles as well as the undispersed or poorly dispersed lipids. The supernatant is carefully removed and gel filtered through a Sephadex G-50 column (40 × 1.5 cm) using PBS as the eluent to separate liposome-associated RFP from free RFP. Liposomes are eluted in the void volume. The liposome-rich fractions are pooled together and concentrated in an Amicon Centriflo CF-25 cone. The outer diameter of liposomes is determined by electron microscopy.

RFP is estimated by measuring its absorbance at 334 nm. The amount of liposome-associated drug is determined after lysing the liposomes with Triton X-100 (1%). The RFP absorbance at 334 nm is found to remain unaffected by the presence of liposomes and the detergent and is linear at least up to 100 μg/ml of RFP. About 28–32% of RFP is found to be incorporated in the unilamellar liposomes.

To determine whether RFP is entrapped or intercalated in liposomes, the above liposome preparation is frozen under liquid nitrogen and immediately thawed. The preparation is then gel filtered again over a Sephadex G-50 column (40 × 1.5 cm). RFP elutes under two different peaks. The one in the void volume corresponds to the liposomal form while the other corresponds to the free form, suggesting that the drug resides in both the internal aqueous space and the lipid bilayer.

Preparation of Liposomes as Drug Carriers in Treatment of Leishmania donovani and Aspergillus Infections

Amphotericin B (Amp B) is a very effective drug for treatment of leishmanial infections that are resistant to treatment with antimonials and fungal infections including aspergillosis. Fungal pathogens, unlike *L. donovani*, do not reside within monocytes and macrophages. It is thought that delivery of Amp-B through tuftsin-bearing liposomes concentrates the drug within macrophages, which migrate to fungally infected lesions and serve as a drug depot, making available high concentrations of the drug in the affected region.

Amp-B-loaded liposomes are prepared from PC and CH (molar ratio 7:3) by probe sonication. Typically, PC (49 μmol), CH (21 μmol), Amp B (1.0 mg), and **I** (7–8% by weight of PC) are dissolved in a round-bottom flask in a minimum volume of chloroform/methanol (1:1, v/v). The solvents are carefully removed under reduced pressure so that a thin film is formed on the wall of the flask. Final traces of solvents are removed by leaving the flask *in vacuo* at 4° overnight. The dried lipid film is hydrated with 150 mM sterile saline (2 ml) under vigorous stirring for 1 h under a nitrogen atmosphere in a bath-type sonicator. The lipid dispersion thus obtained is dialyzed against normal saline for 24 h in the dark at 4°. The dialyzed preparation is centrifuged at 10,000g for 1 h at 4° to remove traces of undispersed lipids. The supernatant is analyzed for both Amp-B and tuftsin.

The amount of liposome-intercalated Amp B is determined by measuring its absorbance at 405 nm. The intercalation efficiencies of Amp B in tuftsin-bearing liposomes and tuftsin-free liposomes are about 95% and 85%, respectively. The extent of tuftsin incorporation is estimated as described above.

Interactions of Tuftsin-Bearing Liposomes with Blood Cells

Specific interactions of tuftsin-bearing liposomes with blood cells are studied by incubating the liposomes with the respective cell population (PMN leukocytes, lymphocytes, erythrocytes, etc.) at 37° and then measuring the transfer/exchange of lipids between the liposomes and cells as well as the solute leakage caused by the interaction of liposomes with cells. These studies show that the tuftsin-bearing liposomes interact specifically with PMN leukocytes, monocytes, and macrophages, but not with lymphocytes or erythrocytes under identical conditions (Singhal *et al.*, 1984).

PMN leukocytes and lymphocytes are isolated from freshly drawn rat blood using a Ficoll-Paque gradient. Erythrocytes are isolated from rat blood by simple centrifugation. The detailed procedures are given below (Singhal *et al.*, 1984).

Isolation of Lymphocytes and PMN Leukocytes

Heparinized rat blood (10 IU/ml blood) is centrifuged at 1000g for 10 min, and the plasma is removed. The cells are suspended in Hanks' balanced salt solution (137 mM NaCl, 37 mM KCl, 1.0 mM CaCl$_2$·2H$_2$O, 0.41 mM MgSO$_4$·7H$_2$O, 0.50 mM MgCl$_2$·6H$_2$O, 0.44 mM KH$_2$PO$_4$, 0.34 mM Na$_2$HPO$_4$·2H$_2$O, 5.55 mM glucose, 4.16 mM NaHCO$_3$, pH 7.4) to 50% hematocrit. One milliliter of this suspension is layered on top of 1 ml Ficoll-Paque (Pharmacia product). It is centrifuged at 300g for 15 min at 20 ± 2°. The top layer containing lymphocytes is removed and washed several times with Hanks' balanced salt solution. Any contaminating erythrocytes are removed by treating one volume of lymphocytes with seven volumes of ammonium chloride (150 mM, pH 7.4) for 5 min at 20°. The mixture is centrifuged, and the cell pellet is washed three times with Hanks' balanced salt solution. The cell viability is determined by Trypan Blue exclusion; usually over 95% of the cells are viable. The purity of the cells is determined in a thin smear stained with Leishmann stain. These cells are contaminated with 5–7% of PMN leukocytes and monocytes. Without any further purification, these cells are used in further experiments after suspending in the above buffer (3.6–4.0×10^7 cells/ml).

The second layer formed in the Ficoll-Paque gradient mainly consists of PMN leukocytes. It is carefully aspirated and washed three times with Hanks' balanced salt solution. Contaminating erythrocytes are removed by the ammonium chloride treatment, but it is not necessary to remove the contaminating lymphocytes (25–30%) from the PMN leukocytes for the study. The viability of these cells is >95%. The cells are suspended in the above buffer (3.6–4.0×10^7 cells/ml) and used within 1 h.

Isolation of Erythrocytes

Heparinized rat blood is centrifuged at $1000g$ for 5 min. Plasma and buffy coat are removed. The pellet is washed several times with sucrose-supplemented Tris-buffered saline. The washed cells are suspended in buffer (3.6–4.0×10^7 cells/ml) and used immediately in further experiments.

Interactions of Liposomes with PMN Leukocytes

Liposomes (0.13–2.5 μmol lipid P/ml) are mixed with PMN leukocytes (1.8–2.0×10^7 cells/ml) in Hanks' balanced salt solution. The mixture is incubated at 37° for 1 h in a shaking waterbath. After incubation is complete, the cells are harvested by centrifugation. The cell pellet is repeatedly washed with Hanks' balanced salt solution until the supernatant is free from radioactivity. The amounts of radioactivity and 6-CF in the cell pellet are assayed after disrupting the cells with Triton X-100 (1% final concentration). The 6-CF fluorescence is monitored by using excitation and emission wavelengths of 490 and 520 nm, respectively. Recovery of cells is about 90% and >95% are viable at the end. To determine the effect of incubation temperature on uptake of liposomes by the leukocytes, liposomes (2 μmol lipid P/ml) are also incubated with PMN leukocytes (1.8–2.0×10^7 cells/ml) at 0° for different periods of time. After incubation is complete, the cells are processed, and the amounts of radioactivity and 6-CF associated with the cell pellet are determined.

Free 6-CF is also incubated with the cells under identical conditions. About 0.01% of the total dye remains bound to the cell pellet.

Interactions of Liposomes with Lymphocytes

Liposomes (2 μmol lipid P/ml) are mixed with lymphocytes (1.8–2.0 × 10^7/ml) in Hanks' balanced salt solution, and the mixture is incubated for different periods of time. After incubation is complete, the cells are washed and assayed for radioactivity and 6-CF.

Interactions of Liposomes with Erythrocytes

Liposomes (2 μmol lipid P/ml) are mixed with erythrocytes (1.8–2.0 × 10^7/ml) in sucrose-supplemented Tris-buffered saline, and the mixture is incubated for different periods of time. After completing the incubation, the cells are washed and assayed for radioactivity. In this case, 6-CF cannot be measured due to quenching of its fluorescence by hemoglobin.

Treatment of Experimental Animals with Tuftsin-Bearing Liposomes to Enhance Their Nonspecific Resistance to Infections

To increase the nonspecific resistance of experimental animals to parasitic infections, tuftsin-bearing liposomes are given intravenously for 3 consecutive days prior to infection (day −3 to day −1). Normally, the dose given is 50 ng tuftsin per day per animal, but this dose may vary depending on the type of infection. Animals are infected on day zero. Under these conditions, animals resist *Plasmodium berghei* (Gupta *et al.*, 1986), *L. donovani* (Guru *et al.*, 1989), and to some extent *Aspergillus fumigatus* (Owais *et al.*, 1993) infection, but no effects are seen on the course of *M. tuberculosis* infection under identical conditions.

Treatment of Leishmanial, Tuberculosis, and Aspergillosis Infections Using Tuftsin-Bearing Liposomes as Drug Carriers

RFP-loaded tuftsin-bearing liposomes are used to treat *M. tuberculosis* $H_{37}Rv$ infections in Swiss albino mice. Various doses of drug-loaded tuftsin-bearing liposomes and drug-loaded but tuftsin-free liposomes and the free drug are delivered intravenously postinfection. The course of infection is determined by measuring the bacilli load in lungs, livers, and spleens, as well as the mean survival times of animals. The numbers of bacilli in lungs, livers, and spleens of mice are quantitated by measuring colony-forming units (cfu). The efficacy of RFP is considerably improved upon its administration in tuftsin-bearing liposomes compared with tuftsin-free liposomes. Consistently, administration of repeated doses of RFP in tuftsin-bearing liposomes shows significantly better antitubercular effects than the administration of a single dose (Agarwal *et al.*, 1994).

AmpB-loaded tuftsin-bearing liposomes are used to treat *L. donovani* infections in golden hamsters and also *A. fumigatus* infections in BALB/c mice. The liposome-associated drug is administered intravenously in both cases (Agrawal *et al.*, 2002; Owais *et al.*, 1993). The drug loaded in tuftsin-bearing liposomes as well as the free drug are given to the control group of animals. While the drug efficacy in leishmanial infections is assessed by measuring the parasite load in spleens of the infected animals, the efficacy of Amp B in *A. fumigatus* infections is determined by measuring both the fungal load (cfu) in various organs and survival times of the treated animals. Antileishmanial and antifungal effects of Amp B are seen to be considerably better if it is administered in tuftsin-bearing liposomes as compared to tuftsin-free liposomes or in the free form (Agrawal *et al.*, 2002; Owais *et al.*, 1993).

Concluding Remarks

Tuftsin is a well-known macrophage activator tetrapeptide that specifically binds monocytes, macrophages, and PMN leukocytes. This activity of tuftsin is further increased upon its incorporation in liposomes, after attaching a sufficiently long hydrophobic anchor at its C-terminus. The tuftsin-bearing liposomes not only enhanced the nonspecific host's resistance against a variety of pathogens but also helped in significantly boosting the chemotherapeutic potential of a number of drugs if delivered after their encapsulation in these liposomes. These studies thus clearly demonstrate the usefulness of this tetrapeptide in augmenting the chemotherapeutic potential of antiparasitic, antitubercular, and antifungal drugs by virtue of its ability to home the drug-loaded nanoparticle-like systems (e.g., liposomes) to monocytes and macrophages and also to simultaneously stimulate the nonspecific killer activity of these cells against pathogens.

References

Agarwal, A., Kandpal, H., Gupta, H. P., Singh, N. B., and Gupta, C. M. (1994). Tuftsin-bearing liposomes as rifampin vehicles in treatment of tuberculosis in mice. *Antimicrob. Agents Chemother.* **38**, 588–593.

Agrawal, A. K., and Gupta, C. M. (2000). Tuftsin-bearing liposomes in treatment of macrophage-based infections. *Adv. Drug Deliv. Rev.* **41**, 135–146.

Agrawal, A. K., Agrawal, A., Pal, A., Guru, P. Y., and Gupta, C. M. (2002). Superior chemotherapeutic efficacy of amphotericin B in tuftsin-bearing liposomes against Leishmania donovani infection in hamsters. *J. Drug Target.* **10**, 41–45.

Fridkin, M., and Gottlieb, P. (1981). Tuftsin, Thr-Lys-Pro-Arg. Anatomy of an immunologically active peptide. *Mol. Cell Biochem.* **41**, 73–97.

Gottlieb, P., Beretz, A., and Fridkin, M. (1982). Tuftsin analogs for probing its specific receptor site on phagocytic cells. *Eur. J. Biochem.* **125**, 631–638.

Guru, P. Y., Agrawal, A. K., Singha, U. K., Singhal, A., and Gupta, C. M. (1989). Drug targeting in Leishmania donovani infections using tuftsin-bearing liposomes as drug vehicles. *FEBS Lett.* **245**, 204–208.

Gupta, C. M., Puri, A., Jain, R. K., Bali, A., and Anand, N. (1986). Protection of mice against Plasmodium berghei infection by a tuftsin derivative. *FEBS Lett.* **205,** 351–354.

Najjar, V. A. (1987). Biological effects of tuftsin and its analogs. *Drugs Future* **12,** 147–160.

Owais, M., Ahmed, I., Krishnakumar, B., Jain, R. K., Bachhawat, B. K., and Gupta, C. M. (1993). Tuftsin-bearing liposomes as drug vehicles in the treatment of experimental aspergillosis. *FEBS Lett.* **326,** 56–58.

Owais, M., Misra-Bhattacharya, S., Haq, W., and Gupta, C. M. (2003). Immunomodulator tuftsin augments antifilarial activity of diethylcarbamazine against experimental brugian filariasis. *J. Drug Target.* **11,** 247–251.

Raleigh, J. W. (1972). Rifampin in treatment of advanced pulmonary tuberculosis. Report of a VA cooperative pilot study. *Am. Rev. Respir. Dis.* **105,** 397–407.

Ralston, E., Hjelmeland, L. M., Klausner, R. D., Weinstein, J. N., and Blumenthal, R. (1981). Carboxyfluorescein as a probe for liposome-cell interactions. Effects of impurities and purification of the dye. *Biochim. Biophys. Acta* **649,** 133–137.

Singh, S. P., Chhabra, R., and Srivastava, V. M. L. (1992). Respiratory burst in peritoneal exudate cells in response to a modified tuftsin. *Experientia* **48,** 994–996.

Singhal, A., Bali, A., Jain, R. K., and Gupta, C. M. (1984). Specific interactions of liposomes with PMN leukocytes upon incorporating tuftsin in their bilayers. *FEBS Lett.* **178,** 109–113.

[17]   Liposomal Polyene Antibiotics

By AGATHA W. K. NG, KISHOR M. WASAN, and
GABRIEL LOPEZ-BERESTEIN

Abstract

Polyene antibiotics (i.e., amphotericin B and nystatin) have been incorporated into lipid-based delivery systems to decrease their toxicity and enhance their therapeutic index, the most common being liposomes. This chapter describes the protocols for preparing liposomal amphotericin B and determining the efficacy and toxicity of the formulations in animals. Furthermore, methods for determining the pharmacokinetics and drug distribution after administration of amphotericin B in lipid-based delivery systems are discussed. Procedures for comparing the toxicity of different amphotericin B formulations in cell culture studies are also elucidated.

Introduction

Amphotericin B and nystatin are polyene antibiotics. They are effective for treatments of both presystemic and systemic fungal infections. These widely used antibiotics are derived from fermentation by *Streptomyces*. Polyene antibiotics are also known to induce renal toxicity (Bolard, 1986; Georgopapadakou and Walsh, 1996; Meyer, 1992).

Polyene antibiotics have been incorporated into lipid-based delivery systems to decrease their toxicity and enhance their therapeutic index, the most common being liposomes. In the 1980s, liposomal amphotericin B, consisting of dimyristoylphosphatidylcholine (DMPC) and dimyristoylphosphatidylglycerol (DMPG) in a lipid-to-drug weight ratio of 12:1 was developed (Wasan et al., 1993). The efficacy of liposomal amphotericin B and amphotericin B was evaluated in mice that had systemic candidiasis. Liposomal amphotericin B was shown to prolong the survival of these mice. This formulation also had a cure rate of 60% (Lopez-Berestein et al., 1983, 1984).

The toxicity of liposomal amphotericin B and amphotericin B is also evaluated in mice with systemic candidiasis. Subjects are injected with free amphotericin B at doses of 0.4, 0.8, 1.0, 1.6, 2.0, 3.0, and 4.0 mg/kg, and the liposomal amphotericin B group is given in doses greater than 10.0 mg/kg in addition. Animals died immediately after receiving amphotericin B at doses higher than 0.8 mg/kg. Even though toxicity was not observed for 21–42 days, autopsies performed at 21 and 42 days showed that there was nephrocalcinoses and diffuse interstitial edema in these animals. On the other hand, liposomal amphotericin B is neither associated with any acute toxic reaction at doses up to 12 mg nor related to any pathological abnormalities at 21 and 42 days (Lopez-Berestein et al., 1983).

The pharmacokinetics of amphotericin B has been investigated in hypercholesterolemic and normolipidemic rabbits after administration of amphotericin B deoxycholate (Doc-AmpB) and amphotericin B lipid complex (ABLC). Animals are administered 1 mg/kg Doc-AmpB or ABLC daily for 7 days. Pharmacokinetics of amphotericin B is altered in hypercholesterolemic rabbits. For example, the areas under the curve (AUCs) for both Doc-AmpB and ABLC are higher in hypercholesterolemic rabbits than in normolipidemic rabbits. ABLC has a prolonged half-life in hypercholesterolemic rabbits. The volume of distribution at steady state (V_{ss}) is lower in hypercholesterolemic rabbits administered Doc-AmpB and ABLC (Ramaswamy et al., 2001).

Hypercholesterolemic and normolipidemic rabbits given doses of 1 mg/kg Doc-AmpB and ABLC daily for 7 days have also been examined for amphotericin B concentrations in different lipoproteins. An increased percent of amphotericin B is recovered in the triglyceride-rich lipoprotein (TRL) fraction when Doc-AmpB was administered to hypercholesterolemic rabbits compared with that in normolipidemic rabbits. Moreover, an increased percentage of amphotericin B is recovered in the low-density lipoprotein (LDL) and TRL fractions when ABLC is administered to hypercholesterolemic rabbits compared with that in normolipidemic rabbits. This finding suggests that an increase in plasma cholesterol level modifies the distribution of amphotericin B in lipoproteins (Ramaswamy et al., 2001; Wasan et al., 1999).

The toxicity of amphotericin B in different formulations can also be investigated in cell culture studies. Preliminary results in our laboratory show that ABLC yields higher protein values and mitochondria transport system (MTS) assay values, indicating that ABLC is less toxic than amphotericin B in its free form.

This chapter describes the protocols for preparing liposomal amphotericin B and determining the efficacy and toxicity of the formulations in animals (Lopez-Berestein *et al.*, 1983, 1984, 1985; Wasan *et al.*, 1993). Furthermore, methods for determining the pharmacokinetics and drug distribution after administration of amphotericin B in a lipid-based delivery system are discussed (Ramaswamy *et al.*, 2001). Procedures for comparing the toxicity of different amphotericin B formulations in cell culture studies are also elucidated.

Preparation of Liposomal Amphotericin B

Amphotericin B and Lipid Stock Solutions

Thirty milligrams of amphotericin B (MW: 924.09 g/mol; Sigma-Aldrich, St. Louis, MO) is dissolved in 750 ml of methanol (40 μg/ml). Seventy milligrams of dimyristoylphosphatidylcholine (DMPC; MW: 677.94 g/mol; Avanti Polar Lipids, Alabaster, AL) and 30 mg of dimyristoylphosphatidylglycerol (DMPG; MW: 688.86 g/mol; Avanti Polar Lipids) are dissolved in 10 ml of chloroform (7:3 w/w) (Wasan *et al.*, 1993).

Preparation of Multilamellar Vesicles

The amphotericin B solution (23.8 ml) is added to 10 ml of the lipid solution in chloroform (DMPC/AmpB molar ratio: 4:1). The organic solvent is evaporated in a rotary evaporator (Rotavapor, Brinkmann Instruments, Westbury, NY), and the residual solvent is dried in a vacuum desiccator for 30 min. Approximately 20 ml 0.9% NaCl is added to the dried mixture. Liposomes are formed by hand shaking. When all drug and lipids are dissolved, the suspension is centrifuged at 2500g. The pellet is resuspended in pyrogen-free saline to achieve a desirable concentration.

Determining the Antifungal Activity of Liposomal Amphotericin B against Candidiasis in Mice

Model for Candida albicans Infection in Mice

Infection is induced in mice (20–25 g) by injecting 0.2 ml of a 0.9% NaCl solution containing 1.75×10^5, 3.50×10^5, or 7.00×10^5 colony forming units (cfu) of *Candida albicans* via the tail vein 4 days prior to the beginning of treatment (Lopez-Berestein *et al.*, 1983, 1984).

Treatment Studies

Single-Dose Treatment: Dose–Response of L-AmpB. Infected mice are injected with a single dose of the treatments listed in Table I. Mice are monitored every day for 25 days, and their survival rate is recorded.

Single-Dose Treatment: Amphotericin B versus Liposomal Amphotericin B. Infected mice are injected with a single dose of the treatments listed in Table II. Mice are monitored every day for 25 days, and their survival rate is recorded.

Single-Dose Treatment: Relation of AmpB/L–AmpB Treatments to the Severity of Infection. Mice are infected with *C. albicans* at the following cfu doses: 1.75×10^5, 3.50×10^5, or 7.00×10^5. Mice are injected with the treatments listed in Table III 3, 4, or 5 days after infection. The mice are

TABLE I
SINGLE-DOSE TREATMENT: DOSE–RESPONSE OF L-AMPB

Groups	Treatments
Untreated mice with candidiasis	N/A
Mice injected with empty liposomes	Empty liposomes (DMPC:DMPG 7:3)
Liposomal amphotericin B	0.8 mg/kg
Liposomal amphotericin B	2.0 mg/kg
Liposomal amphotericin B	3.0 mg/kg
Liposomal amphotericin B	4.0 mg/kg

TABLE II
SINGLE-DOSE TREATEMENT: AMPHOTERICIN B VERSUS LIPOSOMAL AMPHOTERICIN B

Groups	Treatments
Untreated mice with candidiasis	N/A
Mice injected with empty liposomes	Empty liposomes (DMPC:DMPG 7:3)
Amphotericin B	0.8 mg/kg
Liposomal amphotericin B	3.0 mg/kg

TABLE III
SINGLE-DOSE TREATMENT: RELATION OF AMPB/L-AMPB TREATMENTS TO
THE SEVERITY OF INFECTION

Groups	Treatments
Untreated mice with candidiasis	N/A
Mice injected with empty liposomes	Empty liposomes (DMPC:DMPG 7:3)
Amphotericin B	0.8 mg/kg as a single dose
Liposomal amphotericin B	4.0 mg/kg as a single dose

TABLE IV
MULTIPLE DOSES OF L-AMPB

Groups	Treatments
Untreated mice with candidiasis	N/A
Mice injected with empty liposomes	Empty liposomes (DMPC:DMPG 7:3)
Amphotericin B	0.8 mg/kg for 5 days
Liposomal amphotericin B	5.6 mg/kg for 5 days

monitored every day up to 25 days after treatments have begun, and the survival rate of mice is recorded.

Multiple Doses of L-AmpB. This method is used to assess the effects of high-dose L-AmpB on severely infected mice. Mice are injected with the treatments listed in Table IV 5 days after infection. Doses are given every day for 5 days. The mice are monitored every day up to 25 days after treatments have begun, and the survival rate of mice is recorded.

Determining the Toxicity of Liposomal Amphotericin B
Formulations in Mice

Mice (20–25 g) are injected with 0.9% normal saline, empty liposomes, amphotericin B (0.4, 0.8, 1.0, 1.6, 2.0, 3.0, and 4.0 mg/kg), or L-AmpB (0.4, 0.8, 1.0, 2.0, 3.0, 4.0, and 10.0 mg/kg) as a single dose. The acute toxicity is assessed by observing the immediate death following injection of treatments. The subacute toxicity is observed by inspecting mice daily from day 0 to day 21. Serum creatinine, blood urea nitrogen (BUN), and serum glutamic pyruvic transaminase (SGPT) are determined in each mouse during the subacute toxicity phase. Chronic toxicity is evaluated by observing mice up to 42 days after injection (Lopez-Berestein *et al.*, 1983, 1984).

Pharmacokinetics of Amphotericin B Deoxycholate and Amphotericin B Lipid Complex in a Hypercholesterolemic Rabbit Model

Adaptation Period for Rabbits

To induce hypercholesterolemia, rabbits should receive rabbit chow supplemented with 2.5% (w/v) coconut oil and 0.50% (w/v) cholesterol for 7 days prior to the administration of Doc-AmpB or ABLC. At least 3 days are allowed for the normolipidemic rabbits to acclimate to the environment prior to the experiments.

Sample Collection

A dose of Doc-AmpB (Bristol-Myers Squibb, New York, NY) or ABLC (1 mg/kg) is injected intravenously to hypercholesterolemic and normolipidemic rabbits through the jugular vein every day for 7 consecutive days (total of eight doses). Blood samples are collected 24 h after administration of the previous day's dose. Blood samples should be collected prior to and at 5, 15, and 30 min and 1, 2, 4, 8, 12, 24, 48, and 72 h following the administration of the last dose of Doc-AmpB or ABLC. Plasma samples are stored at 4° prior to assay of amphotericin B by high-performance liquid chromatography (HPLC) in each fraction.

Pharmacokinetic Analysis. The concentrations of amphotericin B in plasma are plotted against time on log-scale paper, and the distribution phase and the terminal half-life are determined by the method of residuals. The area under the amphotericin B concentration–time curve (AUC) is estimated from time zero to infinity by the trapezoidal rule. Other pharmacokinetic parameters, such as mean residence time (MRT), total body clearance (CL), and volume of distribution at steady state (V_{ss}), are estimated by compartmental analysis with the WINNONLIN nonlinear estimation program (GraphPad Inc., San Diego, CA) (Ramaswamy *et al.*, 2001).

Serum Distribution of Deoxycholate Amphotericin B and Amphotericin B Lipid Complex in a Hypercholesterolemic Rabbit Model

Primary Salt Solutions

Solution of density 1.006 g/ml: 11.4 g NaCl is dissolved in 1000 ml of distilled water. Solution of density 1.478 g/ml: 78.32 g NaBr is dissolved in 100 ml of 1.006 g/ml density solution.

Secondary Density Solutions

Solution of density 1.019 g/ml: 100 ml of the solution of density 1.006 g/ml is mixed with 2.83 ml of the 1.478 g/ml density solution. Solution of density 1.063 g/ml: 100 ml of the solution of density 1.006 g/ml is mixed with 13.73 ml of 1.478 g/ml density solution. Solution of density 1.21 g/ml: 100 ml of the solution of density 1.006 g/ml is mixed with with 76.10 ml of the solution of density 1.478 g/ml.

Adaptation Period for Rabbits

Hypercholesterolemia is induced in New Zealand White rabbits by feeding rabbit chow supplemented with 2.5% (w/v) coconut oil and 0.50% (w/v) cholesterol for 7 days prior to the administration of Doc-AmpB or ABLC. At least 3 days are allowed for the normolipidemic rabbits to acclimate to the environment prior to the experiments.

Sample Collection

A 1 mg/kg dose of Doc-AmpB or ABLC is injected intravenously to rabbits through the jugular vein every day for 7 consecutive days (total of eight doses). Plasma samples are collected at 5 min following the administration of the last dose of Doc-AmpB or ABLC. Plasma samples are stored at 4° prior to lipoprotein separation and assay of amphotericin B in each fraction. Amphotericin B does not redistribute between lipoprotein fractions at 4°.

Lipoprotein Fraction Separation

Sample Preparation. The sample is thawed in the refrigerator. Three milliliters of each plasma sample is pipetted into six labeled ultracentrifuge tubes (14 × 89 mm Ultra-Clear). NaBr (1.02 g) is added to each plasma sample (final concentration: 0.34 g/ml of plasma), increasing the density of plasma to 1.25 g/ml. The amount of NaBr can actually range from 1.02 to 1.04 g, but, obviously, all samples must be treated identically. The samples are vortexed to dissolve the NaBr in plasma. Glass test tubes (16 × 125 mm) are filled with solutions of density 1.21, 1.063, and 1.006 g/ml. The test tubes should be filled to the very top, because 2.8 ml of each density solution will be needed per gradient. Both the test tubes with density solutions and the plasma tubes are placed in an ice bath and refrigerated for 1–2 h.

Gradient Preparation. The density solutions and plasma samples are removed from the refrigerator. The samples are removed from the ice bath, and the outsides of the tubes are dried. For each plasma sample, 2.8 ml of the 1.21 g/ml density solution is pipetted into a 16 × 100-mm test tube. Using a plastic pipette, this solution is layered onto the plasma sample in the ultracentrifuge tubes; this step is repeated for the solutions of density 1.063 and 1.006 g/ml.

Ultracentrifuge Preparation. The caps and the inside of the buckets are cleaned with Kimwipes. Some vacuum silicon gel is applied onto the cap threads. The outsides of the ultracentrifuge tubes are wiped dry and placed carefully into the buckets of an SW-41 rotor. The caps are screwed on to close the buckets without shaking or disturbing the layers. The buckets are placed into their corresponding places on the rotor, ensuring that both hooks of the bucket are secure. The rotor is placed carefully into the ultracentrifuge (model L8-80M centrifuge; Beckman, Mississauga, Canada). The samples are centrifuged at 40,000 rpm at 15° for 18 h.

Separation of Fractions. At the end of the run, the rotor is removed from the ultracentrifuge, the buckets are removed from the rotor, the caps are unscrewed, and the tubes are removed from the buckets. For each sample, four 16 × 100-mm test tubes are labeled TRL, LDL, HDL, and LPDP. Half-way between the visible layers is marked with a pen. For the TRL layers, a Pasteur pipette is used to scrape carefully the sides of the top layer. The TRL layer is removed by placing the tip just above the surface to get air bubbles. This is continued until close to the mark, and the pipette contents are transferred to the TRL tube. The LDL layer is drawn up slowly from the most concentrated section using a new Pasteur pipette until the mark is reached and transferred to the LDL tubes. The high-density lipoprotein (HDL) layer is removed similarly and transferred to the HDL tube. For the lipoprotein-deficient plasma (LPDP) layer, the bottom of the test tube is scraped, and the deficient fraction is drawn up from the bottom up and transferred to the LPDP tubes (Ramaswamy *et al.*, 2001; Wasan *et al.*, 1999).

Toxicity of Various Amphotericin B Formulations on LLC-PK1 Cells

Preparation of Different Treatments

ABLC. Treatments are prepared to achieve a final concentration of 20 μg/ml. ABLC is prepared at a stock concentration of 5 mg/ml in distilled water with 0.9% NaCl. Seven microliters of ABLC is added to 28 μl of phosphate-buffered saline (PBS) to make 1 mg/ml solution. Twenty-eight microliters of the 1 mg/ml solution is added to 1.372 ml of medium in a 2-ml Eppendorf centrifuge tube and mixed. Two hundred microliters of this "ABLC treatment" is added to five wells of a 96-well plate.

Amphotericin B. Five milligrams of powdered AmpB is dissolved in 300 μl dimethyl sulfoxide (DMSO; Fisher, Scientific, Pittsburgh, PA). Seven hundred milliliters of methanol is added to make a 5 mg/ml solution. To make a 1 mg/ml solution, 14 μl of the 5 mg/ml solution is added to 56 μl methanol. Twenty-eight microliters of this solution is added to 1.372 ml of the medium in a 2-ml Eppendorf tube. To five wells of a 96-well plate is added 200 μl of this "AmpB treatment."

No Treatment Controls. Two hundred microliters of fresh complete medium is added to five wells of a 96-well plate.

Vehicle Controls. Sixty microliters of DMSO is added to 740 μl of methanol to make a 37:3 mixture of methanol/DMSO. After mixing, 28 μl of this mixture is added to 1.372 ml of medium in an Eppendorf tube. After thorough mixing, 200 μl of this "vehicle control" is added to five wells of a 96-well plate.

Cytotoxicity Experiments

Day 1. LLC-PK1 cells are seeded into a 96-well plate at a density of about 40,000 cells/cm^2 (2.46 $\times$ 10^4 cells/well).

Day 2. Treatments or controls are prepared as above. Media and other solutions are warmed to 37°. The wells are examined under the microscope to confirm that the cells are attached and are seeded uniformly. Any wells with suspicious cells are discarded. The medium is aspirated from all chosen wells and is replaced with 200 μl of the relevant treatment. The cells are incubated for 18 h at 37° and 5% CO$_2$.

Day 3. After 18 h of treatment, the media in the treatment groups are aspirated, and the cells are washed twice with 100 μl PBS. Warm medium (100 μl) is added to each well. Twenty microliters of MTS (CellTitre 96 Aqueous One Solution Cell Proliferation Assay Reagent, Fisher, Mississauga, Ontario, Canada) is added to each well. For use as "background" 100 μl medium + 20 μl MTS are added to several wells without cells. The cells are incubated for about 2 h at 37°, without oscillation, and the absorbance is read at 492 nm in a microplate reader (Labsystems Multiscan Ascent). The cells are washed with 150 μl PBS and then with 225 μl PBS. The cells are lysed with 75 μl of 100 m*M* NaOH for 1 h. The plate is oscillated on a plate shaker briefly to mix the contents. Ten microliters is taken from each sample for use in the BioRad Protein Assay (Hercules, CA). The absorbance of the background is subtracted from each absorbance (abs) value (read at 620 nm). The percentage toxicity is calculated in comparison to control cells by using the following formula:

(Abs control cells − Abs treated cells)/Abs control cells × 100%

The results of the protein assay and MTS assay are expected to correlate. Toxic treatments should affect the growth of cells. Therefore, cells given toxic treatments are expected to yield less protein and lower MTS assay values.

References

Bolard, J. (1986). How do the polyene macrolide antibiotics affect the cellular membrane properties? *Biochim. Biophys. Acta* **864**, 257–304.

Georgopapadakou, N. H., and Walsh, T. J. (1996). Antifungal agents: Chemotherapeutic targets and immunologic strategies. *Antimicrob. Agents Chemother.* **40**, 279–291.

Lopez-Berestein, G., Mehta, R., Hopfer, R. L., Mills, K., Kasi, L., Mehta, K., Fainstein, V., Luna, M., Hersh, E. M., and Juliano, R. (1983). Liposome-encapsulated amphotericin B for treatment of disseminated candidiasis in neutropenic mice. *J. Infect. Dis.* **147**, 939–945.

Lopez-Berestein, G., Hopfer, R. L., Mehta, R., Mehta, K., Hersh, E. M., and Juliano, R. L. (1984). Liposome-encapsulated amphotericin for treatment of disseminated candidiasis in neutropenic mice. *B. J. Infect. Dis.* **150**, 278–283.

Lopez-Berestein, G., McQueen, T., and Mehta, K. (1985). Protective effect of liposomal-amphotericin B against *C. albicans* infection in mice. *Cancer Drug Deliv.* **2**, 183–189.

Meyer, R. D. (1992). Current role of therapy with amphotericin B. *Clin. Infect. Dis.* **14**(Suppl. 1), S154–S160.

Ramaswamy, M., Peteherych, K. D., Kennedy, A. L., and Wasan, K. M. (2001). Amphotericin B lipid complex or amphotericin B multiple-dose administration to rabbits with elevated plasma cholesterol levels: Pharmacokinetics in plasma and blood, plasma lipoprotein levels, distribution in tissues, and renal toxicities. *Antimicrob. Agents Chemother.* **45**, 1184–1191.

Wasan, K. M., Brazeau, G. A., Keyhani, A., Hayman, A. C., and Lopez-Berestein, G. (1993). Roles of liposome composition and temperature in distribution of amphotericin B in serum lipoproteins. *Antimicrob. Agents Chemother.* **37**, 246–250.

Wasan, K. M., Cassidy, S. M., Ramaswamy, M., Kennedy, A., Strobel, F. W., Ng, S. P., and Lee, T. Y. (1999). A comparison of step-gradient and sequential density ultracentrifugation and the use of lipoprotein deficient plasma controls in determining the plasma lipoprotein distribution of lipid-associated nystatin and cyclosporine. *Pharm. Res.* **16**, 165–169.

[18]   Drug Delivery by Lipid Cochleates

By LEILA ZARIF

Abstract

Drug delivery technology has brought additional benefits to pharmaceuticals such as reduction in dosing frequency and side effects, as well as the extension of patient life. To address this need, cochleates, a precipitate obtained as a result of the interaction between phosphatidylserine and calcium, have been developed and proved to have potential in encapsulating and delivering small molecule drugs. This chapter discusses the molecules that can be encapsulated in a cochleate system and describes in detail the methodology that can be used to encapsulate and characterize hydrophobic drugs such as amphotericin B, a potent antifungal agent. Some efficacy data in animal models infected with candidiasis or aspergillosis are described as well.

Introduction

Drug delivery technology offers significant benefits to pharmaceuticals such as reduction in dosing frequency and side effects and the extension of patient life (Zarif, 2002). In addition, drug delivery technology allows the use of alternate routes for the administration of drugs that can improve patient compliance and convenience (oral instead of injectable). Multiple drug delivery platforms (Florence and Hussain, 2001; Han and Amidon, 2000; Leone-Bay *et al.*, 2000; Zarif and Perlin, 2002) have therefore been developed to increase oral absorption of drugs. Cochleates are one of the drug delivery platforms that attracted attention recently as a means to enhance and optimize the delivery of pharmaceuticals (Delmas *et al.*, 2002; Santangelo *et al.*, 2000; Segarra *et al.*, 2002; Zarif, 2002; Zarif and Perlin, 2002; Zarif *et al.*, 2000a,b).

This chapter explains, in detail, the use of cochleate delivery technology to formulate small molecule drugs and addresses the different methods that have been developed to date for drug encapsulation. Amphotericin B was taken as an example of a small hydrophobic drug.

Background of the Cochleate System

Cochleates were first described by Papahadjopoulos *et al.* in 1975; they reported that these crystalline structures were formed as intermediates in the preparation of large unilamellar vesicles. Cochleates were obtained as a precipitate after the addition of calcium cations to negatively charged preformed liposomes. Cochleates were investigated as a delivery system in vaccine and gene therapy (Mannino *et al.*, 1998; Zarif and Mannino, 2000). This lipid-based system appears to have potential in the development of future generations of pharmaceuticals (Zarif, 2002; Zarif and Perlin, 2002).

To prepare drug-cochleates three methods can be used. The first method is named trapping/high pH (Zarif *et al.*, 2000a). The second is the trapping/film method, exploiting the hydrophobic property of the drug and its easy inclusion into a lipid bilayer (Zarif *et al.*, 2000a). The third approach uses a biocompatible, hydrogel, two-phase system that leads to the formation of small nanocochleate particles (Zarif *et al.*, 2001). The physicochemical properties of drug cochleates prepared according to these three methods are different (particle size, aggregation) and therefore are expected to result in different biological activities.

Application of Cochleates in Drug Delivery

Cochleates have been used to entrap different kinds of drugs. Preference has been given, however, to hydrophobic moieties, amphipathic molecules that can easily insert into membrane bilayers, and charge-pH–induced molecules such as peptides and aminoglycosides. Figure 1 is a schematic of how hydrophobic (Fig. 1A), amphipathic (Fig. 1B), negatively charged (Fig. 1C), or positively charged (Fig. 1D) molecules could be embedded into a cochleate structure.

The entrapment efficiency depends on the nature of the drug. Hydrophobic drugs would have a high affinity for the hydrophobic interior of the lipid bilayers to minimize the interaction with water (Fig. 1A), as demonstrated by the high entrapment efficiency obtained with amphotericin B. Amphipathic molecules such as doxorubicin, which have hydrophobic regions but are soluble in water, would partition between the bilayers and the external aqueous phase (Fig. 1B). Because calcium induces dehydration of the interbilayer domains, the amount of water in this region is low, and, therefore small hydrophilic molecules are not suitable for encapsulation in the cochleate structures. On the other hand, either small hydrophobic molecules or macromolecules, which have a charge or charge-pH–induced structures such as DNA, peptides, or aminoglycosides, should be easily encapsulated in the interbilayer domains (Fig. 1C and D).

Cochleates are mainly formed of phosphatidylserine (PS) and calcium. PS shows considerable binding affinity for calcium due to the tendency of calcium to lose part of its hydration shell and to displace water upon complex formation (New, 1990; Papahadjopoulos *et al.*, 1975). As a result

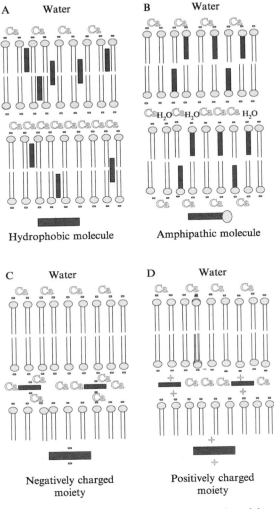

FIG. 1. Drug–lipid bilayer architecture in a cochleate as a function of the nature of the drug: hydrophobic (A), amphipathic (B), negatively charged (C), or positively charged (D) drugs.

of neutralization of the electrostatic charge, calcium causes liposomes composed of PS to aggregate and fuse with each other (Wilschut *et al.*, 1980). Calcium plays a crucial role in bringing bilayers close together through partial dehydration of the membrane surface and cross-linking of opposing molecules of PS (Papahadjopoulos *et al.*, 1990). Therefore, for empty cochleates or cochleates embedding either a hydrophobic molecule such as amphotericin B or an amphipathic molecule such as doxorubicin into their bilayers, very little or almost no water or hydration should be available between the bilayers.

When a charged moiety is included within the interspace between the bilayers of a cochleate, calcium binding to the phospholipid head group or to the charged drug is different, depending on the ionic condition of the drug. For a negatively charged moiety such as DNA, the calcium cation would serve as a bridge between this drug and the phospholipid head located on one bilayer in the region where the drug/calcium coexist (Fig. 1C), and calcium will serve as a bridge between two phospholipid head groups in the region where the drug is excluded. For a positively charged drug such as an aminoglycoside, direct electrostatic attraction between the drug and the negatively charged phospholipid could occur, and the drug could assume the role of calcium by complexing two head groups of two different bilayers, whereas calcium maintains the bridging where the drug is excluded (Fig. 1D). In either case, a higher hydration level would be expected in the interspace between the bilayers and, therefore, a larger interspace compared with an empty cochleate or a cochleate containing a hydrophobic or amphipathic molecule.

A recent feature of cochleates is that the cochleate particle size can be tailored (Zarif et al., 2001). Oral absorption of drugs is known to be closely correlated with the particle size of the vehicle transporting the drug (Florence et al., 1995; O'Hagan et al., 1992; Porter, 1997). Therefore, the cochleate particle size should affect the biological activity of the encapsulated drug.

Preparation of Amphotericin B Cochleates

Multiple methods have been described for cochleate preparation allowing cochleate particle size to be tailored (Zarif et al., 2000a,b, 2001). Amphotericin B (AmpB) was studied, as AmpB represents an ideal model for hydrophobic drugs. The idea was that an oral AmpB cochleate formulation would result in safe systemic delivery of the drug, due to the enhanced stability imparted to AmpB when embedded in the cochleate bilayers. Several AmpB formulations that use the systemic route for delivery of AmpB are available and are used clinically (Adle-Moore and Proffitt, 1993; Guo et al., 1991; Janoff et al., 1993). AmpB cochleates can be prepared using either the trapping/high pH method (Fig. 2), the trapping/film method (Fig. 3), or the aqueous/aqueous emulsion system (Fig. 4). The trapping/high pH method detailed in Fig. 2 uses powdered lipid and AmpB crystals as raw materials; therefore, it has the advantage of the absence of organic solvents. Since this method uses a pH around 11.5,

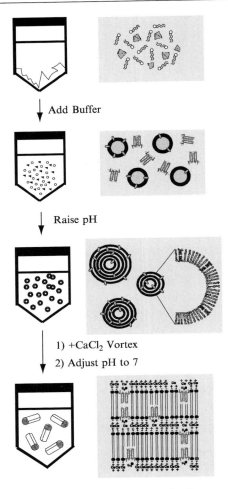

Add Buffer

Raise pH

1) +CaCl$_2$ Vortex
2) Adjust pH to 7

FIG. 2. General procedure to prepare a drug cochleate by the trapping/high pH method.

no sonication is needed to produce small unilamellar vesicles. The use of high-energy sonication combined with a high pH would have a detrimental effect on the lipid integrity and an increase of lipid degradation through hydrolysis (Grit, 1991). This method results in the formation of aggregates of relatively large-sized cochleates and collapsed phospholipid sheets as shown in Fig. 6 (Zarif et al., 2000a). The large size of the cochleate made by the trapping/high pH method is confirmed by freeze-fracture electron microscopy (Fig. 6A), which shows a mixture of rolled-up structures and collapsed sheets.

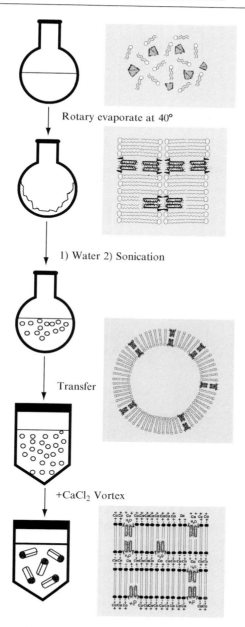

Rotary evaporate at 40°

1) Water 2) Sonication

Transfer

+CaCl$_2$ Vortex

FIG. 3. General procedure to prepare a drug cochleate by the trapping/film method.

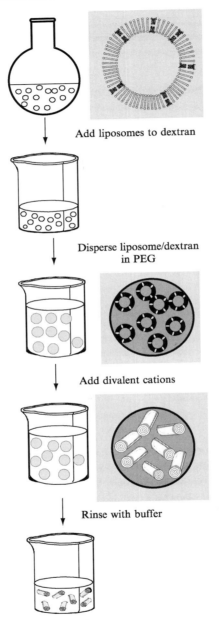

Add liposomes to dextran

Disperse liposome/dextran
in PEG

Add divalent cations

Rinse with buffer

FIG. 4. Hydrogel process to prepare small-sized drug nanocochleates.

Materials

Dioleoylphosphatidylserine (DOPS) can be purchased from Avanti Polar Lipids (Alabaster, Alabama). AmpB used in the following preparations can be purchased from USP (Rockville, MD). All solvents, methanol and acetonitrile, as well as deionized water and normal saline can be purchased from Fisher Scientific. Dextran (500,000), poly(ethylene glycol) (PEG 8000), calcium chloride, ethylenediaminetetraacetic acid (EDTA), and all other chemical reagents (analytical grade) can be obtained from Sigma Chemical Co. (St Louis, MO). Sonication can be performed with a bath sonicator (Laboratory Supplies Co. Inc., Hicksville, NY) model G1125PIG.

Trapping–High pH Method

DOPS powder (91.8 mg) and AmpB crystals (10 mg) are mixed in a lipid/AmpB molar ratio of 10:1 in a sterile polypropylene tube (Fig. 2). TES [N-Tris(hydroxymethyl)-methyl-2-aminomethane sulfonic acid] buffer (10 ml, pH 7.4) is added. Multilamellar liposomes are formed after vortexing for 3 × 1 min. The pH is increased to 11.5 by addition of 140 μl of 1 N NaOH, and AmpB crystals are solubilized. The absence of AmpB crystals and the presence of liposomes are monitored by using phase contrast and polarization optical microscopy. Calcium chloride (1.25 ml of a 0.1 M stock solution) is added slowly to the AmpB liposome suspension at a lipid/calcium molar ratio 2:1; AmpB cochleates are formed. The external pH is adjusted to 7 by the addition of 75 μl of 1 N HCl. The preparation is stored at 4° in the absence of light.

Trapping-Film Method

AmpB (10 mg) is dissolved in methanol at 0.5 mg AmpB/ml with brief sonication and the solution is added to 9.18 ml of DOPS in chloroform at 10 mg lipid/ml in a round-bottom flask (Fig. 3). AmpB is readily soluble in the chloroform/methanol mixture. The mixture is dried to a film using a rotary evaporator and gentle warming at 35–40° under reduced pressure (1 bar). The dried lipid film is hydrated with 10 ml of deionized water and then sonicated in a cooled bath sonicator for 5 × 1 min with 1-min intervals until the suspension becomes clear. The liposome size is measured by laser light scattering (weight analysis); AmpB liposome size should be around 50 nm.

To form AmpB cochleates, a calcium chloride solution (0.1 M) is slowly added to the liposome suspension (lipid/calcium, 2:1 in molar). Yellow AmpB cochleates are obtained. The preparation should be kept at 4° under nitrogen in the absence of light.

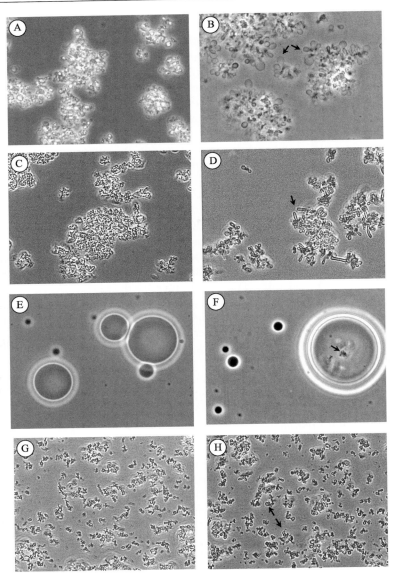

FIG. 5. Phase–contrast microscopy of (A) AmpB cochleates prepared by the trapping/high pH method. (B) AmpB cochleates of (A) after the addition of EDTA; note the conversion of cochleates into liposomes. (C) AmpB cochleates prepared by the trapping/film method. (D) AmpB cochleates of (C) converted to liposomes after addition of EDTA. (E) Phase–contrast microscopic images of a mixture of liposomes in dextran dispersed into PEG gel solution; the

Hydrogel Method

Cochleate particles have been engineered to obtain smaller particle size (<1 μm) and almost no aggregation by using a binary aqueous/aqueous system (Zarif *et al.*, 2001). Biocompatible hydrogels were used to obtain the two-phase system. To obtain the hydrogel-isolated cochleates, a liposome suspension is prepared with or without drug loading (Fig. 4). The liposome suspension is mixed with a hydrogel dextran solution. The dextran phase, which carries the liposomes, is dispersed into another hydrogel (PEG) solution. Under a shear force (vigorous magnetic stirring for small batches), the dextran phase forms microdroplets in which the liposomes are entrapped and isolated due to the partition effect favoring dextran (Fig. 5E). The particle size of the dextran droplets can be controlled by the shear force applied. Addition of calcium ions to the continuous phase (PEG solution) results in the diffusion of calcium into the dextran, collapsing the liposomes into cochleates. The cochleates are recovered by rinsing out the hydrogel.

To prepare AmpB cochleates using the hydrogel method, AmpB in methanol (0.5 mg/ml) is added to DOPS in chloroform (lipid/AmpB, 10:1 in molar) and the mixture is dried into a drug–lipid film using a rotary evaporator (Fig. 4). The film is hydrated with deionized water at a concentration of 10 mg lipid/ml, and the AmpB–lipid suspension is sonicated until small AmpB liposomes are obtained. The AmpB–liposome suspension is mixed with 40% w/w dextran-500,000 in a suspension of 2/1 v/v dextran/liposome. This mixture is then injected using a syringe into 15% w/w PEG-8000 under magnetic stirring (800–1000 rpm). An aqueous–aqueous emulsion of AmpB liposomes/dextran droplets dispersed in a PEG continuous phase is obtained. A CaCl$_2$ solution (100 mM) is added to the emulsion at a final concentration of 1 mM. Stirring is continued for 1 h to allow the slow formation of small-sized AmpB cochleates sequestered into the dextran droplets. The polymer is washed by the addition of a washing buffer containing 1 mM CaCl$_2$ and 150 mM NaCl to the emulsion at a ratio of 10:1 (v/v). The suspension is vortexed and centrifuged at 3000 rpm, 2–4°, for 30 min. After removal of the supernatant, washing buffer is added at a

small black dots are dextran particles formed by the dispersing dextran phase in the PEG phase. (F) Microscopic images of the sample shown in (E) after treatment with CaCl$_2$ solution; the black objects in circles are cochleates formed by the addition of Ca^{2+} ions. (G) Microscopic images of AmpB cochleates at 1 mg AmpB/ml shown in (F) after washing with a buffer containing 1 mM CaCl$_2$ and 100 mM NaCl; aggregates are formed by the cochleate particles. (H) Suspension shown in (G) after addition of EDTA; cochleate particles opened into liposomes with diameter of 1–2 μm, indicating that the intrinsic size of the cochleate particles is in the submicrometer range.

ratio of 5:1 (v/v), followed by centrifugation under the same conditions. Depending on the desired concentration, the resultant AmpB cochleate precipitate can be reconstituted with the same buffer.

Characterization of Cochleates

To analyze cochleates and the drug–cochleate morphology, phase-contrast optical microscopy and freeze-fracture electron microscopy are the ideal methods to be used. The drug content can be analyzed spectrophotometrically or by high-performance liquid chromatography (HPLC).

Optical Microscopy

A microscope equiped with a 100× lens, a digital camera, and a video recording system can be used to assess the morphology of cochleate particles. After recording the images of cochleates, a drop of EDTA solution (100 mM, pH 9) is added to the edge of the coverglass to examine the conversion of cochleates into liposomes.

Figure 5A shows a typical cochleate precipitate. When EDTA is added to chelate the calcium cations, the cochleates open into relatively large liposomes (Fig. 5B). The trapping/film method leads to smaller particle size cochleates (Fig. 5C), which yield liposomes of smaller size than the trapping/ high pH cochleates (Fig. 5D). The trapping method usually results in both rolled-up cochleate cylinders and collapsed sheets (Fig. 6A). This can be attributed to the rate of addition of calcium and the fact that the final structure of cochleates depends on the liposome size, the final lipid concentration, and the rate of addition of calcium. In contrast, the trapping/film method uses small unilamellar vesicles as intermediates for the preparation of cochleates, thus more uniform, cigar-shaped cochleates are obtained.

For the hydrogel method, optical microscopy should be performed at each step of the preparation procedure to detect the formation of small cochleates. Each preparation step is depicted in Fig. 5E–H. When the AmpB/liposome-dextran mixture was dispersed into PEG solution, phase separation occurred as shown in Fig. 5E. Partition of the liposomes favored the dispersed dextran phase as indicated by the yellow color of AmpB. This partitioning ensures that liposomes are isolated in each dextran particle. Addition of calcium ions into the continuous-phase (PEG) resulted in the formation of precipitates in the dextran-dispersed phase (Fig. 5F). Small needle-shaped cochleates were formed and observed under the microscope (Fig. 5G). These cochleates open into unilamellar vesicles of about 1 μm in diameter upon addition of EDTA and chelation of the calcium (Fig. 5H). The needle-shaped morphology is confirmed by electron microscopy after freeze-fracture (Fig. 6B).

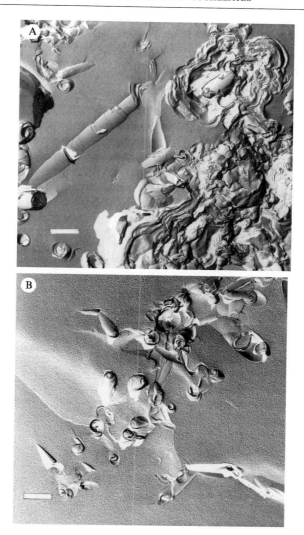

FIG. 6. Freeze-fracture electron micrographs of cochleates prepared either by the trapping (A) or by the hydrogel (B) method.

Freeze-Fracture Electron Microscopy

To analyze the cochleate structure, freeze-fracture can be performed as follows. A thin layer of the sample is deposited on a copper holder, and the sample is then quenched in liquid propane and fractured *in vacuo* (less than

10^{-6} T) with a liquid nitrogen–cooled knife in a Balzers 301 freeze-etching unit. The replication is performed with unidirectional Pt-C shadowing; the mean thickness of the metal deposit should be about 1.5 nm. The replicas are washed with ethanol and water and observed in a Philips EM301 electron microscope (Zarif *et al.*, 2000b). Figure 6 shows a typical morphology of cochleates.

Entrapment Efficiency of a Drug Measured by HPLC: Example of Amphotericin B Cochleates

Extraction of AmpB from Cochleates: Solvent Method (Segarra et al., *2002).* Five aliquots of 100 μl AmpB cochleates are placed in 15-ml polypropylene tubes. Methanol (10 ml) is added to each one of the samples and vortexed for 1 min. Two sets of sonication (1 min) and vortexing (20 min) are performed, and the samples are centrifuged at 4° for 5 min (3000 rpm). Four milliliters of the supernatant is transferred into 15-ml polypropylene tubes, an identical volume of methanol is added, and the mixture is vortexed for 20 s. Around 100 μl is transferred into HPLC inserts, and the vials are sealed. The samples are kept at low temperature (4°) and protected from light until analysis.

Extraction of AmpB from Cochleates: Detergent Method (Zarif et al., *2000b).* One hundred microliters of trapping cochleates is aliquoted into five 15-ml polypropylene centrifuge tubes. To each tube the following substrates are added while vortexing: 60 μl pH 9.5 EDTA and 100 μl of 10% Triton X-100. The resulting solution is clear yellow. Fifty microliters of NaOH (1 *N*) is added with vortexing, and 80–100 μl of the samples is transferred into HPLC inserts. The HPLC vials are kept sealed in a refrigerator protected from light until analysis.

Measurement of AmpB Content by HPLC (Segarra et al., *2002).* To perform the HPLC analysis, the system must be equipped with a photodiode array detector. A 20-μl aliquot of the extracted sample (kept at 5°) is added through an in-line filter into a Nova-Pak C_{18} column (3.9 × 150 mm, 4 μm particle size) at 40°. AmpB is eluted at a flow rate of 0.5 ml/min using a mixture of 29% methanol, 30% acetonitrile, and 41% 2.5 m*M* EDTA and detected at 408 nm. A gradient elution is used as follows: analytical time 0, acetonitrile 30%; analytical time 7.5 min, acetonitrile 30%; analytical time 8.5, acetonitrile 41%; analytical time 9.5, acetonitrile 30%. The methanol percentage is kept constant throughout the run. The column is allowed to re-equilibrate for 4 min. The concentration of AmpB is calculated by comparison to a freshly prepared external standard.

Evaluation of the Efficacy of Amphotericin B Cochleates in a Candidiasis Model

The efficacy of AmpB cochleates has been evaluated in a murine model infected with *Candida albicans*. AmpB cochleates are given parenterally (ip) to ICR mice as a daily dose of 0.1–20 mg/kg/day for 10 days and compared with Fungizone and AmBisome. One hundred percent survival is observed after 30 days even at low doses (0.5 mg/kg/day). In the spleen, cochleates amphotericin B (CAMB) showed enhanced potency compared with Fungizone given ip at 1 mg/kg/day and is equivalent to AmBisome given intravenously. In the kidneys, the three drugs were comparable (Graybill *et al.*, 1999; Zarif *et al.*, 1999b, 2000b).

The efficacy of oral CAMB has been studied in *C. albicans*–infected BALB/c mice. CAMB is given at doses ranging between 0.5 and 20 mg/kg/day for 15 days. All groups show 100% survival. A low dose (2.5 mg/kg/day) succeeds in completely eradicating the *Candida* from the lungs and is as efficient as Fungizone given ip at almost the same dose (2 mg/kg/day) (Santangelo *et al.*, 2000; Zarif *et al.*, 1999a, 2000c).

AmpB cochleates have been described as being safe *in vitro* (Zarif *et al.*, 2000b) and *in vivo* (Santangelo *et al.*, 2000; Zarif and Mannino, 2000; Zarif *et al.*, 1999a, 2000b). This is somewhat expected, since cochleates are a lipid-based system. Lipid systems have been shown to decrease the toxicity of AmpB (Bekersky *et al.*, 1999). In addition, cochleates are formed of natural products, calcium and phosphatidylserine, known for their high biocompatibility and positive health impact (Crook *et al.*, 1991; Villardita *et al.*, 1987) in improving and repairing mental functions.

Evaluation of the Efficacy of Amphotericin B Cochleates in an Aspergillosis Model

AmpB cochleates have been evaluated, additionally, in a murine model of systemic aspergillosis (Delmas *et al.*, 2002). Oral doses of CAMB at 20 and 40 mg/kg/day show 70% survival and more than a 2 log reduction in colony counts in lungs, liver, and kidneys. This can be considered as a promising result given that aspergillosis is a deadly disease for which no ideal treatment is currently available.

Conclusion

Cochleates are a self-assembled lipid system that has shown promising results in delivering orally hydrophobic drugs such as amphotericin B.

328 LIPOSOMAL ANTIBACTERIAL, ANTIFUNGAL, AND ANTIVIRAL AGENTS [18]

The development of this product is underway (website: www.biodelivery sciences.com). Extensive research has led to the development of procedures for the preparation of drug cochleates. Efforts should be directed now toward understand the mechanism by which cochleates deliver drugs to the site of action.

Acknowledgments

The author aknowledges her coworkers whose names are cited in the references. Special acknowledgment to Dr. Chris Lambros from NIAID for his support and to Jina Zarif-Luciani for the artistic work on the figures and help in preparing the manuscript.

References

Adler-Moore, J. P., and Proffitt, R. T. (1993). Development, characterization, efficacy and mode of action of AmBisome, a unilammellar liposomal formulation of amphotericin B. *J. Liposome Res.* **3**, 429–450.

Bekersky, I. I., Fielding, R. M., Buell, D., and Lawrence, I. I. (1999). Lipid-based amphotericin B formulations: From animals to man. *Pharm. Sci. Technol. Today* **2**, 230–236.

Crook, T. H., Tinkenberg, J., Yesavage, J., Petrie, W., Nunzi, M. G., and Massari, D. C. (1991). Effects of phosphatidylserine in age-associated memory impairment. *Neurology* **41**, 644–649.

Delmas, G., Park, S., Chen, Z. W., Tan, F., Kashiwazaki, R., Zarif, L., and Perlin, D. S. (2002). Efficacy of orally delivered cochleates containing amphotericin B in a murine model of aspergillosis. *Antimicrob. Agents Chemother.* **46**, 2704–2707.

Florence, A. T., and Hussain, N. (2001). Transcytosis of nanoparticle and dendrimer delivery systems: Evolving vistas. *Adv. Drug Deliv. Rev.* **50**(Suppl. 1), S69–S89.

Florence, A. T., Hillery, A. M., Hussain, N., and Jani, P. U. (1995). Factors affecting the oral uptake and translocation of polystyrene nanoparticles: Histological and analytical evidence. *J. Drug Target* **3**, 65–70.

Graybill, J. R., Najvar, L. K., Bocanegra, R., Scolpino, A., Mannino, R. J., and Zarif, L. (1999). A new lipid vehicle for amphotericin B. Interscience Conference on Antimicrobial Agents and Chemotherapy (ICAAC), San Francisco, September 26–29, 1999.

Grit, M. (1991). Thesis: Stability of Liposomes, Analytical, Chemical, and Physical Aspects. Department of Pharmaceutics, University of Utrecht, Utrecht, The Netherlands.

Guo, L. S. S., Fielding, R. M., Lasic, D. D., Hamilton, R. L., and Mufson, D. (1991). Novel antifungal drug delivery: Stable amphotericin B-cholesteryl sulfate discs. *Int. J. Pharm.* **75**, 45–54.

Han, H. K., and Amidon, G. L. (2000). Targeted prodrug design to optimize drug delivery. *AAPS Pharm. Sci.* **2**, E6.

Janoff, A. S., Perkins, W. R., Saletan, S. L., and Swenson, C. E. (1993). Amphotericin B lipid complex (ABLC): A molecular rationale for the attenuation of amphotericin B related toxicities. *J. Liposome Res.* **3**, 451–471.

Leone-Bay, A., Paton, D. R., and Weidner, J. J. (2000). The development of delivery agents that facilitate the oral absorption of macromolecular drugs. *Med. Res. Rev.* **20**(2), 169–186.

Mannino, R. J., Canki, M., Feketeova, E., Scolpino, A., Wang, Z., Zhang, F., Kheiri, M., and Gould-Fogerite, S. (1998). Targeting immune response induction with cochleate and liposome-based vaccines. *Adv. Drug Deliv. Rev.* **3**, 273–287.

New, R. R. C. (1990). "Liposomes, a Practical Approach." IRL Press/Oxford University Press, New York.

O'Hagan, D. T., Christy, N. M., and Davis, S. S. (1992). Particulates and lymphatic drug delivery. In "Lymphatic Transport of Drugs" (W. N. Charman and V. J. Stella, eds.), pp. 279–315. CRC Press, Boca Raton, FL.

Papahadjopoulos, D., Vail, W. J., Jacobson, K., and Poste, G. (1975). Cochleate lipid cylinders: Formation by fusion of unilamellar lipid vesicles. Biochem. Biophys. Acta **394,** 483–491.

Papahadjopoulos, D., Nir, S., and Düzgüneş, N. (1990). Molecular mechanisms of calcium-induced membrane fusion. J. Bioenerg. Biomembr. **22,** 157–179.

Porter, C. J. H. (1997). Drug delivery to the lymphatic system. Crit. Rev. Ther. Drug Carrier Syst. **14**(14), 333–393.

Santangelo, R., Paderu, P., Delmas, G., Chen, Z. W., Mannino, R., Zarif, L., and Perlin, D. S. (2000). Efficacy of oral cochleate-amphotericin B in a mouse model of systemic candidiasis. Antimicrob. Agents Chemother. **44,** 2356–2360.

Segarra, I., Movshin, D., and Zarif, L. (2002). Pharmacokinetics and tissue distribution after intravenous administration of a single dose of amphotericin B cochleates, a new lipid-based delivery system. J. Pharm. Sci. **91,** 1827–1837.

Villardita, C., Grioli, S., Salmeri, G., Nicoletti, F., and Pennisi, G. (1987). Multicentre clinical trial of brain phosphatidyl serine in elderly patients with intellectual deterioration. Clin. Trials J. **24,** 84–93.

Wilschut, J., Düzgüneş, N., Fraley, R., and Papahadjopoulos, D. (1980). Studies on the mechanism of membrane fusion: Kinetics of calcium ion induced fusion of phosphatidylserine vesicles followed by a new assay for mixing of aqueous vesicle contents. Biochemistry **19,** 6011–6021.

Zarif, L. (2002). Elongated supramolecular assemblies in drug delivery. J. Control Release **81,** 7–23.

Zarif, L., Graybill, J. R., Perlin, D., and Mannino, R. (2000a). Cochleates: New lipid-based drug delivery system. J. Liposome Res. **10,** 523–538.

Zarif, L., Graybill, J. R., Perlin, D., Najvar, L., Bocanegra, R., and Mannino, R. J. (2000b). Antifungal activity of amphotericin B cochleates against Candida albicans infection in a mouse model. Antimicrob. Agents Chemother. **44,** 1463–1469.

Zarif, L., and Mannino, R. J. (2000). Cochleates. In "Cancer Gene Therapy: Past Achievements and Future Challenge" (N. Habib, ed.), p. 83. Kluwer Academic/Plenum, New York.

Zarif, L., and Perlin, D. (2002). Amphotericin B nanocochleates: From formulation to oral efficacy. Drug Deliv. Technol. **2,** 34.

Zarif, L., Segarra, I., Jin, T., Hyra, D., and Mannino, R. J. (1999a). Amphotericin B cochleates as a novel oral delivery system for the treatment of fungal infections. 26th International Symposium on Controlled Release of Bioactive Materials, Boston, June 20–23, 1999.

Zarif, L., Segarra, I., Jin, T., Hyra, D., Perlin, D., Graybill, J. R., and Mannino, R. J. (1999b). Oral and systemic delivery of amphotericin B mediated by cochleates. American Association of Pharmaceutical Scientists (AAPS) Annual Meeting and Exposition, November 1999.

Zarif, L., Santangelo, R., Mannino, R. J., and Perlin, D. (2000c). Amphotericin B cochleates as a novel oral delivery system for the treatment of fungal infections. The 27th International Symposium on Controlled Release of Bioactive Materials, The Third Consumer and Diversified Products Conference, Paris, France, July 7–13, 2000.

Zarif, L., Jin, T., Segarra, I., and Mannino, R. J. (2001). WO Patent 01/52817 A2.

[19] Lymphoid Tissue Targeting of Anti-HIV Drugs Using Liposomes

By ANDRÉ DÉSORMEAUX and MICHEL G. BERGERON

Abstract

Considering that HIV-1 accumulates and replicates actively within lymphoid tissues, any strategy that will decrease viral stores in these tissues might be beneficial to the infected host. Follicular dendritic cells (FDC), B lymphocytes, antigen-presenting cells like macrophages, and activated $CD4^+$ T cells are abundant in lymphoid tissues, and all express substantial levels of the HLA-DR determinant of the major histocompatibility complex class II (MHC-II). Monocyte-derived macrophages, which are also $CD4^+$ and express HLA-DR, are considered to be the most frequent hosts of HIV-1 in tissues of infected individuals. This chapter describes a method for the generation of sterically stabilized immunoliposomes grafted with anti-HLA-DR antibodies that allows efficient delivery of drugs to lymphoid tissues. The method first involves the production of murine HLA-DR (clone Y-17, IgG_{2b}) and human HLA-DR (clone 2.06, IgG_1) antibodies from hybridomas in mice and their purification from ascites fluids. This step is followed by the production of Fab' fragments of antibodies 2.06 and Y-17 that are grafted at the surface of sterically stabilized immunoliposomes instead of the complete IgG to reduce their immunogenicity. The preparation of sterically stabilized liposomes, the composition of which allows an efficient entrapment and retention of several drugs, by the method of thin lipid film hydratation followed by extrusion through polycarbonate membranes is then described. This step is followed by the removal of unencapsulated drug, when present, by low-speed centrifugation of the liposomal preparation through a Sephadex G-50 column. These liposomes contain a fixed amount of poly(ethylene glycol) chain terminated by a maleimide reactive group for the coupling of Fab' fragments. The procedure for the coupling of Fab' fragments at the surface of sterically stabilized liposomes and the removal of uncoupled fragments of antibodies is described. *In vitro* binding studies of sterically stabilized immunoliposomes to cell lines expressing different surface levels of the mouse or human HLA-DR determinant of MHC-II demonstrate that these liposomes are very specific. When compared with conventional liposomes, the subcutaneous administration in the upper back, below the neck, of mice of anti-HLA-DR immunoliposomes resulted in a 2.9 and 1.6 times greater accumulation in

METHODS IN ENZYMOLOGY, VOL. 391

the cervical and brachial lymph nodes, respectively. The use of sterically stabilized immunoliposomes increases 2 to 4.6 times the concentration of liposomes in all tissues, with a peak accumulation at 240 h in brachial, inguinal, and popliteal lymph nodes and at 360 h or greater in cervical lymph nodes. A single bolus injection of indinavir given subcutaneously to mice results in no significant drug levels in lymphoid organs. Most of the injected drug accumulates in the liver and is totally cleared within 24 h postadministration. In contrast, sterically stabilized immunoliposomes are very efficient in delivering high concentrations of indinavir to lymphoid tissues for at least 15 days postinjection. The drug accumulation in all tissues leads to a 21- to 126-fold increased accumulation when compared with the free agent. Anti-HLA-DR immunoliposomes containing indinavir are as efficient as the free agent in inhibiting HIV-1 replication in PM1 cells that express high levels of cell surface HLA-DR. Sterically stabilized anti-HLA-DR immunoliposomes mostly accumulate in the cortex in which follicles (B cells and FDCs) are located, and in parafollicular areas in which T cells, interdigitating dendritic cells, and other accessory cells are abundant. The delivery of drugs in this area of the lymph nodes could represent a convenient strategy to inhibit more efficiently HIV-1 replication. Although the method described in this chapter is specific to the coupling of anti-HLA-DR antibodies, any antibody fragment or peptide specific for an antigen present in relatively large quantities at the surface of lymphoid cells, that is anchored to the surface of sterically stabilized liposomes with an appropriate coupling method, can be used to concentrate drugs within target tissues and improve the therapeutic effect of drugs.

Introduction

Twenty years after the first clinical evidence of acquired immunodeficiency syndrome (AIDS) was reported, AIDS has become one of the most devastating diseases that the health and scientific community has ever faced. Statistics (as of the end of 2001) from UNAIDS/WHO indicated that since the beginning of the epidemic, more than 60 million people have been infected with human immunodeficiency virus (HIV), the etiological agent of AIDS. An estimated 5 million new cases of HIV-1 infection occurred during the year 2001 (i.e., approximately one infection every 6 s). AIDS deaths totalled 3 million in 2001 alone, giving a cumulative death of over 23 million since the beginning of the epidemic. HIV/AIDS is now the leading cause of death in sub-Saharan Africa and is presently the fourth biggest killer worldwide. As the number of individuals infected with HIV is growing dramatically throughout the world, it is imperative to improve strategies to control the progression of the disease in infected individuals.

A critical event in the initial establishment of HIV infection is the dissemination of viral particles in lymphoid organs that act as major reservoirs for HIV. The high viral load observed in the lymphoid tissues was reported to be associated with trapped HIV particles on the follicular dendritic cells (FDC) located in the germinal centers (Embretson et al., 1993; Fox et al., 1991; Pantaleo et al., 1993; Schrager and D'Souza, 1998). In addition to the extracellular localization of HIV-1 in interdendritic spaces of germinal centers, viral particles were also found within the endosomal and cytoplasmic compartments of FDC. Moreover, viral particles bound onto the FDC remained highly infectious to CD4$^+$ T cells despite the presence of neutralizing antibodies on their surface (Heath et al., 1995; Schrager and Fauci, 1995). Paradoxically, the efficient trapping mechanisms of the lymph nodes, which are ordinarily associated with the initiation of an appropriate immune response, may contribute to the immune deterioration caused by HIV (Fauci, 1993). The persistence of viral particles in the lymphoid tissues results in a chronic stimulation of the immune system that ultimately leads to the destruction of the microenvironment network within the germinal centers. In the advanced stage of the disease, the inability of FDC to retain HIV-1 particles has been postulated to contribute to the increased viral burden in the periphery. Based on the above demonstration, it is essential to reduce or abrogate the production and accumulation of HIV-1 particles in the lymphoid tissues to preserve the architecture and integrity of the latter.

Active antiretroviral therapy, usually consisting of a combination of two nucleoside analogues and one protease inhibitor, has been shown to be effective in controlling infection by reducing the plasma viral load to undetectable levels in HIV-infected individuals and depleting the pools of virus in lymphoid tissues (Wong et al., 1997). Despite the apparent success of antiretroviral therapy, the capacity of HIV-1 to establish latent infection of CD4$^+$ T cells allows viral particles to persist in tissues. Recent studies showed that anti-HIV regimens do not fully eliminate viral replication in secondary lymphoid tissues, and this continued replication of HIV-1 seems to be associated with the presence of drug-sensitive viruses (Zhang et al., 1999). It was shown that replication-competent HIV-1 is routinely isolated from lymphoid organs of patients even after 30 months of therapy (Finzi et al., 1997; Furtado et al., 1999; Zhang et al., 1999). Moreover, the initiation of highly active antiretroviral therapy (HAART) as early as 10 days after the onset of symptoms of primary infection could not prevent the establishment of a pool of latently infected resting CD4$^+$ T cells (Chun et al., 1998). By measuring the decay rate of the latent reservoir in 34 treated adults whose plasma virus levels were undetectable, Finzi *et al.* (1999) estimated that 60 years of therapy would be required for complete eradication of the virus. Latently infected resting $CD4^+$ T cells thus provide a lifelong persistence of HIV-1 and are likely to represent the major barrier to virus eradication in patients on combination antiretroviral therapy. On the other hand, increasing numbers of treatment failures resulting from toxicity, drug-resistant mutants, and poor compliance of patients to the drug regimen are emerging with long-term therapy. Consequently, the development of alternative approaches that specifically target lymphoid tissues, the main reservoir for HIV, remains a high priority to treat this infection more efficiently.

One common feature of retroviruses, as well as of many other enveloped viruses, is the acquisition of host cell surface molecules during the budding process. In addition to its virally encoded structures, HIV-1 has been shown to incorporate a vast array of host proteins while budding out of the infected cell (reviewed in Tremblay *et al.*, 1998). The incorporation of host constituents by HIV-1 during the budding phase seems to be a selective, nonrandom phenomenon. Depending on their nature, these constituents may directly or indirectly participate in early events and putatively affect the pathogenesis of HIV-1 infection. For instance, virion-associated host constituents can modulate steps in the virus life cycle, including the attachment of virions to target cells, binding avidity between virus and host cell, and neutralization susceptibility of virions. FDC, B lymphocytes, antigen-presenting cells like macrophages, and activated $CD4^+$ T cells are abundant in lymphoid tissues, and all express substantial levels of the HLA-DR determinant of the major histocompatibility complex class II (MHC-II). Monocyte-derived macrophages, which are also $CD4^+$ and express HLA-DR, are considered to be the most frequently identified hosts of HIV-1 in tissues of infected individuals. Consequently, the probability that newly formed virions will bear cellular HLA-DR is high. More importantly, it has been demonstrated that plasma HIV-1 isolates from virally infected individuals do carry host-encoded HLA-DR on their surface (Saarloos *et al.*, 1997). The physiological relevance of cellular HLA-DR bound to viral particles is further provided by previous studies, indicating that it is one of the most abundant host-derived molecules carried by HIV-1 (Tremblay *et al.*, 1998).

Considering that HIV-1 accumulates and replicates actively within lymphoid tissues, any strategy that will decrease viral stores in these tissues might be beneficial to the infected host. As liposomes are naturally taken up by cells of the mononuclear phagocytic system (MPS), liposome-based therapy could represent a convenient approach to improve the delivery of anti-HIV agents into infected cells (Désormeaux and Bergeron, 1998).

Previous studies performed in our laboratory demonstrated that the encapsulation of anti-HIV agents into liposomes allows high intracellular penetration of drugs, good *in vitro* antiviral efficacy against HIV-1, efficient targeting of macrophage-rich tissues, and a marked improvement of the pharmacokinetics of drug (Désormeaux *et al.*, 1994; Dusserre *et al.*, 1995; Gagné *et al.*, 2002; Harvie *et al.*, 1995, 1996). On the other hand, the coupling of poly(ethylene glycol) (PEG) on the surface of liposomes (sterically stabilized liposomes) was shown to increase their ability to move through the lymph after subcutaneous injection and to decrease their rate of uptake by the MPS, increasing their residence time in plasma and/or lymph (Allen, 1994; Woodle and Lasic, 1992). Consequently, as the HLA-DR determinant of the MHC-II is abundantly expressed on antigen-presenting cells such as monocytes and macrophages and FDC, attachment of anti-HLA-DR antibodies to PEG-modified liposomes (sterically stabilized immunoliposomes) represents a logical strategy to combine prolonged circulation time and efficient delivery of drugs to specific cell populations and/or pathogen reservoirs.

Given the toxic side effects of antiviral agents actually available for the treatment of HIV infection and their limited ability to target specific cells, strategies aimed at reaching therapeutic levels of drugs into infected cells should be explored. In addition, as suboptimal concentrations of drugs within infected cells can lead to the development of resistance, delivery of high drug concentrations within HIV reservoirs could reduce the frequency of resistance. This chapter describes a method for the generation of sterically stabilized immunoliposomes that allows efficient delivery of drugs to lymphoid tissues. Although the method described in this chapter is specific to the coupling of anti-HLA-DR antibodies, any antibody fragment or peptide, specific for an antigen present in relatively large quantities on the surface of lymphoid cells, that is anchored to the surface of sterically stabilized liposomes with an appropriate coupling method can be used to concentrate drugs within the target tissue and improve the therapeutic effect of drugs.

Preparation and Purification of Monoclonal Antibodies

Hybridomas producing monoclonal antibodies directed against murine HLA-DR (clone Y-17, IgG$_{2b}$) and human HLA-DR (clone 2.06, IgG$_1$) are grown in RPMI 1640 supplemented with 10% fetal bovine serum (FBS), 2 mM L-glutamine, 100 U/ml penicillin G, and 100 mg/ml streptomycin. Cells are cultivated in BALB/c mice (18–20 g), and antibodies are isolated from ascites fluids. Antibodies are purified using a protein-G affinity column (Pharmacia, Baie d'Urfé, QC, Canada) according to the manufacturer's instructions. The total protein concentration in ascites is approximately 20 mg/ml, and the specific antibody concentration is in the range of 1–5 mg/ml, leading to a yield of pure antibody of 5–25%. Antibodies are next sterilized using 0.22-μm low binding protein filters (Millipore, Bedford, MA) and stored at $-20°$ in phosphate-buffered saline (PBS, pH 7.4) until use. Purity of antibodies is assessed using sodium dodecyl sulfate polyacrylamide gel electrophoresis (SDS–PAGE) under nonreducing conditions. The apparent molecular weight is verified using BenchMarker prestained protein ladder (GIBCO BRL, Grand Island, NY). Gel staining is performed with Coomassie brilliant blue (Sigma, St. Louis, MO).

The immunoreactivity of 2.06 and Y-17 antibodies is tested by flow cytometry on RAJI cells, an Epstein–Barr virus-carrying B cell line that expresses high levels of cell surface HLA-DR, -DP, and -DQ proteins and on freshly prepared C3H mouse spleen cells, respectively. In brief, a suspension of cells (10^6 cells/ml) is incubated for 30 min at $4°$ with either 1 μg of 2.06 or biotinylated Y-17 antibodies. Afterward, cells are washed three times with PBS and incubated for 30 min at $4°$ with 1 μg of R-phycoerythrin–conjugated goat antimouse (Southern Biotechnologie Associates, Birmingham, AL) for 2.06 or with 1 μg of R-phycoerythrin–conjugated streptavidin (Jackson ImmunoResearch Laboratories Inc., West Grove, PA) for Y-17. Afterward, cells are washed three times with PBS, fixed with 1% paraformaldehyde, and kept on ice under darkness until assessment of fluorescence by flow cytometry. Irrelevant isotype-matching antibodies are used as control.

Preparation of Fab′ Fragments

Harding *et al.* (1997) previously demonstrated that the immunogenicity of PEG-coated immunoliposomes bearing complete antibodies following repeated subcutaneous administration in mice was almost exclusively associated with the Fc portion of the immunoglobulin. Studies have also demonstrated that the preparations of PEG-grafted immunoliposomes are more immunogenic than the free IgG component, which is of major importance to the antibody-mediated liposomal drug delivery effort. Accordingly, in the present method, anti-HLA-DR Fab′ fragments rather than the entire antibody are used to reduce immunogenicity associated with the Fc portion. Our previous studies have demonstrated that liposomes bearing Fab′ fragments are 2.3-fold less immunogenic than liposomes bearing the entire IgG after once-a-week subcutaneous injections for 4 weeks in rats (Gagné *et al.*, 2002). However, it should be noted that Fab′ fragments still harbor a small part of the Fc portion, which could possibly be responsible for immunological stimulation. Consequently, the use of liposomes bearing Fv fragments or a particular peptide, which constitutes the smallest part of the immunoglobulin that keeps affinity for ligand, could possibly further reduce the induction of an immune response associated with repeated administrations of immunoliposomal preparations.

The $F(ab')_2$ fragments of antibody 2.06 are produced using an Immunopure IgG_1 Fab' and $F(ab')_2$ preparation kit (Pierce, Rockford, IL). In brief, the 2.06 antibody is concentrated with a Centricon-100 (Amicon, Beverly, MA), resuspended in 0.5 ml of PBS, and added to 0.5 ml of Immunopure IgG_1 mild elution buffer containing 1 mM cysteine. The solution is then incubated with an immobilized ficin column for 40 h at 37°. Afterward, the solution is eluted with 4 ml of Immunopure binding buffer, and fragments are separated on an Immunopure protein A column. The column retains Fc fragments and undigested IgG_1, whereas $F(ab')_2$ fragments are collected in 1-ml fractions. Fractions containing $F(ab')_2$ are determined from absorbance readings at 280 nm and pooled together. The $F(ab')_2$ fragments (110 kDa) are then concentrated using Centricon-50 (Amicon, Beverly, MA) and resupended in 1 ml phosphate–ethylenediaminetetraacetic acid (EDTA) buffer (100 mM sodium phosphate and 5 mM EDTA, pH 6.0).

The $F(ab')_2$ fragments of antibody Y-17 are produced following incubation of the antibody (7 mg/ml) with lysyl endopeptidase (Wako Chemicals, Richmond, VA, in 50 mM Tris–HCl, pH 8.5) in an enzyme/substrate molar ratio of 1:50 for 3 h at 37°. IgG_{2b} antibodies are known to be highly resistant to enzymatic digestion involving conventional proteases such as papain or pepsin. However, it was demonstrated that lysyl endopeptidase is very efficient in cleaving the core hinge of IgG_{2b} at Lys 228E/Cys 229 without affecting the inter-heavy chain disulfide bridges that are needed for the coupling of Fab' fragments with the thiol-reactive lipid (Yamagushi et al., 1995). The digestion products contain undigested IgG, $F(ab')_2$ and Fc fragments. The enzyme is removed by gel chromatography on a Sephadex G-25M column (Pharmacia, Baie d'Urfé, QC, Canada), whereas undigested IgG and Fc fragments are removed using a protein A affinity chromatography column. $F(ab')_2$ fragments are resuspended in a phosphate-EDTA buffer (100 mM sodium phosphate and 5 mM EDTA, pH 6.0). The IgG–$F(ab')_2$ percentage yield is approximately 30% for both antibodies.

To obtain Fab' fragments, $F(ab')_2$ fragments are incubated with 6 mg of 2-mercaptoethylamine-HCl (MEA, final concentration of 0.05 M) for 90 min at 37° under a nitrogen atmosphere. MEA cleaves the disulfide bridges between the heavy chains but preserves the disulfide linkages between the heavy and light chains. The solution is then eluted on a

Sephadex G-25 M column preequilibrated with an acetate–EDTA buffer (100 mM anhydrous sodium acetate, 88 mM sodium chloride, and 1 mM EDTA, pH 6.5), and Fab′ fragments are collected in 1-ml fractions. Fractions containing Fab′ fragments are determined using a BCA protein assay reagent kit (Pierce, Rockford, IL) and pooled together. The Fab′ fragments (55 kDa) are concentrated using Centricon-10 (Amicon, Beverly, MA), resuspended in acetate–EDTA buffer (pH 6.5) and kept under a nitrogen atmosphere at 4° until coupling to liposomes. The purity of Fab′ fragments is assessed using SDS–PAGE, and their antigenic specificity is verified by flow cytometry using appropriate cells.

Preparation of (Immuno)liposomes

With a view to develop liposome-based products, it is important to use liposome bilayer characteristics that allow high efficiency of drug encapsulation, as well as reduced leakage of entrapped drug to take advantage of the ability of liposomes to act as a site-specific drug delivery system. In that respect, studies performed in our laboratory showed that best results for entrapment of antivirals (AZT, ddC, ddI, 3TC, foscarnet, indinavir) are obtained with cholesterol-free anionic liposomes, which are composed of synthetic phospholipids having high gel-to-fluid phase transition temperatures. Such phospholipids are physically more stable than those having short or unsaturated hydrocarbon chains and are less prone to release their entrapped drug.

Liposomes composed of dipalmitoylphosphatidylcholine (DPPC)/dipalmitoylphosphatidylglycerol (DPPG)/distearoylphosphatidylethanolamine–[poly(ethylene glycol) 2000]–maleimide (DSPE-PEG-MAL) in a molar ratio of 10:3:0.33 are prepared according to the method of thin lipid film hydration. In brief, the lipid mixture is dissolved in chloroform/methanol (2:1, v/v) in a round-bottom flask, and the organic solvent is evaporated to form a thin lipid film on the wall of the flask. In some experiments, a small proportion of [^{14}C]-DPPC or [^{3}H]-labeled drug is added to the liposome formulation as radioactive tracers. The lipid film is then hydrated with an acetate–EDTA buffer (pH 6.5) with or without the drug to be encapsulated. Multilamellar vesicles (MLVs) are formed by mechanical agitation of the mixture at a temperature higher than the gel-to-liquid phase transition temperature of the phospholipid mixture. MLVs are then extruded through 0.1-μm polycarbonate membranes (Nucleopore, Cambridge, MA) using a stainless-steel extrusion device (Lipex Biomembranes, Vancouver, BC, Canada). The extrusion process provides a convenient method to reduce the mean diameter of liposomes and to produce homogeneously sized unilamellar vesicles. Vesicle size distribution and homogeneity of the large unilamellar vesicles (LUVs) are evaluated by quasielastic light scattering with a submicron particle analyzer. The mean size of the phospholipid vesicles produced by this method is in the range of 100–120 nm. Unencapsulated drug is removed by centrifugation (275g for 15 min at 4°) of the liposomal preparation through a Sephadex G-50 column (Pharmacia, Baie d'Urfé, QC, Canada) and efficiency of drug entrapment is determined by radioactive counting.

For the coupling procedure, sterically stabilized liposomes are incubated with freshly prepared Fab′ fragments (35 μg Fab′ fragments/μmol lipid) overnight at 4° under agitation and under a nitrogen atmosphere. Uncoupled Fab′ fragments are removed using a Sepharose CL-4B size exclusion column (Sigma, St. Louis, MO). The total amount of Fab′ fragments conjugated to the phospholipid vesicles is evaluated using a Coomassie protein assay reagent (Pierce, Rockford, IL). The coupling method of Fab′ fragments to sterically stabilized liposomes is summarized in Fig. 1. Drug retention is evaluated by incubating liposomes in PBS at 4° and 37° and in 80% serum at 37°. At specific time points, 0.1 ml of the dispersion of liposomes is removed and centrifuged (300g for 5 min at 4°) through a 1-ml Sephadex G-50 column. Drug retention is evaluated by radioactive counting of the centrifuged solution relative to that at time zero.

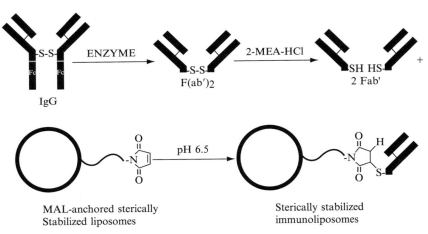

FIG. 1. Schematic representation of the coupling of Fab′ fragments to MPB-DPPE liposomes. F(ab′)$_2$ fragments of IgG$_1$ and IgG$_{2b}$ antibodies were obtained by enzymatic digestion using immobilized ficin and lysyl endopeptidase, respectively. Fab′ fragments were obtained by reduction of F(ab′)$_2$ fragments with 2-mercaptoethylamine-HCl. Immunoliposomes were generated following incubation of Fab′ fragments with freshly prepared MPB-DPPE liposomes. Adapted from Dufresne et al. (1999), with permission from Elsevier Science.

In Vitro Binding and Specificity of Immunoliposomes

The binding and specificity of sterically stabilized liposomes bearing murine HLA-DR (clone Y-17, IgG_{2b}) and human HLA-DR (clone 2.06, IgG_1) Fab' fragments are verified by flow cytometry on freshly prepared C3H mouse spleen cells and RAJI cells, respectively. In brief, a suspension of cells (10^6 cells/ml) is incubated for 30 min with 1.5 μmol of sterically stabilized immunoliposomes labeled with the lipophilic carbocyanine dye 1,1'-dioctadecyl-3,3,3',3'-tetramethylindocarbocyanine perchlorate (DiI, Molecular Probes, Eugene, OR). Cells are washed three times and resuspended in PBS. The specificity of liposomes for cells is verified by flow cytometry analysis from the fluorescence associated to DiI (fluorochrome incorporated into the lipid membrane).

The levels of binding of conventional liposomes and human anti-HLA-DR (IgG_1, clone 2.06) immunoliposomes to cell lines expressing different surface levels of the human HLA-DR determinant of MHC-II have been previously determined by flow cytometry (Dufresne *et al.*, 1999). As expected, anti-HLA-DR immunoliposomes did not bind to SUPT-1 cells, a human $CD4^+$ T lymphoid cell line that does not express the HLA-DR determinant on their surface. In contrast, a very strong binding was observed following incubation of liposomes bearing anti-HLA-DR Fab' fragments with both the HUT-78, a human $CD4^+$ T lymphoid cell line, and RAJI cells, which bear important levels of human HLA-DR on their cellular membrane. The specificity of murine anti-HLA-DR (clone Y-17, IgG_{2b}) immunoliposomes for I-E antigens present on mouse spleen cells has also been confirmed by flow cytometry analysis. However, sterically stabilized liposomes and liposomes bearing an irrelevant nonspecific isotype matching Fab' fragments did not bind to mouse cells. Taken together, these results demonstrate that liposomes bearing human anti-HLA-DR Fab' fragments are very specific to cells expressing the HLA-DR determinant of MHC-II.

Tissue Distribution Studies

The ability of liposomes bearing murine anti-HLA-DR Fab' fragments (clone Y-17, IgG_{2b}) at the end termini of PEG chains to target and concentrate anti-HIV drugs into lymphoid tissues is evaluated following the subcutaneous injection of immunoliposomes in the upper back, below the neck, of female C3H mice. The subcutaneous route of injection is selected, since it constitutes an appropriate route for targeting lymph nodes (Allen *et al.*, 1993; Hawley *et al.*, 1995; Oussoren and Storm, 1997). In addition, in contrast to intravenous administrations of therapeutic agents, it represents a simple route for patient self-administration of liposomal drugs, and it might serve as a depot for the sustained release of drugs *in vivo*, most likely reducing the frequency of liposomal drug administration. Such an approach may also lead to the generation of subcutaneous pumps, which may be worth exploring in the near future for the treatment of HIV-infected individuals. At a specific time after the administration of the liposomal preparations, animals are sacrificed, and blood is collected and separated by centrifugation ($6000g$ for 10 min at $4°$). At the same time, selected tissues (liver, spleen, lung, thymus, and kidney, and cervical, brachial, mesenteric, gluteal, and popliteal lymph nodes) are collected, washed in PBS, and weighed. Afterward, tissues and plasma are treated with a Beckman tissue solubilizer (BTS-450, Beckman Instruments Inc., Irvine, CA) and decolored in H_2O_2. Lipid and drug levels in all samples are then monitored by scintillation counting.

The localization of liposomes (with or without drugs) within tissues is determined by fluorescence microscopy using liposomes labeled with the fluorochrome DiI (2.5 mg/ml in 100% ethanol; 3.75 μg DiI/mg lipid). DiI has low toxicity and is integrated into liposome membranes with high stability and does not leave cells after liposome uptake. In brief, liposomes are incubated with DiI (4 μg DiI/mg lipid) for 60 min at $50°$ with agitation. Unbound DiI is removed by centrifugation ($300g$ for 15 min at $4°$) of the liposomal preparation through a coarse Sephadex G-50 column. Liposomal formulations bearing nonspecific isotype-matching Fab' fragments are used as controls. For the tissue localization studies, a single bolus injection of liposomes is administered subcutaneously in the upper back, below the neck, of female C3H mice. At specific time intervals postinjection, animals are sacrificed, and selected tissues are removed. Tissues are washed in PBS, embedded in OCT Tissues Tek (Bayer, Pointe-Claire, QC, Canada), frozen in liquid nitrogen, and stored at $-20°$. Tissue sections (10 μm thick) are cut using a Jung Frigocut 2800E (Leica Canada Inc., St-Laurent, QC, Canada) and deposited on slides pretreated with 2% aminoalkysilane. Coated slides are immediately observed with a fluorescence microscope. DiI fluorescence is observed with a rhodamine optic excitation filter (ex, 510–560; dichroic mirror, DM 580; and barrier filter, BA 590).

Lymphoid Tissue Targeting of Immunoliposomes

One major advantage in using liposome formulations as a drug delivery system is that the distribution of liposomes can be modulated through variations of their size, lamellarity, lipid composition, charge, and surface-attached molecules such as hydrophilic groups and antibodies. It is thus possible to adapt the physicochemical properties of the liposomal form with the desired therapeutic objective. Oussoren and Storm (2001) recently reviewed the lymphatic absorption and lymph node uptake of liposomes after subcutaneous administration. The most important decisive factor influencing lymph node uptake of subcutaneously administered liposomes was shown to be the size of the phospholipid vesicles. Indeed, it was demonstrated that the degree of lymphatic absorption of liposomes having a diameter of less than 0.1 μm can reach levels up to 70% of the injected dose. In contrast, larger liposomes remained almost completely at the injection site. It was previously demonstrated that the passage of liposomes from the lymphatic vessel fenestration into the lymph nodes occurs via a maximum liposome size having a diameter less than 120 nm (Allen *et al.*, 1993). It was also shown that the presence of PEG on the surface of liposomes represents an important factor for the *in vivo* stability of the liposomal formulations after subcutaneous administrations. Coating the liposome surface with hydrophilic PEG molecules may reduce the nonspecific interactions of the phospholipid vesicles with the interstitial surroundings and inhibit the formation of large particles that could impede lymph node uptake (Oussoren and Storm, 2001). Furthermore, the increased hydrophilicity of liposomes may allow improved migration through the aqueous channels of the interstitium, improving lymphatic absorption.

After their subcutaneous administration, liposomes are known to be taken up by lymphatic capillaries draining the injection site or remain at the site of injection. Once the liposomes have passed the interstitium and entered the lymphatic capillaries, they enter the lymphatic system to be next captured in the regional lymph nodes. Studies on the intranodal fate of liposomes demonstrate that phagocytosis of the liposomal formulations by macrophages constitutes the most important mechanism by which liposomes are taken up by the lymph nodes (Oussoren and Storm, 2001). Similarly, morphological observations on the fate of liposomes in regional lymph nodes suggest that liposomes remain stable within the lymphatic circulation and that they eventually enter cells of the lymph nodes by phagocytosis to end up in the lysosomal apparatus (Velinova *et al.*, 1996). The role of macrophages in lymph node uptake is confirmed by the drastic reduction in the accumulation of liposomes in lymph nodes after the injection of chlodronate-containing liposomes to rodents in which lymph nodes are selectively depleted of macrophages (Oussoren and Storm, 2001). Despite the fact that PEG-coated liposomes are known to reduce the uptake by macrophages, a reduced lymph node accumulation of sterically stabilized liposomes was also observed upon depletion of macrophages chlodronate-containing liposomes. It was hypothesized that the reduced accumulation of PEG-coated liposomes within lymph nodes was due to the slow progression of liposomes through the lymph nodes that give enough time to establish an effective interaction of the sterically stabilized liposomes with phagocytes. In addition, the enhanced vascular permeability coefficient of the PEG-modified liposomes could allow a better penetration through capillaries.

We have previously evaluated the concentration of conventional liposomes and murine anti-HLA-DR immunoliposomes in cervical and brachial lymph nodes after a single subcutaneous administration given below the neck of female C3H mice. Results showed that liposomes bearing murine anti-HLA-DR Fab′ fragments targeted more efficiently the cervical lymph nodes when compared with that of conventional liposomes (Dufresne *et al.*, 1999). When compared with conventional liposomes, the subcutaneous administration of anti-HLA-DR immunoliposomes resulted in a 2.9 and 1.6 times greater accumulation in the cervical and brachial lymph nodes, respectively. These results showed that the presence of anti-HLA-DR Fab′ fragments on the surface of the phospholipid vesicles improved the targeting of regional lymph nodes as evidenced by the higher accumulation of anti-HLA-DR immunoliposomes in these tissues.

The coupling of hydrophilic groups, such as PEG, on the surface of liposome membranes is known to limit their interaction with a variety of blood components, thereby reducing the rapid clearance of phospholipid vesicles from the bloodstream. The reduced interactions with blood proteins prolong the circulation half-lives of these surface-modified liposomes when compared with conventional liposomes. Consequently, attachment of anti-HLA-DR Fab′ fragments to the end termini of PEG-coated liposomes (sterically stabilized immunoliposomes) could further improve their tissue accumulation when compared with PEG-free immunoliposomes. Our previous observations have demonstrated that sterically stabilized immunoliposomes increase by 2 to 4.6 times the concentration of liposomes in all tissues, with a peak of accumulation at 240 h in brachial, inguinal, and popliteal lymph nodes and at 360 h or greater in cervical lymph nodes (Bestman-Smith *et al.*, 2000). The concentration of immunoliposomal formulations was higher in brachial and cervical lymph nodes than in other tissues, confirming that the liposomes given subcutaneously accumulate preferentially in regional lymph nodes. Table I shows the area under the curve of conventional and sterically stabilized immunoliposomes in different tissues. PEG-coated immunoliposomes accumulate much better than conventional immunoliposomes in all tissues, indicating that the presence of the hydrophilic groups on the surface of immunoliposomes has an important effect on their uptake by the lymphatic system.

TABLE I

AREA UNDER THE CURVE OF STERICALLY STABILIZED AND CONVENTIONAL IMMUNOLIPOSOMES
IN DIFFERENT TISSUES AFTER THE ADMINISTRATION OF A SINGLE SUBCUTANEOUS DOSE
(2 μmol) IN C3H MICE[a]

Tissues	Sterically stabilized immunoliposomes	Conventional immunoliposomes	Ratio sterically stabilized immunoliposomes conventional immunoliposomes
Cervical lymph nodes	1514.7	616.1	2.5
Brachial lymph nodes	1693.7	874.7	1.9
Mesenteric lymph nodes	16.0	5.5	2.9
Inguinal lymph nodes	34.8	15.8	2.2
Popliteal lymph nodes	70.8	26.3	2.7
Liver	61.5	25.5	2.4
Spleen	57.4	12.6	4.6
Plasma	3.6	1.7	2.2

[a] Values, expressed in μmol lipids/g tissue or ml of plasma/h, were calculated from the mean values of the tissue distribution profile using the trapezoidal rule.

Lymphoid Tissue Targeting of Anti-HIV Drugs

The efficacy of antivirals is often limited by their poor ability to target and penetrate infected cells. Protease inhibitors entail several disadvantages, such as low bioavailability because of digestion by gut proteases and rapid metabolism, poor intracellular accumulation, and high binding to plasma proteins. Encapsulation of these anti-HIV agents into sterically stabilized immunoliposomes could represent a convenient strategy to protect these antiviral drugs against plasma protein binding, improving their pharmacokinetic parameters and enhancing their accumulation within HIV reservoirs. The delivery of a greater amount of antiviral agents in tissues susceptible to HIV infection and the reduced delivery of drugs at sites where they might be potentially toxic should result in an increased efficacy and reduced toxicity of anti-HIV agents.

Figure 2 shows the tissue distribution of indinavir, either as free or incorporated in sterically stabilized anti-HLA-DR immunoliposomes, after a single bolus injection given subcutaneously to C3H mice. Administration of free indinavir results in no significant drug levels in lymphoid organs (Fig. 2A). In fact, most of the injected drug accumulates in liver and was totally cleared within 24 h postadministration. In contrast, and of major importance, sterically stabilized immunoliposomes are very efficient in delivering high concentrations of indinavir to lymphoid tissues for at least

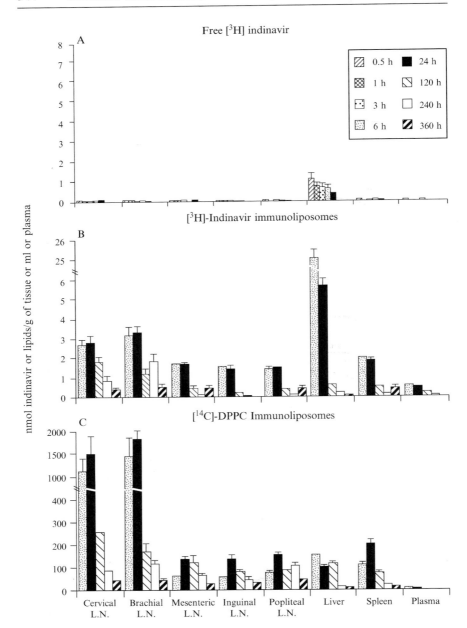

FIG. 2. Tissue and plasma distribution of free indinavir (A) and of sterically stabilized anti-HLA-DR immunoliposomes (B, C) containing indinavir as a function of time after a single bolus subcutaneous administration given in the upper back, below the neck, of C3H mice. Values represent the means (± SEM) obtained from six animals per group per time point. L.N., lymph nodes. From Gagné *et al.* (2002), with permission from Elsevier Science.

15 days postinjection (Fig. 2B). This enhanced drug accumulation is associated with the high specificity of the anti-HLA-DR Fab′ fragments for HLA-DR–expressing cells, since the presence of an irrelevant isotype-matching Fab′ fragment on sterically stabilized liposomes had no effect on the tissue distribution profile of liposomes, being similar to that of nontargeted sterically stabilized liposomes (Bestman-Smith *et al.*, 2000). A high drug accumulation is observed in the cervical and brachial lymph nodes. This could be explained by the fact that when injected subcutaneously, sterically stabilized immunoliposomes are drained by the lymph and pass through several lymph nodes before returning to the thoracic duct. Consequently, after their subcutaneous administration, liposomes will first accumulate within the cervical and brachial lymph nodes, which are the nearest lymphoid tissues from the injection site. Thereafter, if liposomes are not all retained by these tissues, they continue to migrate via the lymph to reach the next nearest lymph nodes. A high concentration of [^{14}C]-labeled lipids is also observed in the cervical and brachial lymph nodes at 6 and 24 h postadministration (Fig. 2C).

Table II shows the corresponding area under the curve of free and immunoliposomal indinavir in tissues. Results clearly demonstrate that the incorporation of indinavir into sterically stabilized anti-HLA-DR immunoliposomes greatly enhances the drug concentration in all tissues, leading to a 21- to 126-fold increased accumulation when compared with the free agent. A high concentration of immunoliposomal indinavir is also observed in the liver of animals, which is most likely due to the small diameter of immunoliposomes (100–120 nm), which allow them to extrav-

TABLE II
AREA UNDER THE CURVE FOR FREE AND IMMUNOLIPOSOMAL INDINAVIR IN TISSUES AFTER A
SINGLE SUBCUTANEOUS ADMINISTRATION IN MICE[a]

Tissues	Immunoliposomal indinavir	Free indinavir	Ratio immunoliposomal/ free indinavir
Cervical lymph nodes	523.2	7.6	68.8
Brachial lymph nodes	617.0	4.9	126.0
Mesenteric lymph nodes	192.8	6.4	30.1
Inguinal lymph nodes	144.5	4.1	35.2
Popliteal lymph nodes	134.2	4.5	29.8
Liver	733.3	35.0	21.0
Spleen	211.3	5.3	39.9
Plasma	77.8	2.3	33.8

[a] Values, expressed in nmol indinavir/g tissue or ml plasma/h, were calculated from the mean values of the tissue distribution profile using the trapezoidal rule.

asate from lymphatic vessels to reach the blood circulation and be taken up by liver macrophages (Allen et al., 1993).

It was previously reported that an HIV protease inhibitor encapsulated in negatively charged multilamellar liposomes was about 10-fold more effective and had a lower EC_{90} than the free drug in inhibiting HIV-1 production in human monocyte–derived macrophages (Düzgüneş et al., 1999; Pretzer et al., 1997). It was also reported that the drug encapsulated in sterically stabilized liposomes was as effective as the free drug. Similarly, we have demonstrated that anti-HLA-DR immunoliposomes containing indinavir were as efficient as the free agent in inhibiting HIV-1 replication in PM1 cells, a clonal derivative of HUT 78 cells, which express high levels of cell surface HLA-DR (Gagné et al., 2002). However, because immuno-liposomes allow efficient drug targeting of HIV reservoirs, the potential therapeutic advantages of immunoliposomes over the free agent for the treatment of HIV infection might be expected to be observed under in vivo situations. Taken together, these results suggest that liposomes may represent a convenient approach to improve the delivery of HIV protease inhibitors with low aqueous solubility and low oral bioavailability and for the targeting of these drugs to lymph nodes.

Lymphoid Tissue Localization of Immunoliposomes

Fluorescence microscopy studies have demonstrated that the presence of anti-HLA-DR Fab′ fragments at the end termini of PEG chains greatly modifies the localization of DiI-labeled PEG-coated liposomes in brachial lymph nodes. Figure 3 compares the localization of fluorescent sterically stabilized liposomes and sterically stabilized immunoliposomes in brachial lymph nodes at 120 h after their subcutaneous administration in mice. Results show that sterically stabilized liposomes are mainly localized in the subcapsular area, probably in the afferent lymphatic vessel and around the afferent area (Fig. 3A and B).

Oussoren et al. (1998) observed the presence of colloidal gold particles in macrophages of regional lymph nodes following the subcutaneous administration of PEG-gold–liposomes to mice. It was hypothesized that because of the slow progression of liposomes through the lymph nodes, enough time was available for efficient interactions of the PEG–liposomal surface with phagocytic cell membranes. In contrast, fluorescent micrographs of brachial lymph nodes show that sterically stabilized anti-HLA-DR immunoliposomes mostly accumulate in the cortex in which follicles (B cells and FDCs) are located and in parafollicular areas in which T cells, interdigitating dendritic cells, and other accessory cells are abundant (Fig. 3C and D). No significant fluorescent signal was observed in the

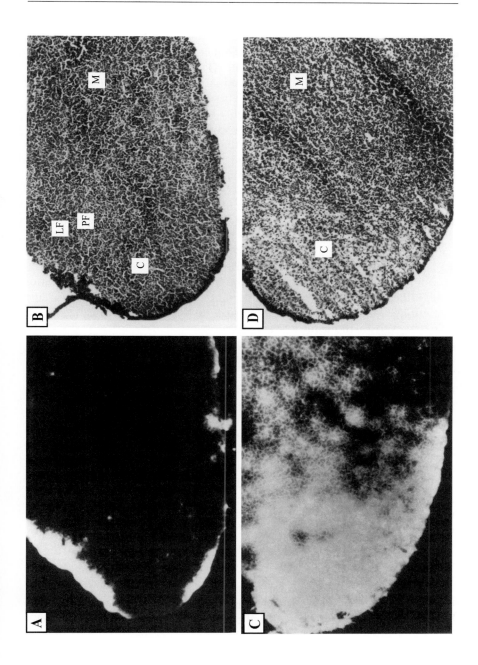

subcapsular area of the lymph nodes when nonspecific isotype-matching Fab' fragments were coupled to sterically stabilized liposomes. In HIV-1 infection, FDCs, which reside in germinal centers, display HIV-1 particles on their surface to selectively activate B cells. $CD4^+$ T cells from the paracortical area migrate to the germinal centers in response to B-cell activation to be eventually infected. These results clearly show that sterically stabilized immunoliposomes highly accumulate in these regions and could represent a convenient strategy to deliver drugs in this area of the lymph nodes, inhibiting more efficiently HIV-1 replication.

The localization of both liposomal formulations has also been determined in the spleen at 48 h after their subcutaneous administration in mice (Bestman-Smith et al., 2000). Results indicate that liposomes bearing PEG hydrophilic groups accumulate in the red pulp and the marginal zone of the white pulp of the spleen. In contrast, sterically stabilized immunoliposomes are concentrated in the follicle of the white pulp and little in the marginal zone. The administration of free DiI does not result in any fluorescent signal in all tissues studied. Furthermore, the injection of DiI-labeled liposomes show a disappearance of the fluorescent signal as a function of time, indicating that the probe is eliminated with time. When nonspecific immunoliposomes are injected into mice, only a weak fluorescence signal was observed in the same area as that observed for sterically stabilized liposomes, suggesting that the presence of irrelevant antibody had no effect on the tissue localization of liposomes.

As the microenvironment of lymphoid tissues is crucial for an effective immune response, it is necessary to develop strategies to target lymphoid tissues and $CD4^+$ T cells to reduce the process of HIV dissemination. A key factor to reduce viral burden in HIV-infected individuals resides in the use of a more selective drug delivery system. The entrapment of antiviral drugs into liposomes represents a convenient strategy to enhance drug accumulation within specific cells targeted by HIV, thereby increasing their efficacy and reducing their toxicity. Such a targeted drug delivery system is expected to reduce the HIV load in lymphoid tissues and preserve the follicular dendritic cell microenvironment that will likely protect the infected host from developing the characteristic immunodeficient state. The use of liposomes may lead to the development of new strategies that

FIG. 3. Fluorescent micrographs of brachial lymph nodes of C3H mouse at 120 h following the administration of a single subcutaneous dose of DiI sterically stabilized liposomes (A) and sterically stabilized immunoliposomes (C). (B, D) The corresponding hematoxylin and eosin coloration of tissues. The cortex area (C), the parafollicular area (PF), the medulla (M), and the lymphoid follicules (LF) are shown. Magnification: 250×. From Bestman-Smith et al. (2000), with permission from Elsevier Science.

may improve the efficacy and safety of drugs used for the treatment of HIV/AIDS (Düzgüneş, 1998).

References

Allen, T. M. (1994). The use of glycolipids and hydrophilic polymers in avoiding rapid uptake of liposomes by the mononuclear phagocyte system. *Adv. Drug Deliv. Rev.* **13**, 285–309.

Allen, T. M., Hansen, C. B., and Guo, L. S. S. (1993). Subcutaneous administration of liposomes: A comparison with the intravenous and intraperitoneal routes of injection. *Biochim. Biophys. Acta* **1150**, 9–16.

Bestman-Smith, J., Gourde, P., Désormeaux, A., Tremblay, M. J., and Bergeron, M. G. (2000). Sterically stabilized liposomes bearing anti-HLA-DR antibodies for targeting the primary cellular reservoirs of HIV-1. *Biochim. Biophys. Acta* **1468**, 161–174.

Chun, T. W., Engel, D., Berrey, M. M., Shea, T., Corey, L., and Fauci, A. S. (1998). Early establishment of a pool of latently infected, resting CD4(+) T cells during primary HIV-1 infection. *Proc. Natl. Acad. Sci. USA* **95**, 8869–8873.

Désormeaux, A., and Bergeron, M. G. (1998). Liposomes as drug delivery system: A strategic approach for the treatment of HIV infection. *J. Drug Targeting* **6**, 1–15.

Désormeaux, A., Harvie, P., Perron, S., Makabi-Panzu, B., Beauchamp, D., Tremblay, M., Poulin, L., and Bergeron, M. G. (1994). Antiviral efficacy, intracellular uptake and tissue distribution of liposome-encapsulated 2′,3′-dideoxyinosine (ddI) in rats. *AIDS* **8**, 1545–1553.

Dufresne, I., Désormeaux, A., Bestman-Smith, J., Gourde, P., Tremblay, M. J., and Bergeron, M. G. (1999). Targeting lymph nodes with liposomes bearing anti-HLA-DR Fab′ fragments. *Biochim. Biophys. Acta* **1421**, 284–294.

Dusserre, N., Lessard, C., Paquette, N., Perron, S., Poulin, L., Tremblay, M., Beauchamp, D., Désormeaux, A., and Bergeron, M. G. (1995). Encapsulation of foscarnet in liposomes modifies drug intracellular accumulation, *in vitro* anti-HIV-1 activity, tissue distribution, and pharmacokinetics. *AIDS* **9**, 833–841.

Düzgüneş, N. (1998). Treatment of human immunodeficiency virus, *Mycobacterium avium*, and *Mycobacterium tuberculosis* infections by liposome-encapsulated drugs. *In* "Medical Applications of Liposomes" (D. D. Lasic and D. Papahadjopoulos, eds.), pp. 189–219. Elsevier Sciences B. V., Amsterdam.

Düzgüneş, N., Pretzer, E., Simoes, S., Slepushkin, V., Konopka, K., Flasher, D., and Pedroso de Lima, M. C. (1999). Liposome-mediated delivery of antiviral agents to human immunodeficiency virus-infected cells. *Mol. Memb. Biol.* **16**, 111–118.

Embretson, J., Zupancic, M., Ribas, J. L., Burke, A., Racz, P., Tenner-Racz, K., and Haase, A. (1993). Massive covert infection of helper T lymphocytes and macrophages by HIV during the incubation period of AIDS. *Nature* **362**, 359–362.

Fauci, A. S. (1993). Multifactorial nature of human immunodeficiency virus disease: Implications for therapy. *Science* **262**, 1011–1017.

Finzi, D., Blankson, J., Siliciano, J. D., Margolick, J. B., Chadwick, K., Pierson, T., Smith, K., Lisziewicz, J., Lori, F., Flexner, C., Quinn, T. C., Chaisson, R. E., Rosenberg, E., Walker, B., Gange, S., Gallant, J., and Siliciano, R. F. (1999). Latent infection of CD4+ T cells provides a mechanism for lifelong persistence of HIV-1, even in patients on effective combination therapy. *Nature Med.* **5**, 609–611.

Finzi, D., Hermandova, M., Pierson, T., Carruth, L. M., Buck, C., Chaisson, R. E., Quinn, T. C., Chadwick, K. M. J., Brookmeyer, R., Gallant, J., Markowitz, M., Ho, D. D., Richman, D. D., and Siliciano, R. F. (1997). Identification of a reservoir for HIV-1 in patients on highly active antiretroviral therapy. *Science* **278**, 1295–1300.

Fox, C. H., Tenner-Rácz, K., Rácz, P., Firpo, A., Pizzo, P. A., and Fauci, A. S. (1991). Lymphoid germinal centers are reservoirs of human immunodeficiency virus type 1 RNA. *J. Infect. Dis.* **164,** 1051–1057.

Furtado, M. R., Callaway, D. S., Phair, J. P., Kunstman, K. J., Stanton, J. L., Macken, C. A., Perelson, A. S., and Wolinsky, S. M. (1999). Persistence of HIV-1 transcription in peripheral-blood mononuclear cells in patients receiving potent antiretroviral therapy. *N. Engl. J. Med.* **340,** 1614–1622.

Gagné, J. F., Désormeaux, A., Perron, S., Tremblay, M. J., and Bergeron, M. G. (2002). Targeted delivery of indinavir to HIV-1 primary reservoirs with immunoliposomes. *Biochim. Biophys. Acta* **1558,** 198–210.

Harding, A. J., Engbers, C. M., Newman, M. S., Goldstein, N. I., and Zalipsky, S. (1997). Immunogenicity and pharmacokinetic attributes of poly(ethyleneglycol)-grafted immunoliposomes. *Biochim. Biophys. Acta* **1327,** 181–192.

Harvie, P., Désormeaux, A., Bergeron, M. C., Tremblay, M., Beauchamp, D., Poulin, L., and Bergeron, M. G. (1996). Comparative pharmacokinetics, distribution in tissues, and interactions with blood proteins of conventional and sterically stabilized liposomes containing 2′,3′-dideoxyinosine. *Antimicrob. Agents Chemother* **40,** 225–229.

Harvie, P., Désormeaux, A., Gagné, N., Tremblay, M., Poulin, L., Beauchamp, D., and Bergeron, M. G. (1995). Lymphoid tissues targeting of liposome-encapsulated 2′,3′-dideoxyinosine (ddI). *AIDS* **9,** 701–707.

Hawley, A. E., Davis, S. S., and Illum, L. (1995). Targeting of colloids to lymph nodes: Influence of lymphatic physiology and colloidal characteristics. *Adv. Drug Deliv. Rev.* **17,** 129–148.

Heath, S. L., Tew, J. G., Szakal, A. K., and Burton, G. F. (1995). Follicular dendritic cells and human immunodeficiency virus infectivity. *Nature* **377,** 740–744.

Oussoren, C., and Storm, G. (1997). Subcutaneous administration of liposomes for lymphatic targeting. *J. Lipos. Res.* **7,** 227–240.

Oussoren, C., and Storm, G. (2001). Liposomes to target the lymphatics by subcutaneous administration. *Adv. Drug Deliv. Rev.* **50,** 143–156.

Oussoren, C., Velinova, M., Scherphof, G., van der Want, J. J., van Rooijen, N., and Storm, G. (1998). Lymphatic uptake and biodistribution of liposomes after subcutaneous injection. IV. Fate of liposomes in regional lymph nodes. *Biochim. Biophys. Acta* **1370,** 259–272.

Pantaleo, G., Graziosi, C., Demarest, J. F., Butini, L., Montoni, M., Fox, C. H., Orenstein, J. M., Kotler, D. P., and Fauci, A. S. (1993). HIV infection is active and progressive in lymphoid tissue during the clinically latent stage of disease. *Nature* **362,** 355–358.

Pretzer, E., Flasher, D., and Düzgüneş, N. (1997). Inhibition of human immunodeficiency virus type-1 replication in macrophages and H9 cells by free or liposome-encapsulated L-689,502, an inhibitor of the viral protease. *Antiviral Res.* **34,** 1–15.

Saarloos, M. N., Sullivan, B. L., Czerniewski, M. A., Parameswar, K. D., and Spear, G. T. (1997). Detection of HLA-DR associated with monocytropic, primary, and plasma isolates of human immunodeficiency virus type 1. *J. Virol.* **71,** 1640–1643.

Schrager, L. K., and D'Souza, M. P. (1998). Cellular and anatomical reservoirs of HIV-1 in patients receiving potent antiretroviral combination therapy. *JAMA* **280,** 61–71.

Schrager, L. K., and Fauci, A. S. (1995). Trapped but still dangerous. *Nature* **377,** 680–681.

Tremblay, M. J., Fortin, J. F., and Cantin, R. (1998). The acquisition of host-encoded proteins by nascent HIV-1. *Immunol. Today* **19,** 346–351.

Velinova, M., Read, N., Kirby, C., and Gregoriadis, G. (1996). Morphological observations on the fate of liposomes in the regional lymph nodes after footpad injection into rats. *Biochim. Biophys. Acta* **1299,** 207–215.

Wong, J. K., Gunthard, H. F., Havlir, D. V., Zhang, Z. Q., Haase, A. T., Ignacio, C. C., Kwok, S., Emini, E., and Richman, D. D. (1997). Reduction of HIV-1 in blood and lymph nodes following potent antiretroviral therapy and the virologic correlates of treatment failure. *Proc. Natl. Acad. Sci. USA* **94,** 12574–12579.

Woodle, M. C., and Lasic, D. D. (1992). Sterically stabilized liposomes. *Biochim. Biophys. Acta* **1113,** 171–199.

Yamagushi, Y., Kim, H., Kato, K., Masuda, K., Shimada, I., and Arata, Y. (1995). Proteolytic fragmentation with high specificity of mouse immunoglobulin G. *J. Immun. Methods* **181,** 259–267.

Zhang, L., Ramratnam, B., Tenner-Racz, K., He, Y., Vesanen, M., Lewin, S., Talal, A., Racz, P., Perelson, A. S., Korber, B. T., Markowitz, M., and Ho, D. D. (1999). Quantifying residual HIV-1 replication in patients receiving combination antiretroviral therapy. *N. Engl. J. Med.* **340,** 1605–1613.

[20]   Delivery of Antiviral Agents in Liposomes

By Nejat Düzgüneş, Sergio Simões, Vladimir Slepushkin, Elizabeth Pretzer, Diana Flasher, Isam I. Salem, Gerhard Steffan, Krystyna Konopka, and Maria C. Pedroso de Lima

Abstract

The intracellular activity of certain antiviral agents, including antisense oligonucleotides, acyclic nucleoside phosphonates, and protease inhibitors, is enhanced when they are delivered in liposome-encapsulated form. In this chapter we describe the preparation of pH-sensitive liposomes encapsulating antisense oligonucleotides, ribozymes, and acyclic nucleoside phosphonate analogues and their effects on HIV replication in macrophages. We outline the use of liposomal HIV protease inhibitors in infected macrophages. We present two methods for the covalent coupling of soluble CD4 to liposomes and show the association of these liposomes with HIV-infected cells. We also describe the synthesis of a novel antiviral agent based on cyclodextrin and its incorporation into liposomes.

Introduction

The treatment of human immunodeficiency virus (HIV) infection presents a global health challenge. Drugs that inhibit the viral reverse transcriptase and protease and one fusion inhibitor constitute the current armamentarium against HIV. These drugs have been effective in many cases in reducing the viral load to undetectable levels, particularly when used in combination. The high mutation rate of HIV, however, has enabled the virus to develop mutants that are resistant to these drugs. Noncompliance due to the toxic side effects of these drugs is an additional factor that facilitates the emergence of drug-resistant strains. New inhibitors against drug-resistant viral enzymes and alternative viral targets need to be developed continuously, a process that is likely to take long periods of time. As an alternative, it may be relatively straightforward to identify viral gene sequences that can serve as targets for antisense oligonucleotides (ODN), ribozymes, or short interfering RNAs (Akhtar and Rossi, 1996; Düzgüneş et al., 2001a,b; Dykxhoorn et al., 2003; Stephens and Rivers, 2003; Stevenson, 2003; Wagner and Flanagan, 1997). This appears to be a simpler approach to the development of new drugs against multidrug-resistant viral strains. The problem of the emergence of any viral strains resistant to these drugs may be addressed by the introduction of corresponding changes in the sequences of the drugs (Düzgüneş, 1998; Düzgüneş et al., 1999). Furthermore, the generation of mutants resistant to antisense oligonucleotides is likely to be slower than the emergence of those resistant to viral enzymes, since a single mutation may not be sufficient for escape from antisense inhibition. This would be due to the complementarity of a large number of base pairs between the oligonucleotide and its target sequence (Lisziewicz et al., 1994), while a single mutation can confer resistance to reverse transcriptase or protease inhibitors. Despite the potential advantages of antisense ODN and ribozymes, the intracellular delivery of these macromolecular drugs in vivo presents an important challenge (Akhtar and Rossi, 1996; Shi and Hoekstra, 2004; Wagner and Flanagan, 1997). These agents are most likely taken up by receptor-mediated endocytosis, a relatively inefficient route requiring high concentrations of the drugs to be active. Although the complexation of antisense ODN or ribozymes with cationic liposomes can enhance their intracellular delivery (Bennett et al., 1992; Konopka et al., 1998), intravenous injection of the complexes can result in their rapid uptake by the liver, making delivery to other organs problematic (Litzinger et al., 1996).

Liposomes that destabilize and deliver their contents into the cytoplasm at the mildly acidic pH achieved in endosomes (Straubinger et al., 1985; Düzgüneş et al., 1991; Torchilin et al., 1993; Simões et al., 2001) have been utilized as carriers for the intracellular delivery of oligonucleotides (Düzgüneş et al., 1995, 2001a; Ropert et al., 1992). Although first-generation pH-sensitive liposomes are cleared rapidly from the bloodstream following intravenous administration, the inclusion of a low mole fraction of poly(ethylene glycol)-phosphatidylethanolamine (PEG-PE) in the membrane of these liposomes results in prolonged circulation, without compromising their ability to deliver charged fluorescent molecules into macrophage-like cells (Slepushkin et al., 1997, 2004). In this chapter, we describe the use of antisense oligonucleotides and ribozymes encapsulated in pH-sensitive liposomes to inhibit virus production in HIV-infected macrophages derived from human peripheral blood monocytes. We also describe the delivery in pH-sensitive liposomes of acyclic nucleoside phosphonate analogues (Balzarini et al., 1991) that are taken up by cells relatively slowly via an endocytosis-like process (De Clercq, 1995, 2003). We outline the use of liposomes for enhancing the therapeutic effect of HIV protease inhibitors and methods to target liposomes to HIV-infected cells. Finally we describe the synthesis of a novel antiviral agent based on cyclodextrin and its incorporation into liposomes.

Antiviral Drugs in pH-Sensitive Liposomes

Macromolecular Antiviral Drugs

pH-sensitive liposomes are composed of cholesteryl hemisuccinate (CHEMS; Sigma-Aldrich, St. Louis, MO) and dioleoylphosphatidylethanolamine (DOPE; Avanti Polar Lipids, Alabaster, AL) at a molar ratio of 4:6. Sterically stabilized pH-sensitive liposomes are composed of CHEMS, DOPE, and poly(ethylene glycol) (2000)-distearoylphosphatidylethanolamine (PEG-PE; originally obtained from Sequus Pharmaceuticals, now ALZA, Menlo Park; now available from Avanti Polar Lipids) at a molar ratio of 4:6:0.3). Control, non-pH-sensitive liposomes are composed of phosphatidylglycerol from egg (PG) and DOPE at a molar ratio of 4:6 and are prepared by reverse-phase evaporation (Düzgüneş, 2003; Düzgüneş et al., 1983; Szoka and Papahadjopoulos, 1978). Lipid mixtures dried from chloroform are dissolved in 780 μl of diethyl ether in a high-quality screw-cap tube with a Teflon-lined cap. The diethyl ether is prewashed with water by vigorous shaking to enable any peroxidation products to dissolve in the water phase; the upper organic phase is then used to dissolve lipids. An ODN or ribozyme solution is prepared at a concentration of 0.2 mM in 100 mM HEPES buffer, pH 7.5, made isotonic with NaCl. An aliquot of this preparation (260 μl) is added to the lipid dissolved in diethyl ether, and the tube is sealed with Teflon tape, followed by the screw-cap, under a stream of argon. The mixture is sonicated briefly in a bath-type sonicator (Laboratory Supply Co., Hicksville, NY) to form a stable emulsion. The ether is removed at 32–34° under controlled vacuum in a rotary evaporator (Büchi, Flawil, Switzerland) to avoid excessive bubbling. After a gel is formed, an additional 240 μl of ODN solution is added, and the gel is broken by vortexing. Evaporation is continued for 30 min to remove any residual ether. HEPES buffer without ODNs is used to prepare control "empty" liposomes. Liposomes are extruded 21 times through two polycarbonate filters of 100-nm pore diameter, using a LiposoFast device (Avestin, Inc., Ottawa, Canada), to obtain a uniform size distribution.

Unencapsulated ODN is removed, and the external buffer exchanged by different procedures: (1) Liposomes are subjected to dialysis, using Spectra/Por (Spectrum, Houston, TX) dialysis bags (50,000 Da cutoff), against isotonic saline containing 10 mM HEPES buffer, pH 7.4 (with two changes of 4 liters of buffer, 20 h dialysis each, at 4°). (2) Liposomes are subjected to size-exclusion chromatography using a Sepharose CL-4B column (1 × 20 cm) equilibrated in 25 mM HEPES and 140 mM NaCl (pH 7.4). (3) Liposomes are ultracentrifuged (three times, 150,000g, 40 min each, 4°). Liposomes are sterilized by filtration through 0.45-μm syringe filters (MSI, Westboro, MA).

An anti-REV-responsive element (RRE) 15-mer phosphorothioate ODN (5′-F-GTGCTTCCTGCTGCOT-3′) and a 15-mer ODN of nonspecific sequence (5′-F-CCTATCAGGCAGTAOT-3′), where F and O represent fluorescein and biotin, respectively, are synthesized, purified by high-performance liquid chromatography (HPLC), and provided by Lynx Therapeutics, Inc. (Hayward, CA). A 38-mer, 5′ fluorescein-labeled, chimeric DNA–RNA hammerhead ribozyme (5′-ACACAACAcugau-gaGTCCGTGAGGACgaaa-CGGGC*A*C-3′, where an asterisk indicates a phosphorothioate linkage, capital letters are deoxyribonucleotides, and lower case letters are ribonucleotides) targeted to the HIV-1 5′ LTR is synthesized using an automated Applied Biosystems 394 RNA/DNA instrument (courtesy of P. Swiderski, City of Hope, Duarte, CA). A modified ribozyme lacking catalytic activity (5′-CAAACAACcugaugaGTCCGT-GAGGA-CgaaaACCGG*G*C-3′) is used as a control.

The amount of ODN or ribozyme associated with liposomes after removal of the unencapsulated material is assessed by either of two methods: (1) If the ODNs or ribozymes are labeled with fluorescein, the fluorescence of an aliquot of the liposome suspension is measured. This is then converted to concentration by using a standard curve obtained with free, labeled ODN. Small volumes (up to 5 μl) of the liposome suspension are added to 1.5–2 ml of HEPES-buffered saline containing a detergent ($C_{12}E_8$, final concentration 1 mg/ml), and the fluorescence is measured at 20° using an excitation wavelength of 490 nm and an emission wavelength of 520 nm. (2) If the ODNs or ribozymes are not labeled, the optical density is measured at 260 nm after addition of a detergent ($C_{12}E_8$, final concentration 1 mg/ml), using the following equation:

$$[\text{ODN}](\mu g/ml) = A_{260} \times \text{Extinction coefficient} (\mu g/ml) \times \text{Dilution factor}$$

The lipid–phosphate concentration is measured according to the Bartlett (1959) colorimetric assay (see also Düzgüneş, 2003, for a protocol).

Encapsulation efficiency (E.E., %) is defined by the ratio of the drug (D)/lipid (L) concentrations after (D_f/L_f), and before (D_i/L_i), removal of the unencapsulated material:

$$E.E.(\%) = (D_f/L_f)/(D_i/L_i) \times 100$$

This parameter eliminates the effect of lipid loss during the liposome preparation steps.

In addition, the dose of ODN or ribozyme that is loaded into a certain amount of lipid (loading capacity, L. C.) can be calculated according to:

$$L.C.(\mu g/\mu mol) = \text{Mass of encapsulated ODN/Amount of lipid}$$

Nucleoside Phosphonate Antivirals

The acyclic nucleoside phosphonate analogues 9-(2-phosphonylmethoxyethyl)adenine (PMEA) and 9-(2-phosphonylmethoxypropyl)adenine (PMPA) are encapsulated in pH-sensitive liposomes in a similar manner. PMEA is dissolved at 5.4 mg/ml in 100 mM HEPES, pH 7.8, made isotonic with NaCl, and PMPA is dissolved at 5.8 mg/ml in the same buffer. The amount of encapsulated PMEA or PMPA is determined by fluorescence after lysing the liposomes with detergent and reacting the drugs with chloracetaldehyde (Naesens et al., 1992).

Isolation, HIV-1 Infection, and Treatment of Human Macrophages

Human macrophages are isolated from buffy coats seronegative for HIV, hepatitis C, and hepatitis B by centrifugation on Ficoll–Hypaque (Histopaque 1077, Sigma, Columbia, MD) and adherence to plastic, as described previously (Pretzer et al., 1997). HIV-1$_{BaL}$ is purchased from Advanced Biotechnologies (Rockville, MD) and propagated in macrophages (Pretzer et al., 1997). The number of cells remaining at the end of the experiment is estimated by counting the nuclei after staining the cells with Naphthol Blue Black (Nakagawara and Nathan, 1993). TSQ1 cell viability is ascertained by a modified Alamar Blue assay (Konopka et al., 1996). Macrophages are plated in 48-well plates and infected with HIV-1$_{BaL}$ at a multiplicity of infection of 0.1 on day 7 after isolation, by incubation with 140–165 μl of virus-containing culture medium for 2 h at 37°. The cells are then washed with fresh Dulbecco's modified Eagle's medium-high glucose (DME-HG) (Irvine Scientific, Santa Ana, CA) medium with 20% heat-inactivated fetal bovine serum (FBS, Sigma), and incubation at 37° is continued. Various dilutions of ODN or ribozyme preparations are made in the same culture medium and added to the cells 24 h after infection. Fresh dilutions of treatments are added with two medium replacements until day 8 following infection. On this day, fresh medium without treatments is added to the cells, and supernatants are saved for subsequent analysis. Viral p24 levels are monitored in cell culture supernatants collected every 2–3 days, using an enzyme-linked immunosorbent assay (ELISA), as described (Konopka et al., 1990; Pretzer et al., 1997). ELISA kits obtained from Coulter (Miami, FL) or the AIDS Vaccine Program (NCL-Fredrick Cancer Research and Development Center, Fredrick, MD) are also utilized in some experiments. Supernatants of uninfected cells are used as controls for the p24 determinations by ELISA.

Inhibition of HIV Replication in Macrophages by Antisense Oligonucleotides, Ribozymes, and Nucleoside Phosphonate Analogues Encapsulated in pH-sensitive Liposomes

While the free ODN up to 3 μM is inactive against HIV infection in macrophages, the ODN encapsulated in pH-sensitive CHEMS/DOPE liposomes inhibits p24 production by 42% and 91% at 1 and 3 μM, respectively (Düzgüneş et al., 2001a; Slepushkin et al., 2004). It is interesting to note that the ODN is also effective when delivered in sterically stabilized pH-sensitive CHEMS/DOPE/PEG-PE liposomes. This observation should be contrasted with the highly inhibitory effect of PEGylated lipids in cationic lipid-mediated delivery of ODN (Shi et al., 2002; Song et al., 2002). The ODN encapsulated in conventional, non-pH-sensitive liposomes is not effective. Although the nonspecific ODN encapsulated in pH-sensitive liposomes has no effect at 1 μM, it inhibits HIV infection by 53% at 3 μM. Nonspecific inhibition of HIV infection of lymphocyte cell lines by phosphorothioate ODN has been reported (Lisziewicz et al., 1992; Weichold et al., 1995; Zelphati et al., 1994). The effect may be due to the inhibition of virus binding to the CD4 receptor (Stein et al., 1991) or the nonspecific inhibition of viral reverse transcriptase (Zelphati et al., 1994). Since the treatments in our experiments are started 24 h after the initial infection step, and since the ODN are encapsulated, virus entry is most likely not inhibited. Control, buffer-loaded, pH-sensitive and sterically stabilized pH-sensitive liposomes affect HIV infection to some extent, but their effect is usually inconsistent from experiment to experiment.

Delivery of the 38-mer chimeric ribozyme complementary to HIV 5'-LTR in pH-sensitive liposomes inhibits HIV production in macrophages by 88%, while the free ribozyme causes decrease of only 10% (Fig. 1) (Düzgüneş et al., 2001a). Virus production is inhibited by 73% by a non-catalytic, modified ribozyme delivered in pH-sensitive liposomes. This inhibition is probably the result of the antisense-like action of the ribozyme rather than the cleavage of the target mRNA sequence.

The acyclic nucleoside phosphonates PMEA (adefovir) and PMPA (tenofovir) (Balzarini et al., 1991; De Clercq, 1995, 2003) inhibit reverse

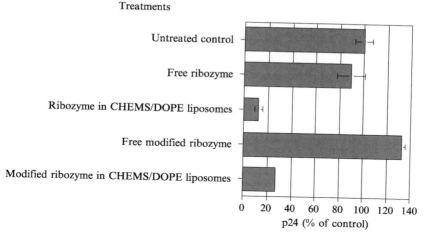

Fig. 1. The effect of an anti-HIV 5′-LTR 38-mer chimeric (RNA–DNA) hammerhead ribozyme encapsulated in pH-sensitive liposomes on HIV-1 production in human macrophages. The modified ribozyme lacks catalytic activity, but possibly acts through an antisense mechanism. The data are expressed as the percentage of viral p24 produced by untreated macrophages. Reproduced from Düzgüneş *et al.* (2001a), with permission.

transcription. The intracellular phosphorylation of these compounds is not a limiting step for their antiviral activity, unlike nucleoside analogues. The disadvantages of these antiviral drugs are their slow cellular uptake and their poor oral bioavailability. When PMEA and PMPA are delivered to HIV-infected macrophages in pH-sensitive liposomes, the antiviral effect of the drugs is enhanced (Fig. 2) (Düzgüneş *et al.*, 2001a). The encapsulated drugs are more effective in inhibiting HIV production by macrophages throughout the concentration range tested. The EC_{50} of the liposome-encapsulated PMEA is about 10-fold lower than that of the free antiviral.

Protease Inhibitors in Conventional Liposomes

Processing of the HIV-1 Gag–Pol precursor polyprotein by the viral protease is essential for the production of infectious virions by host cells (Kay and Dunn, 1990; Robins and Plattner, 1993). Numerous inhibitors with high specificity for the viral protease over cellular proteases have been developed and are in clinical use. Since macrophages are a major reservoir of HIV-1 in infected individuals (Aquaro *et al.*, 2002; Meltzer *et al.*, 1990), effective delivery of protease inhibitors to these cells is likely to facilitate the reduction of the viral burden. Liposomes are targeted naturally to

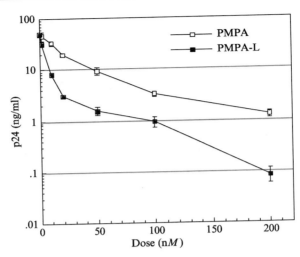

Fig. 2. Inhibition of virus production in HIV-1$_{BaL}$–infected macrophages by free or liposome-encapsulated (R)-9-(2-phosphonylmethoxypropyl)adenine (PMPA). The macrophages were infected with HIV-1$_{BaL}$ for 2 h, and after washing, further incubated for 24 h. They were then treated for 8 days, and viral p24 in culture supernatants was determined on day 15 after infection. Data from Düzgüneş et al. (2001a).

macrophages and may be an effective carrier for protease inhibitors. Since these drugs are highly hydrophobic, they partition readily in the membrane phase of liposomes. Since some protease inhibitors can associate with blood proteins and become unavailable, liposomal targeting to infected macrophages may present advantages over the free drug. Liposome incorporation also provides a means of delivery for protease inhibitors that are too hydrophobic to be amenable for oral administration.

Control multilamellar vesicles (MLV) are prepared at a 1:1:1 molar ratio of PC/PG/chol. MLV containing the protease inhibitor L-689,502 (PI) (Thompson et al., 1992) are prepared at a molar ratio of 0.4:1:1:1 (PI/PC/PG/chol). Briefly, chloroform solutions of the lipids are mixed, with or without PI, and evaporated dry under vacuum in a rotary evaporator. The dried mixture is hydrated by vortexing in argon-saturated buffer (140 mM NaCl, 10 mM KCl, 10 mM HEPES, pH 7.4, 290–300 mOsm). Liposomes are prepared under sterile conditions, within a week before the start of each experiment. Following centrifugation, 96% of the PI copellets with the liposomes. Phase-contrast microscopy indicates that the liposomes are greater than 3 μm in diameter and have a heterogeneous size distribution.

To evaluate the antiviral effect of liposomal protease inhibitors, macrophages in 48-well plates (7–15 days after isolation) are infected with 5 ng p24/well of HIV-1$_{BaL}$, and in 96-well plates with 1.4 ng/well, in a volume of medium sufficient to cover the cells (100–200 μl/well in 48-well plates, 28 μl in 96-well plates). The cells and virus are incubated 2–2.5 h at 37°, then washed three times with medium, and incubated in DME-HG/20% FBS with or without treatments, with medium replacements three times a week (Pretzer et al., 1997). For overnight treatment, medium is removed on the following day, the cells are washed three times, fresh medium is placed in the wells, and the cells are cultured without further treatment. For continuous treatment, the medium replacement on the following day is omitted, and fresh dilutions of treatments are added with each medium replacement. For treatment of cells with free PI as a control, serial dilutions are made in dimethly sulfoxide (DMSO), and the final dilution into the medium results in a concentration of 0.4% DMSO in the well. In addition to the untreated control, each experiment includes a control containing 0.4% DMSO and empty liposomes at the same concentration of lipid as that of the highest PI-liposome concentration. Macrophages treated overnight with PI-MLV produce very low levels of viral p24 (at or below 0.1 ng/ml), an order of magnitude below that produced by the cells treated with free PI (Fig. 3) (Pretzer et al., 1997).

Liposome Targeting to HIV-Infected Cells

Cells infected with HIV-1 express the viral envelope glycoprotein gp120/gp41 on their surface when they are producing new virions. The transmembrane glycoprotein CD4, expressed mainly on the surface of helper T lymphocytes and monocyte/macrophages, is the primary high-affinity receptor for HIV-1. The recombinant ectodomain of CD4 (soluble CD4 or sCD4), which retains high affinity binding to gp120, can block infection and syncytium formation by laboratory isolates of HIV-1. Although toxins have been coupled to sCD4 to kill virally infected cells, the potential nonspecific toxicity of such constructs in vivo may limit their eventual use. An alternative approach is to couple sCD4 to liposomes as a targeting ligand (Flasher et al., 1994). A potential advantage of targeted liposomes over toxin conjugates is that they can encapsulate a diverse array of molecules, including cytotoxic drugs. Certain antiviral drugs that have limited ability to reach infected tissues, or that have limited access to the cytoplasm of infected cells, may be delivered via certain types of liposomes, such as sterically stabilized pH-sensitive liposomes.

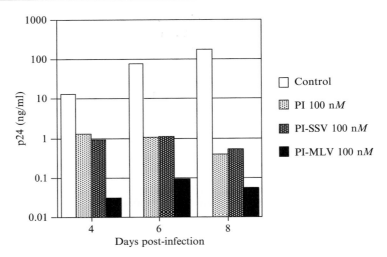

Fɪɢ. 3. Effect of free and liposome-encapsulated L-689,502 on virus production by macrophages infected with HIV-1$_{BaL}$. The protease inhibitor was added in free form (dissolved in DMSO and diluted) or encapsulated in either multilamellar (PI-MLV) or sterically stabilized (PI-SSV) vesicles. Data from Pretzer *et al.* (1997).

sCD4 Coupling to Liposomes with N-Succinimidylacetylthioacetate (SATA)

Soluble CD4, obtained from Genentech (South San Francisco, CA), is coupled to liposomes using a modification of the method of Martin and Papahadjopoulos (1982). The protein is thiolated using *N*-succinimidylace-tylthioacetate (SATA; Pierce, Rockford, IL) (Derksen and Scherphof, 1985; Duncan *et al.*, 1983) and coupled to liposomes containing the thiol-reactive lipid 4-(*p*-maleimidophenyl) butyryl-phosphatidylethanolamine (MPB-PE). A 54.1 m*M* solution of SATA in dimethyl formamide (DMF, Pierce) is prepared immediately before use and added to 6 mg/ml sCD4 (in 20 m*M* HEPES, 188 m*M* NaCl, pH 7.0) at room temperature, under argon, at molar ratios of 2.25:1 or 1.125:1 (SATA/sCD4). The mixture is stirred gently for 1 h, and the unreacted SATA is removed by dialysis (Spectrapor 2, Spectrum Medical Industries, Los Angeles, CA) against a 660-fold volume of Buffer A [20 m*M* morpholinoethanesulfonic acid (MES), 20 m*M* 4-morpholinopropanesulfonic acid (MOPS), 125 m*M* NaCl, 1 m*M* ethylenediaminetetraacetic acid (EDTA), pH 6.7] at 4°, with three changes of buffer over 24 h. Sulfhydryl residues are deprotected by reaction with 10 m*M* hydroxylamine at room temperature under argon. Activated thiol groups are quantified by reaction of the thiol-modified protein with 5,5′-dithiobis(2-nitrobenzoic acid) (DTNB; Sigma) against a cysteine standard (FitzGerald, 1987; Riddles *et al.*, 1983).

The lipids (obtained from Avanti Polar Lipids) are mixed at a molar ratio of 9.7:5:0.15:0.2 [PC/chol/N-(lissamine-rhodamine B sulfonyl) dioleoylphosphatidylethanolamine (Rh-PE):MPB-PE] in chloroform and dried in a rotary evaporator. Liposomes are prepared by reverse-phase evaporation (Düzgüneş *et al.*, 1983; Szoka and Papahadjopoulos, 1978), as described in detail in Düzgüneş (2003), using Buffer A saturated with argon. To obtain a uniform size distribution, the liposomes are extruded four times through double polycarbonate membranes of 0.1-μm pore diameter (Poretics, Pleasanton, CA) under argon pressure in a high-pressure stainless-steel extruder. MPB-PE–containing liposomes are prepared immediately before the coupling reaction. An aliquot of the liposomes is set aside to be used as the uncoupled liposome control.

Thiolated sCD4 is incubated with liposomes at final concentrations of 3.1 mg/ml sCD4 and 8 mM lipid (2.25:1 molar ratio derivative), or 2.9 mg/ml sCD4 and 8 mM lipid (1.125:1 molar ratio derivative), at room temperature, under argon, with gentle stirring overnight.

Protein A Coupling to Liposomes with SATA

Protein A can be used as an anchor for the Fc region of "immunoadhesin" (CD4-IgG), a recombinant protein combining sCD4 and the crystallizable region of IgG (Capon *et al.*, 1989). Protein A (purified from *Staphylococcus aureus*, Cowan I strain; Zymed Laboratories, South San Francisco, CA) is modified with SATA as described above for the derivatization of sCD4, except that 54.1 mM SATA in DMF is added to 5 mg/ml protein A in phosphate-buffered saline (PBS)/1 mM EDTA at a molar ratio of 9:1 SATA/protein A. Unreacted SATA is removed from modified protein A by dialysis against two changes of 640-fold volumes of Buffer B (10 mM HEPES, 140 mM NaCl, 10 mM KCl, 1 mM EDTA, pH 7.0). Thiolated protein A (2.2 mg/ml) is incubated with liposomes at a final concentration of 8 mM lipid.

sCD4 Coupling to Liposomes with 4-(4-N-Maleimidophenyl) Butyric Acid Hydrazide (MPBH)

For liposome coupling, sCD4 can also be derivatized by the oligosaccharide-directed heterobifunctional reagent MPBH (Chamow *et al.*, 1992; Flasher *et al.*, 1994). sCD4 (5.4 mg/ml in 0.1 M Na-acetate buffer, pH 5.5) is oxidized in the presence of 10 mM sodium periodate (NaIO$_4$, Sigma) for 30 min at room temperature. Excess NaIO$_4$ is removed by gel filtration on Sephadex G-25 equilibrated with 0.1 M Na-acetate buffer, pH 5.5. MPBH is dissolved in DMF and added to oxidized sCD4 to produce final concentrations of 1 mM and 2.7 mg/ml, respectively. The reaction proceeds for

2 h at room temperature, and the unreacted MPBH is removed by gel filtration on Sephadex G-25 equilibrated with 10 mM Na-acetate/150 mM NaCl, pH 5.5. The sCD4–MPBH conjugate is concentrated to 10 mg/ml by ultrafiltration in a Centricon 10 concentrator (Amicon, Beverly, MA).

Liposomes composed of PC/chol/Rh-PE/N-[3-(2-pyridyldithio)propionyl] dipalmitoyl phosphatidylethanolamine (PDP-PE) at a molar ratio of 9.7:5:0.15:0.2 are prepared by reverse-phase evaporation as described above. PDP-PE–containing liposomes are reduced to expose free thiol groups, using a 20-fold excess (12 mM) of dithiothreitol (DTT) at room temperature under argon for 1 h. Excess DTT is removed by gel filtration on Sephadex G-75 (equilibrated with buffer A under a constant stream of argon). The reduction of sulfhydryl groups on the liposomes is measured by their reaction with DTNB (Sigma). Right before mixing with thiol-containing liposomes, the pH of the sCD4–MPBH solution is adjusted to 7.0 by titration with 500 mM MES/500 mM MOPS, pH 8.0. sCD4 is coupled to liposomes, using final concentrations of 2.4 mg/ml and 10 mM lipid, respectively, under the same conditions as described for SATA. To confirm that the coupling reaction of maleimide-derivatized sCD4 with thiol-containing liposomes is specific, an aliquot of sCD4 is not oxidized by periodate prior to incubation with MPBH, and this mock-derivatized sCD4 is incubated subsequently with thiol-containing liposomes.

Purification of Protein–Liposome Conjugates

The protein-coupled liposomes are purified by flotation on a discontinuous metrizamide (Sigma) gradient, as described by Heath (1987), to remove free protein, followed by overnight dialysis (Spectrapor 2) against ≥ 1000 volumes of either Buffer A or Buffer B, both without EDTA, at 4°, under argon. Conjugates are sterilized by passage through a 0.22-μm filter (Millipore, Billerica, MA) and assayed for lipid concentration by a phosphate assay (Bartlett, 1959; Düzgüneş, 2003) and for protein concentration by the Lowry et al. (1951) assay. At a SATA/sCD4 ratio of 2.25:1, the amount of protein coupled per micromole phospholipid was 142 μg, corresponding to approximately 280 molecules of CD4 per liposome (Flasher et al., 1994).

Binding of gp120 to Liposomes

Binding of gp120 to CD4-liposomes is a test of the functionality of the coupled CD4 and is determined by radioimmunoassay (Ashkenazi et al., 1990; Chamow et al., 1990). Microtiter wells are coated with anti-IgG antibody to immobilize CD4–IgG. Then, ^{125}I-labeled gp120 (HIV-1$_{IIIB}$) is added simultaneously with free CD4–IgG or liposome-coupled CD4 to determine the ability of the latter two constructs to compete with the immobilized CD4–IgG for binding to gp120.

Association of CD4-Liposomes with HIV-Infected Cells

Specific binding of the CD4-liposomes to HIV-infected cells is determined by using chronically infected cell lines. THP-1 cells are obtained from the American Type Culture Collection. A chronically HIV-infected cell line (designated THP-1/HIV-1$_{IIIB}$) has been developed in our laboratory by infecting THP-1 cells with HIV-1$_{IIIB}$ (Konopka et al., 1993b). Chronically infected H9/HTLV-IIIB cells used in our experiments have been obtained from J. Mills and T. El-Beik (San Francisco General Hospital, San Francisco, CA). These cells can also be obtained from the AIDS Research and Reference Reagent Program (McKesson BioServices Corporation, Germantown, MD). Uninfected H9 cells are obtained from R. C. Gallo, through the AIDS Research and Reference Reagent Program. Human lymphoblastoid A3.01 cells are provided by T. M. Folks (CDC, Atlanta, GA). All cells are maintained at 37°, under 5% CO_2 in RPMI 1640 medium supplemented with 10% (v/v) heat-inactivated fetal bovine serum (FBS) (both from Irvine Scientific), penicillin (50 units/ml), streptomycin (50 μg/ml), and L-glutamine (2 mM). The cells are passaged 1:5 or 1:6 every 3–4 days. The percentage of viable cells in each culture is measured by Trypan Blue dye exclusion.

An amount of liposomes containing 10 μg/ml CD4 is incubated with uninfected H9 or THP-1 cells, or the chronically infected H9/HTLV-IIIB or THP-1/HIV-1$_{IIIB}$ cells, for 1 h at 37° [10^6 viable cells in a final volume of 100 μl of RPMI/0.2% bovine serum albumin (BSA)] (Fig. 4). An equivalent lipid concentration of uncoupled liposomes is used as a control. The effect of 10 μg/ml free sCD4 on CD4-liposome association to cells is also determined to assess the specificity of the interaction. In the case of protein A-liposomes, H9/HTLV-IIIB cells are first preincubated for 1 h on ice, either in the presence or absence of 10 μg/ml CD4–IgG. Cells are then washed three times with phosphate-buffered saline (PBS)/0.2% BSA and incubated for 30 min on ice and 30 min at 37° in the presence of either protein A-liposomes (100 μg protein/ml) or an equivalent lipid concentration of uncoupled liposomes.

Cells are washed four times with cold PBS/0.2% BSA to remove unassociated liposomes and are then fixed in 2% paraformaldehyde. Flow cytometric analysis is performed using a custom-designed flow cytometer composed of Becton Dickinson FACS IV parts (Lebo et al., 1993), equipped with three argon lasers (Spectra Physics). The system is driven by a Consort 40 computer and software (Becton Dickinson). Obviously,

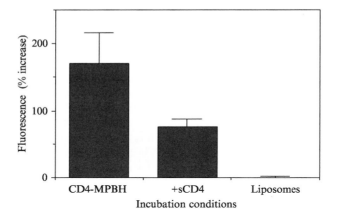

FIG. 4. Association of CD4-coupled and control liposomes with chronically infected H9/HTLV-IIIB cells and the effect of free sCD4. Liposomes containing rhodamine-phosphatidylethanolamine were incubated with the cells for 1 h at 37°, washed, and fixed. Cell association is expressed as the percentage increase in mean fluorescence relative to the autofluorescence of the cells. Data from Flasher *et al.* (1994).

any flow cytometer equipped for rhodamine fluorescence can be used for these experiments. Samples are analyzed for lissamine rhodamine, with excitation at 528 nm and resulting emission at 575 nm, using a bandpass filter of 575/25 nm. Forward scatter energy is used to trigger the signal collection. Ten thousand events are recorded for each sample. Forward and side scatter signals are used to gate the cell subset of interest and to eliminate debris and cell aggregates.

Association of CD4-liposomes with HIV-1–infected cells is ascertained by a significant shift in fluorescence relative to cell autofluorescence, as well as a widening of the fluorescence peak, suggesting some degree of heterogeneity with respect to the level of gp120 expression on the surface of the cells (Flasher *et al.*, 1994). Control liposomes without CD4 do not associate significantly with infected cells (Fig. 4) and neither do CD4-liposomes with uninfected cells. Free sCD4 inhibits the association of sCD4-SATA liposomes with H9/HTLV-IIIB cells by about 76% and that of CD4-MPBH liposomes by about 55%, indicating the specificity of the interaction of the liposomes with the cells. The multivalency of the liposomes is the most likely explanation for the incomplete inhibition of liposome–cell association by free sCD4. In experiments utilizing CD4–IgG, the incubation of protein A-liposomes with H9/HTLV-IIIB cells preincubated with CD4-IgG results in a significant fluorescence shift, indicating the association of the liposomes with the infected cells.

Effect of CD4-Liposomes on HIV-1 Infectivity

The viral isolate HIV-1$_{LAV-BRU}$ (initially designated as the LAV strain, and reidentified as LAI) is obtained from Dr. F. Barré-Sinoussi, propagated in CEM cells and purified as described in Larsen *et al.* (1993). The virus preparation is then propagated in A3.01 cells, and harvested at times of peak p24 production. The latter virus preparation is designated as HIV-1$_{LAV-BRU/P}$ (Konopka *et al.*, 1993a) and used for the experiments on inhibition of infectivity by CD4-liposomes. The infectious titer per 100 ng of p24 antigen is 4.6×10^5 tissue culture infectious dose, 50% end point (TCID$_{50}$), as determined in both H9 and A3.01 cells by the method of Johnson and Byington (1990).

The virus (HIV-1$_{LAV-BRU/P}$; 0.4 μg p24/ml) is incubated with 0.1, 0.3, 1.0, or 3.0 μg sCD4, the same amount of CD4 in liposome-coupled form, or uncoupled liposomes (corresponding to the lipid concentration of 1.5 μg of CD4-MPBH liposomes) in a final volume of 50 μl RPMI/10% FBS for 2 h at 37°. Ten microliters (4 ng viral p24) of each treated virus sample is added to 90 μl of A3.01 cells in RPMI/10% FBS at a final infection density of 2.0×10^7 cells/ml (i.e., 2 ng viral p24/10^6 cells). Cells are incubated with virus for 2 h at 37° to allow for viral adsorption and fusion, washed three times with 2 ml RPMI/10% FBS to remove unbound virus, and plated at a cell density of 5×10^5 cells/ml (4 ml/well) in 12-well cell culture plates (Corning, Corning, NY). Three days after infection, 2 ml of the culture medium is removed for analysis and replaced with fresh medium until day 7. Cell-free supernatants from day 3 and day 7 cultures are analyzed for virus production by an ELISA for viral p24 (Konopka *et al.*, 1990), and the plates are read in a Molecular Devices (Menlo Park, CA) V_{max} microplate reader (Table I).

Peptide Coupling to Liposomes

The complementarity determining region 2 (CDR2)-like domain of CD4 (residues 40–60) is thought to be involved in binding to gp120 (Brand *et al.*, 1995; Jameson *et al.*, 1988). We have shown that a CDR2-derived synthetic peptide coupled to liposomes can mediate the targeting of the liposomes to HIV-infected cells (Slepushkin *et al.*, 1996). The peptide MV-647 (SLWDQGNFPLGGGC-Acm) is coupled to the carboxyl groups on CHEMS/PC (4:6) liposomes via the amino-terminus of the peptide, using 1-ethyl-3(3-dimethylaminopropyl)carbodiimide (EDC) and *N*-hydroxysulfosuccinimide (Sulfo-NHS), according to the two-step coupling protocol provided by Pierce. The initial peptide/CHEMS ratio is chosen to be 1:1. Alternatively, the peptide is coupled to PC/chol/*N*-glutarylphosphatidylethanolamine (G-PE) (2:1:0.1) liposomes using the same procedure. The

TABLE I
THE EFFECT OF LIPOSOME-COUPLED CD4 AND sCD4 ON THE INFECTIVITY OF
HIV-1$_{BRU}$ IN A3.01 CELLS[a,b]

Reagent	CD4 concentration (μg/ml)	Viral p24 (ng/ml)	Inhibition (%)
Control	0	52.8 ± 7.1	0
CD4-MPBH liposomes	0.1	12.5 ± 1.0	76.3
	0.3	9.8 ± 1.3	81.4
	1	1.8 ± 0.1	96.6
	3	0	100
CD4-SATA liposomes	0.1	32.1 ± 6.2	39.2
	0.3	16.9 ± 3.2	68
	1	5.1 ± 0.6	90.3
	3	0	100
sCD4	0.1	2.7 ± 1.3	94.8
	0.3	0.22 ± 0.13	99.6
	1	0	100
	3	0	100
Plain liposomes[c]	0	23.2 ± 8.4	56.1

[a] Data from Flasher et al. (1994).
[b] HIV-1$_{BRU}$ (0.4 μg p24/ml) was preincubated with liposome-coupled CD4 or sCD4 at the indicated concentrations for 2 h at 37°. CD4$^+$ A3.01 cells were infected at 2 ng p24/10^6 cells. The results obtained at 7 days postinfection are shown. Data represent the mean and standard deviation of p24 values obtained from duplicate determinations from duplicate wells.
[c] The lipid concentration of the liposomes (10 μM) corresponds to that of CD4-MPBH liposomes containing 1.5 μg/ml coupled CD4.

liposomes encapsulating 80 mM calcein (Molecular Probes, Eugene, OR), pH 7.4, 290 mOsm, are prepared by reverse-phase evaporation followed by extrusion through 100-nm-pore-diameter polycarbonate membranes (Düzgüneş, 2003; Slepushkin et al., 1996). Unencapsulated calcein is removed by dialysis overnight at 4° with two changes of 4 liters of 150 mM NaCl, 10 mM TES (Sigma), pH 7.4, and then against 4 l of 140 mM NaCl, 20 mM MES (Sigma), pH 6.0. After the coupling reaction, uncoupled peptide is removed by dialysis against two changes of 150 mM NaCl, 10 mM TES, pH 7.4, using Spectra/Por dialysis membranes with a molecular weight cutoff of 300 kDa. The concentration of the peptide in the liposome preparation is determined by tryptophan fluorescence. Following incubation of these liposomes (containing 6.6 μM peptide) with chronically infected H9/IIIB cells (10^6 cells in 1 ml), the cells are subjected to flow cytometry. Significant binding of liposomes coupled to MV-647 is observed, while liposomes coupled to various CDR-3 peptides do not bind significantly to these cells (Slepushkin et al., 1996).

Cholesteryl β-Cyclodextrin Sulfate as an Antiviral Agent

Sulfated dextran was identified as an anti-HIV agent in the early days of the search for antiviral drugs against AIDS (De Clercq, 1989). We have synthesized sulfated cyclodextrins coupled to cholesterol as a means to generate liposomes with antiviral activity (Steffan *et al.*, 2000).

Synthesis of Sulfated β-Cyclodextrin-NH$_{prim}$-CO-CH$_2$-CH$_2$-CO-O- Cholesteryl

The synthesis of mono-6-deoxy-6-amino-β-cyclodextrin is described in detail by Henke *et al.* (1996). As described by Bellanger and Perly (1992) for fatty acids, CHEMS (Sigma) is reacted in ethyl acetate with equimolar amounts of *N*-hydroxysuccinimide and dicyclohexylcarbodiimide at room temperature overnight. Precipitated dicyclohexyl urea is filtered off, and ethyl acetate is removed by applying vacuum. The raw product (activated CHEMS) is recrystallized in ethanol. The product is obtained in theoretical yield.

Activated CHEMS and mono-6-deoxy-6-amino-β-cyclodextrin are dissolved under stirring at 60° in an amount of dimethyl formamide, which is just sufficient to dissolve the reactants. The progress of the reaction is monitored by thin-layer chromatography (TLC) on silica gel, using an ethyl acetate/isopropanol/water/concentrated ammonia solution (7:7:10:0.5) as the solvent and monitoring development by spraying with ethanol/concentrated sulfuric acid. After quantitative reaction, the product is precipitated with a mixture of ethanol and water, washed with ethanol, and dried in a vacuum oven. The product is obtained in theoretical yield.

The β-cyclodextrin-NH$_{prim}$-CO-CH$_2$-CH$_2$-CO-O-cholesteryl is sulfated with trimethylamine/sulfurtrioxide in dimethyl formamide under stirring at 60° overnight. Per 1 mol of β-cyclodextrin-NH$_{prim}$-CO-CH$_2$-CH$_2$-CO-O- cholesteryl (which equals 20 mol of hydroxyl groups), about 60 mol of trimethylamine/sulfurtrioxide is used. The raw product is precipitated in an excess of ethanol, redissolved in water, and precipitated in ethanol again. The trimethyl ammonium salt of the product is dissolved in an amount of aqueous sodium acetate solution (30% by weight), which is just sufficient. The solution is laced with activated charcoal and heated in a water bath at 80° for 5 min and subsequently filtered through diatomaceous earth. The sodium salt of the product is obtained by adding ethanol in a sufficient amount without coprecipitating sodium acetate. The sodium salt of sulfated β-cyclodextrin-NH$_{prim}$-CO-CH$_2$-CH$_2$-CO-O-cholesteryl is obtained in theoretical yield. The degree of sulfation is about 14.5 ± 3 sulfate groups per molecule, as determined by LSIMS. For storage, the product is dissolved in water, filtered aseptically through a 200-nm sterile filter, and freeze-dried.

Syncytium-Inhibition Assay

The cyclodextrin derivatives obtained are tested for their anti-HIV activity using a syncytium-inhibition assay. CD4$^+$ H9 cells are used as receptor cells, and TF228.1.16 cells are used as viral gp120/gp41-expressing cells. The syncytia assay is carried out following Lifson (1993) and Jonak *et al.* (1993), but using commercially available DME medium containing 16% FBS. After an 18-h incubation, syncytium formation is assessed. At a concentration as low as 8 μg/ml (approx. 2.6 μM) of the cyclodextrin derivative, no syncytia formation can be observed, indicating complete inhibition of gp120/gp41-mediated membrane fusion (N. Düzgüneş, E. Pretzer, and G. Steffan, unpublished data). No cytotoxic effect is observed up to a concentration 50-fold higher than the effective concentration, using the Trypan Blue and Alamar Blue tests (Konopka *et al.*, 1996).

Commercially available β-cyclodextrin, sulfated β-cyclodextrin, and sulfated β-cyclodextrin-NH$_{prim}$-CO-C$_{15}$H$_{31}$ have also been tested in the above assay. Only the *N*-palmitoylated cyclodextrin derivative (sulfated β-cyclodextrin-NH$_{prim}$-CO-C$_{15}$H$_{31}$) is able to inhibit syncytia formation within the investigated range of concentration [up to 200 μg/ml of cyclodextrin (derivative)], but it is necessary to use 40 μg/ml to completely inhibit syncytia formation.

Conclusions

The methods and studies presented in this chapter indicate that liposomes may be useful in the intracellular delivery of antiviral agents. We have reviewed previously the *in vitro* and *in vivo* studies with liposomal antivirals (Düzgüneş, 1998; Düzgüneş *et al.*, 2001b). The necessity for intravenous or subcutaneous administration of liposomal anti-HIV agents over a long period of time, potentially excessive localization in the liver and spleen, and the unknown long-term effects of chronic administration are disadvantages of this system. Nevertheless, the potential of relatively infrequent administration of liposomal antivirals, their enhanced localization in lymph nodes, the ability of liposomes to deliver macromolecular drugs against multidrug-resistant HIV strains, and protection of the drugs from inactivation may counter these disadvantages. Although phosphorothiate antisense ODN can be delivered to target cells *in vivo* following administration in the free form, the clinical use of these molecules may be limited due to the potential toxicity associated with non-sequence-specific effects. Delivery of such ODN, or ribozymes, in sterically stabilized pH-sensitive liposomes may decrease or eliminate this complication. Targeting of these liposomes to infected cells or potential host cells may further enhance the therapeutic effect of these macromolecular drugs.

Current anti-HIV-1 drugs, as well as ODN and ribozymes, are intended to keep virus production at low levels. While their use is essential, these drugs cannot cure HIV infection, since the virus is integrated in the genome of a very large number of cells in infected individuals. Some of these cells are infected latently and do not produce virions, making it impossible to detect them. These cells can be activated by certain compounds to produce virions, however. Liposomes containing cytotoxic compounds, such as doxorubicin, targeted to HIV envelope glycoproteins, would then be able to recognize and enter these cells, thereby eliminating a reservoir of virus. In an alternative approach, it may be possible to develop macromolecular drugs that can detect and inactivate integrated proviral DNA (Düzgüneş et al., 2001b). Triple helix-forming oligonucleotides (TFOs) linked with psoralen are a prototype of such macromolecular drugs (Giovannangeli et al., 1997; Praseuth et al., 1999). TFOs can also induce mutations in target genes in vivo, without the need for psoralen (Vasquez et al., 2000). As we have indicated previously (Düzgüneş et al., 2001b), the next generation of TFOs and peptide nucleic acids (Pooga et al., 2001) may be able to recognize, cleave out, and degrade proviral DNA and reseal the chromosomal DNA, possibly in conjunction with other macromolecules (Seidman and Glazer, 2003). The delivery of such macromolecular drugs into infected cells may be achieved by the use of specifically designed liposomes.

Acknowledgments

Some of the experimental work presented in this chapter was supported by Grant AI32399 from the National Institute of Allergy and Infectious Diseases.

References

Akhtar, S., and Rossi, J. J. (1996). Anti-HIV therapy with antisense oligonucleotides and ribozymes: Realistic approaches or expensive myths? *J. Antimicrob. Chemother.* **38,** 159–165.

Aquaro, S., Calio, R., Balzarini, J., Bellocchi, M. C., Garaci, E., and Perno, C. F. (2002). Macrophages and HIV infection: Therapeutical approaches toward this strategic virus reservoir. *Antiviral Res.* **55,** 209–225.

Ashkenazi, A., Presta, L. G., Marsters, S. A., Camerato, T. R., Rosenthal, K. A., Fendly, B. M., and Capon, D. J. (1990). Mapping the CD4 binding site for human immunodeficiency virus by alanine-scanning mutagenesis. *Proc. Natl. Acad. Sci. USA* **87,** 7150–7154.

Balzarini, J., Perno, C.-F., Schols, D., and De Clercq, E. (1991). Activity of acyclic nucleoside phosphonate analogues against human immunodeficiency virus in monocyte/macrophages and peripheral blood lymphocytes. *Biochem. Biophys. Res. Commun.* **178,** 329–335.

Bartlett, G. R. (1959). Phosphorus assay in column chromatography. *J. Biol. Chem.* **234,** 466–468.

Bellanger, N., and Perly, B. (1992). NMR investigations of the conformation of new cyclodextrin-based amphiphilic transporters for hydrophobic drugs: Molecular lollipops. *J. Mol. Struct.* **273,** 215–226.

Bennett, C. F., Chiang, M. Y., Chan, H. C., Shoemaker, J. E. E., and Mirabelli, C. K. (1992). Cationic lipids enhance cellular uptake and activity of phosphorothioate antisense oligonucleotides. *Mol. Pharmacol.* **41**, 1023–1033.

Brand, D., Srinivasan, K., and Sodroski, J. (1995). Determinants of human immunodeficiency virus type 1 entry in the CDR2 loop of the CD4 glycoprotein. *J. Virol.* **69**, 166–171.

Capon, D. J., Chamow, S. M., Mordenti, J., Marsters, S. A., Gregory, T., Mitsuya, H., Byrn, R. A., Lucas, C., Wurm, F. M., Groopman, J. E., Broder, S., and Smith, D. H. (1989). Designing CD4 immunoadhesins for AIDS therapy. *Nature* **337**, 525–531.

Chamow, S. M., Peers, D. H., Byrn, R. A., Mulkerrin, M. G., Harris, R. J., Wang, W.-C., Bjorkman, P. J., Capon, D. J., and Ashkenazi, A. (1990). Enzymatic cleavage of a CD4 immunoadhesin generates crystallizable, biologically active Fd-like fragments. *Biochemistry* **29**, 9885–9891.

Chamow, S. M., Kogan, T. P., Peers, D. H., Hastings, R. C., Byrn, R. A., and Ashkenazi, A. (1992). Conjugation of soluble CD4 without loss of biological activity via a novel carbohydrate-directed cross-linking reagent. *J. Biol. Chem.* **267**, 15916–15922.

De Clercq, E. (1989). Potential drugs for the treatment of AIDS. *J. Antimicrob. Chemother.* **23**(Suppl. A), 35–46.

De Clercq, E. (1995). Antiviral therapy for human immunodeficiency virus infections. *Clin. Microbiol. Rev.* **8**, 200–239.

De Clercq, E. (2003). Clinical potential of the acyclic nucleoside phosphonates cidofovir, adefovir, and tenofovir in treatment of DNA virus and retrovirus infections. *Clin. Microbiol. Rev.* **16**, 569–596.

Derksen, J. T. P., and Scherphof, G. L. (1985). An improved method for the covalent coupling of proteins to liposomes. *Biochim. Biophys. Acta* **814**, 151–155.

Duncan, J. S., Weston, P. D., and Wrigglesworth, R. (1983). A new reagent which may be used to introduce sulfhydryl groups into proteins, and its use in the preparation of conjugates for immunoassay. *Anal. Biochem.* **132**, 68–73.

Düzgüneş, N. (1998). Treatment of human immunodeficiency virus, *Mycobacterium avium* and *Mycobacterium tuberculosis* infections by liposome-encapsulated drugs. *In* "Medical Applications of Liposomes" (D. D. Lasic and D. Papahadjopoulos, eds.), pp. 189–219. Elsevier Science B. V., Amsterdam.

Düzgüneş, N. (2003). Preparation and quantitation of small unilamellar liposomes and large unilamellar reverse-phase evaporation liposomes. *Methods Enzymol.* **367**, 23–27.

Düzgüneş, N., Wilschut, J., Hong, K., Fraley, R., Perry, C., Friend, D. S., James, T. L., and Papahadjopoulos, D. (1983). Physicochemical characterization of large unilamellar vesicles prepared by reverse-phase evaporation. *Biochim. Biophys. Acta* **732**, 289–299.

Düzgüneş, N., Straubinger, R. M., Baldwin, P. A., and Papahadjopoulos, D. (1991). pH-sensitive liposomes: Introduction of foreign substances into cells. *In* "Membrane Fusion" (J. Wilschut and D. Hoekstra, eds.), pp. 713–730. Marcel Dekker, New York.

Düzgüneş, N., Flasher, D., Pretzer, E., Konopka, K., Slepushkin, V. A., Steffan, G., Salem, I. I., Reddy, M. V., and Gangadharam, P. R. J. (1995). Liposome-mediated therapy of human immunodeficiency virus type-1 and Mycobacterium infections. *J. Liposome Res.* **5**, 669–691.

Düzgüneş, N., Pretzer, E., Simões, S., Slepushkin, V., Konopka, K., Flasher, D., and Pedroso de Lima, M. C. (1999). Liposome-mediated delivery of antiviral agents to human immunodeficiency virus-infected cells. *Mol. Membrane Biol.* **16**, 111–118.

Düzgüneş, N., Simões, S., Slepushkin, V., Pretzer, E., Rossi, J. J., De Clercq, E., Antao, V. P., Collins, M. L., and Pedroso de Lima, M. C. (2001a). Enhanced inhibition of HIV-1 replication in macrophages by antisense oligonucleotides, ribozymes and acyclic oligonucleotide phosphonate analogs delivered in pH-sensitive liposomes. *Nucleosides Nucleotides Nucleic Acids* **20**, 515–523.

Düzgüneş, N., Simões, S., Konopka, K., Rossi, J. J., and Pedroso de Lima, M. C. (2001b). Delivery of novel macromolecular drugs against HIV-1. *Expert Opin. Biol. Ther.* **1,** 949–970.

Dykxhoorn, D. M., Novina, C. D., and Sharp, P. A. (2003). Killing the messenger: Short RNAs that silence gene expression. *Nat. Rev. Mol. Cell. Biol.* **4,** 457–467.

FitzGerald, D. J. (1987). Construction of immunotoxins using *Pseudomonas* exotoxin A. *Methods Enzymol.* **151,** 139–145.

Flasher, D., Konopka, K., Chamow, S., Ashkenazi, A., Dazin, P., Pretzer, E., and Düzgüneş, N. (1994). Liposome targeting to human immunodeficiency virus type 1-infected cells via recombinant soluble CD4 and CD4-immunoadhesin (CD4-IgG). *Biochim. Biophys. Acta* **1194,** 185–196.

Giovannangeli, C., Diviacco, S., Labrousse, V., Gryaznov, S., Charneau, P., and Hélène, C. (1997). Accessibility of nuclear DNA to triplex-forming oligonucleotides: The integrated HIV-1 provirus as a target. *Proc. Natl. Acad. Sci. USA* **94,** 79–84.

Heath, T. D. (1987). Covalent attachment of proteins to liposomes. *Methods Enzymol.* **149,** 111–119.

Henke, C., Steinem, C., Janshoff, A., Steffan, G., Luftmann, H., Sieber, M., and Galla, H.-J. (1996). Self-assembled monolayers of monofunctionalized cyclodextrins onto gold: A mass spectrometric characterization and impedance analysis of host-guest interaction. *Anal. Chem.* **68,** 3158–3165.

Jameson, B. A., Rao, P. E., Kong, L. I., Hahn, B. H., Shaw, G. M., Hood, L. E., and Kent, S. B. H. (1988). Location and chemical synthesis of a binding site for HIV-1 on the CD4 protein. *Science* **240,** 1335–1339.

Johnson, V. A., and Byington, R. E. (1990). Infectivity assay (virus yield assay). *In* "Techniques in HIV Research" (A. Aldovoni and B. D. Walker, eds.), pp. 71–76. Stockton Press, New York.

Jonak, Z. L., Clark, R. K., Matour, D., Trulli, S., Craig, R., Henri, E., Lee, E. V., Greig, R., and Debouck, C. (1993). A human lymphoid recombinant cell line with functional human immunodeficiency virus type 1 envelope. *AIDS Res. Hum. Retroviruses* **9,** 23–32.

Kay, J., and Dunn, B. M. (1990). Viral proteinases: Weakness in strength. *Biochim. Biophys. Acta* **1048,** 1–18.

Konopka, K., Davis, B. R., Larsen, C. E., Alford, D. R., Debs, R. J., and Düzgüneş, N. (1990). Liposomes modulate human immunodeficiency virus infectivity. *J. Gen. Virol.* **71,** 2899–2907.

Konopka, K., Davis, B. R., Larsen, C. E., and Düzgüneş, N. (1993a). Anionic liposomes inhibit human immunodeficiency virus type 1 (HIV-1) infectivity in CD4+ A3.01 and H9 cells. *Antiviral Chem. Chemother.* **4,** 179–187.

Konopka, K., Pretzer, E., Plowman, B., and Düzgüneş, N. (1993b). Long-term noncytopathic productive infection of the human monocytic leukemia cell line THP-1 by human immunodeficiency virus type 1 (HIV-1$_{IIIB}$). *Virology* **193,** 877–887.

Konopka, K., Pretzer, E., Felgner, P. L., and Düzgüneş, N. (1996). Human immunodeficiency virus type-1 (HIV-1) infection increases the sensitivity of macrophages and THP-1 cells to cytotoxicity by cationic liposomes. *Biochim. Biophys. Acta* **1312,** 186–196.

Konopka, K., Rossi, J. J., Swiderski, P., Slepushkin, V. A., and Düzgüneş, N. (1998). Delivery of anti-HIV-1 ribozyme into HIV-infected cells via cationic liposomes. *Biochim. Biophys. Acta* **1372,** 55–68.

Larsen, C. E., Nir, S., Alford, D. R., Jennings, M., Lee, K.-D., and Düzgüneş, N. (1993). Human immunodeficiency virus type 1 (HIV-1) fusion with model membranes: Kinetic analysis and the effects of pH and divalent cations. *Biochim. Biophys. Acta* **1147,** 223–236.

Lebo, R. V., Bruce, B. D., Dazin, P. F., and Payan, D. G. (1993). Design and application of a versatile triple-laser cell and chromosome sorter. *Cytometry* **8,** 71–82.

Lifson, J. D. (1993). Fusion of human immunodeficiency virus-infected cells with uninfected cells. *Methods Enzymol.* **221,** 3–12.

Lisziewicz, J., Sun, D., Klotman, M., Agrawal, S., Zamecnik, P., and Gallo, R. (1992). Specific inhibition of human immunodeficiency virus type 1 replication by antisense oligonucleotides: An *in vitro* model for treatment. *Proc. Natl. Acad. Sci. USA* **89,** 11209–11213.

Lisziewicz, J., Sun, D., Weichold, F. F., Thierry, A. R., Lusso, P., Tang, J., Gallo, R. C., and Agrawal, S. (1994). Antisense oligodeoxynucleotide phosphorothioate complementary to Gag mRNA blocks replication of human immunodeficiency virus type 1 in human peripheral blood cells. *Proc. Natl. Acad. Sci. USA* **91,** 7942–7946.

Litzinger, D. C., Brown, J. M., Wala, I., Kaufman, S. A., Van, G. Y., Farrell, C. L., and Collins, D. (1996). Fate of cationic liposomes and their complex with oligonucleotide *in vivo. Biochim. Biophys. Acta* **1281,** 139–149.

Lowry, O. H., Rosebrough, N. J., Farr, A. L., and Randall, R. J. (1951). Protein measurement with the Folin phenol reagent. *J. Biol. Chem.* **193,** 267–275.

Martin, F. J., and Papahadjopoulos, D. (1982). Irreversible coupling of immunoglobulin fragments to preformed vesicles. An improved method for liposome targeting. *J. Biol. Chem.* **257,** 286–288.

Meltzer, M. S., Skillman, D. R., Gomatos, P. J., Kalter, D. C., and Gendelman, H. E. (1990). Role of mononuclear phagocytes in the pathogenesis of human immunodeficiency virus infection. *Annu. Rev. Immunol.* **8,** 169–194.

Naesens, L., Balzarini, J., and De Clercq, E. (1992). Acyclic adenine nucleoside phosphonates in plasma determined by high-performance liquid chromatography with fluorescence detection. *Clin. Chem.* **38,** 480–485.

Nakagawara, A., and Nathan, C. F. (1983). A simple method for counting adherent cells: Application to cultured human monocytes, macrophages and multinucleated giant cells. *J. Immmunol. Methods* **56,** 261–268.

Pooga, M., Land, T., Bartfai, T., and Langel, Ü. (2001). PNA oligomers as tools for specific modulation of gene expression. *Biomol. Eng.* **17,** 183–192.

Praseuth, D., Guieysse, A. L., and Hélène, C. (1999). Triple helix formation and the antigene strategy for sequence-specific control of gene expresssion. *Biochim. Biophys. Acta* **1489,** 181–206.

Pretzer, E., Flasher, D., and Düzgüneş, N. (1997). Inhibition of human immunodeficiency virus type-1 replication in macrophages and H9 cells by free or liposome-encapsulated L-689,502, an inhibitor of the viral protease. *Antiviral Res.* **34,** 1–15.

Riddles, P. W., Blakeley, R. L., and Zerner, B. (1983). Reassessment of Ellman's reagent. *Methods Enzymol.* **91,** 49–60.

Robins, T., and Plattner, J. (1993). HIV protease inhibitors: Their anti-HIV activity and potential role in treatment. *J. Acquir. Immune Defic. Syndr.* **6,** 162–170.

Ropert, C., Lavignon, M., Dubernet, C., Couvreur, P., and Malvy, C. (1992). Oligonucleotides encapsulated in pH sensitive liposomes are efficient toward Friend retrovirus. *Biochem. Biophys. Res. Commun.* **183,** 879–885.

Seidman, M. M., and Glazer, P. M. (2003). The potential for gene repair via triple helix formation. *J. Clin. Invest.* **112,** 487–494.

Shi, F., and Hoekstra, D. (2004). Effective intracellular delivery of oligonucleotides in order to make sense of antisense. *J. Control. Release* **97,** 189–209.

Shi, F., Wasungu, L., Nomden, A., Stuart, M. C., Polushkin, E., Engberts, J. B., and Hoekstra, D. (2002). Interference of poly(ethylene glycol)-lipid analogues with cationic-lipid-mediated delivery of oligonucleotides; role of lipid exchangeability and non-lamellar transitions. *Biochem. J.* **366,** 333–341.

Simões, S., Slepushkin, V., Düzgüneş, N., and Pedroso de Lima, M. C. (2001). On the mechanisms of internalization and intracellular delivery mediated by pH-sensitive liposomes. *Biochim. Biophys. Acta* **1515,** 23–37.

Slepushkin, V. A., Salem, I. I., Andreev, S. M., Dazin, P., and Düzgüneş, N. (1996). Targeting of liposomes to HIV-1 infected cells by peptides derived from the CD4 receptor. *Biochem. Biophys. Res. Commun.* **227,** 827–833.

Slepushkin, V. A., Simões, S., Dazin, P., Newman, M. S., Guo, L. S., Pedroso de Lima, M. C., and Düzgüneş, N. (1997). Sterically stabilized pH-sensitive liposomes: Intracellular delivery of aqueous contents and prolonged circulation *in vivo. J. Biol. Chem.* **272,** 2382–2388.

Slepushkin, V., Simões, S., Pedroso de Lima, M. C., and Düzgüneş, N. (2004). Sterically stabilized pH-sensitive liposomes. *Methods Enzymol.* **387,** 134–147.

Song, L. Y., Ahkong, Q. F., Rong, Q., Wang, Z., Ansell, S., Hope, M. J., and Mui, B. (2002). Characterization of the inhibitory effect of PEG-lipid conjugates on the intracellular delivery of plasmid and antisense DNA mediated by cationic lipid liposomes. *Biochim. Biophys. Acta* **1558,** 1–13.

Steffan, G., Galla, H. -J., Düzgüneş, N., and Henke, C. (2000). Sterinverknüpfte, anionische cyclodextrinderivative und deren verwendung in arzneimittelformulierungen. German Patent DE 198 14 815 C2.

Stein, C. A., Neckers, L. M., Nair, B. C., Mumbauer, S., Hoke, G., and Pal, R. (1991). Phosphorothioate oligodeoxycytidine interferes with binding of HIV-1 gp1/20 to CD4. *J. Acquir. Immune Defic. Synd.* **4,** 686–693.

Stephens, A. C., and Rivers, R. P. (2003). Antisense oligonucleotide therapy in cancer. *Curr. Opin. Mol. Ther.* **5,** 118–122.

Stevenson, M. (2003). Dissecting HIV-1 through RNA interference. *Nat. Rev. Immunol.* **3,** 851–858.

Straubinger, R. M., Düzgüneş, N., and Papahadjopoulos, D. (1985). pH-Sensitive lioposomes mediate cytoplasmic delivery of encapsulated macromolecules. *FEBS Lett.* **179,** 148–154.

Szoka, F., Jr., and Papahadjopoulos, D. (1978). Procedure for preparation of liposomes with large internal aqueous space and high capture by reverse-phase evaporation. *Proc. Natl. Acad. Sci. USA* **75,** 4194–4198.

Thompson, W. J., Fitzgerald, P. M. D., Holloway, M. K., Emini, E. A., Darke, P. L., McKeever, B. M., Schleif, W. A., Quintero, J. C., Zugay, J. A., Tucker, T. J., Schwering, J. E., Homnick, C. F., Nunberg, J., Springer, J. P., and Huff, J. R. (1992). Synthesis and antiviral activity of a series of HIV-1 protease inhibitors with functionality tethered to the P1 or P1′ phenyl substituents: X-ray crystal structure assisted design. *J. Med. Chem.* **35,** 1685–1701.

Torchilin, V. P., Zhou, F., and Huang, L. (1993). pH-sensitive liposomes. *J. Liposome Res.* **3,** 201–255.

Vasquez, K. M., Narayanan, L., and Glazer, P. M. (2000). Specific mutations induced by triplex-forming oligonucleotides in mice. *Science* **290,** 530–533.

Wagner, R. W., and Flanagan, W. M. (1997). Antisense technology and prospects for therapy of viral infections and cancer. *Mol. Med. Today* **3,** 31–38.

Weichold, F. F., Lisziewicz, J., Zeman, R. A., Nerurkar, L. S., Agrawal, S., Reitz, M. S., Jr., and Gallo, R. C. (1995). Antisense phosphorothioate oligodeoxynucleotides alter HIV type 1 replication in cultured human macrophages and peripheral blood mononuclear cells. *AIDS Res. Hum. Retroviruses* **11,** 863–868.

Zelphati, O., Imbach, J.-L., Signoret, N., Zon, G., Rayner, B., and Leserman, L. (1994). Antisense oligonucleotides in solution or encapsulated in immunoliposomes inhibit replication of HIV-1 by several different mechanisms. *Nucleic Acids Res.* **22,** 4307–4314.

Section III

Miscellaneous Liposomal Therapies

[21]   Liposomal Vasoactive Intestinal Peptide

By VARUN SETHI, HAYAT ÖNYÜKSEL, and ISRAEL RUBINSTEIN

Abstract

Liposomes have been investigated as drug carriers since first discovered in the 1960s. However, the first-generation, so-called classic liposomes found relatively limited therapeutic utility. Nonetheless, the advent in the 1980s of the second-generation sterically stabilized liposomes (SSL) that evade uptake by the host's reticuloendothelial system greatly enhanced their utility as drug carriers because of their prolonged circulation half-life and passive targeting to injured and cancerous tissues. Over the past decade, our work focused on exploiting the bioactivity of vasoactive intestinal peptide (VIP), a ubiquitous 28-amino acid, amphipathic and pleiotropic mammalian neuropeptide, as a drug. To this end, the peptide expresses distinct and unique innate bioactivity that could be harnessed to treat several human diseases that represent unmet medical needs, such as pulmonary hypertension, stroke, Alzheimer's disease, sepsis, female sexual arousal dysfunction, acute lung injury, and arthritis. Unfortunately, the bioactive effects of VIP last only a few minutes due to its rapid degradation and inactivation by enzymes, catalytic antibodies, and spontaneous hydrolysis in biological fluids. Hence, our goal was to develop and test stable, long-acting formulations of VIP using both classic and SSL as platform technologies. We found that spontaneous association of VIP with phospholipid bilayers leads to a transition in the conformation of the peptide from random coil in an aqueous environment to α-helix, the preferred conformation for ligand–receptor interactions, in the presence of lipids. This process, in turn, protects VIP from degradation and inactivation and amplifies its bioactivity *in vivo*. Importantly, we discovered that the film rehydration and extrusion technique is the most suitable to passively load VIP onto SSL at room temperature and yields the most consistent results. Collectively, these attributes indicate that VIP on SSL represents a suitable formulation that could be tested in human disease.

Introduction

Since the discovery of classic liposomes (CL) or first-generation liposomes in the 1960s (Bangham *et al.*, 1965), CL have been extensively investigated as potential carriers for biologically active molecules and drugs *in vivo* (Düzgüneş *et al.*, 1988; Heath *et al.*, 1986; Huang and Rorstad,

1987; Matthay *et al.*, 1986; Mayhew and Rustum, 1985; Papahadjopoulos *et al.*, 1985; Straubinger *et al.*, 1985). However, their short half-life in the circulation due to rapid uptake by the reticuloendothelial system (RES) has limited their potential use as drug carrier systems. In the 1990s, development of second-generation sterically stabilized liposomes (SSL) greatly advanced the ability to utilize liposomes as drug carrier systems (Klibanov *et al.*, 1990, 1991; Maruyama *et al.*, 1991a). SSL were found to have increased circulation half-life because of decreased RES uptake through "steric hindrance" of opsonins that would otherwise adsorb on their surface, thereby tagging them for RES uptake.

Over the past few years, our laboratory has been investigating the delivery of vasoactive intestinal peptide (VIP) using various formulations of liposomes. VIP is a 28-amino acid ubiquitous amphipathic neuropeptide that displays a broad range of biological activities (Gomariz *et al.*, 2001; Said, 1986). It is widely distributed in the central and peripheral nervous systems, where it functions as a nonadrenergic, noncholinergic neurotransmitter and neuromodulator (Gao *et al.*, 1994). VIP is localized in the gastrointestinal tract, heart, lung, thyroid, kidney, urinary bladder, genital organs, and brain (Gozes *et al.*, 1994, 1996). It is also localized in perivascular nerves and upon its release elicits a potent, though short-lived, endothelium-dependent and -independent vasodilation (Brizzolara and Burnstock, 1991; Gao *et al.*, 1994; Gaw *et al.*, 1991; Huang and Rorstad, 1987; Ignarro *et al.*, 1987; Suzuki *et al.*, 1995, 1996a). VIP has important actions on the respiratory tract, including relaxation of airway smooth muscle and attenuation of inflammation (Said, 1992). Recently, VIP has been shown to have several immunomodulatory functions and is secreted by several immune-competent cells in response to various immune signals (Gomariz *et al.*, 2001). Deficiency in the release and/or degradation of VIP has been implicated in the pathogenesis of several diseases, such as cystic fibrosis, impotence, congenital megacolon in Hirschsprung's disease, and achalasia of the esophagus (Gozes *et al.*, 1994, 1996; Ollerenshaw *et al.*, 1989). The bioactive effects of VIP are mediated via high-affinity transmembrane receptors on target cells (Harmar *et al.*, 1998). These receptors are overexpressed in various cancer cells and thus can be actively targeted for tumor imaging and therapeutics using VIP as the ligand (Dagar *et al.*, 2001; Thakur *et al.*, 2000). VIP could be used to treat essential and pulmonary hypertension, congestive heart failure, bronchial asthma, erectile dysfunction, acute food impaction in the esophagus, acute lung injury, and arthritis (Davis *et al.*, 1981; Delgado *et al.*, 2001; Refai *et al.*, 1999).

However, as seen with other peptides, the bioactive effects of VIP are short-lived, lasting only a few minutes after intravenous administration

(Domschke *et al.*, 1978; Hassan *et al.*, 1994). This short-term efficacy of VIP is most likely due to its degradation by enzymes, catalytic antibodies, and by spontaneous hydrolysis. Several attempts have been made to synthesize more stable and potent analogues of VIP for the treatment of asthma (Bolin *et al.*, 1995), impotence (Gozes *et al.*, 1994), neurodegenerative diseases (Gozes *et al.*, 1996), and inadequate regional blood flow (Refai *et al.*, 1999). However, none of these analogues has reached the clinic. In fact, a recent clinical study using a cyclic VIP analogue in patients with asthma has been terminated prematurely. This often implies that high concentrations of VIP are required to observe significant physiological effects, thus causing severe side effects such as reduced systemic arterial pressure (Thakur *et al.*, 2000).

Clearly, for VIP to be considered in therapeutics, its half-life in biological fluid should be prolonged appreciably. To this end, this chapter describes work in our laboratories using liposomes as a novel delivery system for VIP.

VIP on Classical Liposomes

Preparation

Synthetic VIP can either be purified by successive reverse-phase high-performance liquid chromatography (HPLC) on a preparative C_{18} column in triethylamine phosphate/acetonitrile and trifluoacetic acid (TFA)/acetonitrile solvent systems or purchased from commercial sources such as American Peptide Company (Sunnyvale, CA). Radioiodination and purification of VIP can be performed as previously described (Paul *et al.*, 1989). Classic liposomes can be prepared by several well-established procedures.

Reverse-Phase Evaporation. In this method, a mixture of egg yolk phosphatidylcholine (ePC), egg yolk phosphatidylglycerol (ePG), and cholesterol in a molar ratio of 1:4:5 is used (Sigma Chemical Co., St. Louis, MO) (Gao *et al.*, 1994). The phospholipid solution (12 mM each) is mixed with 3 ml diethyl ether containing 3 mM synthetic VIP, a mixture of 0.2 mM labeled VIP and 3 mM unlabeled VIP, or 0.2 mM radioactive peptide alone with 1 ml of 50 mM HEPES buffer, pH 7.3. Sonication of the mixture and subsequent evaporation of the ether *in vacuo* form vesicles. The resulting dispersion is diluted with 10 ml 50 mM HEPES buffer and centrifuged (12,500g, 7 min). The supernatant is discarded, and the pellet is washed three times with buffer containing 0.15 M NaCl. The larger vesicles are removed by passing the dispersion through 1-μm polycarbonate filters (Nuclepore, Pleasonton, CA). The inorganic phosphate content of the liposomes is measured by a microtiter Bartlett assay (Gao *et al.*, 1994; Suzuki *et al.*, 1995). This method of preparation yields liposomes of ~600 nm. The amount of VIP encapsulated in these liposomes is estimated by using radioimmuno assay (RIA) specific toward VIP (Gao *et al.*, 1994; Suzuki *et al.*, 1995). A VIP/lipid mole ratio of ~0.008 is anticipated using this method of preparation.

Film Rehydration and Extrusion Method. This method provides a more controlled method to prepare CL of the desired size compared with reverse-phase evaporation. For the film rehydration and extrusion method, the phospholipids are dissolved in chloroform, and the solution is dried to a film. The dried lipid film is hydrated with 100 μl of 0.15 M NaCl containing 0.7 mg VIP by vortex mixing followed by sonication of the dispersion. The dispersion obtained is frozen and thawed five times using a dry ice–acetone bath. The resulting liposomes are extruded nine times through two polycarbonate filters (3 μm) using a Liposofast apparatus (Avestin Inc., Ottawa, Canada). The liposomes are passed through a gel filtration column to separate free VIP from liposome-encapsulated VIP. The liposomes are recovered in the void volume and stored at 4° prior to use. Size of liposomes is determined by dynamic laser light scattering. A liposome size of ~600 nm is obtained for liposomes prepared by this method. The phosphate content is estimated by modified Bartlett assay as previously mentioned. The VIP/lipid mole ratio is assessed by an enzyme-linked immunosorbent assay (ELISA) specific for VIP. A VIP/lipid mole ratio of ~0.008 is anticipated using this preparation method as well (Gao *et al.*, 1994; Suzuki *et al.*, 1995). Both these methods have been successful in forming liposome-encapsulating VIP that showed improved *in vitro* and *in vivo* activities compared with aqueous VIP (Gao *et al.*, 1994; Gozes *et al.*, 1996).

Classic liposomes are also prepared in the gel state to study the interaction between VIP and phospholipid bilayers (Ikezaki *et al.*, 1999; Önyüksel *et al.*, 2003). Phospholipids with a higher transition temperature such as dipalmitoylphosphatidyl choline (DPPC) and ePG in a molar ratio of 9:1 are used to prepare gel state CL using the film rehydration and extrusion method. Different sizes of liposomes (30 and 100 nm) having the same composition as above are used to study the effects of rigidity and curvature of the bilayer on their interaction with VIP (Ikezaki *et al.*, 1999; Önyüksel *et al.*, 2003).

In Vitro *Effects*

Encapsulation of VIP on CL using either the reverse-phase evaporation or film rehydration/extrusion method was successful in preventing the trypsin-catalyzed degradation of the peptide over 60 min (Gao *et al.*,

1994; Gololobov *et al.*, 1998). This was clearly evident as liposomal VIP was three to four orders of magnitude less susceptible to the trypsin-catalyzed reaction than was aqueous VIP (Gololobov *et al.*, 1998). Moreover, in an autoantibody-catalyzed reaction for VIP hydrolysis, VIP encapsulated in CL showed a 5-fold increase in the K_m values without any significant increase in the V_{max} values compared with aqueous VIP. This suggested that the change in K_m was due to poor recognition of the substrate by antibodies and not to a change in the enzyme rate constant (Gololobov *et al.*, 1998). Stability experiments conducted in the presence of human plasma showed that only 10% of VIP encapsulated in CL was released after incubation for 8 days at 37°, thereby improving the stability of the peptide *in vitro* (Fig. 1).

Secondary structure of liposomal VIP evaluated using circular dichroism (CD) showed that the enhanced *in vitro* activity of the VIP encapsulated in CL was due to change in the conformation of the peptide from predominantly random coil in aqueous solution to α-helix in the presence of phospholipids (Gololobov *et al.*, 1998). These data corroborated the results obtained from other groups, demonstrating the increase in the α-helicity of the peptide in the presence of phospholipids compared with VIP in aqueous solution (Robinson *et al.*, 1982; Rubinstein *et al.*, 2000). The increase in the α-helicity of VIP can be attributed to the interaction and subsequent partitioning of the VIP into the phospholipid bilayer (Noda *et al.*, 1994). However, recent data from our laboratory have

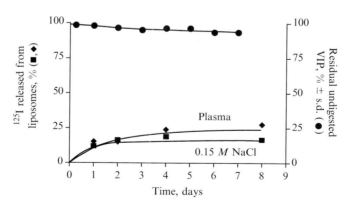

Fig. 1. VIP release from and stability in liposomes. [Tyr$^{10-125}$I] VIP-containing liposomes (4 nmol VIP/1.4/μmol Pi) were incubated for 8 days at 37° in the absence and presence of 25% human plasma in 0.15 *M* NaCl. Total available [Tyr$^{10-125}$I]VIP per assay tube was 10^4 cpm. Released VIP was separated from liposomal VIP by passage through Sepharose 4B columns. Residual undigested VIP in the liposome fractions was estimated by the TCA precipitation method. Reproduced with permission from Gololobov *et al.* (1998).

shown that the interaction of VIP with the phospholipid bilayer is dependent on the biophysical state of the bilayer. A change in the bilayer from liquid-crystalline to the gel state was associated with a marked decrease in α-helicity of the peptide (Önyüksel et al., 2003).

In Vivo Effects

To study the effect of the peptide in situ, a well-established hamster cheek pouch model to assess the vasoactive effects of neuropeptides was used (Davis et al., 1981; Gao et al., 1994; Rubinstein and Mayhan, 1995; Suzuki et al., 1995). The check pouch microcirculation can be visualized using a method previously described in our laboratory (Suzuki et al., 1995, 1996a). The effect of VIP either as aqueous solution or liposomal formulation on arteriolar diameter and its duration of action are summarized in

TABLE I

EFFECTS OF VASOACTIVE INTESTINAL PEPTIDE (VIP) AND LIPOSOMAL VIP ON ARTERIOLAR DIAMETER IN THE INTACT HAMSTER CHEEK POUCH

| | Observed effects | | | |
| | | Arteriolar diameter | | |
Formulation	Size of liposomes (nm)	(% change from baseline)	Duration of effect (min)	Reference
VIP in saline[a]	N/A	13	4	Gao et al., 1994
VIP on classical liposomes[a]				
Reverse-phase evaporation	600	35	16	Gao et al., 1994
Film rehydration and extrusion	600	38	13	Suzuki et al., 1995
VIP on sterically stabilized liposomes[a]				
Freeze–thaw extrusion	224	10	5	Séjourné et al., 1997a
Extrusion/dehydration–rehydration	250	38	16	Séjourné et al., 1997a
Optimized extrusion/ dehydration–rehydration method[b]	89	22	15	Patel et al., 1999
VIP in gel-state classic liposomes[a]	100	10	5	Ikezaki et al., 1999

[a] VIP 1.0 nmol was used in all experiments.
[b] VIP 0.1 nmol was used in this group only.

Table I. These data show that encapsulation of VIP in CL prolongs both the duration of action and magnitude of vasodilation compared with VIP in aqueous solution (Suzuki *et al.*, 1995). Importantly, encapsulation of VIP in CL restored vasoreactivity in spontaneously hypertensive hamsters and rats (Gao *et al.*, 1994; Suzuki *et al.*, 1996b) (Fig. 2). However, no prolongation in duration of vasodilation was observed in these animals due to possible elaboration of vasoconstrictory substances in the microcirculation.

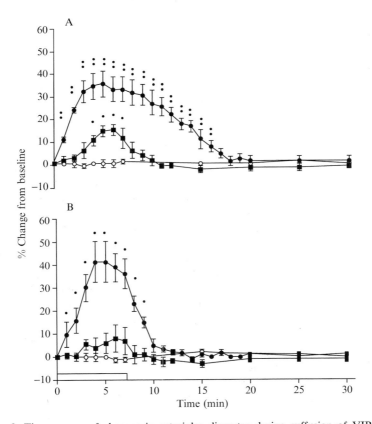

FIG. 2. Time course of changes in arteriolar diameter during suffusion of VIP alone (0.1 nmol; ■), empty liposomes (○), and VIP (0.1 nmol) encapsulated into liposomes (●) in normotensive (A) and hypertensive (B) hamsters. Values are means ± SE; each group, $n =$ four animals. *$p < 0.05$ in comparison with baseline. **$p < 0.05$ in comparison with VIP alone. Open bar indicates duration of suffusion. Reproduced with permission from Suzuki *et al.* (1996a).

In vivo studies using mice (Gololobov *et al.*, 1998) and rats (Refai *et al.*, 1999) have shown a prolonged half-life for VIP encapsulated in CL. Encapsulation of VIP in liposomes increases the half-life of the peptide by ~8-fold (Refai *et al.*, 1999). This is a marked improvement from the previous estimated half-life of 1.0 min in humans and 0.4 min in rats (Domschke *et al.*, 1978; Hassan *et al.*, 1994). Although the half-life of VIP encapsulated in CL was markedly improved relative to VIP in aqueous solution, the use of VIP encapsulated in CL for therapeutic purposes is limited due to rapid uptake of CL by the RES. This problem was overcome by using SSL as an improved delivery system for VIP.

VIP in Sterically Stabilized Liposomes

Preparation

Three successive methods of preparing SSL encapsulating/incorporating VIP were developed in our laboratory. These are the (1) freeze–thaw extrusion, (2) extrusion/dehydration–rehydration, and (3) optimized extrusion/dehydration–rehydration methods (Patel *et al.*, 1999; Séjourné *et al.*, 1997a). All three methods use the same composition of phospholipids, distearoylphosphatidylethanolamine-poly(ethylene glycol) (DSPE–PEG)$_{2000}$, ePC, ePG, and cholesterol, at a molar ratio of 0.5:5:1:3.5 with total phospholipid content of 17 μmol. Phospholipids are dissolved in chloroform in a round-bottom flask. The chloroform is evaporated in a rotary evaporator (Labconco, Kansas City, MO) at 45°. The film is used to prepare SSL using the various methods outlined later.

Freeze–Thaw Extrusion. The dried lipid film is hydrated with 250 μl 0.15 *M* saline containing 0.4 mg VIP. The dispersion is vortexed and then sonicated for 5 min in a waterbath sonicator (Fisher Scientific, Itasca, IL). Finally, the dispersion is frozen and thawed five times in an acetone–dry ice bath. Following freeze–thaw, the dispersion is extruded through stacked polycarbonate filters (200 nm) using the Liposofast apparatus (Avestin Inc., Ottawa, Canada). Liposomes with an average diameter of 224 nm are obtained with this method. The VIP-associated liposomes are separated from the free VIP by passing the extruded dispersion through a gel filtration column (Bio-Rad, A-5m, Bio-Rad Labs, Richmond, CA). The separated fraction, which is normally eluted in the void volume of the column at 4°, is stored for further *in vitro* testing (Séjourné *et al.*, 1997a). The problem with this method is the high stress associated with extrusion of the SSL in the presence of the peptide leading to partial sloughing of the peptide, and some loss in bioactivity (Séjourné *et al.*, 1997a).

Extrusion/Dehydration–Rehydration. The dried lipid film is hydrated with 250 μl of 0.15 M NaCl. The dispersion is vortexed rigorously and sonicated for 5 min. The dispersion is extruded through stacked polycarbonate filters of pore size 200, 100, and 50 nm to form liposomes (Séjourné *et al.*, 1997a). Then 0.4 mg VIP and 30 mg of trehalose (cryoprotectant) in powder form, are added to the extruded dispersion. The mixture is frozen in an acetone–dry ice bath and lyophilized at $-46°$ under $\sim$5 $\times$ 10^{-3} mBar pressure overnight using the Lab-Scale freeze-dryer (Labconco, Kansas City, MO). The freeze-dried cake is resuspended in 250 μl deionized water. Liposomes of $\sim$250 nm in size are anticipated after resuspending the freeze-dried cake. The dispersion is subsequently separated as in method 1 and stored at 4° until use (Séjourné *et al.*, 1997a). This method avoids extrusion of SSL after incubation with the peptide. However, freeze drying the formulation following extrusion changes the liposome size following rehydration of the freeze-dried cake due to the fusion of the preformed liposomes (Szucs and Tilcock, 1995). This problem is avoided by optimizing the formulation, such that the freeze-dry step is eliminated from the preparation method.

Optimized Extrusion/Dehydration–Rehydration Method. This new and optimized method to prepare VIP-SSL avoids the freeze-drying step in method 2, thereby decreasing the preparation time, and conserves liposome size. In this optimized preparation, the dried phospholipid film (same composition as in methods 1 and 2) is hydrated with 250 μl 0.15 M NaCl. The dispersion is vortexed thoroughly and sonicated for 5 min in a bath sonicator. The dispersion is extruded through stacked polycarbonate filters to yield an average size of 80 nm liposomes (using sequential filters of size 200, 100, and 50 nm). VIP (0.3 mg) is added to the extruded suspension, and the mixture is incubated at room temperature (25°) for 2 h. Free VIP is separated from liposome-associated VIP by gel filtration chromatography as described above. The final amount of VIP associated with liposomes is quantified by ELISA (Peninsula Laboratories, Santa Clara, CA). Liposome size is measured by quasielastic light scattering to confirm the size (Nicomp 270, Particle Sizing System, Santa Barbara, CA). The inorganic phosphate content of the liposomes is measured using a modified Bartlett assay, as mentioned above. A phospholipid yield of $\sim$50% is anticipated using this method. The VIP-SSL are stored at 4° until use (Patel *et al.*, 1999). All three methods outlined above yield a VIP/phospholipid molar ratio of $\sim$0.004.

VIP on SSL prepared by method 3 was evaluated as a passively targeted system for breast cancer (Dagar *et al.*, 1999). The results suggested that VIP dissociates from the surface of liposomes in the vicinity of VIP receptors, due to stronger affinity of VIP for its cognate receptors compared with SSL (Dagar *et al.*, 1999). To overcome this problem, VIP was actively conjugated to the distal end of an activated DSPE–PEG$_{3400}$–NHS moiety (Dagar *et al.*, 2001). The reaction mixture needs to be maintained at pH 6.6 such that the histidine (pK_a = 7–8) group of the peptide will be involved in the formation of the conjugate. This is important because using a different pH exposes multiple lysine groups in the peptide, which are a critical component of the α-helical domain, involved in VIP receptor recognition. SSL are prepared as previously described in method 3, without adding VIP. These preformed SSL are incubated with the conjugated mixture (DSPE–PEG$_{3400}$–VIP) overnight at 4° to insert them into the liposomes (Dagar *et al.*, 2001). VIP-SSL prepared by this method are more stable, and VIP does not dissociate from the SSL upon interaction with its receptors (Dagar *et al.*, 2001).

In Vitro *Studies*

As shown above, VIP on liposomes overcomes trypsin-, autoantibody-, and plasma-induced inactivation of VIP *in vitro* (Gao *et al.*, 1994; Gololobov *et al.*, 1998). Studies conducted using VIP on SSL significantly potentiated DNA synthesis in oral keratinocytes *in vitro* compared with synthesis in the presence of VIP alone (Fig. 3) (Rubinstein *et al.*, 2001). The effect on DNA synthesis was not dependent on empty SSL thus indicating that the increase in activity was due to the marked improvement in stability and activity of VIP on SSL. Conformation studies conducted in the presence of PEGylated phospholipid aggregates (sterically stabilized micelles, SSM) have shown an increase in α-helicity of VIP associated with SSM (Rubinstein *et al.*, 2000). We have used SSM instead of SSL to study the change in conformation of VIP in the presence of phospholipids, because experiments conducted with SSL expressed a large signal-to-noise ratio, thereby rendering data analysis impractical. Moreover, α-helicity of VIP associated with SSM is potentiated in the presence of low concentrations (10^{-11}–10^{-10} M) of calmodulin (CaM), a ubiquitous extracellular membrane-bound and intracellular protein (Table II; Rubinstein *et al.*, 2000). This interaction requires the presence of preformed α-helix VIP, i.e., phospholipid-associated VIP. This strategy could be used to amplify the effect of other therapeutic amphipathic peptides with limited bioavailability.

VIP (alone or on SSL) did not modulate the chemotactic response of human neutrophils to N-formyl-methionylleucyl-phenylalanine (FMLP) and zymosan, two potent chemotaxis agents, *in vitro* (Hatipoğlu *et al.*, 1998). On the contrary, chemotaxis was attenuated by DSPE–PEG$_{2000}$

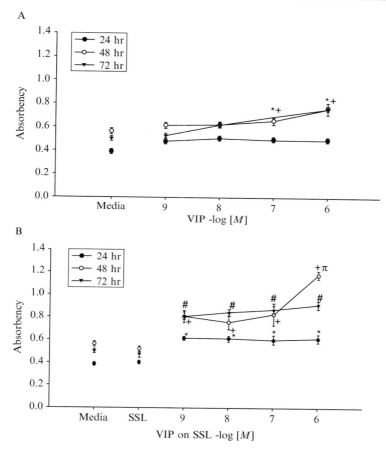

Fig. 3. (A) Incorporation of bromodeoxyuridine (BrdU) into DNA of HCPC-1 cells incubated in the absence and presence of vasoactive intestinal peptide (VIP) for 24, 48, and 72 h. Values are means ± SEM; each group, n = four experiments in triplicate. $*p < 0.05$ in comparison to media. $^+p < 0.05$ in comparison to 24 h. (B) BrdU incorporation into DNA of HCPC-1 cells incubated in the absence and presence of vasoactive intestinal peptide on sterically stabilized liposomes (VIP on SSL) and empty SSL for 24, 48, and 72 h. Values are means ± SEM; each group, n = four experiments in triplicate. $*p < 0.05$ in comparison to media. $^+p < 0.05$ in comparison to 24 h. $^\#p < 0.05$ in comparison to 24 h. $^\pi p < 0.05$ in comparison to 72 h. Reproduced with permission from Rubinstein et al. (2000).

and empty SSL (Hatipoğlu et al., 1998). Stability studies of VIP on SSL showed ~80% peptide activity even after storage for 30 days at 4°. Although studies detailing stability of the peptide in SSL in the presence of serum and trypsin were not conducted, we anticipate its stability to be

TABLE II
EFFECT OF PHOSPHOLIPIDS ON α-HELICITY OF VASOACTIVE INTESTINAL PEPTIDE
(VIP) AT 25° AND 37°[a]

	α-Helix (%)	
	Saline	Micelles[b]
VIP at room temperature	5 ± 1	27 ± 2[*, **]
VIP at 37°	2 ± 1	67 ± 3[***]
Calmodulin at room temperature	0	0
VIP and calmodulin at room temperature	1 ± 1	42 ± 2[*, **, ****]
Arginine[z]-vasopressin at room temperature	8 ± 1	16 ± 2[*]
Arginine[z]-vasopressin and calmodulin at room temperature	8 ± 1	16 ± 2[*]

[a] α-Helix conformation of VIP and arginine[z]-vasopressin in saline and sterically stabilized micelles by circular dichroism.
[b] Values are means $\pm$ SEM of four experiments (each experiment is an average of nine continuous scans).
* $p < 0.05$ in comparison to saline.
** $p < 0.05$ in comparison to arginine[z]-vasopressin.
*** $p < 0.05$ in comparison to VIP in micelles at room temperature.
**** $p < 0.05$ in comparison to VIP in micelles at room temperature.
Reproduced with permission from Rubinstein et al. (2000).

similar or even better than what is observed for VIP encapsulated in CL (Gololobov et al., 1998).

In Vivo Studies

The bioactivity of VIP are in either in CL or SSL using the in situ cheek pouch model hamster summarized in Table I. The effects on both arteriolar diameter and duration of action of both formulations are very similar, suggesting no change in the in situ bioactivity of the peptide. However, the effect of VIP on SSL on mean arterial pressure following intravenous administration in hamsters is markedly different from that observed with aqueous VIP (Fig. 4) (Séjourné et al., 1997a). Intravenous administration of VIP on SSL to normotensive hamsters has no significant effect on systemic arterial pressure, as SSL cannot extravasate through normal blood vessel wall (Fig. 5). Moreover, it has been previously shown that VIP receptors are expressed on the abluminal side of blood vessel wall (Fahrenkrug et al., 2000); thus VIP would need to extravasate across the luminal cavity to exert its action. VIP on SSL extravasates preferentially in areas with damaged endothelium, as evident in hypertensive animals (Rubinstein and Mayhan, 1995; Thomas et al., 1997). Figure 5 shows preferential

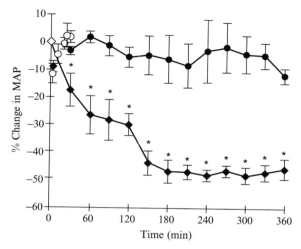

Fig. 4. Percentage change in mean arterial pressure after bolus intravenous injection of 0.1 nmol VIP on SSL (method B, ◆), native VIP (○), and empty SSL (●) in hamsters with spontaneous hypertension. Values are mean ± SEM; each group n = four animals. *$p < 0.05$ compared to baseline. Reproduced with permission from Séjourné *et al.* (1997a).

lowering of systemic arterial pressure after intravenous administration of a low concentration of VIP on SSL in hypertensive animals. Moreover, clinical studies from other laboratories have previously shown that free VIP express adverse effects, most likely due to extravasation of the free peptide followed by subsequent intraction with VIP receptors (Thakur *et al.*, 2000). Collectively, these data show that VIP on SSL can be used selectively in systemic hypertension circumventing adverse effects seen when using higher concentrations of aqueous VIP.

We have previously shown that VIP on SSL activates VIP receptors in the peripheral microcirculation (Séjourné *et al.*, 1997b). However, receptor-independent pathways are also involved (Ikezaki *et al.*, 1998; Séjourné *et al.*, 1997b).

Formulations of VIP on SSL elicit a significant concentration-dependent increase in arteriolar diameter in the intact hamster cheek pouch. Suffusion of anti-VIP antibody had no significant effects on VIP on SSL-induced responses. Arteriolar diameter increased by 22% from baseline during VIP-SSL (0.1 nmol) suffusion alone and by 21% and 19% during suffusion of 0.1 nmol VIP-SSL together with 0.02 and 0.04 mg anti-VIP antibody, respectively, whereas suffusion of VIP alone (1.0 nmol) in the presence of anti-VIP antibody (0.02 and 0.04 mg) shows significantly attenuated vasodilation (Ikezaki *et al.*, 1998). These results confirm the

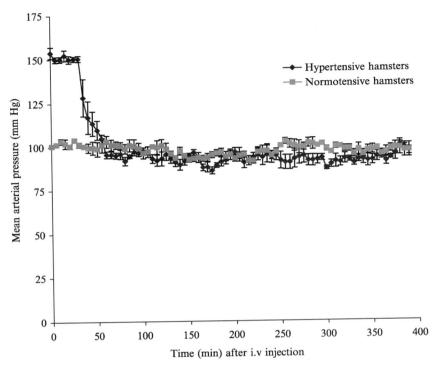

FIG. 5. Effects of intravenous VIP on SSC (0.1 nmol; bolus) on mean arterial pressure in normotensive and spontaneously hypertensive hamsters (data are means ± SEM; each $n =$ four animals).

in vitro stability improvement of VIP in either CL or SSL formulations. Collectively, interaction of VIP with these liposomes significantly increases biological activity, *in vivo* half-life, stability, and targetability of the peptide.

Conclusions and Future Directions

Several methods to prepare liposomal VIP have been described. However, SSL prepared by method 3 (optimized film rehydration and extrusion) are the most practical for preparation of liposomal VIP. There are several reasons for this conclusion; first, method 3 involves the incorporation of the peptide on preformed SSL, thereby avoiding possible loss of VIP by extrusion during liposome preparation in the presence of the peptide (method 1). Second, method 3 provides control over the size of the liposomes, unlike freeze-drying the formulation and then rehydrating (method 2), where the size of liposomes cannot be controlled. Third, the formulation according to method 3 is simple to prepare, as the peptide needs only to be incubated for only 2 h at room temperature with preformed liposomes before use. These factors favor the use of method 3 in preparation of liposomal VIP.

Association of VIP with liposomes, either CL or SSL, stabilizes the peptide and improves its bioactivity. Liposomal VIP elicits significant potentiation and prolongation of the vasoactive effects of VIP, when VIP is either encapsulated in liposomes (CL) or associated with SSL. VIP interactions with liposomes prevent VIP degradation by trypsin and autocatalytic antibodies, thus prolonging the molecular half-life of the peptide. Moreover, association of VIP with liposomes converts the conformation of the peptide from predominantly random coil in aqueous solution to α-helix in the presence of phospholipids. These factors contribute to a longer half-life of VIP when associated with liposomes. Although the in vivo half-life of VIP encapsulated in CL is ~8-fold higher than that of aqueous VIP, preliminary experiments from our laboratory indicate that the half-life of VIP on SSL would be much longer due to their longer circulation time. This is in agreement with other groups that have shown longer circulation half-lives for other proteins and peptides using SSL (Liu et al., 1992; Maruyama et al., 1991b; Oku et al., 1994). Development of second-generation liposomal VIP using SSL is a major breakthrough, because VIP is now associated with SSL and its RES uptake is reduced, resulting in a longer circulation half-life of the peptide in vivo. Evidently, association of VIP with either CL or SSL also significantly increases peptide stability of the VIP in vitro to almost ~15 days when stored at 4° compared with a few hours for aqueous (Gololobov et al., 1998; Patel et al., 1999).

Recent work in our laboratory involving conjugation of VIP to the distal end of phospholipids has expanded the use of the liposomal VIP formulation. This third-generation VIP on SSL could be used for active targeted delivery of compounds, either therapeutic or diagnostic/imaging, to specific sites where VIP receptors are overexpressed, thereby circumventing systemic adverse reactions (Dagar et al., 2001). In this regard, SSL have been shown to extravasate preferentially in tumors and inflamed tissues due to the leakiness of the local microvasculature (Carr et al., 2001; Dvorak et al., 1991; Ezaki et al., 2001; Gabizon and Martin, 1997; Oku and Namba, 1994; Yuan et al., 1994). These disease states thus provide a suitable opportunity to use VIP-SSL as a delivery system for both passive and active targeting.

In summary, developments in pharmaceutical technologies have made it possible to bring liposomal formulations to the clinic. Various methods have been designed to produce liposomes on a large scale, making it possible to move this field from the bench to the market. This trend is evident as more liposome-based drugs have been made available to patients. We propose that liposomal-VIP is an appropriate candidate for such a product.

Acknowledgments

Supported, in part, by VA Merit Review grant, NIH Grants R01A G024026 and R01 HL72323, and Department of Defense Grant DAMD17-02-1-0415. We thank the students, research associates, and postdoctoral fellows who contributed to this research.

References

Bangham, A. D., Standish, M. M., and Watkins, J. C. (1965). Diffusion of univalent ions across the lamellae of swollen phospholipids. *J. Mol. Biol.* **13**, 238–252.

Bolin, D. R., Michalewsky, J., Wasserman, M. A., and O'Donnell, M. (1995). Design and development of a vasoactive intestinal peptide analog as a novel therapeutic for bronchial asthma. *Biopolymers* **37**, 57–66.

Brizzolara, A. L., and Burnstock, G. (1991). Endothelium-dependent and endothelium-independent vasodilatation of the hepatic artery of the rabbit. *Br. J. Pharmacol.* **103**, 1206–1212.

Carr, D. B., McDonald, G. B., Brateng, D., Desai, M., Thach, C. T., and Easterling, T. R. (2001). The relationship between hemodynamics and inflammatory activation in women at risk for preeclampsia. *Obstet. Gynecol.* **98**, 1109–1116.

Dagar, S., Sekosan, M., Blend, M., Rubinstein, I., and Önyüksel, H. (1999). Optimized formulation of vasoactive intestinal peptide on sterically stabilized liposome. *In Proc. 26th Int. Symp. Control. Rel. Bioactive Mater.*, pp. 22–23.

Dagar, S., Sekosan, M., Lee, B. S., Rubinstein, I., and Önyüksel, H. (2001). VIP receptors as molecular targets of breast cancer: Implications for targeted imaging and drug delivery. *J. Control. Rel.* **74**, 129–134.

Davis, M. J., Joyner, W. L., and Gilmore, J. P. (1981). Responses of pulmonary allograft and cheek pouch arterioles in the hamster to alterations in extravascular pressure in different oxygen environments. *Circ. Res.* **49**, 125–132.

Delgado, M., Abad, C., Martinez, C., Leceta, J., and Gomariz, R. P. (2001). Vasoactive intestinal peptide prevents experimental arthritis by downregulating both autoimmune and inflammatory components of the disease. *Nat. Med.* **7**, 563–568.

Domschke, S., Domschke, W., Bloom, S. R., Mitznegg, P., Mitchell, S. J., Lux, G., and Strunz, U. (1978). Vasoactive intestinal peptide in man: Pharmacokinetics, metabolic and circulatory effects. *Gut* **19**, 1049–1053.

Düzgüneş, N., Perumal, V. K., Kesavalu, L., Goldstein, J. H., Debs, R. J., and Gangadharam, P. R. J. (1988). Enhanced effect of liposome-encapsulated amileacin on *Mycobacterium avium-intraccellulare* complex in beige mice. *Antimicrob. Agents Chemother.* **27**, 1404–1411.

Dvorak, H. F., Nagy, J. A., and Dvorak, A. M. (1991). Structure of solid tumors and their vasculature: Implications for therapy with monoclonal antibodies. *Cancer Cells* **3**, 77–85.

Ezaki, T., Baluk, P., Thurston, G., La Barbara, A., Woo, C., and McDonald, D. M. (2001). Time course of endothelial cell proliferation and microvascular remodeling in chronic inflammation. *Am. J. Pathol.* **158**, 2043–2055.

Fahrenkrug, J., Hannibal, J., Tams, J., and Georg, B. (2000). Immunohistochemical localization of the VIP1 receptor (VPAC1R) in rat cerebral blood vessels: Relation to PACAP and VIP containing nerves. *J. Cereb. Blood Flow Metab.* **20**, 1205–1214.

Gabizon, A., and Martin, F. (1997). Polyethylene glycol-coated (pegylated) liposomal doxorubicin. Rationale for use in solid tumours. *Drugs* **54**(Suppl. 4), 15–21.

Gao, X.-P., Noda, Y., Rubinstein, I., and Paul, S. (1994). Vasoactive intestinal peptide encapsulated in liposomes: Effects on systemic arterial blood pressure. *Life Sci.* **54**, 247–252.

Gaw, A. J., Aberdeen, J., Humphrey, P. P., Wadsworth, R. M., and Burnstock, G. (1991). Relaxation of sheep cerebral arteries by vasoactive intestinal polypeptide and neurogenic stimulation: Inhibition by L-N^G-monomethyl arginine in endothelium-denuded vessels. *Br. J. Pharmacol.* **102**, 567–572.

Gololobov, G., Noda, Y., Sherman, S., Rubinstein, I., Baranowska-Kortylewicz, J., and Paul, S. (1998). Stabilization of vasoactive intestinal peptide by lipids. *J. Pharmacol. Exp. Ther.* **285**, 753–758.

Gomariz, R. P., Martinez, C., Abad, C., Leceta, J., and Delgado, M. (2001). Immunology of VIP: A review and therapeutical perspectives. *Curr. Pharm. Des.* **7**, 89–111.

Gozes, I., Reshef, A., Salah, D., Rubinraut, S., and Fridkin, M. (1994). Stearylnorleucine-vasoactive intestinal peptide (VIP): A novel VIP analog for noninvasive impotence treatment. *Endocrinology* **134**, 2121–2125.

Gozes, I., Bardea, A., Reshef, A., Zamostiano, R., Zhukovsky, S., Rubinraut, S., Fridkin, M., and Brenneman, D. E. (1996). Neuroprotective strategy for Alzheimer disease: Intranasal administration of a fatty neuropeptide. *Proc. Natl. Acad. Sci. USA* **93**, 427–432.

Harmar, A. J., Arimura, A., Gozes, I., Journot, L., Laburthe, M., Pisegna, J. R., Rawlings, S. R., Robberecht, P., Said, S. I., Sreedharan, S. P., Wank, S. A., and Waschek, J. A. (1998). International Union of Pharmacology. XVIII. Nomenclature of receptors for vasoactive intestinal peptide and pituitary adenylate cyclase-activating polypeptide. *Pharmacol. Rev.* **50**, 265–270.

Hassan, M., Refai, E., Andersson, M., Schnell, P. O., and Jacobsson, H. (1994). *In vivo* dynamical distribution of 131I-VIP in the rat studied by gamma-camera. *Nucl. Med. Biol.* **21**, 865–872.

Hatipoğlu, U., Gao, X., Verral, S., Séjourné, F., Pitrak, D., Alkan-Önyüksel, H., and Rubinstein, I. (1998). Sterically stabilized phospholipids attenuate human neutrophils chemotaxis *in vitro*. *Life Sci.* **63**, 693–699.

Heath, T. D., Lopez, N. G., Piper, J. R., Montgomery, J. A., Stern, W. H., and Papahadjopoulos, D. (1986). Liposome-mediated delivery of pteridine antifolates to cells *in vitro*: Potency of methotrexate, and its alpha and gamma substituents. *Biochim. Biophys. Acta* **862**, 72–80.

Huang, M., and Rorstad, O. P. (1987). VIP receptors in mesenteric and coronary arteries: A radioligand binding study. *Peptides* **8**, 477–485.

Ignarro, L. J., Byrns, R. E., Buga, G. M., and Wood, K. S. (1987). Mechanisms of endothelium-dependent vascular smooth muscle relaxation elicited by bradykinin and VIP. *Am. J. Physiol.* **253**, H1074–H1082.

Ikezaki, H., Paul, S., Alkan-Önyüksel, H., Patel, M., Gao, X. P., and Rubinstein, I. (1998). Vasodilation elicited by liposomal VIP is unimpeded by anti-VIP antibody in hamster cheek pouch. *Am. J. Physiol.* **275**, R56–R62.

Ikezaki, H., Patel, M., Önyüksel, H., Akhter, S. R. A., Gao, X. P., and Rubinstein, I. (1999). Exogenous calmodulin potentiates vasodilation elicited by phospholipid-associated VIP *in vivo*. *Am. J. Physiol.* **276**, R1359–R1365.

Klibanov, A. L., Maruyama, K., Torchilin, V. P., and Huang, L. (1990). Amphipathic polyethyleneglycols effectively prolong the circulation time of liposomes. *FEBS Lett.* **268**, 235–237.

Klibanov, A. L., Maruyama, K., Beckerleg, A. M., Torchilin, V. P., and Huang, L. (1991). Activity of amphipathic poly(ethylene glycol) 5000 to prolong the circulation time of liposomes depends on the liposome size and is unfavorable for immunoliposome binding to target. *Biochim. Biophys. Acta* **1062**, 142–148.

Liu, D., Mori, A., and Huang, L. (1992). Role of liposome size and RES blockade in controlling biodistribution and tumor uptake of GM1-containing liposomes. *Biochim. Biophys. Acta* **1104**, 95–101.

Maruyama, K., Yuda, T., Okamoto, A., Ishikura, C., Kojima, S., and Iwatsuru, M. (1991a). Effect of molecular weight in amphipathic polyethyleneglycol on prolonging the circulation time of large unilamellar liposomes. *Chem. Pharm. Bull.* **39**, 1620–1622.

Maruyama, K., Mori, A., Bhadra, S., Subbiah, M. T., and Huang, L. (1991b). Proteins and peptides bound to long-circulating liposomes. *Biochim. Biophys. Acta* **1070**, 246–252.

Matthay, K. K., Heath, T. D., Badger, C. C., Bernstein, I. D., and Papahadjopoulos, D. (1986). Antibody-directed liposomes: Comparison of various ligands for association, endocytosis, and drug delivery. *Cancer Res.* **46**, 4904–4910.

Mayhew, E., and Rustum, Y. M. (1985). The use of liposomes as carriers of therapeutic agents. *Prog. Clin. Biol. Res.* **172B**, 301–310.

Noda, Y., Rodriguez-Sierra, J., Liu, J., Landers, D., Mori, A., and Paul, S. (1994). Partitioning of vasoactive intestinal polypeptide into lipid bilayers. *Biochim. Biophys. Acta* **1191**, 324–330.

Oku, N., and Namba, Y. (1994). Long-circulating liposomes. *Crit. Rev. Ther. Drug Carrier Syst.* **11**, 231–270.

Oku, N., Naruse, R., Doi, K., and Okada, S. (1994). Potential usage of thermosensitive liposomes for macromolecule delivery. *Biochim. Biophys. Acta* **1191**, 389–391.

Ollerenshaw, S., Jarvis, D., Woolcock, A., Sullivan, C., and Scheibner, T. (1989). Absence of immunoreactive vasoactive intestinal polypeptide in tissue from the lungs of patients with asthma. *N. Engl. J. Med.* **320**, 1244–1248.

Önyüksel, H., Ashok, B., Dagar, S., Sethi, V., and Rubinstein, I. (2003). Interactions of VIP with rigid phospholipid bilayers: Implications for vasoreactivity. *Peptides* **24**, 281–286.

Papahadjopoulos, D., Heath, T., Bragman, K., and Matthay, K. (1985). New methodology for liposome targeting to specific cells. *Ann. N.Y. Acad. Sci.* **446**, 341–348.

Patel, M., Önyüksel, H., Ikezaki, H., Dagar, S., Akhter, S. R., and Rubinstein, I. (1999). *Proc. Int. Symp. Control. Rel. Bioactive Mater* **26**, #6238.

Paul, S., Said, S. I., Thompson, A. B., Volle, D. J., Agrawal, D. K., Foda, H., and de la Rocha, S. (1989). Characterization of autoantibodies to vasoactive intestinal peptide in asthma. *J. Neuroimmunol.* **23**, 133–142.

Refai, E., Jonsson, C., Andersson, M., Jacobsson, H., Larsson, S., Kogner, P., and Hassan, M. (1999). Biodistribution of liposomal 131I-VIP in rat using gamma camera. *Nucl. Med. Biol.* **26**, 931–936.

Robinson, R. M., Blakeney, E. W., Jr., and Mattice, W. L. (1982). Lipid-induced conformational changes in glucagon, secretin, and vasoactive intestinal peptide. *Biopolymers* **21**, 1271–1274.

Rubinstein, I., and Mayhan, W. G. (1995). L-arginine dilates cheek pouch arterioles in hamsters with hereditary cardiomyopathy but not in controls. *J. Lab. Clin. Med.* **125**, 313–318.

Rubinstein, I., Patel, M., Ikezaki, H., Dagar, S., and Önyüksel, H. (2000). Conformation and vasoreactivity of VIP in phospholipids: Effects of calmodulin. *Peptides* **20**, 1497–1501.

Rubinstein, I., Dagar, S., Sethi, V., Krishnadas, A., and Önyüksel, H. (2001). Surface-active properties of vasoactive intestinal peptide*. *Peptides* **22**, 671–675.

Said, S. I. (1986). Vasoactive intestinal peptide. *J. Endocrinol. Invest.* **9**, 191–200.

Said, S. I. (1992). Vasoactive intestinal polypeptide (VIP) in lung function and disease. *Biomed. Res.* **13**, 257–268.

Séjourné, F., Rubinstein, I., Suzuki, H., and Alkan-Önyüksel, H. (1997a). Development of a novel bioactive formulation of vasoactive intestinal peptide in sterically stabilized liposomes. *Pharm. Res.* **14**, 362–365.

Séjourné, F., Suzuki, H., Alkan-Önyüksel, H., Gao, X. P., Ikezaki, H., and Rubinstein, I. (1997b). Mechanisms of vasodilation elicited by VIP in sterically stabilized liposomes *in vivo. Am. J. Physiol.* **273**, R287–292.

Straubinger, R. M., Düzgüneş, N., and Papahadjopoulos, D. (1985). pH-sensitive liposomes mediate cytoplasmic delivery of encapsulated macromolecules. *FEBS Lett.* **179**, 148–154.

Suzuki, H., Noda, Y., Paul, S., Gao, X. P., and Rubinstein, I. (1995). Encapsulation of vasoactive intestinal peptide into liposomes: Effects on vasodilation *in vivo. Life Sci.* **57**, 1451–1457.

Suzuki, H., Noda, Y., Gao, X. P., Séjourné, F., Alkan-Önyüksel, H., Paul, S., and Rubinstein, I. (1996a). Encapsulation of VIP into liposomes restores vasorelaxation in hypertension *in situ. Am. J. Physiol.* **271**, H282–287.

Suzuki, H., Gao, X. P., Olopade, C. O., and Rubinstein, I. (1996b). Aqueous smokeless tobacco extract impairs endothelium-dependent vasodilation in the oral mucosa. *J. Appl. Physiol.* **81**, 225–231.

Szucs, M., and Tilcock, C. (1995). Lyophilization and rehydration of polymer-coated lipid vesicles containing a lipophilic chelator in the presence of sucrose: Labeling with 99mTc and biodistribution studies. *Nucl. Med. Biol.* **22**, 263–268.

Thakur, M. L., Marcus, C. S., Saeed, S., Pallela, V., Minami, C., Diggles, L., Pham, H. L., Ahdoot, R., Kalinowski, E. A., and Moody, T. (2000). 99mTc-labeled vasoactive intestinal peptide analog for rapid localization of tumors in humans. *Ann. N.Y. Acad. Sci.* **921**, 37–44.

Thomas, C. L., Artwohl, J. E., Suzuki, H., Gao, X., White, E., Saroli, A., Bunte, R. M., and Rubinstein, I. (1997). Initial characterization of hamsters with spontaneous hypertension. *Hypertension* **30**, 301–304.

Yuan, F., Salehi, H. A., Boucher, Y., Vasthare, U. S., Tuma, R. F., and Jain, R. K. (1994). Vascular permeability and microcirculation of gliomas and mammary carcinomas transplanted in rat and mouse cranial windows. *Cancer Res.* **54**, 4564–4568.

[22] Liposomal Superoxide Dismutases and Their Use in the Treatment of Experimental Arthritis

By M. Eugénia M. Cruz, M. Manuela Gaspar,
M. Bárbara F. Martins, and M. Luísa Corvo

Abstract

It has long been suggested that superoxide dismutase (SOD) be used for antioxidant therapy on the basis of its ability to catalyze the dismutation of superoxide radicals involved in the pathogenesis of several inflammatory disorders such as rheumatoid arthritis. However, the administration of SOD in free form has some disadvantages, most importantly, the low accumulation of SOD in inflamed areas due to its reduced half-life in the bloodstream and its rapid renal excretion. To overcome this, SOD can be incorporated either in highly loaded conventional liposomes (SA-liposomes) or long circulating liposomes (PEG-liposomes). After an appropriate formulation of SOD in SA-liposomes, the therapeutic effect is strongly increased, as indicated by a reduction of about 40% of inflammation edema compared with treatment with nonencapsulated enzyme. Compared with SA-liposomes, PEG-liposomes show superior therapeutic activity. A second approach consists of the construction of a hydrophobic SOD derivative (Ac-SOD) that can be partially inserted within the lipid matrix of liposomes and that expresses enzymatic activity to the external medium. This hydrophobic enzyme, Ac-SOD, associated with liposomes (so called Ac-SOD-enzymosomes), is able to exert its therapeutic activity while circulating in the organism, regardless of the integrity of the liposomes. Ac-SOD-enzymosomes have a more rapid antiinflammatory effect than SOD liposomes, confirming that the release of Ac-SOD from liposomes is no longer required to achieve dismutation. Different methodologies for the preparation of SOD and Ac-SOD liposomal formulations (conventional and long circulating) have been established and are described in detail here.

Introduction

Rheumatoid arthritis is an inflammatory pathology characterized by an overproduction of reactive oxygen species, namely superoxide radical, which accumulates and damages the peripheral joints (Dowling et al., 1993). It has long been suggested that superoxide dismutase (SOD) be used for antioxidant therapy on the basis of its ability to catalyze the dismutation of superoxide radicals (Jadot et al., 1995). Several attempts have been made to use SOD as a therapeutic agent against oxygen radical–related diseases (Henrotin et al., 1992; Starkebaum, 1993; Tanswell and Freeman, 1987). However, the therapeutic success of SOD as an antiarthritic drug has been hindered by its short residence time and consequent low accumulation in inflamed areas (Dowling et al., 1993; Petkau et al., 1987; Takakura et al., 1994). One strategy to overcome these problems is to incorporate SOD in liposomes, thereby increasing the circulation time and targeting it to the site of action (Corvo et al., 1997). Early attempts to treat rheumatoid arthritis with liposomal SOD were not very successful due to formulation problems, namely low enzyme encapsulation efficiency, resulting in low concentrations at the inflammation site (Baker et al., 1992; Hariton et al., 1988; Michelson et al., 1981; Niwa et al., 1985). After an appropriate formulation of SOD in highly loaded conventional liposomes (SA-liposomes), the therapeutic effect was strongly increased, as indicated by a reduction of 40–45% of inflammation edema compared with treatment with nonencapsulated enzyme (Corvo *et al.*, 1997). Long circulating liposomes (PEG-liposomes) showed advantages compared with SA-conventional liposomes. Small-sized PEG-liposomes (0.1 μm) accumulate to a larger extent at the inflammatory focus in rats with adjuvant arthritis (Corvo *et al.*, 1999). Comparative tests have shown that the higher specific targeting ability of PEG-liposomes is translated into a higher therapeutic activity of these formulations compared with SA-liposomes. In any of the above-mentioned liposomal formulations, the therapeutic activity of SOD is restricted by its release from liposomes localized at the inflamed site. In fact, SOD, a hydrophilic enzyme, is likely to be encapsulated in the internal aqueous space of the liposomes, being unable to exert its activity in the intact liposomal form (Corvo *et al.*, 1997). To improve *in vivo* SOD activity independent of liposome disruption, another approach was explored, consisting of the construction of a hydrophobic SOD derivative able to be partially inserted in the lipid matrix of liposomes and partially exposed to the external medium (Cruz *et al.*, 1994; Martins *et al.*, 1992). Theoretically, this entity (hydrophobic enzyme) associated with liposomes would be able to exert its therapeutic activity while circulating in the organism. Figure 1 illustrates SOD (I) and Ac-SOD

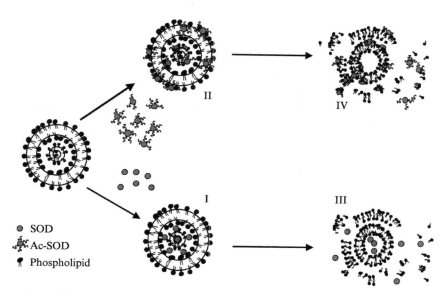

○ SOD
Ac-SOD
Phospholipid

Fig. 1. Schematic representation of SOD (I) and Ac-SOD (II) localization in liposomes and correspondent release (III, IV).

(II) localization in liposomes in the aqueous space and in the lipid matrix, respectively (Cruz, 1994, 1995). Accordingly, a different behavior in regard to enzymatic activity of SOD and Ac-SOD liposomes either in intact form (I and II) or during liposome disruption (III and IV) is expected. Hydrophobic derivatives of SOD, Ac-SOD, were constructed by covalent linkage of fatty acid residues to accessible amino groups of the enzyme (Martins *et al.*, 1990, 1996). They were incorporated in conventional and long circulating liposomes and extensively characterized, in terms of incorporation parameters, ζ-potential, and enzymatic activity (Cruz *et al.*, 1999; Gaspar *et al.*, 2000, 2003). As expected, in contrast to SOD liposomes, Ac-SOD liposomes are able to express enzymatic activity while in the intact liposomal form. Therefore, they are called Ac-SOD-enzymosomes, which means liposomes with surface-exposed SOD (Cruz *et al.*, 1999). Enzymosomes have been shown to possess therapeutic activity in experimental arthritis, irrespective of the type of liposomes used. Long circulating enzymosomes had antiinflammatory activity superior to conventional ones. Also, it was demonstrated that enzymosomes have a faster antiinflammatory effect than SOD liposomes, confirming that the release of Ac-SOD from liposomes is no longer required to achieve dismutation (Gaspar *et al.*, 2000).

Based on the physicochemical properties of SOD and Ac-SOD, different methodologies for the preparation of corresponding liposomal formulations (conventional and long circulating) have been established and are described in detail here.

Methods

Preparation of SOD-Liposomes

Preparation of Multilamellar SOD Liposomes. Multilamellar liposomes are prepared by the simple dehydration–rehydration method (sDRV) (Fig. 2) (Cruz *et al.*, 1989, 1993). Mixtures in a molar ratio of 7:2:1 of phosphatidylcholine (PC) (Lipoid, Ludwigshafen, Germany), cholesterol (Chol), and stearylamine (SA) (Sigma Chemical Co., St. Louis, MO) for SA-liposomes or 1.85:1:0.15 of PC, Chol, and distearoylphosphatidylethanolamine–poly(ethylene glycol) 2000 (DSPE-PEG; Avanti Polar Lipids Inc., Alabaster, AL) for long circulating liposomes (PEG-liposomes) in chloroform, in a total lipid concentration of 32 μmol/ml, are dried under a nitrogen stream until a homogeneous film is formed. This film is dispersed in 1.0 mg/ml or 5.0 mg/ml bovine erythrocytes Cu-Zn superoxide dismutase (SOD; Sigma Chemical Co., St. Louis, MO) in deionized water,

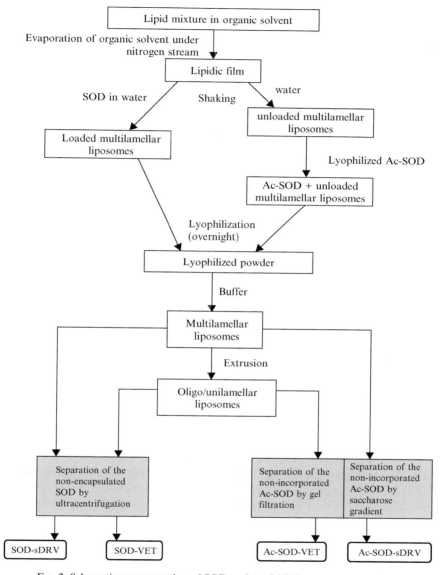

FIG. 2. Schematic representation of SOD and Ac-SOD liposome preparation.

respectively, for SA- or PEG-liposomes, frozen in liquid nitrogen and lyophilized overnight. A 0.28 M mannitol/10 mM citrate buffer, pH 5.6, is then added to the lyophilized powder, up to one-tenth of the volume of the original dispersion. This hydration step lasts 30 min and, subsequently, 0.145 M NaCl/10 mM citrate buffer, pH 5.6, is added up to the starting volume. Except for the dispersions that are extruded (see below), nonencapsulated protein is separated from the liposome dispersion by ultracentrifugation at 300,000g for 20 min at 4° in a Beckman LM-80 ultracentrifuge. Finally, liposomes are dispersed in the desired volume of 0.145 M NaCl/10 mM citrate buffer, pH 5.6. The encapsulation efficiency of this method is 45–70% depending on the lipid composition used.

Preparation of Extruded SOD Liposomes (VET). Multilamellar liposomes obtained by the previously described method are extruded sequentially through polycarbonate filters of appropriate pore size until the desired vesicle size is reached (i.e., 0.6, 0.4, 0.2, 0.1, and 0.05 μm) under a nitrogen pressure of 100–500 lb/in.[2] with an extruder device (Lipex, Biomembranes Inc., Vancouver, British Columbia, Canada). To guarantee the homogeneity of the preparation, at the last filter size, the liposomal suspension is collected and reextruded until a polydispersity lower than 0.2 is obtained. Nonencapsulated protein is separated from the liposome dispersion by 26-fold dilution with 0.145 M NaCl/10 mM citrate buffer, pH 5.6, followed by ultracentrifugation at 300,000g for 120 min at 4° in a Beckman LM-80 ultracentrifuge. Finally, liposomes are dispersed in the desired volume of 0.145 M NaCl/10 mM citrate buffer, pH 5.6.

Preparation of Ac-SOD-Liposomes

Preparation of Ac-SOD. The Ac-SOD is prepared by acylation of the accessible ε-NH$_2$ groups of the enzyme with acyl chlorides dispersed in micellar medium, according to Torchilin *et al.* (1980) with some modifications (Martins *et al.*, 1990, 1996). The percentage of the amino groups covalently linked to hydrophobic acyl chains is controlled by the molar ratio between the acyl chloride and the protein and is also a function of the acyl chain length (Martins *et al.*, 1992).

As an example for the modification of 30–35% of the ε-NH$_2$ groups of SOD by linkage to fatty acid residues with an acyl chain of 16 carbon atoms, a molar ratio of palmitoyl chloride/protein of 140 is adequate. In brief, an aliquot of the palmitoyl chloride (2 μl) is added to a test tube containing 3 ml of 50 mM carbonate buffer, pH 9.4, with 1% sodium cholate. The tube is immediately immersed in an ultrasonic waterbath,

Sonorex, RK 156, for 90 s. An adequate dispersion is achieved when the preparation becomes opalescent. If it does not occur, the dispersion procedure must be checked. To aid in this step, special care should be taken to place the test tube in the regions of the waterbath where some slight turbulence is noticed. The opalescent preparation is immediately transferred to a test tube containing 1500 μg of the enzyme. The tube containing the mixture must be vigorously shaken and placed in a rotating mixer for tubes, at room temperature, for 2 h. The modified enzyme is then recovered from the reaction medium by two centrifugations, with a dialysis step in between. In brief, to facilitate the separation of the excess palmitoyl chloride (hydrolyzed to palmitic acid and remaining in suspension because of the presence of the sodium cholate micelles), the preparation is diluted with an equal volume of cold water and placed in a refrigerator to achieve a temperature of 5°. The palmitic acid is then removed by centrifugation at 20,000g, at 5°, in a Sigma 202MK centrifuge. The supernatant is assayed for the percentage of blocked ε-NH$_2$ groups of the enzyme and dialyzed against water using a 12,000 D cut-off membrane. The dialysis is carried out for 72 h, with efficient stirring of the outside solution. The sample is carefully removed from the dialysis bag and again centrifuged at 20,000g, at 5°. The final preparation of Ac-SOD is recovered in the supernatant. The preparation is assayed for total protein, transferred to appropriate vials, and lyophilized.

Recommendations: If considering scaling-up the acylation procedure, it is important to know that it is very difficult to achieve a good dispersion for a volume of buffer higher than 3 ml and for a volume of acyl chloride higher than 3 μl, especially for acyl chlorides with more than 14 carbon atoms. The acyl chlorides are very unstable in the presence of humidity and that should be considered during the manipulation and storage of these reagents. The test tube for dispersion must have a diameter of 18–20 mm, and the aliquot of acyl chloride must be dispensed near the surface of the buffer to avoid its retention at the test tube walls. The enzyme concentration in the reaction medium must be 16 μM or lower. To achieve a good preservation of the catalytic activity of the enzyme, substrate must be present in the reaction medium (Martins et al., 1996). In this case, the test tube must have a capacity higher than the volume of the reaction medium.

As a reference for all the assays of characterization of the modified enzyme, native enzyme is previously solubilized in 50 mM carbonate buffer, pH 9.4, with 1% sodium cholate, mixed, and centrifuged, as described for the modification. Aliquots are taken and used as a reference for quantifying of ε-NH$_2$ groups. The enzyme solution is then dialyzed against water, assayed for protein, and lyophilized.

Preparation of Ac-SOD Multilamellar Liposomes. For the incorporation of Ac-SOD in multilamellar liposomes, vesicles without protein are
first prepared and then mixed with lyophilized Ac-SOD (Jorge *et al.*, 1994)
(Fig. 2). The lipid mixtures and the procedure to obtain the lipidic film are
the same as described for the SOD liposomes. One milliliter of water is
added per 32 μmol of lipid to form empty liposomes. These liposomes
are added to 1 or 3 mg of lyophilized Ac-SOD per ml of liposomes for
SA-liposomes or PEG-liposomes, respectively. The mixture is frozen and
lyophilized overnight. The rehydration of the lyophilized powder to obtain
Ac-SOD multilamellar vesicles is performed as for SOD multilamellar liposomes. Except for the dispersions that are extruded (see below),
nonincorporated protein is separated from the liposome dispersion by a
discontinuous saccharose gradient from 0 to 40 mg/ml. In a 5-ml ultracentrifuge tube the gradient is made by the sequential addition, at the bottom
of the tube, of 350 μl of water and 800 μl of 4, 8, 12, 16, and 40 mg/ml
saccharose. Finally, 0.35 ml of the liposomal suspension is added at the top
of the gradient and is ultracentrifuged for 30 min at 170,000g in a Beckman
swinging bucket rotor. The liposomes are collected at the middle of the
gradient, and the nonincorporated protein at the bottom of the tube is
neglected. To wash the saccharose out of the liposome medium, collected
liposomes are diluted 20 times and ultracentrifuged, at 300,000g for 20 min
at 4° in a Beckman LM-80 ultracentrifuge, and the precipitated liposomes
are finally ressuspended in the desired volume of 0.145 M NaCl/10 mM
citrate buffer, pH 5.6. The incorporation efficiency of this method is
40–60%, depending on the lipid composition used.

Preparation of Ac-SOD Extruded Liposomes (VET). Multilamellar
liposomes obtained by the previous method are extruded as described
for the preparation of SOD extruded liposomes, except for the separation
of the nonincorporated Ac-SOD. To remove the nonincorporated Ac-SOD
from liposomes a gel filtration technique on Sephadex G-200 (Pharmacia,
Uppsala, Sweden) is used. The gel is allowed to swell in the presence
of 0.145 M NaCl/10 mM citrate buffer, pH 5.6, at 90° for 5 h, and columns
of 18 × 1 cm (Bio-Rad Laboratories, Hercules, CA) with the obtained
gel are prepared. In each column, 1 ml of liposomes is applied to the top of
the column and eluted with the same buffer. Samples of 1 ml are collected
and the elution profile is analyzed spectrophotometrically at 280 and
450 nm for protein and vesicle detection, respectively. After the void
volume (5 ml) and based on spectrophotometric data, the fractions containing liposomes are eluted first followed by the fractions corresponding
to the nonincorporated Ac-SOD. The fractions containing the eluted
liposomes are pooled and ultracentrifuged for concentration at 250,000g
for 3 h at 15° in a Beckman LM-80 ultracentrifuge. The pellet is finally

resuspended in a desired volume of 0.145 M NaCl/10 mM citrate buffer, pH 5.6.

Characterization of Ac-SOD

The modified enzyme is characterized for degree of modification, retention of enzymatic activity, partition coefficient, isoelectric point, and ζ-potential according to the following procedures.

Degree of Modification. The percentage of modified ε-NH$_2$ groups is quantified by a fluorometric assay (Böhlen *et al.*, 1973). The Ac-SOD is assayed for unblocked ε-NH$_2$ groups, and the native enzyme is assayed for the total accessible ε-NH$_2$ groups of the protein. Briefly, aliquots of 5–125 μl, containing 0.25–25 μg of protein, are diluted with 50 mM phosphate buffer, pH 8, to a final volume of 750 μl. A fluorescamine solution, freshly prepared by solubilization of 30 mg of fluorescamine in 100 ml of dioxane (the presence of traces of humidity in dioxane should be avoided), is added in a volume of 250 μl. This addition has to be performed under constant stirring. Since fluorescamine can react with trace amines, test tubes must be carefully cleaned and a reagent blank must be run routinely. The intensity of fluorescence emission (I_f) of bound fluorescamine at 475 nm is measured for excitation at 390 nm in a Hitachi, F 3000 spectrofluorimeter. The degree of modification is calculated as follows:

$$[1 - (I_f \text{ per mg of Ac-SOD})/(I_f \text{ per mg of SOD})] \times 100$$

Catalytic Activity Retention Assay. The Ac-SOD is assayed for catalytic activity measuring the ability of the enzyme to decrease the rate of autoxidation of epinephrine to adrenochrome (Misra and Fridovich, 1972; Sun and Zigman, 1978). The extent of inhibition is taken as a measure of the catalytic activity of the enzyme. The catalytic assays are performed including 1% cholic acid in the reaction medium. The volume of epinephrine that results in a ΔAbs (480 nm)/min of 0.025 must be determined, following the autoxidation reaction for 6 min. The reaction is initialized by the addition of 40–60 μl of 0.01 M epinephrine in 10^{-2} M hydrochloric acid to 0.05 M carbonate buffer, pH 10.2, with 10^{-4} M ethylenediametetraacetic acid (EDTA) at 30°, in a final volume of 1.5 ml. A curve of the inhibition (%) of the autoxidation of epinephrine as a function of the amount of Ac-SOD is plotted. The amount (μg) of Ac-SOD required to inhibit 50% of the epinephrine autoxidation is calculated. An equivalent assay is performed for native SOD. The retention of catalytic activity is calculated as follows:

$$[(\mu\text{g SOD})/(\mu\text{g Ac-SOD})] \times 100$$

Partition Coefficient Determination. Water (2.5 ml) and octanol (2.5 ml) are mixed and vortexed until an emulsion is formed. The mixture is added to 0.25 mg of protein and the emulsion stirred for 2 h. The preparation is centrifuged at 5000g for 5 min. The two phases formed after centrifugation are analyzed for protein content measuring the intrinsic emission of SOD or Ac-SOD fluorescence. Equivalent preparations without protein are used as reference. The partition coefficient in octanol/water is determined as the ratio between the protein present in the octanol and water phases.

Isoelectric Point and ζ-Potential Determination. The isoelectric point (p*I*) of Ac-SOD is quantified by microelectrophoresis combined with Doppler laser anemometry and photon correlation spectroscopy, using a Zeta-Sizer 2000 (Malvern, UK). In brief, the electrophoretic mobility of Ac-SOD or the native enzyme is monitored as a function of the pH of the electrolyte. Different buffers as electrolytes are used for pH values in the range 2.5–8.5, with a constant and low ionic strength ($I = 0.01$), to promote electrostatic interactions either for the modified or the native enzyme. The concentration of enzyme is 500 μg/ml. The experiments are performed at 20°. The isoelectric point corresponds to the pH of null electrophoretic mobility. At each pH, the ζ-potentials of the modified and native macromolecules are also obtained.

Characterization of SOD and Ac-SOD Liposomes

The liposomal formulations are characterized by lipid composition, lipid and protein concentration, enzymatic activity retention, ζ-potential, and mean diameter. These data allow us to evaluate the encapsulation/incorporation parameters.

Encapsulation/Incorporation Parameters. Encapsulation efficiency (EE) for a hydrophilic protein (SOD) is calculated as follows:

$$[(Final\,(Prot/Lip))/(Initial\,(Prot/Lip))] \times 100$$

The incorporation efficiency (IE) for hydrophobic protein (Ac-SOD) is calculated as follows:

$$[(Final\,(Prot/Lip))/(Initial\,(Prot/Lip))] \times 100$$

The EE and IE, being ratios between final to initial (protein/lipid), are the measure of efficiency of an available lipid mixture to encapsulate hydrophilic SOD or incorporate hydrophobic SOD in the final liposomal form.

Protein Quantification. The method for protein quantification in liposomes is based on the Lowry method (Lowry *et al.*, 1951) with previous disruption of the vesicles with Triton X-100 and sodium dodecyl sulfate (SDS) (Wang and Smith, 1975). Samples, in triplicate, containing a protein amount between 7 and 35 μg (maximum volume, 0.5 ml) are pipetted into 15-ml tubes. In parallel, a calibration curve with 0.1 mg/ml bovine serum albumin (BSA) stock solution is prepared. In triplicate, 0, 5, 10, 20, 30, and 40 μg of BSA stock solution per tube are pipetted. Water is then added to obtain a total volume in each tube of 0.5 ml. To disrupt liposomes, 0.4 ml of SDS [20% (w/v)] and 0.5 ml of Triton X-100 [2% (w/v)] are added to all samples (including the calibration curve). After the addition of 0.1 ml of 1 N NaOH, all tubes are vortexed and heated at 60° for 20 min or until a clear solution is observed that corresponds to liposome disruption. After cooling samples to room temperature, 1.0 ml of Lowry solution is added to all tubes, vortexed, and incubated at room temperature for 10 min. Then, 0.1 ml of the diluted Folin-Ciocalteau (1:1 with water) solution is added. After 45 min, the absorbance of all samples is recorded at 750 nm, against the blank of the calibration curve, in a spectrophotometer (UV 160, Shimadzu, Tokyo, Japan). The colored complex formed is stable for about 2 h.

Phospholipid Quantification. The method for phospholipid quantification is based on the colorimetric determination of PO_4^{3-}. The inorganic phosphate is converted to phosphomolybdic acid, which is quantitatively converted to a blue color due to reduction of ascorbic acid during heating, using the method described by Rouser et al. (1970). Briefly, samples in triplicate containing a phosphate amount between 10 and 80 nmol (sample volume, less than 100 μl) are pipetted into 15-ml glass tubes. In parallel, a calibration curve with a 0.5 mM phosphate solution is prepared. In triplicate, phosphate amounts of 20, 30, 40, 50, 60, and 80 nmol are pipetted into glass tubes. All the tubes are heated (180°) in a heating block until dryness. After cooling, 0.3 ml of perchloric acid (70–72%) is added to all tubes. Marbles are placed on the top of all tubes and heated in the heating block (180°) for 45 min to convert to inorganic form all the organic lipid phosphate and until a clear solution is obtained. After cooling samples to room temperature, 0.4 ml of hexaammoniumheptamolybdate solution [1.25% (w/v)] followed by 0.4 ml of ascorbic acid solution [5% (w/v)] are added to all glass tubes. A blue color is obtained due to the reduction of ascorbic acid during heating in a boiling waterbath for 5 min. After cooling the tubes, the absorbance at 797 nm of all samples, against the blank of the calibration curve, is recorded in a UV 160 Spectrophotometer (Shimadzu). The amount of phosphate in the samples is obtained through the calibration curve with the aid of linear regression. The calibration curve is linear up to absorbance values of at least 1.000.

Enzymatic Activity of SOD. The retention of total enzymatic activity (Total Act.) of SOD is defined as the percentage of the enzymatic activity of the final liposomes related to a control SOD solution, obtained after disruption of samples with Triton X-100. The percentage of the enzymatic activity exposed to the external surface (Partial Act.) is quantified in intact liposomes (without disruption with Triton X-100) and related to the Total Act.

To determine total enzymatic activity, SOD or Ac-SOD encapsulated/incorporated in liposomes must be released from the vesicles by the addition of 20% Triton X-100 in a ratio 1:1. Appropriate dilutions must be done to prepare a final protein concentration of 6 μg/ml, resulting in a maximum of 0.3% Triton X-100. To determine the enzymatic activity exposed to the external surface (Partial Act.) of SOD or Ac-SOD, appropriate dilutions with 0.145 M NaCl/10 mM citrate buffer (pH 5.6) must be made to prepare a final protein concentration of 12 μg/ml. The retention of Total Act. is calculated as follows:

$$(\mu\text{g SOD or Ac-SOD (Control solution))}/ (\mu\text{g SOD or Ac-SOD (Final preparation))} \times 100$$

The Partial Act. is calculated as follows:

$$(\mu\text{g SOD or Ac-SOD (Intact liposomes))}/ (\mu\text{g SOD or Ac-SOD (Total Act.))} \times 100$$

Liposome Size Measurements. Liposome mean diameter is determined by dynamic light scattering in a ZetaSizer 1000 HS$_A$ (Malvern, UK). This technique is based on Brownian motion and is related to the size of the particles. For viscosity and refractive index, the values of pure water are used. As a measure of particle size distribution of the dispersion, the system reports the polydispersity index, (PI). The index ranges from 0.0 for an entirely monodisperse sample up to 1.0 for a polydisperse suspension. To determine the mean diameter and polydispersity index of the liposomal preparations, in an appropriate cell for the ZetaSizer, 25 μl of the liposomal suspension is added to 1 ml of the working buffer. To ensure that appropriate mean diameter and polydispersity index are achieved, besides the measurements done for final liposomal preparations, these parameters are also determined during the extrusion procedure.

ζ-Potential Determination. Before determination of the ζ-potential of liposomal formulations, an initial check of the apparatus is made with a standard of a well-known ζ-potential value (standard DTS5050, Malvern Instruments, Ltd., UK). Dilutions of liposome formulations are made with the filtered (0.1 μm) working buffer to achieve 3 μmol/ml lipid concentration in all samples. Samples (5 ml) are introduced into the capillary cell via a syringe. The ζ-potential of the samples at a temperature of 25° is recorded. Three independent dilutions are prepared for each liposomal formulation under study. The scattering angle is 12° and the electric field intensity ranges from 18.5 to 19.6 V/cm.

Biological Evaluation of SOD and Ac-SOD Liposomes

Male Wistar rats aged 3 months old and weighing 250–300 g are kept under standard hygiene conditions, fed with commercial chow, and given acidified drinking water *ad libitum*. All experimental procedures are carried out with the permission of the local laboratory animal committee.

Rat Adjuvant Arthritis Model. For the induction of inflammation, male Wistar rats are injected with a single intradermal injection of 0.1 ml of a suspension of *Mycobacterium butiricum* killed and dried in incomplete Freund's adjuvant (Difco Laboratories, Detroit, MI) at 10 mg/ml, into the subplantar area of the right hind paw. In this animal model, the inflammation is established 7 days after the induction.

According to the treatment schedule, rats receive 1, 3, 6, or 11 intravenous injections of SOD or Ac-SOD formulations during 11 days. The administered dose ranges from 33 to 363 μg per rat. Treatments begin 7 days after induction of inflammation.

Evaluation of Antiinflammatory Activity. The evaluation of the antiinflammatory activity is assessed by changes of the physical swelling of the hind paw. The ankle circumference of the inflamed paw is measured before the induction of inflammation, 7 days after induction, and during the 11-day treatment period. The regression of the edema is calculated according to the following formula:

$$[(\text{Ankle day } X - \text{Ankle 7 days after ind.})/ (\text{Ankle bef. ind.} - \text{Ankle 7 days after ind.})] \times 100(\%)$$

where ankle day X is ankle circumference measured at day X during the 11-day treatment period, ankle 7 days after ind. is ankle circumference measured 7 days after the induction of inflammation, and ankle bef. ind. is ankle circumference measured before the induction of inflammation.

In addition to the measure of the ankle circumference of the inflamed paw, the evaluation of paw edema is also monitored by changes in the volume of the affected paw before inflammation induction, 7 days after induction, and during the 11-day treatment period. Inflamed paw volumes are determined using a plethysmometer (7140 Plethysmometer, Ugo, Basile, Italy), equipment for accurate measurement of the rat paw swelling. The regression of the volume of inflamed paw is calculated according to the following formula:

$$[(\text{Vol day } X - \text{Vol. after ind.})/(\text{Vol. bef. ind.} - \text{Vol. after ind.})] \times 100(\%)$$

where vol. day X is the volume of the inflamed paw measured at day X during the 11-day treatment period, vol. after ind. is the volume of the inflamed paw measured 7 days after induction of inflammation, and vol. bef. ind. is the volume of the paw measured before induction of inflammation.

Results

Properties of Ac-SOD

For Ac-SOD conjugated with palmitic chains a degree of modification of 30% is achieved, with 90% retention of the catalytic activity. A minimized perturbation for the structure of Ac-SOD occurred, considering the results of the retention of catalytic activity. An alteration of surface charge is indicated by the electrophoretic mobility curves and the ζ-potential values. The isoelectric point of Ac-SOD is 4.83 and of SOD is 4.96. The decrease of pI is expected due to the blockage of 30% of the accessible ε-NH$_2$ [pK_a of ε-NH$_2$ is in the range 9.3–9.5 (Means and Feeney, 1971)]. Also, a deviation between the profiles of electrophoretic mobility, as a function of pH, for both forms of the enzyme is noticed. For example, at pH 5.6, ζ-potentials of -16.7 mV for Ac-SOD and -5.7 mV for SOD allow us to preview a different interaction of Ac-SOD with charged liposomal bilayers. The hydrophobicity of Ac-SOD evaluated by the octanol/water partition coefficient indicates a 20-fold increase in the affinity of the bioconjugate to octanol, without accumulation of Ac-SOD at the interface. This points to a significant increase in the affinity of Ac-SOD to hydrophobic media. Considering these results, both hydrophobic and electrostatic interactions can contribute to the association between Ac-SOD and the bilayers of liposomes. Consistent with the purpose of the construction of a modified enzyme with affinity for the lipid matrix of liposomes, Ac-SOD can be a candidate to replace the native enzyme in processes by which a hydrophobic interaction with the liposome bilayer is required.

Properties of SOD and Ac-SOD Liposomes

Table I shows the typical parameters of SOD and Ac-SOD liposomes of two different types (SA-liposomes and PEG-liposomes). The incorporation efficiencies for Ac-SOD liposomes are substantially higher than those obtained for SOD liposomes. To reach a similar enzyme loading capacity [final (Prot/Lip)] of about 14 μg/μmol, a 2- to 6-fold higher initial protein concentration is needed for SOD encapsulation in SA- and PEG-liposomes, respectively, compared with the acylated derivative. The higher

TABLE I
TYPICAL PARAMETERS OF SOD AND Ac-SOD LIPOSOMAL FORMULATIONS[a]

Formulations	Initial protein (mg/ml)	IE/EE (%)	Partial activity (%)
Ac-SOD-PEG-liposomes	0.6–0.8[b]	43–54[b]	43–47[b]
SOD-PEG-liposomes	3.2–4.8[c]	10–18[c]	<2[c]
Ac-SOD-SA-liposomes	0.6–0.7[b]	58–60[b]	40–45[b]
SOD-SA-liposomes	1–1.45[c]	35–45[c]	<9

[a] Partial activity: percentage of the enzymatic activity exposed on the enzymosome surface in relation to total activity. Liposomes presented a mean diameter of 0.1 μm with a polydispersity index lower than 0.12. Final (protein/lipid) is about 14 $\mu g/\mu$mol.
[b] From Gaspar et al. (2000).

IE of Ac-SOD compared with EE of SOD seems to reflect the higher affinity of the acylated enzyme for the hydrophobic region of the liposomes. This is in agreement with the higher partition coefficient value of Ac-SOD.

Liposomes prepared with Ac-SOD can express enzymatic activity before disruption of the vesicles typically around 40–47% of the total activity. This property of SA- and PEG-liposomes incorporating Ac-SOD indicates that Ac-SOD is exposed on the external surface of the liposomes and is accessible to the substrate. SOD liposomes do not show relevant enzymatic activity before disruption (lower than 9%), indicating that this enzyme is encapsulated in the internal space of liposomes. This assumption is confirmed by ζ-potential measurements of SOD and Ac-SOD liposomes (Gaspar et al., 2003).

Therapeutic Activity of SOD and Ac-SOD Liposomes

SOD encapsulated in either SA- or PEG-liposomes shows therapeutic activity in the rat adjuvant arthritis model. In this model, for untreated animals a mean increase in edema of 40% is observed. Any result showing a regression of the edema compared with the control group means that effective therapeutic activity is observed. Different therapeutic profiles can be obtained according to different formulations, schedules, doses, and site of injection. Figure 3 shows an example of therapeutic activity obtained with several SOD liposomal formulations. Free SOD shows a therapeutic effect but without regression of edema. In contrast, all SOD liposomal formulations show a higher therapeutic activity with an effective regression of the edema varying from 20–45%. No significant differences are observed for therapeutic efficacy with daily injections during the 11-day treatment

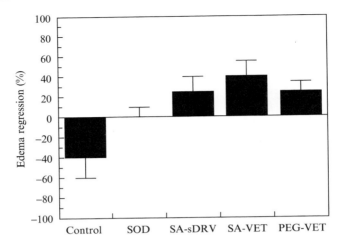

FIG. 3. Typical results obtained with liposomal SOD treatments in a daily treatment schedule with a dose of 33 μg SOD per animal in the rat adjuvant arthritis model. Control, control animals without treatment; SOD, animals treated with free SOD administered iv; SA-sDRV, animals treated with SOD encapsulated in sDRV SA-liposomes administered iv; SA-VET, animals treated with SOD encapsulated in VET SA-liposomes administered iv; PEG-VET, animals treated with SOD encapsulated in VET PEG-liposomes administered sc.

period either with SOD SA-liposomes or SOD PEG-liposomes. However, SOD PEG-liposomes are superior to SA-liposomes when one dose of a highly concentrated formulation (363 μg per rat) is used. For each type of liposome, no significant therapeutic differences are observed between SOD and Ac-SOD. Nevertheless, the antiinflammatory effect can be achieved earlier (4 days after the 11-day treatment period) with the Ac-SOD liposomes than with the SOD liposomes (Gaspar *et al.*, 2000).

Conclusions

We have developed methodologies to efficiently encapsulate/incorporate either the native (SOD) or the acylated form (Ac-SOD) of superoxide dismutase in conventional and long circulating liposomes. We have prepared a new entity, Ac-SOD, that, irrespective of having different physicochemical properties, retains the enzymatic activity of SOD, which is the most relevant characteristic for therapeutic purposes. We have established the experimental conditions to prepare SOD and Ac-SOD liposomes with similar final protein loading, allowing the comparison of respective therapeutic activities. Both types of liposomes, encapsulating/

incorporating either SOD or Ac-SOD, have been shown to be efficient therapeutic agents against experimental arthritis. A higher efficiency was found for PEG-liposomes compared with SA-liposomes, while a faster therapeutic effect of Ac-SOD enzymosomes was found. We can envisage that enzymosomes can be therapeutically used either for passive targeting to inflammation sites (conventional form) or long circulating microreservoirs (PEG form) for the treatment of other diseases, where enzymatic activity independent of liposome disruption is required.

We challenge researchers to explore novel applications of SOD liposomes and enzymosomes and to apply the encapsulation/incorporation methodologies described in this chapter to other proteins.

Acknowledgments

The authors would like to thank Prof. J. G. Morais for help with the pharmacokinetics studies, Prof. Gert Storm and Prof. Daan Crommelin for their contribution to the development of long circulating liposomes and for accepting M. L. Corvo and M. M. Gaspar at the Department of Pharmaceutics, University of Utrecht. This work was partially supported by research grants from JNICT/Portugal (contract numbers PBIC/C/SAU/1551/92 and PECS/C/SAU/140/95) and PRAXIS XXI/Portugal (contract number 2/2.1/SAU/1360/95). M. L. Corvo was the recipient of a personal grant from Fundação Calouste Gulbenkian and Comissão INVOTAN (Grant 3/A/95/PO).

References

Baker, R. R., Czopf, L., Jilling, T., Freeman, B. A., Kirk, K. L., and Matalon, S. (1992). Quantitation of alveolar distribution of liposome-entrapped antioxidant enzymes. *Am. J. Physiol.* **263,** L585–L594.

Böhlen, P., Stein, S., Dairman, W., and Undenfriend, S. (1973). Fluorimetric assay of proteins in the nanogram range. *Arch. Biochem. Biophys.* **155,** 213–220.

Corvo, M. L., Martins, M. B., Francisco, A. P., Morais, J. G., and Cruz, M. E. M. (1997). Liposomal formulations of Cu,Zn-superoxide dismutase: Physico-chemical characterization and activity assessment in an inflammation model. *J. Control. Rel.* **43,** 1–8.

Corvo, M. L., Boerman, O. C., Oyen, W. J. G., Bloois, L. V., Cruz, M. E. M., Crommelin, D. J. A., and Storm, G. (1999). Intravenous administration of superoxide dismutase in long circulating liposomes.II. *In Vivo* fate in a rat model of adjuvant arthritis. *Biochim. Biophys. Acta* **1419,** 325–334.

Corvo, M. L., Jorge, J. C., Ron van't Hof, Cruz, M. E. M., Crommelin, D. J. A., and Storm, G. (2002). Superoxide dismutase entrapped in long-circulating liposomes; formulation design and therapeutic activity in rat adjuvant arthritis. *Biochim. Biophys. Acta* **1564,** 227–236.

Cruz, M. E. M. (1994). Liposomal Delivery of Proteins. International Workshop on New Drug Delivery Systems, Coimbra, Portugal.

Cruz, M. E. M. (1995). Native and Hydrophobized Proteins Immobilized in Liposomes. 1st European Intensive Course on New Forms and New Routes of Administration for Drugs, Coimbra, Portugal.

Cruz, M. E. M., Corvo, M. L., Jorge, J. C. S., and Lopes, F. (1989). Liposomes as carrier systems for proteins: Factors affecting protein encapsulation. *In* "Liposomes in the Therapy of Infectious Diseases and Cancer" (G. Lopez-Berestein and I. Fidler, eds.), p. 417. A. R. Liss, New York.

Cruz, M. E. M., Gaspar, M. M., Lopes, F., Jorge, J. S., and Perez-Soler, R. (1993). Liposomal L-asparaginase: *In Vitro* evaluation. *Int. J. Pharmaceut.* **96**, 67–77.

Cruz, M. E. M., Martins, M. B., Corvo, M. L., Jorge, J. C. S., and Gaspar, M. M. (1994). Native and Lipophilic Derivatives of L-Asparaginase and Superoxide Dismutase and Respective Liposomal Forms. *In* "Proceedings of the International Symposium on Controlled Release Bioactive Materials" Vol. 21, pp. 346–347. Controlled Release Society, Inc. Deerfield, IL, USA.

Cruz, M. E. M., Gaspar, M. M., Martins, M. B., Corvo, M. L., and Storm, G. (1999). SOD enzymosomes for the treatment of adjuvant arthritis. *In* "Proceedings of the 4th International Conference on Liposome Advances—Progress in Drug and Vaccine Delivery." London.

Dowling, E. J., Chander, C. L., Claxson, A. W., Lillie, C., and Blake, D. R. (1993). Assessment of human recombinant manganese superoxide dismutase in models of inflammation. *Free Radic. Res. Commun.* **18**, 291–298.

Gaspar, M. M., Corvo, M. L., and Cruz, M. E. M. (2000). Acylated SOD liposomal formulations (SOD enzymosomes). *In* "Proceedings of the 4th Spanish-Portuguese Conference on Controlled Drug Delivery," Vitoria, Spain.

Gaspar, M. M., Martins, M. B., Corvo, M. L., and Cruz, M. E. M. (2003). Design and characterization of enzymosomes with surface-exposed superoxide dismutase. *Biochim. Biophys. Acta* **1609**, 211–217.

Hariton, C., Jadot, G., Michelson, A. M., and Mandel, P. (1988). Superoxide Dismutase treatment reduces [3H]flunitrasepam affinity in cortex and hippocampus of the rat. *Neurosci. Lett.* **102**, 313–318.

Henrotin, Y., Deby-Dupont, G., Deby, C., Franchimont, P., and Emerit, I. (1992). Active oxygen species, articular inflammation and cartilage damage. *In* "Free Radicals and Aging" (I. Emerit and B. Chance, eds.), pp. 309–322. Birkhauser Verlag, Basel.

Jadot, G., Vaille, A., Maldonado, J., and Vanelle, P. (1995). Clinical pharmacokinetics and delivery of bovine superoxide dismutase. *Clin. Pharmacokinetic* **28**, 17–25.

Jorge, J. C. S., Perez-Soler, R., Morais, J. G., and Cruz, M. E. M. (1994). Liposomal palmitoyl-L-asparaginase: Characterization and biological activity. *Cancer Chemother. Pharmacol.* **34**, 230–234.

Lowry, O. H., Rosebrough, N. J., Farr, A. L., and Randall, R. T. (1951). Protein measurement with the folin phenol reagent. *J. Biol. Chem.* **193**, 266–275.

Martins, M. B. F., Jorge, J. C. S., and Cruz, M. E. M. (1990). Acylation of L-asparaginase with total retention of enzymatic activity. *Biochimie* **72**, 671–675.

Martins, M. B. F., Corvo, M. L., and Cruz, M. E. M. (1992). Lipophilic derivatives of Cu, Zn-superoxide dismutase: characterization and immobilization in liposomes. *In* "Proceedings of the International Symposium on Controlled Release Bioactive Materials" (J. Kopeček, ed.), Vol. 19, pp. 524–525. Controlled Release Society, Inc. Deerfield, IL, USA.

Martins, M. B. F., Gonçalves, A. P. V., and Cruz, M. E. M. (1996). Biochemical Characterization of an L-asparaginase bioconjugate. *Bioconjugate Chem.* **4**, 430–435.

Means, G. E., and Feeney, R. E. (1971). "Chemical Modifications of Proteins." Holden-Day Inc., San Francisco.

Michelson, A. M., Puget, K., and Durosay, P. (1981). Studies of liposomal superoxide dismutase in rats. *Mol. Physiol.* **1**, 85–96.

Misra, H. P., and Fridovich, I. (1972). The role of superoxide anion in the autoxidation of epinephrine and simple assay for superoxide dismutase. *J. Biol. Chem.* **247**, 3170–3175.

Niwa, Y., Somiya, K., Michelson, A. M., and Puget, K. (1985). Effect of liposomal encapsulated superoxide dismutase on active oxygen related human disorders. A preliminary study. *Free Radic. Res. Commun.* **1**, 137–153.

Petkau, A. (1987). Role of superoxide dismutase in modification of radiation injury. *Br. J. Cancer* **55**, 87–95.

Rouser, G., Flusher, S., and Yamamoto, A. (1970). Two dimensional thin layer chromatographic separation of polar lipids and determination of phospholipids by phosphorus analysis of spots. *Lipids* **5**, 494–496.

Starkebaum, G. (1993). Review of rheumatoid arthritis. Recent developments. *Autoimmune Dis.* **13**, 273–289.

Sun, M., and Zigman, S. (1978). An improved spectrophotometric assay for superoxide dismutase based on epinephrine autoxidation. *Anal. Biochem.* **90**, 81–87.

Takakura, Y., Masuda, S., Tokuda, H., Nishikawa, M., and Hashida, M. (1994). Targeted delivery of superoxide dismutase to macrophages via mannose receptor-mediated mechanism. *Biochem. Pharmacol.* **47**, 853–858.

Tanswell, A. K., and Freeman, B. A. (1987). Liposome-entrapped antioxidant enzymes prevent lethal O_2 toxicity in newborn rat. *J. Appl. Physiol.* **63**, 347–352.

Torchilin, V. P., Omel'Yanenko, V. G., Klibanov, A. L., Mikhailov, A. I., Gol'Danskii, V. I., and Smirnov, V. N. (1980). Incorporation of hydrophilic protein modified with hydrophobic agent into liposome membrane. *Biochim. Biophys. Acta* **602**, 511–521.

Wang, C. H., and Smith, R. L. (1975). Lowry determination of protein in thye presence of triton X-100. *Anal. Biochem.* **63**, 414–417.

[23]   The Use of Sterically Stabilized Liposomes to Treat Asthma

By Kameswari S. Konduri, Sandhya Nandedkar, David A. Rickaby, Nejat Düzgüneş, and Pattisapu R. J. Gangadharam[†]

Abstract

Asthma is characterized by airway hyperresponsiveness, chronic inflammation, and airway remodeling, which may lead to progressive, irreversible lung damage. Liposomes have been used for the delivery of aerosolized asthma medications into the lungs. This method could facilitate sustained action of steroids while using only a fraction of the dosage and a less frequent dosing interval than conventional therapy. We describe the evaluation of the effect of budesonide encapsulated in sterically stabilized liposomes on lung inflammation and airway hyperreactivity in a mouse model of asthma. We outline the determination of markers implicated in the progression of asthma, including histopathology, eosinophil peroxidase activity in bronchoalveolar

[†] Deceased.

METHODS IN ENZYMOLOGY, VOL. 391

lavage, and airway hyperresponsiveness to methacholine. Weekly administration of budesonide in sterically stabilized liposomes results in a significant reduction in the total lung inflammation score, peripheral blood eosinophil counts, and the total serum IgE level, similar to that obtained with daily budesonide. Airway hyperresponsiveness to methacholine challenge decreases significantly in the group treated with weekly budesonide in sterically stabilized liposomes, while it does not decrease in the daily budesonide group.

Introduction

Asthma is the most common chronic illness in childhood and affects 5–10% of the population in North America. It accounts for the most hospitalizations and missed school and parent workdays, with an estimated cost of $12 billion per year (Weiss *et al.*, 1992). Asthma is characterized by airway hyperresponsiveness, chronic inflammation, and airway remodeling, which may lead to progressive, irreversible lung damage. It is primarily an inflammatory disease that occurs after a triggering agent (allergen) induces the release of histamine from mast cells. Histamine and other mediators released from the mast cells attract numerous inflammatory cells (i.e., lymphocytes, eosinophils) to the bronchial epithelium along with their proinflammatory cytokines and mediators (Bousquet *et al.*, 1990; Broide *et al.*, 1991; Kips *et al.*, 1995; Robinson *et al.*, 1992). Regular antiinflammatory medication use is crucial in preventing airway remodeling and the irreversible lung damage that occurs in asthma (Agertoft and Pedersen, 1994; Haahtela *et al.*, 1994; Overbeck *et al.*, 1996; USPHS NHLBI, 2002).

The mainstay of asthma therapy is directed at reducing the pulmonary inflammation with the use of anti-inflammatory drugs. Although the current inhaled steroids are very effective in preventing the significant inflammation that occurs in asthma, they have major drawbacks. They require daily administration for the drugs to be effective, which may lead to patient non-compliance and treatment failure. Since compliance in the patient is critical in interrupting the chronic inflammation that occurs in the asthmatic, this becomes a significant issue for therapy. The inhaled steroids are delivered to the lungs either with use of dry powdered inhalers, metered does inhalers or via nebulization. Even with the best technique, it is estimated that only 8–15% of a given dose is delivered to the lungs via a dry powdered inhaler and only about 3–8% via a metered dose inhaler (O'Callaghan *et al.*, 1994). In addition, only a fraction of the drug reaches the lower or peripheral airways. Inhaled steroids have a short half-life *in vivo* and potential toxicity when using higher doses (Brown *et al.*, 1991; O'Bryne and Pedersen, 1998; Pedersen and O'Bryne, 1997; Toogood *et al.*, 1994; Wong and Black, 1992).

Liposomes can encapsulate a variety of drugs and provide a repository for the slow release of the drugs, thereby providing a prolonged therapeutic effect. Liposome-encapsulated antibiotics show increased efficacy in a variety of infectious diseases (Düzgüneş, 1998; Gangadharam *et al.*, 1995; Wasan and Lopez-Berestein, 1998). Liposomes have also been used for the delivery of aerosolized asthma medications such as cromolyn sodium and albuterol sulfate (Gonda, 1990; Taylor *et al.*, 1989). Delivery in liposomes into the lungs could facilitate a sustained action of steroids while using only a fraction of the dosage and a less frequent dosing interval than conventional therapy.

Liposomes may be characterized by their lipid composition, surface charge, steric interactions, and number of lamellae. "Conventional liposomes" used for drug delivery are usually composed of naturally occurring phospholipids, such as phosphatidylcholine or phosphatidylglycerol with or without cholesterol. Although conventional liposomes can encapsulate a variety of drugs, they are recognized *in vivo* by the cells of the reticuloendothelial system (RES), are characterized by a nonspecific reactivity with the biological milieu, and are cleared rapidly from the circulation (Gregoriadis, 1995; Kamps and Scherphof, 2004). In addition, incorporation of triamcinolone or beclomethasone into conventional liposomes results in their rapid redistribution and leakage from liposomes into the medium (Farr *et al.*, 1989; Gonzalez-Rothi *et al.*, 1996; Schreier *et al.*, 1994; Vidgren *et al.*, 1995; Waldrep *et al.*, 1994, 1997).

Unlike conventional liposomes, "sterically stabilized liposomes" are relatively nonreactive to the environment due to surface coating of the liposomes with moieties, such as poly(ethylene glycol) (PEG) (Allen *et al.*, 1991; Gabizon and Papahadjopoulos, 1988). These liposomes have enhanced stability and decreased immunogenicity due to their surface coating with PEG. PEG derivatives can be prepared and purified inexpensively, and many have already been approved for pharmaceutical use, such as PEG adenine deaminase (ADA), and liposomes containing these derivatives make them ideal for therapeutic application. In addition, it has been reported that empty liposomes (without drug) can decrease inflammation (Huang *et al.*, 1995; Konduri *et al.*, 2003), and this may be an additional benefit of using liposomes as a delivery system for inhaled steroids.

This chapter describes the evaluation of the effect of budesonide encapsulated in sterically stabilized liposomes on lung inflammation and airway hyperreactivity in a mouse model of asthma. Our experiments have shown that budesonide encapsulated in sterically stabilized liposomes provides an effective means to decrease lung inflammation in experimental asthma (Konduri *et al.*, 2003). Levels of immunological markers implicated in the progression of asthma, including bronchoalveolar lavage (BAL), eosinophil peroxidase (EPO) activity, and peripheral blood eosinophil

(EOS) counts, have decreased significantly with this approach. Eosinophil peroxidase, one of the major eosinophil cytotoxins, has been identified as an important mediator of airway inflammation. It has ribonuclease activity and is a potent helminthotoxin and neurotoxin (Bousquet *et al.*, 1990; Broide *et al.*, 1991; Wardlaw and Kay, 1987).

Ovalbumin Sensitization of Animals

Mice are anesthetized on day 0 with methoxyflurane given by inhalation. A fragmented, heat-coagulated ovalbumin (OVA; Sigma Chemical Co., St. Louis, MO) implant is inserted subcutaneously on the dorsal aspect of the cervical region. On days 14–24, the mice are given a 30 min aerosolization of a 6% OVA solution on alternate days for a period of 10 days. This method of sensitization has led to significant elevations in EPO activity, peripheral blood (PB) eosinophils, and serum IgE levels, along with lung inflammation on histopathological examination by day 24 (Nandedkar *et al.*, 2001).

Treatment Groups

Therapy can be initiated on day 25, 1 day after the OVA sensitization is completed. Sensitized animals receive nebulized treatments for 4 weeks as follows: (1) budesonide (20 μg; Sigma) encapsulated in sterically stabilized liposomes, once a week; (2) budesonide (20 μg) without liposome encapsulation given daily (standard therapy); (3) budesonide (20 μg) encapsulated in conventional liposomes, once a week; (4) buffer-loaded (empty) sterically stabilized liposomes, once a week; and (5) budesonide (20 μg) without liposome encapsulation, once a week. The nebulization doses are all given at a volume of 1 ml for 2 min, using a chamber in which the mouse is allowed to breathe freely. All treatment groups are compared with either the untreated sensitized or unsensitized (normal) mice.

Each study group consists of 20 mice and is followed over a 4-week period. Five animals from each treatment and control group [untreated sensitized and unsensitized (normal)] are sacrificed by overdose of methoxyflurane by inhalation 24 h after the first treatments are given and then at weekly intervals for 4 weeks. At each time point, measurements of EPO in BAL, PB eosinophils, and total serum IgE level, and histopathological examination of the lung tissues are obtained.

Pulmonary mechanics are studied using a protocol described by Thakker *et al.* (1999). Measurements to evaluate the effect of drug or drug-liposome therapy on airway responses to methacholine challenge are determined using spontaneously breathing mice that have been intubated tracheally. The treatment groups are compared with sensitized, untreated mice and with healthy, normal mice undergoing the same procedures and receiving the same doses of methacholine. Methacholine challenges are determined every 2 weeks for 16 weeks on all experimental groups. As an antigen challenge and to demonstrate sensitization, an aerosolized dose of 6% OVA is given to each animal 24 h before the evaluation of the pulmonary mechanics.

The amounts of lipids used for the empty liposome control group are based on the amount of lipid nebulized for each of the liposome formulations encapsulating budesonide. The dose of budesonide chosen is based on preliminary dose–response studies with 5–50 μg of budesonide. Budesonide (5, 10, 15, 20, or 50 μg) is administered via nebulization daily to a group of sensitized mice, and the dose-dependent effects on the inflammatory parameters are evaluated. These data are compared with a group of either untreated sensitized or unsensitized (normal) mice. The 20-μg dose of budesonide was shown to effectively decrease EPO activity in BAL, PB eosinophils, and inflammation on histopathological examination of the lung tissues, along with other inflammatory parameters, without evidence of toxicity to the spleen, liver, bone marrow, or gastrointestinal tract. There were also no granulomas or tissue abnormalities in any of the tissues evaluated.

Drugs and Reagents

Budesonide for daily therapy is diluted from premixed vials (0.25 mg/ml) commercially available from Astra Pharmaceuticals (Wayne, PA) and is administered via a Salter Aire Plus Compressor (Salter Labs, Irvine, CA). Budesonide for encapsulation and N-2-hydroxethylpiperzine-N'-2-ethanesulfonic acid (HEPES) can be purchased from Sigma. Phosphatidylcholine, phosphatidylglycerol, and poly(ethylene glycol)-distearoylphosphatidylethanolamine are obtained from Avanti Polar Lipids (Alabaster, AL). Cholesterol is purchased from Calbiochem (La Jolla, CA). NaCl and KCl is purchased from Fisher Scientific (Pittsburgh, PA).

Liposome Preparation

Budesonide is encapsulated into either sterically stabilized liposomes [phosphatidylglycerol/phosphatidylcholine/poly(ethylene glycol)-distearoylphosphatidylethanolamine/cholesterol] or conventional liposomes (phosphatidylglycerol/phosphatidylcholine/cholesterol) using a protocol modified from that described by Gangadharam *et al.* (1995). A portion of the cholesterol used in control liposomes is replaced by budesonide during the preparation of the lipid mixture. Budesonide is dissolved in chloroform/methanol (2:1). Lipids in chloroform (with or without budesonide) are dried onto the sides of a round-bottom glass flask or glass tube by rotary evaporation in a Büchi evaporator. The lipid is dried further in a vacuum oven at room temperature. The dried film is then hydrated by adding sterile 140 mM NaCl, 10 mM HEPES (pH 7.4) at a concentration of 20 μmol lipid/ml, and vortexing. The resulting multilamellar liposome preparation is extruded 21 times through polycarbonate membranes (either 0.2- or 0.8-μm pore diameter; Nuclepore, Pleasanton, CA) using an Avestin (Toronto, Canada) extrusion apparatus.

Histopathology

Histopathological examination is performed on lungs that are removed and fixed with 10% formalin in phosphate buffer. The tissues are embedded in paraffin, sectioned at 5-μm thickness and stained with hematoxylin and eosin, and analyzed using light microscopy at 100× magnification. Examples of lung tissues from the six treatment groups are shown in Fig. 1. A significant reduction in the total lung inflammation score (Fig. 2) is noted with weekly treatments of budesonide encapsulated in sterically stabilized liposomes ($p = 0.020$) compared with the untreated sensitized mice. This reduction is similar to that obtained with daily budesonide therapy ($p = 0.030$). Similar decreases are not observed with the weekly administration of budesonide encapsulated in conventional liposomes, free budesonide, or empty sterically stabilized or conventional liposome treatments. There is also a significant decrease in lung inflammation in the animals given weekly budesonide encapsulated in sterically stabilized liposomes compared with treatments with weekly empty sterically stabilized liposomes ($p = 0.0009$) or weekly budesonide ($p = 0.05$). There is no significant difference between the daily budesonide and the weekly budesonide encapsulated in sterically stabilized liposomes treatment groups. The untreated unsensitized (normal) mice do not exhibit any inflammatory changes during the 4-week period. In contrast, the lung tissues from the untreated sensitized mice have persistent and significant inflammation (Konduri et al., 2003).

Determination of EPO Activity in BAL and PB Eosinophils

At the time of sacrifice, the trachea is cannulated with a ball-tipped 24-gauge needle. The lungs are lavaged three times with 1 ml phosphate-buffered saline. All the washings are pooled and the samples are frozen at

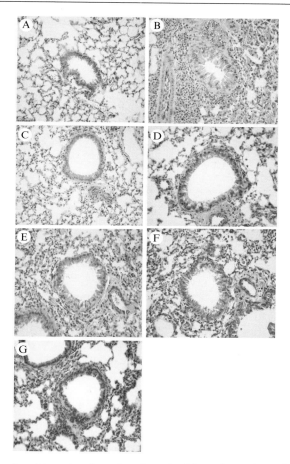

FIG. 1. Examples of hematoxylin and eosin–stained lung tissues from the six treatment groups. (A) Normal; (B) sensitized; (C) weekly budesonide in sterically stabilized liposomes; (D) daily budesonide; (E) weekly budesonide in conventional liposomes; (F) weekly empty sterically stabilized liposomes; (G) weekly budesonide. (See color insert.)

$-70°$. The samples can be thawed later and assayed for determining EPO activity. Eosinophil peroxidase activity in the BAL can be assessed using a modified approach of the method described by Strath *et al.* (1985). Substrate solution consisting of 0.1 M Na-citrate, O-phenylenediamine, and H_2O_2 (3%), at pH 4.5, can be mixed with BAL supernatants at a volume ratio of 1:1. The reaction mixture should be incubated at $37°$, and the reaction can be stopped by adding 4 N H_2SO_4. Horseradish peroxidase

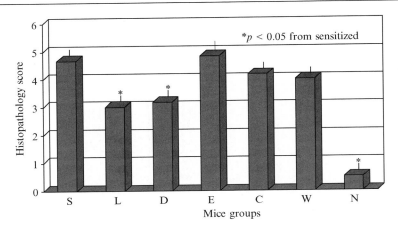

FIG. 2. Histopathology score in the different treatment groups. N, normal; S, sensitized; L, weekly budesonide encapsulated in sterically stabilized liposomes; D, daily budesonide; E, weekly empty sterically stabilized liposomes; C, weekly budesonide encapsulated in conventional liposomes; W, weekly unencapsulated budesonide.

can be used as a standard. EPO activity (ng/ml) can be measured by spectrophotometric analysis at 490 nm. The percentages of eosinophils can be obtained by counting the number of eosinophils in 100 white blood cells under a high-power field (100×) from the peripheral blood smears stained with Wright–Giemsa stain.

Eosinophil Peroxidase Activity (EPO)

Weekly treatment with budesonide encapsulated in sterically stabilized liposomes significantly decreases EPO activity ($p = 0.019$) in BAL when compared with the untreated sensitized mice and is comparable to daily budesonide therapy ($p = 0.015$) (Fig. 3). Weekly budesonide ($p = 0.419$), empty sterically stabilized liposomes ($p = 0.213$) or budesonide encapsulated in conventional liposomes ($p = 0.366$) treatment groups do not show a significant decrease in EPO activity (Konduri *et al.*, 2003). Our preliminary data have shown that our method of sensitization produces, by day 24, significant elevations in EPO activity, peripheral blood EOS count, and serum IgE levels; significant lung inflammation on histopathological examination and increased airway hyperresponsiveness to methacholine challenge. We have also noted continued lung inflammation and elevations in EPO activity, EOS, serum IgE levels, and airway hyperresponsiveness upon rechallenge with ovalbumin 2, 4, and 5 months after sensitization was completed.

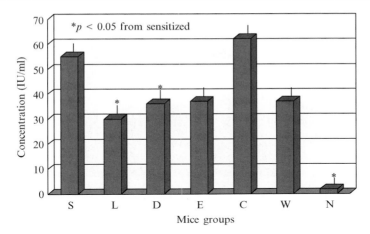

FIG. 3. Eosinophil peroxidase activity in the different treatment groups. N, normal; S, sensitized; L, weekly budesonide encapsulated in sterically stabilized liposomes; D, daily budesonide; E, weekly empty sterically stabilized liposomes; C, weekly budesonide encapsulated in conventional liposomes; W, weekly unencapsulated budesonide.

Peripheral Blood EOS Count

Therapy with weekly budesonide encapsulated in sterically stabilized liposomes ($p = 0.007$) and daily budesonide ($p = 0.001$) significantly decreases EOS compared with the sensitized, untreated group. None of the other treatment groups, including the weekly budesonide encapsulated in conventional liposomes, decreases EOS significantly compared with the sensitized, untreated group (Fig. 4).

Determination of Total Serum IgE

Ninety-six-well flat-bottom plates (Fisher Scientific) are coated with 100 μl/well of 2 μg/ml rat antimouse IgE monoclonal antibody (BD PharMingen, San Diego, CA) and incubated overnight at 4°. Serum is added at 1:50 dilution and incubated overnight at 4°. Purified mouse IgE (k isotype, small b allotype anti-TNP; BD PharMingen) can be used as the standard for total IgE. The samples are incubated for 1 h with biotin-conjugated rat antimouse IgE (detection antibody is purchased from Southern Biotechnology, Birmingham, AL).

Treatment with the weekly budesonide encapsulated in sterically stabilized liposomes and daily budesonide significantly lowers the total serum IgE level ($p = 0.016$ and $p = 0.005$, respectively) (Fig. 5). The

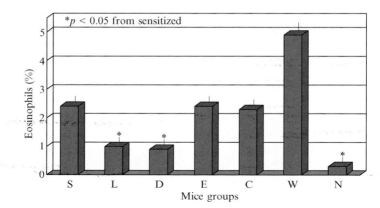

Fig. 4. Peripheral blood eosinophil as a percentage of total cells in the different treatment groups. N, normal; S, sensitized; L, weekly budesonide encapsulated in sterically stabilized liposomes; D, daily budesonide; E, weekly empty sterically stabilized liposomes; C, weekly budesonide encapsulated in conventional liposomes; W, weekly unencapsulated budesonide.

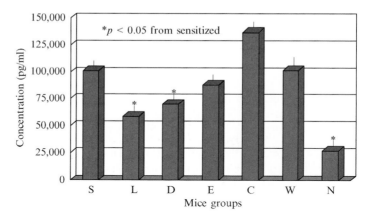

Fig. 5. Serum IgE levels in the different treatment groups. N, normal; S, sensitized; L, weekly budesonide encapsulated in sterically stabilized liposomes; D, daily budesonide; E, weekly empty sterically stabilized liposomes; C, weekly budesonide encapsulated in conventional liposomes; W, weekly unencapsulated budesonide.

total serum IgE level is not reduced significantly in the weekly budesonide encapsulated in conventional liposomes treatment group, or any of the other treatment groups, compared with the untreated sensitized group.

Measurement of Airway Responsiveness to Methacholine

Pulmonary mechanics are studied using a modified protocol described by Thakker *et al.* (1999). Measurements to evaluate the effect of drug or drug-liposome therapy on airway responses to methacholine challenge are determined using spontaneously breathing mice that have been intubated tracheally. The treatment groups are compared with sensitized, untreated mice and with healthy, normal mice undergoing the same procedures and receiving the same doses of methacholine. Methacholine challenges are determined every 2 weeks for 16 weeks on all experimental groups. As an antigen challenge, and to demonstrate sensitization, an aerosolized dose of 6% OVA is given to each animal 24 h before the evaluation of the pulmonary mechanics.

The animals are anesthetized with an intraperitoneal injection of a solution of ketamine and xylazine (40 mg/kg body weight for each drug) (Sigma). A 20 mg/kg body weight maintenance dose of pentobarbital sodium (Sigma) is given before placement in the body plethysmograph. These doses were noted to maintain a steady level of anesthesia without causing significant respiratory depression in our preliminary studies. A tracheotomy is performed, followed by the placement of a tracheostomy tube connected to a tube through the wall of the body plethysmograph chamber, allowing the animal to breathe room air spontaneously. A saline-filled polyethylene tube with side holes is placed in the esophagus and connected to a pressure transducer for measurements of flow, volume, and pressure. A screen pneumotachometer and a Valadyne differential pressure transducer are used to measure flow in and out of the plethysmograph.

Signals from the pressure transducer and the pneumotachometer are processed, using a Grass polygraph (model 7) recorder. The flow signal is integrated using a Grass polygraph integrator (model 7P10) to measure corresponding changes in pulmonary volume. Pressure, flow, and volume signal outputs are digitized and stored on a computer using an analogue-to-digital data acquisition system (CODAS-Dataq Instruments, Inc., Akron, OH). The pressure and volume signals are also displayed to verify catheter placement and monitor the animal during the experiment. The digitized data are analyzed for dynamic pulmonary compliance, pulmonary resistance, tidal volume, respiratory frequency, and minute ventilation from about 6–10 consecutive breaths at each recording event. After baseline measurements are obtained, methacholine is injected intraperitoneally, at 3-min intervals, in successive cumulative doses of 30, 100, 300, 1000, and 3000 μg.

Airway hyperresponsiveness to methacholine challenge decreases significantly in the group that receives weekly budesonide encapsulated

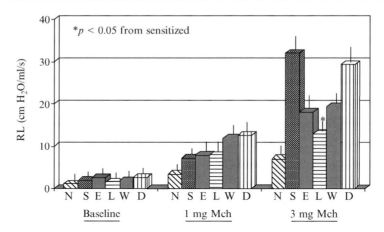

Fig. 6. Airway hyperresponsiveness to methacholine challenge (Mch) in the different treatment groups. N, normal; S, sensitized; L, weekly budesonide encapsulated in sterically stabilized liposomes; D, daily budesonide; E, weekly empty sterically stabilized liposomes; W, weekly unencapsulated budesonide.

in sterically stabilized liposomes ($p = 0.001$) (Fig. 6) but does not decrease in any of the other treatment groups, including the daily budesonide group.

Conclusions

Eosinophils have a primary role in the inflammatory phase of asthma as they secrete cytotoxins that directly damage lung mucosa and epithelium. In our preliminary experiments, peripheral blood eosinophil counts were significantly decreased. This is consistent with previous reports of a decrease in proinflammatory cytokine production with inhaled steroids, which, in turn, inhibits bone marrow production or release of eosinophils (Bousquet *et al.*, 1990; Broide *et al.*, 1991; Lonnkvist *et al.*, 2001; Wardlaw and Kay, 1987; Woolley *et al.*, 1994). Consistent with these previous studies, we have noted decreased levels of interleukin (IL)-4 and IL-5 in BAL and splenocyte culture supernatants (performed by ELISA, R & D Systems, Minneapolis, MN) in only the weekly budesonide encapsulated in sterically stabilized liposomes and daily budesonide treatment groups.

The histopathological analysis of lung tissue shows a significant decrease in lung inflammation in the weekly budesonide encapsulated in sterically stabilized liposomes and daily budesonide treatment groups.

These studies have shown that lung inflammation is significantly decreased using this novel drug delivery system.

Bronchoconstriction also plays an important role in the pathophysiology of asthma. Studies evaluating the effect on airway hyperresponsiveness to methacholine challenge of weekly therapy with budesonide encapsulated in sterically stabilized liposomes also reveal significant decreases in airway hyperresponsiveness to methacholine using this method of delivery.

It is crucial to obtain safety data on long-term therapy. Studies conducted with weekly treatments of budesonide encapsulated in sterically stabilized liposomes for a 6-week period do not reveal any toxicity to the animals using this drug delivery system. This unique drug delivery system may provide a valuable alternative to daily inhaled steroid therapy, with the potential to reduce toxicity and improve compliance for inhaled steroid therapy in asthma. The prevalence of asthma has been increasing, and it affects a significant portion of the population. The cost to treat asthma has also been rising at an alarming rate. Thus, it is imperative to find efficient mechanisms to lessen both the morbidity and the financial burden of this chronic disease. The results from these studies may enable us to develop a new therapy for the most common chronic disease in children. Furthermore, it may be possible to develop encapsulation in sterically stabilized liposomes as a new method to deliver other drugs to the airways.

References

Agertoft, L., and Pedersen, S. (1994). Effects of long-term treatment with an inhaled corticosteroid on growth and pulmonary function in asthmatic children. *Respir. Med.* **88,** 373–381.

Allen, T. M., Hansen, C., Martin, F., Redemann, C., and Yau-Young, A. (1991). Liposomes containing synthetic lipid derivatives of poly(ethylene glycol) show prolonged circulation half-lives *in vivo. Biochim. Biophys. Acta* **1066,** 29–36.

Bousquet, J., Chanez, P., Lacoste, J. Y., Barneon, G., Ghavamian, N., Enander, I., Venge, P., Ahlestedt, S., Simony-Lafontaine, J., Godard, P., *et al.* (1990). Eosinophilic inflammation in asthma. *N. Engl. J. Med.* **323,** 1033–1039.

Broide, D. H., Gleich, G. J., Cuomo, A. J., Coburn, D. A., Federman, E. C., Schwartz, L. B., and Wasserman, S. I. (1991). Evidence of ongoing mast cell and eosinophil degranulation in symptomatic asthma airway. *J. Allergy Clin. Immunol.* **88,** 637–648.

Brown, P. H., Blundell, G., Greening, A. P., and Crompton, G. K. (1991). Screening for hypothalamic-pituitary-adrenal axis suppression in asthmatics taking high dose inhaled corticosteroids. *Respir. Med.* **85,** 511–516.

Düzgüneş, N. (1998). Treatment of human immunodeficiency virus, *Mycobacterium avium,* and *Mycobacterium tuberculosis* infections by liposome-encapsulated drugs. *In* "Medical Applications of Liposomes" (D. D. Lasic and D. Papahadjopoulos, eds.), pp. 189–219. Elsevier Science BV, Amsterdam.

Farr, S. J., Kellaway, I. W., and Carman-Meakin, B. (1989). Comparison of solute partitioning and efflux of liposomes formed by a conventional and aerosolized method. *Int. J. Pharmaceut.* **51,** 39–46.

Gabizon, A., and Papahadjopoulos, D. (1988). Liposome formulations with prolonged circulation time in blood and enhanced uptake by tumors. *Proc. Natl. Acad. Sci. USA* **85,** 6949–6953.

Gangadharam, P. R. J., Ashtekar, D. R., Flasher, D., and Düzgüneş, N. (1995). Therapy of *Mycobacterium avium* complex infection in beige mice with streptomycin encapsulated in sterically stabilized liposomes. *Antimicrob. Agents Chemother.* **39,** 725–730.

Gonda, I. (1990). Aerosols for delivery of therapeutic and diagnostic agents to the respiratory tract. *CRC Crit. Rev. Ther. Drug Carrier Syst.* **6,** 272–313.

Gonzalez-Rothi, R. J., Suarez, S., Hochhaus, G., Schreier, H., Lukyanov, A., Derendorf, H., and Dalla Costa, T. (1996). Pulmonary targeting of liposomal triamcinolone acetonide. *Pharm. Res.* **13,** 1699–1703.

Gregoriadis, G. (1995). Fate of liposomes *in vivo* and its control: A historical perspective. *In* "Stealth Liposomes" (D. Lasic and F. Martin, eds.), pp. 7–12. CRC Press, Boca Raton, FL.

Haahtela, T., Jarvinen, M., Kava, T., Kiviranta, K., Koskinen, S., Lehtonen, K., Nikander, K., Persson, T., Selroos, O., Sovijarvi, A., Stenius-Aarnila, B., Svahn, T., Tammivaara, R., and Laitinen, L. A. (1994). Effects of reducing or discontinuing inhaled budesonide in patients with mild asthma. *N. Engl. J. Med.* **331,** 700–705.

Huang, S. K., Martin, F. J., Friend, D. S., and Papahadjopoulos, D. (1995). Mechanism of stealth liposome accumulation in some pathological tissues. *In* "Stealth Liposomes" (D. Lasic and F. Martin, eds.), pp. 119–125. CRC Press, Boca Raton, FL.

Kamps, J. A., and Scherphof, G. L. (2004). Biodistribution and uptake of liposomes *in vivo*. *Methods Enzymol.* **387,** 257–266.

Kips, J. C., Brusselle, G. G., Joos, G. F., Peleman, R. A., Devos, R. R., and Tavernier, J. H. (1995). Importance of interleukin-4 and interleukin-12 in allergen-induced airway changes in mice. *Int. Arch. Allergy Immunol.* **107,** 115–118.

Konduri, K. S., Nandedkar, S. D., Düzgüneş, N., Suzara, V., Artwohl, J., Bunte, R., and Gangadharam, P. R. J. (2003). Efficacy of liposomal budesonide in experimental asthma. *J. Allergy Clin. Immunol.* **111,** 321–327.

Lonnkvist, K., Hellman, C., Lundahl, J., Hallden, G., and Hedlin, G. (2001). Eosinophil markers in blood, serum, and urine for monitoring the clinical course in childhood asthma: Impact of budesonide treatment and withdrawal. *J. Allergy Clin. Immunol.* **107,** 812–817.

Nandedkar, S. D., Konduri, K. S., Artwohl, J., Bunte, R., and Gangadharam, P. R. J. (2001). A novel approach to ovalbumin sensitization in C57bl/6 mice. *J. Allergy Clin. Immunol.* **107,** S176(Abst. #580).

O'Byrne, P. M., and Pedersen, S. (1998). Measuring efficacy and safety of different inhaled corticosteroid preparations. *J. Allergy Clin. Immunol.* **102,** 879–886.

O'Callaghan, C., Cant, M., and Robertson, C. (1994). Delivery of beclomethasone diproprionate from a spacer device: What dose is available for inhalation? *Thorax* **49,** 961–964.

Overbeck, S. E., Kerstjens, H. A. M., Bogaard, J. M., Mulder, P. G. H., Postma, D. S., and Dutch Chronic Non-specific Lung Disease Study Group (1996). Is delayed introduction of inhaled corticosteroids harmful in patients with obstructive airways disease (asthma and COPD)? *Chest* **110,** 35–41.

Pedersen, S., and O'Byrne, P. M. (1997). A comparison of the efficacy and safety of inhaled corticosteroids in asthma. *Allergy* **52,** 1–34.

Robinson, D., Hamid, Q., Ying, S., Tsicopoulos, A., Barkans, J., Bentley, A., Corrigan, C., Durham, S., and Kay, B. (1992). Predominant Th2-like bronchoalveolar T-lymphocyte population in atopic asthma. *N. Engl. J. Med.* **326,** 298–304.

Schreier, H., Lukyanov, A. N., Hochhaus, G., and Gonzalez-Rothi, R. J. (1994). Thermodynamic and kinetic aspects of the interaction of triamcinolone acetonide with liposomes. *Proc. Int. Symp. Cont. Rel. Bioact. Mater.* **21,** 228–229.

Strath, M., Warren, D. J., and Sanderson, C. J. (1985). Detection of eosinophils using an eosinophil peroxidase assay. Its use as an assay for eosinophil differentiation factors. *J. Immunol. Methods* **83,** 209–215.

Taylor, K. M. G., Taylor, G., Kellaway, I. W., and Stevens, J. (1989). The influence of liposomal encapsulation on sodium cromoglycate pharmacokinetics in man. *Pharm. Res.* **6,** 633–636.

Thakker, J., Xia, J.-Q., Rickaby, D., Krenz, G., Kelly, K., Kurup, V., and Dawson, C. (1999). A murine model of latex allergy induced airway hyperreactivity. *Lung* **177,** 89–100.

Toogood, J. H., Sorva, R., and Puolijoki, H. (1994). Review of the effects of inhaled steroids therapy on bone. *Int. J. Risk. Saf. Med.* **5,** 1–14.

U.S. Public Health Service. National Heart, Lung, and Blood Institute (2002). National asthma education and prevention program. Expert panel report 2: Guidelines for the diagnosis and management of asthma. NIH publication No. 97–4051.

Vidgren, M., Waldrep, J. C., Arppe, J., Black, M., Rodarte, J. A., Cole, W., and Knight, V. (1995). A study of [99]technetium-labeled beclomethasone diproprionate dilauroyl-phosphatidylcholine liposome aerosol in normal volunteers. *Int. J. Pharmaceut.* **115,** 209–216.

Waldrep, J. C., Keyhani, K., Black, M., and Knight, V. (1994). Operating characteristics of 18 different continuous-flow jet nebulizers with beclomethasone dipropionate liposome aerosol. *Chest* **105,** 106–110.

Waldrep, J. C., Gilbert, B. E., Knight, C. M., Black, B. M., Scherer, P., Knight, V., and Eschenbacher, W. (1997). Pulmonary delivery of beclomethasone liposome aerosol in volunteers. *Chest* **111,** 316–323.

Wardlaw, A. J., and Kay, A. B. (1987). The role of the eosinophil in the pathogenesis of asthma. *Allergy* **42,** 321–335.

Wasan, K. M., and Lopez-Berestein, G. (1998). The development of liposomal amphotericin B: An historical perspective. *In* "Medical Applications of Liposomes" (D. Lasic and D. Papahadjopoulos, eds.), pp. 165–180. Elsevier Science BV, Amsterdam.

Weiss, K. B., Gergen, P. J., and Hodgson, T. A. (1992). An economic evaluation of asthma in the United States. *N. Engl. J. Med.* **326,** 862–866.

Wong, J., and Black, P. (1992). Acute adrenal insufficiency associated with high dose inhaled steroids. *Br. Med. J.* **304,** 1415–1416.

Woolley, M. J., Denburg, J. A., Ellis, R., Dahlback, M., and O'Byrne, P. M. (1994). Allergen-induced changes in bone marrow progenitors and airway responsiveness in dogs and the effect of inhaled budesonide on these parameters. *Am. J. Respir. Cell Mol. Biol.* **11,** 600–606.

Section IV

Electron Microscopy of Liposomes

[24]   Cryoelectron Microscopy of Liposomes

By PETER M. FREDERIK and D. H. W. HUBERT

Abstract

A thin aqueous film of suspended lipid vesicles/micelles is the object of choice for vitrification and subsequent study by cryoelectron microscopy. Just prior to vitrification, a thin film (compare with a soap film) is vulnerable to heat and mass exchange. Preparation of thin films in a temperature- and humidity-controlled environment is essential to prevent osmotic and temperature-induced alterations of the lipid structure, as will be explained in this chapter. Further automation of the preparative procedure by automatic blotting and PC control over the timing of critical steps (including vitrification) may further assist in the reproducible throughput of high-quality specimens. By cryotomography, taking a tilt series under low-dose conditions, a three-dimensional reconstruction of the specimen can be analyzed.

Introduction

Cryoelectron microscopy provides images of thin vitrified specimens. Because a vitrified specimen is relatively transparent to the electron beam, a "through vision" of the objects suspended in the vitrified matrix is obtained. By using physical fixation (rapid cooling) as the only fixation method for cryoelectron microscopy (cryo-EM), the specimen remains chemically unmodified, providing a "true vision" of the suspended material. Recent advances in cryoelectron microscopy have had a great impact on colloidal chemistry, including the structure of molecular assemblies of lipids. Advances in cryo-EM concern improved specimen preparation, a major issue to be treated in more detail in this chapter. Further advances in cryo-EM rely on developments in microscope automation, which have improved both the speed and consistency of the performance of the electron microscope and thus contributed to the throughput of high-quality data. Data handling and analysis, as well as microscope operation, have all benefited from advances in microelectronics and software development. For the study of the structure of lipid molecular assemblies by cryo-EM, specimen preparation is a key event in the chain of procedures that starts with the preparation of a suspension and ends with an image and/or the interpretation of image data. One of the claims of cryo-EM is that the images represent the unperturbed specimen as it exists in aqueous

METHODS IN ENZYMOLOGY, VOL. 391

suspension. As long as specimen preparation does not interfere with the original structure this may hold to a certain extent. Specimen preparation normally will not take long; vitrification by itself only takes 10^{-5} s, and it is the period just prior to vitrification that deserves attention. There are in principle two ways to obtain a specimen thin enough for cryo-EM: (1) prepare a thin film (compare with a soap film) and vitrify this thin film, or (2) section a vitrified suspension at low temperature.

The first option is a straightforward procedure and therefore a first choice when investigating a lipid suspension. This chapter will be confined to thin film procedures for vitrification and cryo-observation. To understand the critical steps in cryopreparation and observation, a concise outline is given of the physical chemistry involved in thin film formation and some of the thermodynamics involved in the stability of thin films. Cryo-EM of vitrified thin films is still in development for dynamic ("time-resolved") applications. For "static" applications some standard procedures will be presented after a brief introduction to the relevant background.

Thin Film Formation

In the course of the formation of a thin film from a suspension, the spatial relations between the suspended materials change from the three-dimensional relations in the suspension toward a (pseudo) two-dimensional relationship within a thin film. The presence and proximity of air–water interfaces are experienced throughout the thin film, whereas for the relations within a suspension the air–water interface plays only a minor role. Thus, a thin film represents a suspension in a number of aspects; it has, however, some characteristics of its own that have to be considered when cryo-EM images are interpreted and extrapolated as characteristic for the parent suspension. Preparing a thin specimen from a suspension thus involves certain transformations and modifications of the suspension. In the first place, this has to do with dynamic aspects of thin film formation: creating air–water interfaces and the competition of molecules from the suspension to positions at the air–water interface. When a fresh interface is created, e.g., by taking a drop from a suspension into the air, a "clean" air–water interface may be present for which molecules from the suspension are competing. Small molecules with an amphiphilic character (having hydrophilic and hydrophobic parts within one molecule) will get at the interface at first. Larger molecules with a strong interface affinity will be slower at the interface and may replace some of the early occupants. From the soap literature it is known that sodium dodecyl sulfate (SDS) molecules occupy the surface within a few tenths of a second (room temperature and SDS concentration above the critical micelle concentration; CMC).

Furthermore, it is known that the composition of the interface does not reflect the composition of the suspension; in foam, the concentration of free fatty acids may far exceed the concentration in the solution. Thus, once an interface is formed, it can be remodeled and interact with material in the underlying suspension.

How much material is involved in creating an interfacial layer? To obtain a crude estimate for a phospholipid suspension, let us assume an average molecular area of 0.5 nm^2. Two interfaces of a cryo-EM specimen grid (diameter 3 mm) have a surface of approximately 14 mm^2 (14×10^{12} nm^2) and may thus accommodate 28×10^{12} phospholipid molecules. In a routine experiment, we apply 3 μl of a suspension containing 10 mg phospholipid/ ml on a specimen grid. This droplet contains 3×10^{16} phospholipid molecules, and this is in large excess (about a factor 10^5) of the amount needed to create the interfacial layers. Upon blotting, the situation changes drastically; a 3-μl droplet creates a cylinder of about 0.5 mm thick on a grid and by blotting away excess liquid this cylinder is reduced to a layer of about 100 nm (some 1/5000 of the applied volume), a thin layer that contains only 6×10^{12} phospholipid molecules, not enough to cover two interfacial layers. Apparently, blotting leaves some of the interfacial layers intact, since the presence of intact/continuous interfacial layers in the final cryo-EM image is all too obvious, as will be discussed later in this chapter.

When a thin film has attained a certain thickness, both interfacial layers come to attract each other. Gravity and capillary forces initially dominate the thinning behavior of a thin film; later when the interfacial layers approach each other, molecular interactions (London–van der Waals attractions) gradually take over to become the dominating attracting/ thinning force. The mutual attraction of the interfaces may become counterbalanced by electrostatic repulsion, and eventually also by steric repulsion, and come to an equilibrium structure when the interfacial layers come close. The suspended material may act as a spacer, keeping the interfacial layers apart in an equilibrium configuration with the electrostatic and steric repulsion between particles and interfaces balancing out attractive forces. Such an equilibrium configuration can be considered as an optimal configuration for electron microscopy, since image noise (related to the thickness of the specimen) is minimal. In such an ideal specimen an intact and undisturbed specimen is captured in a minimum amount of noise-generating matrix (i.e., vitrified suspending medium and interfacial layers).

It is in this configuration that the three-dimensional relations of the suspension are reduced to (pseudo) two-dimensional relations of an optimal thin film. With micelle-forming lipids (Moschetta et al., 2002), the thickness of such a film is about 12 nm; the micelle diameter (5–6 nm) plus two interfacial monolayers (total 4 nm) and a few water layers form a hydration mantle. Such a thin film has an extreme surface/volume ratio. This implies that surface effects, including heat and mass exchange, have a profound effect on the bulk properties. To achieve fast cooling (10^7 K/s) such a surface/volume ratio is a prerequisite for rapid heat exchange. Only by extremely fast cooling do aqueous solutions (without "cryoprotectants") vitrify, whereas at slower cooling rates crystallization of water is inevitable. Formation of ice crystals deteriorates the ultrastructure and may have osmotic effects as well. Dealing with a thin specimen with a large surface/volume ratio also implies that such a thin specimen is highly vulnerable and sensitive to environmental conditions. Heat and mass exchange with the environment may interfere with specimen preparation when considering the time scale and magnitude of these exchange processes, as will be outlined in the next section. When the vitrified specimen has to represent the parent suspension faithfully, the environmental conditions (notably temperature and humidity) just prior to vitrification are key parameters.

Thin Films and Heat and Mass Exchange with the Environment

During its formation, a thin film will interact with its environment by heat and mass exchange. This can be considered as a dew point effect; the specimen will attain the temperature that is related to the relative humidity. At the dew point, temperature evaporation and condensation of water are in equilibrium when there is no further heat input in the system and at the dew point the partial vapor pressure of water in the environment equals the saturated vapor pressure at the dew point temperature. From the tables of the saturated vapor pressure of water (e.g., Lide, 2003) these temperatures and vapor pressures can be found for a relevant set of conditions applicable to specimen preparation for cryo-EM. More important than this rather obvious description is the rate of this temperature change. We will first treat this problem in a thermodynamic model, and from these results conclusions can be drawn that allow experimental verification.

First, let us assume a stationary situation: a thin film standing in a temperature- and humidity-controlled environment. We take film thicknesses of 5000, 1000, and 500 nm as starting values for our calculations, assuming that these thicknesses represent various stages in thin film formation. With a stationary film, we assume that this thin film is bordered on both sides by a stagnant layer of air with a thickness of 10 μm. At the surface of the thin film, the water vapor pressure equals the saturated vapor pressure of the film temperature, and 10 μm from this surface vapor pressure equals the relative humidity of the environment. Over 10 μm there will be diffusion of water from the film to the environment, as long as the environment is not saturated with water. As long as there is net water transport from the film, heat will be extracted from the film, and a temperature gradient will be established. This temperature gradient will result in a heat flow toward the film, and heat diffusion follows the same geometric model (Fig. 1).

The time to reach dew point is one of the results that can be plotted on the basis of the outcome of the thermodynamic calculations, and this is represented in Fig. 2. As an example, we calculated for the various film thicknesses the time needed to reach dew point, and this is even for a "relatively" thick film in the order of a few tenths of a second and for real cryo-EM specimens (thinner than 500 nm) less than one-tenth of a second in "normal" laboratory conditions. We assume for these calculations that 25° and 40% relative humidity are typical for a laboratory environment. The outcome shows that dew point temperature will be attained during preparation of a thin film. This implies that environmental humidity and

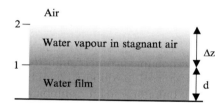

FIG. 1. Geometry of thin films as used for thermodynamic calculations.

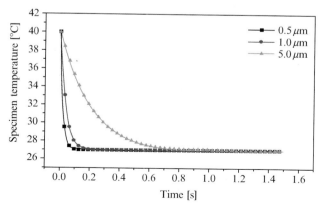

FIG. 2. Relation between film thickness and time to reach "dew point."

temperature are equally important for the control of the specimen temperature prior to vitrification.

The close relation between humidity and specimen temperature has to be considered when investigating temperature-sensitive specimens such as phospholipids by cryo-EM. The relative humidity has a profound effect when studying the thermotropic phase behavior of phospholipids as is illustrated with dipalmitoylphosphatidylcholine (DPPC). With an environment at 50° and a relative humidity of 12%, one obtains the $P_{\beta'}$ structure; only raising the humidity to 100% in the environmental chamber preserves the L_α structure associated with the temperature (Fig. 3).

When the temperature of a thin specimen is lower than the environmental temperature, heat influx from the environment toward the specimen will occur. Roughly the same geometric considerations for the outflow of water vapor apply to the influx of heat. Once the specimen has reached (approximately) its dew point, the heat influx will be practically constant and counterbalanced by further evaporation of the specimen. In thin films there is hardly any temperature gradient over the thickness of the film, and the evaporation from a thin film is (for the range of cryo-EM specimens) independent of the film thickness. In Fig. 4, the evaporation velocities as a function of temperature and relative humidity in a range of relevant conditions for the specimen are plotted.

Under our laboratory conditions (20°, relative humidity 40%), the evaporation velocity is in the order of 40–50 nm/s.

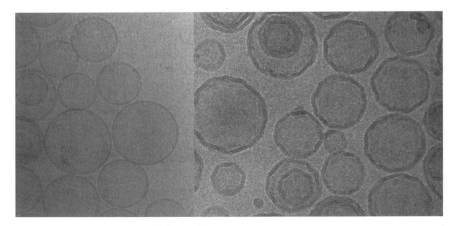

FIG. 3. Extruded vesicles prepared from DPPC (in distilled water) and vitrified from 50° at 100% relative humidity (left) and 12% relative humidity (right). At low relative humidity the lipid adopts the $P_{\beta'}$ (ripple) phase as a consequence of the dew point effect (right); the L_α phase (left) is associated with the temperature of the environment.

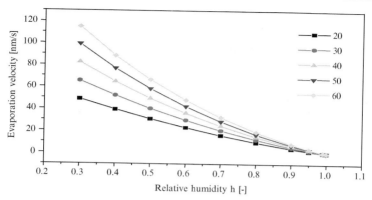

FIG. 4. Rate of evaporation of a thin film subjected to several temperatures (20–60°) and environmental humidities. Relative humidity is scaled between 30% and 100% (0.3 to 1.0).

At this rate of thinning, by evaporation of "pure" water, the solute concentration in the remaining thin film increases. On a time scale of seconds, osmotic effects are to be expected since the osmolarity may double in this period. For osmotically active specimens such as liposomes, preparation at nonsaturating vapor pressure of water has a dual effect. Osmotic changes (e.g., loss of internal volume, Fig. 5) are induced concomitant with a temperature drop (eventually involving a thermotropic phase change). Furthermore, it should be noted that the concentration of solutes also implies that the density of the suspended material increases, e.g., the number of liposomes per field of view. In the old literature on cryo-EM, there is some confusion about the relation between the number of particles in suspension (viruses, liposomes) on the one hand and the number of particles observed in the vitrified thin film on the other.

Small variations in preparation time may result in a range of particle densities when the evaporation velocity (typically 40 nm/s or higher) is of the same order of magnitude as the thickness of the vitrified film (50–200 nm). Only by film preparation in a reproducible way in a temperature-controlled, high-humidity environment can artifacts be prevented. Based on our previous experience (Frederik *et al.*, 1989, 1991) and the work of the group of Talmon (Bellare *et al.*, 1988, Talmon *et al.*, 1990), we decided to further improve specimen preparation by constructing a vitrification robot. A prototype was developed and tested and modified over a period of 5 years (1995–2000). In the course of this development, it became clear that such an instrument was in demand in the scientific community, and the prototype instrument was completely redesigned to build a new series of instruments, initially for colleagues. The vitrification

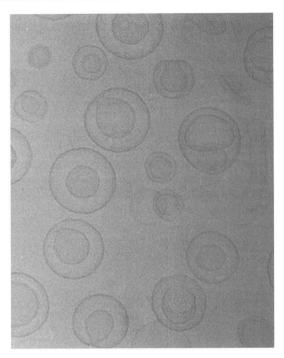

FIG. 5. Osmotic collapse of spherical vesicles into "vase"-like structures (two concentric vesicles connected by an orifice/neck). Of the internal volume, 50% was lost by preparation of liposomes (DOTAP/cholesterol in glucose/300 mosm) at 25° and 40% relative humidity.

robot, now called the Vitrobot (patent applied), was more compact and "user-friendlier" than the prototype and became the basis for a commercial version marketed by FEI (Hillsboro, OR; also see http://www.vitrobot.com/). The description of procedures recommended for the cryo-EM preparation of liposomes are based on the use of a Vitrobot (or any other instrument that meets the same specifications).

Vitrobot: The Vitrification Robot

The major design philosophy behind our vitrification system was to consistently vitrify samples with minimal manual interference. We took the thus far "established" procedures (e.g., Burger *et al.*, 1993; Dubochet *et al.*, 1988, 1992) as a starting point for automation. The established vitrification procedure is usually considered as a three- or four-step procedure:

1. A drop of the suspension is applied to a specimen grid (dipping the grid into a vial or "one sided" application by using a pipette).
2. The excess liquid is blotted away with filter paper.
3. The aqueous thin film is allowed to drain/thin (the often neglected step).
4. The thin film is dropped or shot into a suitable coolant (e.g., melting ethane).

A guillotine-like instrument is used to hold a rod with forceps clamping a grid; all preparative steps are carried along the traveling axis of the grid using a release mechanism for a guided drop/shooting into coolant at the end of the processing. Blotting is often considered as the tricky part of the procedure, successful only in the hands of the experienced. Automation thus implies, first, the designing and testing of an automatic blotting system with the aim to eliminate a major source of variability in the preparation procedure. A tested and approved blotting system could later be housed in a "closed" temperature- and humidity-controlled chamber. The blotting action should be gentle and reproducible, preferentially with an adjustable blotting pressure to meet the requirements of various specimens differing in composition and viscosity. The blotting principle chosen was based on a "hand-clapping action" of two (rotating) disks holding the two blot papers on a foam pad. By symmetric movement of these disks toward a centered grid, a blotting action is achieved (Fig. 6). The blot pads rotate between two blotting actions over a 22.5°-arc to expose fresh filter paper on every blotting action. A spring-lever and cam system (16 cams on a full turn) is used to propel the blot pads. Pneumatics are used to keep the blot pads apart, and an adjustable coil spring is used to generate the force for the blotting action. Initially we started with perfectly parallel blotting pads meeting the grid in a vertical plane. The filter papers had to be firmly attached to the blot pads to prevent them from falling off when processing. Therefore, the blot pads were given a slight inclination to the vertical plane, thereby reducing the contact area between the filter papers to the specimen and its immediate vicinity. After testing and improving the blotting device as described, a humidity- and temperature-controlled chamber was built around the blot mechanism with a guiding system to move the specimen in different positions. The guiding mechanism is driven by a stepping motor for accurate vertical positioning under software control. For the vitrification action we have chosen an accelerated drop principle, using air pressure to accelerate the specimen over a 7-cm trajectory from its blotting position toward the surface of the melting ethane. The forceps holding the grid is mounted on a rod with an accurate clamping and positioning mechanism at its lower end that does not affect the grid-holding action of the forceps.

FIG. 6. Specimen grid held by forceps in the blotting position within the environmental chamber.

At the end of the process of sample application, movement, and blotting (eventually including waiting and equilibration periods around these actions), the rod with tweezers is disengaged from the stepping motor and is shot into the ethane. The melting ethane is placed in a container separated by a shutter from the humidity- and temperature-controlled chamber. The entry velocity of the specimen in the ethane is in the order of 2 m/s. This speed is obtained by adding an additional force to gravitation by using a blowpipe principle. Pressurized air is introduced through a needle valve into the inner tube of two concentric tubes. The outer tube holds the specimen and is secured to the inner tube by a leaky septum. By regulating the air flow, the weight of the outer tube can be tripled (as measured by standard electronic scales). A mechanical damping device is used to brake the specimen movement without bouncing to secure good vitrification without spilling the ethane. With the ethane container in its upper position, a position as close as possible to the controlled environment for optimal vitrification, we needed space for manipulation of the forceps after the vitrification action. This room is created by placing the ethane container on a lift driven by a pneumatic cylinder bringing the ethane close to the chamber just prior to the vitrification action. Upon completion of the vitrification action, the lift is coupled to the rod bearing the forceps, and by moving this assembly downward, the forceps is exposed for further manipulation of the vitrified specimen.

In a full process cycle, the movement of parts relative to each other is achieved, which can be summarized as follows:

A new grid is entered in the procedure: The stepping motor is connected to the rod, and this rod is driven through the shutter at the bottom of the chamber. Foreceps holding a grid can be placed (or removed) from outside by the operator.

The grid is moved into the chamber: The forceps is driven up by the stepping motor, and the shutter closes when the assembly has entered the chamber. On the discretion of the operator, time is allowed for thermal equilibration of the grid.

During further processing, the forceps with the grid travels to several positions in the chamber, and the timing and other process parameters are selected by the operator from a dialogue screen/user interface. The time needed for sample application and the position of the forceps during application are among the parameters that can be selected. The sample can be applied through a port using a pipette from the outside, taking the sample from a vial inside the controlled environment. This is preferred for small sample volumes (3 μl per grid is our standard) and special support films (e.g., lacy carbon films, Quantifoils). Alternatively, the sample can be collected by dipping/withdrawing the grid into a vial; in a typical Eppendorf vial, a minimum volume of 50 μl is required to submerge the grid without mechanical damage. For volumes between 50 and 1500 μl, the instrument can be handle grids to become just submerged, thereby preventing excessive wetting of the forceps.

Blotting Action

The grid is moved upward by the stepping motor and positioned between the blotting pads. The number of blotting actions is chosen from the menu as well as the blotting time—the time the blotting pads come together with the grid in between. The blotting pressure can be selected by accurate positioning ("blot offset") of the grid between the wedge of the blot pads. For a firm blotting action the grid is positioned deep in the wedge; for a gentle blot (e.g., for a grid with continuous carbon) the grid is positioned higher and touches the blot paper only with its edge. In a gentle blotting action, a prolonged contact with the blot papers is required to remove enough liquid. After blotting, the grid may remain for some time in the controlled environment (drain time) or can be dropped immediately into the ethane.

Vitrification Action

The shutter at the bottom of the chamber opens pneumatically, and at the same time the rod bearing the forceps is pneumatically disengaged from the stepping motor and propelled down with additional air pressure. Directly after vitrification, the specimen rod is coupled to the lift of the ethane container, and this assembly moves downward to allow handling of the forceps and collecting the vitrified specimen using conventional cryotransfer tools.

These complex movements are controlled by software (stored on a hard-disk incorporated in the instrument together with a PC-104 card) and user control is exerted over all relevant steps in the procedure. In this setup, a user-friendly system is created that has an inherent flexibility to allow the implementation of new and/or special applications. In the prototype phase, we worked with a PC-controlled system, and in this configuration we had problems when programs were running in the background that could interfere (in a rather unpredictable way) with process control. To exclude this type of priority conflicts, we opted for a dedicated software/hardware solution for reliable process control.

The development of sufficient environmental control deserves some special attention. We learned and appreciated how critical environmental conditions are (see preceding section) only during the development and testing of the instrument. By eliminating variability in mechanical handling, we became aware of other variables, notably the effects of environmental temperature and humidity, which affect the quality of cryofixation. For our work the environmental chamber was primarily designed for temperatures ranging from room temperature to high temperatures (up to 70°): high temperatures to study the thermotropic phase behavior of phospholipids and around 37° to study living cells/organisms. The back of the chamber holds a Peltier type of heating/cooling device and in the cooling option 4° can be reached under normal laboratory conditions (18–25°). A fan blows air through a funnel at the back of the chamber, and the Peltier stage is connected to a back-plate with many ribs to improve heat exchange in the funnel. By choosing the proper type of Peltier element, the entire temperature range (4–60°) can be reached in a reasonable time (15–45 min).

A sensor (PT 100) placed near the blot pads monitors the temperature for readout and control, a location we considered representative for the thermal history of the specimen.

Humidity control went through several stages in the development. Initially we used wetted sponges placed inside the chamber with the idea that after enough equilibration time the environment would become saturated with water vapor. Only when we built in an electronic humidity sensor did we become aware that equilibration took a long time [starting from 40% relative humidity (rH) at room temperature it took some hours to attain 85% rH and more than half a day to attain above 95% rH]. These equilibration times are hardly compatible with a working routine that requires opening of the chamber to place blot papers and/or a vial in the chamber. Such actions cause an immediate drop in the relative humidity, and equilibration almost has to start all over again.

Therefore, we experimented with forced humidification, initially by circulating the air of the chamber through a microporous stone placed in a beaker with water. The results were not impressive, and we considered heating the water while at the same time we came across an ultrasonic device used to create a magic mist in a household environment. This ultrasonic device placed into the chamber gave an immediate response; at room temperature, the rH can be raised to about 100% within a minute. In fact, it rains inside the chamber, and by choosing a proper cover over the beaker with the ultrasonic device, only small droplets (mist) enter the chamber and the fans carry the microdroplets through the entire chamber. It should be borne in mind that at higher temperatures more water is needed to saturate the atmosphere, and it takes some time to attain a saturated environment; during this time additional heating power is required to maintain a constant temperature. A side effect of working under conditions of (over)saturated humidity is that minor temperature differences are eliminated. On relatively cold spots water will condense and locally deliver heat of condensation, and on hot spots water will evaporate, locally extracting heat of evaporation.

We have come to consider the controlled humidification as a cornerstone to successful cryofixation, another essential element being controlled blotting.

Liposome Processing for Cryo-EM

The size of liposomes is important when liposome shape and shape changes are to be studied. The resolution of the electron microscope is related to the thickness of the object; the thinner the object the better the signal-to-noise ratio in the image. Since high resolution requires a specimen as thin as possible, small liposomes are preferred. Small liposomes form thin films that are in the optimal parts slightly thicker than the diameter of the vesicles. Small vesicles (sonicated vesicles; SUV) are thus excellent objects for cryo-EM in this respect, but there are a number of characteristics of SUVs, such as a restricted shape repertoire and an anomalous thermotropic phase behavior, virtually lacking a $P_{\beta'}$ phase. In such a perspective, sonicated vesicles may be considered as "atypical" liposomes. Therefore, we prefer to work with extruded vesicles (hand-shaken vesicles extruded through a filter with 100- or 200-nm pores (see Mui *et al.*, 2003). The size distribution is rather narrow when a 100-nm filter is used (e.g., Anotop 10, 0.1 μm, Whatman; to be attached to a standard syringe); with a 200-nm filter (Anotop 20, 0.2 μm, Whatman), the range of sizes is somewhat broader, but still useful.

Thermotropic Phase Behavior of DPPC

An aliquot of 10 mg DPPC (Avanti Polar Lipids, Alabaster, AL) is taken from a stock solution of DPPC in chloroform. It is evaporated to dryness under a stream of nitrogen gas, and 1 ml of buffer (or distilled water) is added. The tube is sealed and heated well above the phase transition temperature, e.g., placed in a water bath or incubator at 50°. For 1 h the sample is left at 50° with intermittent handshaking of the suspension. Extrusion by three or more passes through a filter is done well above the phase transition (filter and syringe are placed in an incubator at 50° long before use). If the temperature is too low, the filter will be clogged immediately, making it impossible to pass the suspension through the filter by hand operation of the syringe. By at least three passes through the filter, a rather monodisperse suspension is obtained that is suitable for morphological studies by cryo-EM.

Just prior to sample application, a Quantifoil (R 2/2 Quantifoil Gmbh Jena/BRD) is glow discharged to make the surface hydrophilic. Note that the surface characteristics of this support film with a cubic array of 2-μm-sized holes change within minutes. Directly after glow discharge, an applied phospholipid suspension behaves as one film; later on, the film has the tendency to break up into many small films, each small film filling one of the 2-μm-sized holes in the Quantifoil. With a repellant surface, the water film brakes into small droplets after blotting, and these droplets appear as positive lenses (biconvex) in the cryoimage. Because these lenses are unable to drain, they are usually too thick for useful imaging. With a hydrophilic support, the whole surface remains wetted, the holes in the Quantifoil now appearing as flat areas or slightly negative lenses (biconcave) by the mutual attraction of the interfaces. This biconcave geometry may result in sorting by size of the suspended particles; larger particles are forced into the thicker part of the film, while the thinner part of the film accommodates only those particles that fit into the thickness of the liquid layer.

The DPPC suspension is applied as a 3-μl drop using a pipette. For application we prefer pipettes with a plunger tip; other types may have a lot of air between the solution/suspension and the plunger. The thermal expansion of air may result in spontaneous discharge when elevated temperatures are employed.

After application of a drop of suspension, blotting is initiated (one blot for 1 s in the examples shown in Figs. 3, 5, and 7), and immediately after blotting, the specimen is shot into melting ethane. From the melting ethane the specimen grid is carried over to the liquid nitrogen in the peripheral ring of the cooling vial and stored in a transfer box. The vitrified specimen is transferred to a cryoholder of a microscope (e.g., using a Gatan 626 transfer system and cryoholder). After a thermal equilibration period (15–30 min), images are taken at 120 kV and low-dose conditions in a Philips CM 12 microscope. Defocus values of about 900 nm ($\pm$ 300 nm) are employed to reveal the bilayer with optimal contrast (critical defocus values relate to the contrast transfer function of this microscope). The image data presented in Figs. 3 and 5 again illustrate the importance of humidity control just prior to vitrification of critical specimens such as liposome suspensions.

Concluding Remarks

In this chapter we have described important aspects of thin film formation and the interaction of thin films with the environment. Formations of interfacial layers, heat, and mass exchange have typical time constants in the subsecond range and are therefore important to cryofixation. The relationship between environmental humidity and evaporation from the specimen affecting both temperature and solute concentration is analyzed in some detail, sufficient to understand and interpret results of cryofixation. This section of the chapter is probably significant for cryofixation and cryo-EM of colloids in general. Only by working under strictly controlled conditions over a period of several years (and reporting results in life science and colloid chemistry) have we become fully aware of critical variables, such as blotting and humidity control. We think that automatic and controlled vitrification is a rapidly evolving field with a potential impact on various aspects of lipid structure and structural dynamics. Our group has contributed with cryo-EM to various fields of applied "liposomology": drug-loaded vesicles (e.g., Fig. 7, Burger et al., 2002, Lasic et al.,

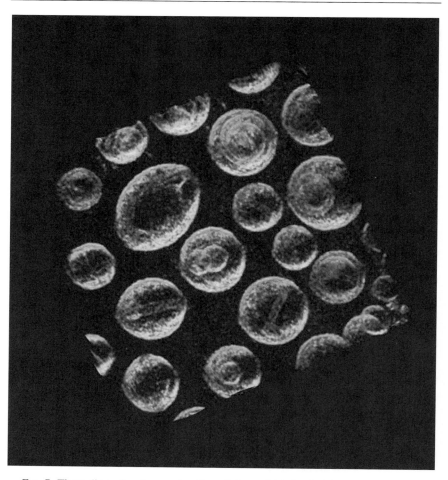

Fig. 7. Three-dimensional reconstruction of doxorubicin-loaded liposomes. A cryo-EM specimen was tilted over +70° to −70° at 2° increments. At every tilt angle a low-dose image was taken and by back projection of this tomographic series a three-dimensional reconstruction of the original specimen was made. Doxil/Caelyx specimen courtesy of Alza corporation (Menlo Park, CA). (See color insert.)

1992, 1995), lipid–DNA complexes for gene delivery (Hui *et al.*, 1999; Lasic *et al.*, 1997; Perrie *et al.*, 2001; Templeton *et al.*, 1997; Xu *et al.*, 2002), polymerization in/on vesicles (Hubert *et al.*, 2000; Jung *et al.*, 1997, 2000), bile action (Moschetta *et al.*, 2002), as well as in fundamental aspects of vesicle formation/fusion (Burger *et al.*, 1993; Lasic *et al.*, 2001; Malinin *et al.*, 2002) and colloid chemistry (Butter *et al.*, 2003; Mladenovic *et al.*, 2003).

The development of a vitrification robot for cryo-EM, as outlined in this chapter, was driven by liposome-related research but has opened a wider perspective. A (tailored) vitrification robot such as the Vitrobot can be considered an instrument that prepares thin slices (in the nanometer range) from a suspension. The vitrified sample offers through vision of the native specimen in the electron microscope and thus enables three-dimensional reconstruction of the individual suspended particles, for instance, by tomography (e.g., Fig. 7, also see McEwen and Koster, 2002). A vitrification robot taking slices from suspensions is likely to become standard equipment in cryo-EM laboratories. As a "liquid slicer," it has its counterpart in the ultramicrotome slicing solid (embedded) materials for almost half a century. We expect that automatic and controlled specimen preparation will boost a new era in cryo-EM, an era in which a wealth of reliable data are generated related to bilayer structures and (bio)colloids in general.

References

Bellare, J. R., Davis, H. T., Scriven, L. E., and Talmon, Y. (1988). Controlled environment vitrification system: An improved sample preparation technique. *J. Electron Microsc. Tech.* **10,** 87–111.

Burger, K. N., Calder, L. J., Frederik, P. M., and Verkleij, A. J. (1993). Electron microscopy of virus-liposome fusion. *Methods Enzymol.* **220,** 362–379.

Burger, K. N., Staffhorst, R. W., de Vijlder, H. C., Velinova, M. J., Bomans, P. H., Frederik, P. M., and de Kruijff, B. (2002). Nanocapsules: Lipid-coated aggregates of cisplatin with high cytotoxicity. *Nat. Med.* **8,** 81–84.

Butter, K., Bomans, P. H. H., Frederik, P. M., Vroege, G. J., and Philipse, A. P. (2003). Direct observation of dipolar chains in iron ferrofluids by cryogenic electron microscopy. *Nat. Mater.* **2,** 88–91.

Dubochet, J., Adrian, M., Chang, J. J., Homo, J. C., Lepault, J., McDowall, A. W., and Schultz, P. (1988). Cryo-electron microscopy of vitrified specimens. *Q. Rev. Biophys.* **21,** 129–228.

Dubochet, J., Adrian, M., Dustin, I., Furrer, P., and Stasiak, A. (1992). Cryoelectron microscopy of DNA molecules in solution. *Methods Enzymol.* **211,** 507–518.

Frederik, P. M., Stuart, M. C., Bomans, P. H., and Busing, W. M. (1989). Phospholipid, nature's own slide and cover slip for cryo-electron microscopy. *J. Microsc.* **153**(Pt. 1), 81–92.

Frederik, P. M., Stuart, M. C., Bomans, P. H., Busing, W. M., Burger, K. N., and Verkleij, A. J. (1991). Perspective and limitations of cryo-electron microscopy. From model systems to biological specimens. *J. Microsc.* **161**(Pt. 2), 253–262.

Hubert, D. H. W., Jung, M., Frederik, P. M., Bomans, P. H. H., Meuldijk, J., and German, A. L. (2000). Vesicle-directed growth of silica. *Adv. Mater.* **12,** 1286–1290.

Hui, S. W., Frederik, P., and Szoka, F. C., Jr. (1999). Physicochemical characterization and purification of cationic lipoplexes. *Biophys. J.* **77,** 341–353.

Jung, M., Hubert, D. H. W., Bomans, P. H. H., Frederik, P. M., Meuldijk, J., van Herk, A. M., Fischer, H., and German, A. L. (1997). New vesicle-polymer hybrids: The parachute architecture. *Langmuir* **13,** 6877–6880.

Jung, M., Hubert, D. H. W., Bomans, P. H. H., Frederik, P., van Herk, A. M., and German, A. L. (2000). A topology map for novel vesicle-polymer hybrid architectures. *Adv. Mater.* **12,** 210–213.

Lasic, D. D., Frederik, P. M., Stuart, M. C., Barenholz, Y., and McIntosh, T. J. (1992). Gelation of liposome interior. A novel method for drug encapsulation. *FEBS Lett.* **312,** 255–258.

Lasic, D. D., Ceh, B., Stuart, M. C., Guo, L., Frederik, P. M., and Barenholz, Y. (1995). Transmembrane gradient driven phase transitions within vesicles: Lessons for drug delivery. *Biochim. Biophys. Acta* **1239,** 145–156.

Lasic, D. D., Strey, H., Stuart, M. C. A., Podgornik, R., and Frederik, P. M. (1997). The structure of DNA-liposome complexes. *J. Am. Chem. Soc.* **119,** 832–833.

Lasic, D. D., Joannic, R., Keller, B. C., Frederik, P. M., and Auvray, L. (2001). Spontaneous vesiculation. *Adv. Colloid Interface Sci.* **89–90,** 337–349.

Lide, D. R. (ed.) (2003). "Handbook of Chemistry and Physics," 84th ed. CRC Press, Boca Raton, FL.

Malinin, V. S., Frederik, P., and Lentz, B. R. (2002). Osmotic and curvature stress affect PEG-induced fusion of lipid vesicles but not mixing of their lipids. *Biophys. J.* **82,** 2090–2100.

McEwen, B. F., and Koster, A. J., eds. (2002). Academic colloquium on electron tomography. Amsterdam, the Netherlands, October 17–20, 2001. *J. Struct. Biol.* **138,** 1–2.

Mladenovic, I. L., Kegel, W. K., Bomans, P., and Frederik, P. M. (2003). Observation of equilibrium, nanometer-sized clusters of silver iodide in aqueous solutions. *J. Phys. Chem. B* **107,** 5717–5722.

Moschetta, A., Frederik, P. M., Portincasa, P., vanBerge-Henegouwen, G. P., and van Erpecum, K. J. (2002). Incorporation of cholesterol in sphingomyelin-phosphatidylcholine vesicles has profound effects on detergent-induced phase transitions. *J. Lipid Res.* **43,** 1046–1053.

Mui, B., Chow, L., and Hope, M. J. (2003). Extrusion technique to generate liposomes of defined style. *Methods Enzymol.* **367,** 3.

Perrie, Y., Frederik, P. M., and Gregoriadis, G. (2001). Liposome-mediated DNA vaccination: The effect of vesicle composition. *Vaccine* **19,** 3301–3310.

Talmon, Y., Burns, J. L., Chestnut, M. H., and Siegel, D. P. (1990). Time-resolved cryotransmission electron microscopy. *J. Electron Microsc. Tech.* **14,** 6–12.

Templeton, N. S., Lasic, D. D., Frederik, P. M., Strey, H. H., Roberts, D. D., and Pavlakis, G. N. (1997). Improved DNA: Liposome complexes for increased systemic delivery and gene expression. *Nat. Biotechnol.* **15,** 647–652.

Xu, L., Frederik, P., Pirollo, K. F., Tang, W. H., Rait, A., Xiang, L. M., Huang, W., Cruz, I., Yin, Y., and Chang, E. H. (2002). Self-assembly of a virus-mimicking nanostructure system for efficient tumor-targeted gene delivery. *Hum. Gene Ther.* **13,** 469–481.

Author Index

A

R

Raad, I., 212
Raber, M. N., 98, 104
Rácz, P., 332
Rader, J. S., 98
Radford, J. A., 9
Radhakrishnan, B., 9
Rahman, A., 9, 72, 146, 179
Rainov, N. G., 202
Rait, A., 446
Raleigh, J. W., 298
Ralston, E., 187, 297
Ram, Z., 200, 202
Ramaswamy, M., 305, 306, 309, 311
Ramirez, M. J., 9
Ramratnam, B., 332
Randall, R. J., 214, 215, 217, 223, 362
Randall, R. T., 404
Rao, P. E., 365
Rarick, M. U., 10
Ratain, M. J., 128
Rawlings, S. R., 378
Rawstron, A. C., 10
Rayner, B., 356
Raz, A., 262, 268
Read, N., 341
Reboiras, M. D., 217
Reddy, M. V., 262, 275, 276, 352
Redelmeier, M. J., 79
Redelmeier, T. E., 9, 10, 11, 14, 18, 19,
    51, 52, 78, 79, 285
Redemann, C., 9, 44, 60, 75, 147, 230,
    236, 272, 415
Refai, E., 378, 379, 384
Regts, D., 262
Regts, J., 262, 268
Reich, S. D., 128
Reig, F., 285
Reimann, H. J., 100
Reingold, A. L., 262
Reinish, L. W., 9, 10, 11, 18, 19, 78
Reitz, M. S., Jr., 356
Remick, S. C., 9
Renard, N., 139
Rentsch, K. M., 62, 67, 68
Reshef, A., 378, 379, 380
Reszka, R., 201
Reynolds, C. W., 130
Reynolds, J. A., 11

Ribas, J. L., 332
Rich, R., Jr., 42
Richards, F., 42, 176
Richards, F. II, 42
Richman, D. D., 332
Rickaby, D., 416, 422
Riddles, P. W., 360
Rijsbergen, Y., 130
Ringdén, O., 229
Ringel, I., 99, 100
Rinkes, I. H., 134
Riondel, J., 103
Rios, A., 128
Rivers, R. P., 352
Rivoltini, L., 30
Robberecht, P., 378
Robert, J., 177
Roberts, D. D., 446
Roberts, W. G., 71
Robertson, C., 426
Robins, T., 357
Robinson, A. M., 212, 216, 226
Robinson, D., 414
Robinson, R. M., 381
Rodarte, J. A., 415
Rodriguez, G., 100, 101
Rodriguez, J., 98
Rodriguez, M. A., 43
Rodriguez-Sierra, J., 381
Rodvold, K. A., 278
Roerdink, F. H., 43, 230, 262, 268, 271, 272
Romaguera, J., 43
Roman, L., 9
Rong, Q., 356
Roovers, D. J., 146
Ropert, C., 352
Rorstad, O. P., 377, 378
Rosales, R., 9
Rose, P. G., 9
Rose, W. C., 100
Rosebrough, N. J., 362, 404
Roseburgh, N. J., 214, 215, 217, 223
Rosenberg, E., 333
Rosenberg, S. A., 134
Rosenblum, M. G., 128
Rosenthal, K. A., 362
Ross, M. E., 10, 261
Rossi, J. J., 284, 352, 356, 357, 358, 368, 369
Rottenberg, H., 19, 51
Rouser, G., 121, 122, 405

Subject Index

A

479

PET

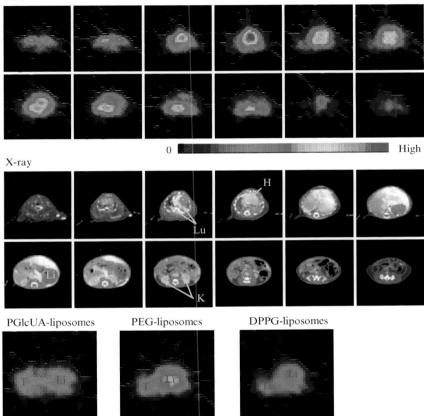

OKU AND NAMBA, CHAPTER 8, FIG. 5. Biodistibution of [2-^{18}F]FDG-labeled PGlcUA liposomes imaged by PET. PGlcUA liposomes composed of DPPC, cholesterol, and PGlcUA (4:4:1 as a molar ratio) were labeled with [2-^{18}F]FDG and extruded through a polycarbonate membrane with 100-nm pores. The [2-^{18}F]FDG-labeled liposomes were injected intravenously into 7-week-old BALB/c male mice, and the emission scan was performed immediately after injection. PET images (upper panel) and corresponding X-ray images (middle panel) are shown where the slice aperture is 3.25 mm from the head (upper left) to the tail (lower right). Lu, lung; H, heart; Li, liver; K, kidney. PGlcUA-and PEG-modified liposomes, as well as DPPG liposomes, were labeled with [2-^{18}F]FDG and injected into BALB/c male mice bearing Meth A sarcoma. The tumor accumulation of these liposomes is shown in the lower panel.

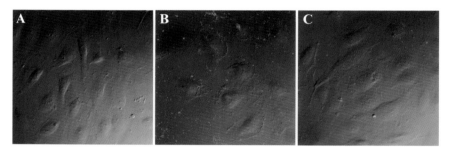

Asai and Oku, Chapter 9, Fig. 5. Specific binding of APRPG-modified liposomes to VEGF-stimulated human umbilical endothelial cells. (B) Confocal microscopic observation indicates that the binding pattern of APRPG-modified liposomes to VEGF-stimulated HUVECs is specific, but not (A) that of control liposomes. (C) Binding of PRP-Lip was inhibited in the presence of excess APRPG, suggesting the presence of specific molecule(s) on the VEGF-stimulated HUVECs that have affinity to APRPG.

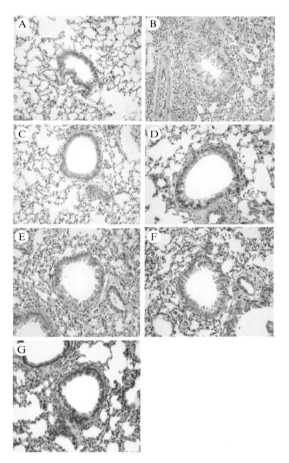

Konduri, *et al.*, Chapter 23, Fig. 1. Examples of hematoxylin and eosin–stained lung tissues from the six treatment groups. (A) Normal; (B) sensitized; (C) weekly budesonide in sterically stabilized liposomes; (D) daily budesonide; (E) weekly budesonide in conventional liposomes; (F) weekly empty sterically stabilized liposomes; (G) weekly budesonide.

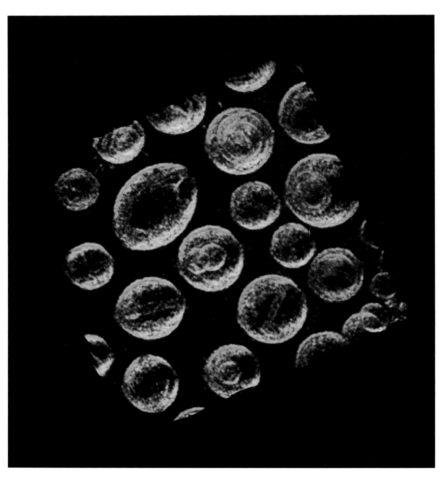

FREDERIK AND HUBERT, CHAPTER 24, FIG. 7. Three-dimensional reconstruction of doxorubicin-loaded liposomes. A cryo-EM specimen was tilted over $+70°$ to $-70°$ at $2°$ increments. At every tilt angle a low-dose image was taken and by back projection of this tomographic series a three-dimensional reconstruction of the original specimen was made. Doxil/Caelyx specimen courtesy of Alza corporation (Menlo Park, CA).